Astrolabes in Medieval Cultures

Astrolabes in Medieval Cultures

Edited by

Josefina Rodríguez-Arribas
Charles Burnett
Silke Ackermann
Ryan Szpiech

BRILL

LEIDEN | BOSTON

Originally published, in part, as Volume 23, No. 1–5 (2017) of Brill's journal *Medieval Encounters*.

Library of Congress Cataloging-in-Publication Data

Names: Rodríguez Arribas, Josefina, editor. | Burnett, Charles (Charles S. F.), editor. | Ackermann, Silke, editor. | Szpiech, Ryan, editor.
Title: Astrolabes in Medieval cultures / edited by Josefina Rodriguez-Arribas, Charles Burnett, Silke Ackermann, Ryan Szpiech.
Other titles: Medieval encounters.
Description: Leiden ; Boston : Brill, [2019] | Originally published, in part, as Volume 23, No. 1–5 (2017) of Brill's journal Medieval encounters. | Includes index.
Identifiers: LCCN 2018043238 (print) | LCCN 2018044134 (ebook) | ISBN 9789004387867 (Ebook) | ISBN 9789004383807 (hardback : alk. paper)
Subjects: LCSH: Astrolabes—History. | Astronomical instruments—History. | Nautical instruments—History.
Classification: LCC QB85 (ebook) | LCC QB85 .A88 2019 (print) | DDC 522/.20902—dc23
LC record available at https://lccn.loc.gov/2018043238

Typeface for the Latin, Greek, and Cyrillic scripts: "Brill". See and download: brill.com/brill-typeface.

ISSN 978-90-04-38380-7 (paperback)
ISSN 978-90-04-38786-7 (e-book)

This book is printed on acid-free paper and produced in a sustainable manner.

Contents

Preface to the New Edition

Ryan Szpiech
University of Michigan
szpiech@umich.edu

This volume was first published in Medieval Encounters 23.1–5 (2017) and was guest edited by Josefina Rodríguez-Arribas, Charles Burnett, and Silke Ackermann. It is now my pleasure as Editor-in-Chief of the journal to publish that special issue in updated form as a book. The contents of this volume are mostly identical with those of the journal issue, but this book also includes a number of corrections and a few important additions.

Among the additions are brief postscripts added by John Davis and Günther Oestmann to their articles that discuss new findings and clarifications of their original points. In addition, images and diagrams in some articles have been resized and improved, a few passages rewritten and updated, and some errata corrected. Most significantly, a new article by Flora Vafea (whose own original contribution also reappears here) has been appended to the original issue as an epilogue. Vafea's epilogue provides a response to the article by Johannes Thomann, "Astrolabes as Eclipse Computers," included in the original publication and reproduced here. Vafea proposes that two of the texts considered by Thomann as independent sources about the plate of eclipses, according to ʿAlī ibn ʿĪsā's description, should be considered instead as two parts of the same work. Vafea provides a text and translation of the proposed single text and attempts a reconstruction of the instrument described therein.

With this new edition of the Medieval Encounters issue "Astrolabes in Medieval Culture," we four editors aim to disseminate this important research to a wider body of readers and scholars and to offer them a corrected and up-to-date version of the research it contains. I am confident that this volume will provide a touchstone for ongoing research in medieval astronomy and serve as a foundation for future studies.

Preface to the First Edition

Astrolabes in Medieval Cultures

*Josefina Rodríguez-Arribas**
Institut für Jüdische Studien Westfälische Wilhelms-Universität, Münster
**Corresponding author, e-mail: rodriguezarribas@gmail.com*

Charles Burnett
Warburg Institute, Woburn Square, London WC1H 0AB, UK
charles.burnett@sas.ac.uk

Silke Ackermann
Museum of the History of Science, Broad St, Oxford OX1 3AZ, UK
silke.ackermann@mhs.ox.ac.uk

This book, first published as a Special Issue of *Medieval Encounters*, is based on the papers of the conference on "Astrolabes in Medieval Cultures" held at the Warburg Institute, University of London, on 24–25 April 2014, under the aegis of a three-year research project on "Astrolabes in Jewish Culture." This project was supported by the British Arts and Humanities Research Council (AH/I003800) and based at the Warburg Institute (Josefina Rodríguez-Arribas and Charles Burnett) with the participation of the British Museum (in its initial stages, Silke Ackermann) and the Museum of the History of Science, Oxford University (Stephen Johnston). The aims of the project were to produce a monograph on the place of the astrolabe in medieval Jewish culture, to draw up an illustrated catalogue of the instruments, and to edit and translate at least one of the Hebrew treatises on the use and construction of astrolabes. All three aims are nearing completion. In this context it seemed opportune to bring together scholars and students interested in the instrument, its history, and its literature, in which experienced and new scholars could update the knowledge so far available in this field. We were interested in all periods and cultures, and in all kinds of approaches, but especially those that have received less attention. This meeting was an occasion for scholarly and friendly exchanges, and a source of inspiration for future research. Several participants brought

 DOI:10.1163/9789004387867_002

instruments they possessed or had made themselves, which were admired, displayed, discussed—and used! The general feeling was that astrolabe is still very much alive and there is still much to learn about its subtleties. After the conference, further articles were commissioned to fill out the subject matter of the book and it was decided to include David King's list of European astrolabes (to ca. 1500 arranged by category) as a useful tool to place the astrolabes referred to in this volume in context.

The editors would like to express their gratitude to the Warburg Institute, their staff and fellows, and to the British Museum and the Oxford Museum of the History of Science and their staffs, for supporting this research project and this conference in different ways. We would particularly like to thank Stephen Johnston for his part in preparing the conference and for the unstinted help and advice that he has always given us. We would like to extend our gratitude to Ryan Szpiech, the editor of *Medieval Encounters*, for accepting the proceedings of this conference into the journal. Last but not least we also thank all the participants in the conference "Astrolabes in Medieval Cultures" for sharing their knowledge and their enthusiasm for this intelligent, beautiful, and emblematic object and making this scholarly event a very special occasion.

The Editors

Introduction

∴

Hic Sunt Dracones—Astrolabe Research Revisited

Silke Ackermann
Museum of the History of Science, Broad Street, Oxford OX1 3AZ, UK
silke.ackermann@mhs.ox.ac.uk

Hic sunt dracones (Here be dragons) is said to be the phrase commonly used on medieval maps to indicate blank areas that are awaiting further exploration and are thus somewhat dangerous. In fact it is not—there appears to be only one globe that bears this phrase on the eastern coast of Asia,[1] but nevertheless the expression has taken on a life of its own, understood in a variety of contexts.

Hic sunt leones (Here be lions) is used with a similar connotation by brother Jorge, the evil monk in *The Name of the Rose* when he is referring to dangerous knowledge and again, now in the ambiguous sense, by the main character and hero of the book, Franciscan brother William of Baskerville when he finds Africa on a map in the library.[2]

It is this same William of Baskerville who, portrayed by Sean Connery in the Hollywood movie, brought an astrolabe to the attention of a wider public—a sort of outreach, if you like. Because it is he who in one of the key scenes of the movie furtively reaches for an astrolabe in his bag to take some readings (what those are we are never told),[3] anxious not be seen to use this instrument introduced from the Islamic—and thus heretical, according to the church authorities—world.

1 Lenox globe, c. 1503–1507, New York Public Library, New York, NY, USA.

2 Umberto Eco, *The Name of the Rose*, trans. William Weaver (San Diego, CA: Harcourt Brace Jovanovich, 1983), 421.

3 "*The Name of the Rose*" (Columbia Pictures, 1986) by Jean-Jacques Annaud.

 | DOI:10.1163/9789004387867_003

Astrolabes have lions and dragons aplenty, in fact a whole menagerie of them—the lion as part of the zodiac, amongst the star-pointers and in astrological contexts; the dragon equally appears amongst the constellations, often figuratively on the rete, in astrological contexts and to explain eclipses.

But equally research on the astrolabe has lots of lions and dragons in the non-tangible sense. In spite of more than 150 years of scholarly work there remain many unchartered territories or those where only the coasts have been explored. A commented and illustrated inventory of all surviving astrolabes that goes beyond a mere list[4] is still outstanding, in spite of a number of long ongoing projects.[5] In the second decade of the twenty-first century one would wish for an online astrolabe portal that enables the user to search an always increasing database by period and region, maker and date, current (or last known) location, with hyperlinks to photos, descriptions, relevant publications and ongoing scholarly discussions and research projects. A database that facilitates interdisciplinary research in a wide variety of contexts in respect to one of the most fascinating instruments that has ever existed—be it art or philology, craftsmanship or technology, ethnology or sociology, economy or law, politics or religion—or a mixture of any of the above.

The more we know, the more we are aware how little we know. What were the legal implications of using an astrolabe in the second half of the thirteenth century? What prices were paid for astrolabes in Renaissance Italy and was it possible to make a living based on this? How are astrolabes deployed in films and literature in the second half of the twentieth century and how do the audiences "read" these? What did a Nuremberg citizen know about astrolabes in the first half of the sixteenth century? Did women make, use, and teach the use of astrolabes in the Islamic World?

This volume, a slightly revised and updated version of the special issue of *Medieval Encounters* 23 (2017), cannot answer all these questions, of course, but it intends to fill some of the blank areas—to tame at least a few dragons (or lions), so to speak.

4 Derek de Solla Price and Sharon Gibbs, *Computerized Checklist of Astrolabes* (New Haven, CT: Yale University, 1973).

5 Alain Brieux and Francis R. Maddison, *Repertoire des facteurs d' astrolabes et de leurs oeuvres* (forthcoming; at the time of writing and following the death of both authors there are indications that the Repertoire may actually soon be published); David A. King, *A Catalogue of Medieval Astronomical Instruments to c. 1500* (forthcoming, the 1996 version is available as PDF on http://davidaking.academia.edu/research#catalogueofmedievalastronomicalinstruments). See also DA King's article in this volume.

Petra Schmidl's article is clear evidence once again (if it was needed) how useful it would be to have the aforementioned astrolabe portal to be able to (re)examine the instruments in detail and to (re)evaluate earlier research claims. She shows convincingly that close examination of the actual evidence can lead to surprising conclusions—and that astrolabes do not only have the obvious moving parts, but do move themselves (and evidence for this motion is evident on the astrolabes) as knowledge is in motion between the Arabic and Latin speaking worlds.

The necessity of learning to "read" objects in the same way as textual and pictorial sources is demonstrated admirably by John Davis. Based on comparisons with two illuminated manuscripts, the "Milemete Treatise" and a copy of the "Secretum Secretorum," Davis puts forward an intriguing historical contextualisation of the famous "Sloane Astrolabe" in the British Museum, the largest and most important of all medieval English instruments, as a teaching tool for Prince Edward of Windsor who was soon to become King Edward III, likely to have been commissioned, c. 1326, by his tutor Richard de Bury. He suggests that two other medieval astrolabes, now in Liège and in the London Science Museum, are believed to be closely associated with the Sloane instrument and derived from it. A tour de force after which quatrefoils, wyverns, or saints' days on astrolabes will never again be allowed to be seen as mere decoration.

The importance of textual evidence and the innumerable sources that have not even been tapped yet, let alone edited and translated, is demonstrated clearly by three articles in this issue. Johannes Thomann discusses, edits, and translates four hitherto unpublished Arabic texts of the ninth century CE on the construction and use of astrolabe-like instruments for showing the conditions of eclipses. These texts appear to be the earliest technical descriptions of eclipse computers in any language. As Flora Vafea proposes in her epilogue at the end of this volume, two of those texts dealing with the plate of eclipses as described by ʿAlī ibn ʿĪsā are, in her view, actually part of the same text. She thus provides an edition merging the two texts into one and attempts a reconstruction of the instrument described.

Josefina Rodríguez Arribas deploys her excellent command of Jewish sources on the astrolabe to make accessible Bonetus de Latis's (Jacob ben Immanuel Provenzale) treatise on the astrolabe ring (1492/1493), an intriguing and apparently unique object that, according to at least one eye-witness account, was not just a mind game, but did, indeed, exist in physical form. The treatise was highly regarded among Bonetus's contemporaries and the instrument depicted in Bonetus's booklet can be considered, as she states, to be one of the last contributions of Jewish culture to the history of the astrolabe.

By comparing Kūshyār ibn Labbān's (tenth century CE) *Book of the Astrolabe* with Athīr al-Dīn al-Abharī's (d. 1262 or 1265) *Treatise on Knowing the Astrolabe*, Taro Mimura shows that both presupposed their own astrolabes when writing. He concludes that many Arabic treatises on the astrolabe had their novelty at least guaranteed by the uniqueness of a specific astrolabe that each author used as his basis.

False certainties can be more damaging than no certainty at all. This is especially true in the context of the dating of astrolabes by "scientific" methods—methods, that is, that are a historical *contradictio in se* with the data available at the time of production. Giorgio Strano competently examines, and then dismisses these methods, reminding scholars that "measuring the star positions on the rete requires one to adopt the same coordinate system used during the Middle Ages." Indeed, it does, and Strano thus introduces an alternative approach tested on six medieval astrolabes that has shown remarkable results, but simultaneously opens up a new set of questions. But those "and many other questions reveal that the new approach to the astrolabes data is not a dead end, but a long path of which only the first few steps have been taken." Face your dragons, one might say.

Knowledge of stereographic projection is at the heart of astrolabe construction. Based on her thesis, Flora Vafea shows that the stereographic projection of the celestial globe onto the plane of its equator is an essential prerequisite of the construction of the planispheric astrolabe in order to preserve and transfer all the functions of the globe. Whilst the representation of the celestial coordinates on an astrolabe is more abstract and further removed from the three-dimensional model of the celestial sphere that a globe represents, the resulting instrument can be used for more accurate measurements. Its functionality thus goes far beyond that of the more didactic role of the globe.

The relationship between the less common stereographic projection from the North Pole (resulting in so-called southern astrolabes) and early astronomical clocks showing time confusingly in an anti-clockwise motion, is explored by Günther Oestmann, who is able to confirm that the knowledge of variations of the stereographic projection and special astrolabe constructions must have reached Europe from Islamic areas (probably Spain) in the twelfth century, thus providing essential knowledge for the construction of astrolabe clock dials.

Four papers in this issue focus on the rich culture of scientific exchange and instrument production in medieval Spain. How little we know even about seemingly basic workshop practices and division of labour in the Middle Ages is lamented by Azucena Hernández. She discusses the only signed and dated Latin astrolabe known to have survived from medieval Spain and proposes an identification for "Petrus Raimundi, a member of the House of the

King of Aragon" in the context of a whole dynasty of kings supporting scientific research, focusing mostly on its practical dimension: the construction of scientific instruments. But whether Petrus Raimundi was involved in the construction of the instrument or in its commission remains, for now, an enigma.

Emilia Calvo investigates some features of the treatise on the *lámina universal*, an astronomical instrument devised by ʿAlī ibn Khalaf, an eleventh-century Andalusī mathematician and astronomer who belonged to the scientific circle of the *qāḍī* Ṣāʿid al-Andalusī. The text is preserved only in the thirteenth-century Old Castilian Alfonsine translation.

Miguel Forcada explores a number of texts written in al-Andalus between the twelfth and the early fourteenth centuries that mention Ibn al-Zarqālluh or the "Toledan astronomers." These texts express scholarly reactions to the contributions of the circle of astronomers of eleventh-century Toledo—and the textual evidence reveals attitudes ranging from a somewhat naïve admiration to harsh criticism.

Not the physical instruments themselves, but "astrolabes on parchment" are the focus of Laura Fernández Fernández's study of four texts relating to the construction and use of astrolabes in the *Libro del Saber de Astrología* (LSA) commissioned by Alfonsio el Sabio in the second half of the thirteenth century with a particular focus on the fine and ample illustrations. She concludes that the drawings were designed in close contact with real instruments as well as the use of illustrated manuscripts. Here, too, we find that we are nowhere near completely chartered territory due to the lack of a catalogue of illuminated astrolabe manuscripts that would facilitate this research, especially in the light of the use of earlier sources from the Islamic world and the collaboration of Muslim and Jewish scholars in the creation of the astrolabe chapter of LSA.

The study of astrolabes has tended to be largely centred on the role, manufacture, and use of the instrument in the classical world, the Middle East, and Europe, with a focus on the Mediterranean. That the trajectory did not only go west around the first Millennium, but also east towards India is a far less well understood story. This has been Raja Sarma's research focus for many years and few can claim to have such a thorough understanding of this part of the world. Here he argues that a huge astrolabe, which was completely reworked at a later stage with Sanskrit legends, was in fact originally produced between 1648 and 1658 by Ḍiyā' al-Dīn Muḥammad of Lahore for the Mughal Emperor Shāh Jahān. One could hardly imagine a more stunning manifestation of the dynamic development of a much earlier tradition.

The present volume endeavours to shed further light on what remains one of the most intriguing areas of instrument research. It is also meant to encourage others to tackle those areas where we would still claim *hic sunt dracones*.

CHAPTER 1

Astrolabes as Eclipse Computers: Four Early Arabic Texts on Construction and Use of the *Ṣafīḥa Kusūfiyya*

Johannes Thomann
Department of Islamic Studies, Institute of Asian and Oriental Studies, University of Zurich, Rämistrasse 59, 8001 Zürich, Switzerland
johannes.thomann@aoi.uzh.ch

Abstract*

The main topic of this article comprises four unpublished Arabic texts on astrolabe-like instruments for showing the conditions of eclipses. They are the earliest technical descriptions of eclipse computers in any language. The first text is a treatise by ʿAlī ibn ʿĪsā (ninth century) on a special astrolabe for lunar phases and eclipses. The second text is an anonymous redaction of the first text with some omissions and additions. The third text describes a similar instrument with a different design, which was invented by Muḥammad ibn Muḥammad Nasṭūlus al-Asṭurlābī in the year 893/894 CE. The fourth anonymous text describes a plate for a graphical solution of the size of a lunar eclipse. In the concluding part, later Arabic descriptions of eclipse computers are summarized, some traces of these texts on real astrolabes are mentioned, and finally some comparable medieval Latin texts are referred to. Four Appendices contain an edition of the Arabic texts and English translations.

Keywords

Alī ibn ʿĪsā – Arabic treatises – astrolabe – eclipse computer – Muḥammad ibn Muḥammad Nasṭūlus al-Asṭurlābī

---

* Editors' note: See also Flora Vafea's epilogue to this volume for a response to this study proposing that two of the four texts actually derive from the same text.

 | DOI:10.1163/9789004387867_004

Introduction

Eclipses of the sun and moon are exceptional events and have had a great impact on human society throughout history. The prediction of eclipses was one of the primary goals of ancient astronomy.[1] The report about Thales of Miletus having predicted the solar eclipse during the battle at the river Halys in the year 585 BCE became a paradigm of the triumph of rational science over blind superstition.[2] Eclipses play an important role in the reports of Babylonian astronomers to the king, and in their astronomical records. The discovery of the eclipse cycle allowed for an easy way to predict possible dates of solar and lunar eclipses. Such cycles might have been in use among illiterate cultures and there are prominent objects that have been presented as candidates for encoding them. The most famous and the most disputed example is the megalithic monument at Stonehenge.[3] More recently, the so-called "Berliner Goldhut," one of four known conical golden hats known from Bronze Age Europe so far, has also been proposed. Wilfried Menghin, the late director of the "Museum für Vor- und Frühgeschichte," published an elaborate interpretation of the patterns on the hat as lunisolar cycles, and, adding to his analysis one may say that the cycle of 235 synodic months must be considered a viable candidate for an eclipse cycle, since 235 synodic months are almost equal to 255 draconitic months.[4] More recently in 2012 a dish found in the palace of Knossos has been interpreted as a Minoan eclipse calculator.[5]

---

1 Clemency Montelle, *Chasing Shadows. Mathematics, Astronomy, and the Early History of Eclipse Reckoning* (Baltimore, MD: Johns Hopkins University Press, 2011).

2 Johannes Thomann, "Zur Nachrechnung antiker Sonnenfinsternisse," *Antike Naturwissenschaft und ihre Rezeption* 9 (1999): 103–109; F.R. Stephenson and Louay J. Fatoohi, "Thales' Prediction of a Solar Eclipse," *Journal for the History of Astronomy* 28.4 (1997): 279–282; Dimitri Panchenko, "Thales's Prediction of a Solar Eclipse," *Journal for the History of Astronomy* 25.4 (1994): 275–288; Thomas Worthen, "Herodotos's Report on Thales' Eclipse," *Electronic Antiquity* 3.7 (1997) no pagination; D.W. Roller, "Some Thoughts on Thales' Eclipse," *Liverpool Classical Monthly* 8 (1983): 58–59; A.A. Mosshammer, "Thales' Eclipse," *Transactions of the American Philological Association* 109 (1981): 145–155; D. O'Brien, "Derived Light and Eclipses in the Fifth Century," *Journal of the History of Astronomy* 88 (1968): 114–127.

3 John David North, *Stonehenge. A New Interpretation of Prehistoric Man and the Cosmos* (New York, NY: The Free Press, 1996).

4 Wilfried Menghin, *Der Berliner Goldhut. Macht, Magie und Mathematik in der Bronzezeit* (Berlin: Staatliche Museen zu Berlin–Stiftung preussischer Kulturbesitz, 2010).

5 M. Tsikritsis, E. Theodossiou, V.N. Manimanis, P. Mantarakis and D. Tsikritsis, "A Minoan Eclipse Calculator," in *5th International Conference of the European Society for the History of Science. Scientific Cosmopolitanism and Local Cultures: Religions. Book of Abstracts*, ed. G. Katsiampoura (Athens: Institute of Historical Research/National Hellenic Research Foundation, 2012), 166.

Greek astronomers became aware that solar eclipses were produced by the lunar body and lunar eclipses by the shadow of the earth, and constructed three-dimensional models of the solar system. Cicero described such a model in Rome, thought to be made by Archimedes, which could perform solar and lunar eclipses in the true sequence of the orbital periods.[6] The proof that such complex devices existed in Antiquity exists in the famous Antikythera mechanism. In a recently undertaken new reading of the inscription, glyphs were identified as symbols for solar and lunar eclipses.[7]

Despite the celebrity such devices had in Antiquity, no technical text on the construction or use of them seems to have existed. There are some side-remarks in Ptolemy's *Almagest* on physical models, but these are far from the sort of detailed descriptions of instruments for astronomical observations that make precise modern reconstructions of them possible.[8] Independent works on the construction of astronomical instruments are scarce; works by Theon, Philoponos, and others on the plane astrolabe are the only exceptions.[9]

Things changed significantly in the Baghdadian renaissance of the sciences in the eighth and ninth centuries. The majority of astronomical texts also dealt with the description of astronomical instruments. A number of works of al-Khwārizmī, Ḥabash al-Ḥāsib, al-Farghānī, and other authors of the early Abbasid period are extant.[10] It was not only a time of acquisition of ancient techniques but also a time of innovation. Ḥabash wrote his treaty on the melon-shaped astrolabe based on a non-stereographic projection with equidistant declination mapping.[11]

---

6 Cicero, *On the Republic. On the Laws*, trans. by Clinton W. Keyes (Cambridge, MA: Harvard University Press, 1928), 1.22, p. 40.

7 Tony Freeth, Alexander Jones, John M. Steele, and Yanis Bitsakis, "Calendars with Olympiad Display and Eclipse Prediction on the Antikythera Mechanism," *Nature* 454 (2008): 614–617.

8 Ptolemy, "Almagest XIII.2," in *Claudii Ptolemaei opera quae exstant omnia*, ed. J.L. Heiberg (vols. 1–2), F. Boll and E. Boer (vol. 3), 3 vols. (Leipzig: B.G. Teubneri, 1898–1954), vols. 1.1 and 1.2 (1898, 1903), 533; instruments of observation: see Almagest V.12.

9 Jean Philopon, *Traité de l'astrolabe* (Paris: Belles Lettres, 2015).

10 François Charette and Petra Schmidl, "Al-Khwārizmī and Practical Astronomy in Ninth-Century Baghdad. The Earliest Extant Corpus of Texts in Arabic on the Astrolabe and Other Portable Instruments," *SCIAMVS* 5 (2004): 101–198; Richard P. Lorch and Paul Kunitzsch, "Ḥabash al-Ḥāsib's Book on the Sphere and its Use," *Zeitschrift für Geschichte der arabisch-islamischen Wissenschaften* 2 (1985): 68–98; Richard Lorch, *Al-Farghānī. On the Astrolabe* (Stuttgart: Steiner, 2005).

11 Edward Stewart Kennedy, Paul Kunitzsch and Richard Lorch, *The Melon-Shaped Astrolabe in Arabic Astronomy* (Stuttgart: Steiner, 1999).

ʿAlī ibn ʿĪsā's Epistle on the Eclipse Plate (I)

In the context of this illustrious circle in Baghdad at the time of al-Ma'mūn the sources mention a certain ʿAlī ibn ʿĪsā al-Asṭurlābī as a member of the team formed to measure one degree of geographical latitude.[12] He is a somewhat shadowy figure and not much is known about his biography.[13] A treatise on the astrolabe that bears his name was edited in 1913 and translated into German.[14] According to al-Qabīsī (died 967 CE) ʿAlī ibn ʿĪsā wrote a *Risāla fī ibṭāl ṣināʿat aḥkām al-nujūm* (Letter on the invalidation of judicial astrology), against which al-Qabīsī wrote his *Kitāb fī ithbāt ṣināʿat aḥkām al-nujūm wa-naqd Risālat ʿAlī ibn ʿĪsā fī ibṭālihā* (Book on the validation of judicial astrology and the shortcomings of ʿAlī ibn ʿĪsā's letter on the invalidation of it).[15] Neither of the two works has survived.

Besides that, there is another treatise ascribed to him with the title *Risāla fī maʿrifat al-ʿamal bi-l-ṣafīḥa l-qamariyya wa-l-ḥuqq*[*a*] *al-kusūfiyya* (Epistle on the knowledge how to use the lunar plate and the eclipse box).[16] As the title indicates, devices for the display of the lunar phases and of eclipses were closely related tasks. While the first task of lunisolar devices has received some scholarly attention already a century ago, the study of the second task—the display of the eclipse—is still in its infancy.[17] The lunisolar device

12 David King, "Too Many Cooks ... A New Account of the Earliest Geodetic Measurements," *Suhayl* 1 (2000): 207–241 (210, 215, 218, 223); Y. Tzvi Langermann, "The Book of Bodies and Distances of Ḥabash al-Ḥāsib," *Centaurus* 28 (1985): 108–128 (122).

13 Marvin Bolt, "ʿAlī Ibn ʿĪsā al-Asṭurlābī," in *The Biographical Encyclopedia of Astronomers*, ed. Thomas Hockey (New York, NY: Springer, 2007), 34; Boris A. Rosenfeld and Ekmeleddin İhsanoğlu, *Mathematicians, Astronomers, and Other Scholars of Islamic Civilization and Their Works (7th–19th c.)* (Istanbul: IRCICA, 2003), 28.

14 Louis Cheikho, "Kitāb al-ʿamal bi-l-aṣṭurlāb li-ʿAlī Ibn ʿĪsā," *al-Mashriq* 16 (1913): 29–46; Carl Schoy and Joseph Drecker, "ʿAlî ibn ʿÎsâ, Das Astrolab und sein Gebrauch," *Isis* 9 (1927): 239–254.

15 Charles Burnett, Keiji Yamamoto, and Michio Yano, *Al-Qabīṣī (Alcabitius): The Introduction to Astrology* (London: The Warburg Institute, 2004), 18–19.

16 MS Istanbul, Topkapı Saray Kütüphanesi, Ahmet III 3509 ff. 304r–309v, written in 1277 CE. Descriptions: Fuat Sezgin, *Geschichte des arabischen Schrifttums: Band VI Astronomie bis ca. 430 H.* (Leiden: Brill, 1978) 144; Fehmi Edhem Karatay, *Arapça yazmalar katalŏgu* (Istanbul: Topkapı Sarayı Müzesi, 1962–1969), 3:740; Max Krause, "Stambuler Handschriften islamischer Mathematiker," *Quellen und Studien zur Geschichte der Mathematik, Astronomie und Physik; Abt.B: Studien* 3 (1936): 437–532 (447).

17 For the eclipse plate see Flora Vafea, "Al-Bīrūnī: The Plate of the Eclipses," *Suhayl* 15 (2017): forthcoming; Johannes Thomann and Matthias Vogel, *Schattenspur. Sonnenfinsternisse in Wissenschaft, Kunst und Religion* (Basel: Naturhistorisches Museum, 1999), 40 and

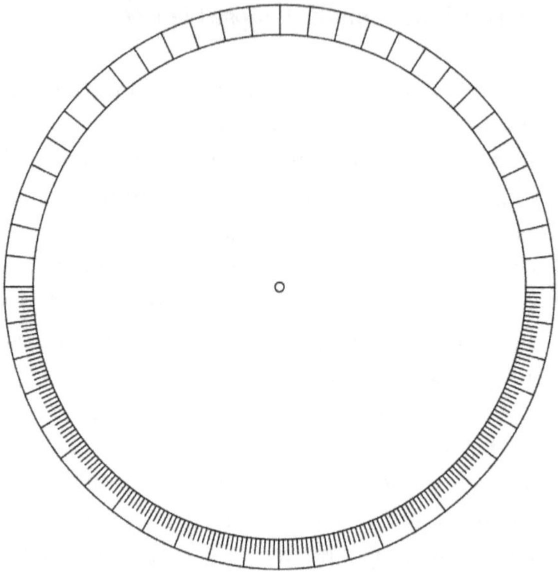

FIGURE 1.1 *The mater of ʿAlī ibn ʿĪsā.*

described by ʿAlī ibn ʿĪsā is different from later examples of geared mechanisms. Its parts are standard parts of an astrolabe, but have a different design. The mater, which is called *ḥujra* ("limb"), is divided into four quadrants. The rim of the upper half is divided into twenty-eight parts. They correspond to the twenty-eight days during which the moon is visible during a lunar month. The rim of the lower right quarter is divided into twelve parts, which correspond to the twelve daily hours. The hours are marked with *abjad* letters. The rim of the lower left quadrant is divided accordingly, representing the twelve hours of the night. The scale in the lower half of the limb is designed to indicate the time of the setting of the moon (see Figure 1.1).

The rete, called *ʿankabūt* ("rete"), has a circular hole in one half. Its dimensions are not indicated in the text but, in order to fulfil the described task, the ratio of its radius to the distance of its centre from the centre of its rete must be one to the square-root of two. In this hole the phases of the Moon are shown if the lunar plate is used, and the magnitude of the lunar eclipse, if the eclipse plate is used.

49–50. For lunar plates and mechanisms, see Donald R. Hill, "Al-Bīrūnī's Mechanical Calendar," *Annals of Science* 42 (1985): 89–163; Eilhard Wiedemann, "Ein Instrument, das die Bewegung von Sonne und Mond darstellt, nach al-Bîrûnî," *Der Islam* 4 (1913): 5–13.

The reconstruction of the lunar plate is somewhat problematic, since the text in the Istanbul manuscript seems to contain a lacuna in a crucial passage. Moreover, at some points there seems to be a confusion of diacritical points. Arabic *muḥraq* ("burned") is confused with *mukhriq* ("perforated"). In the later *Istīʿāb al-wujūh al-mumkina fī ṣanʿat al-aṣṭurlāb* of al-Bīrūnī, the Persian expression *sīm-i sūkht* ("burned silver") is used for the black coloring.[18] An analogous Arabic expression is *abār muḥraq* ("burnt lead") for a compound of lead.[19] Therefore, the reading *al-fiḍḍa al-muḥraqa* in the text of ʿAlī Ibn ʿĪsā seems preferable. The expression might have meant a silver nitrate or another compound of silver. Even so, through comparison with later texts, a tentative reconstruction can be given. On the lunar plate are two symmetrically arranged circles with the same disposition as the circular hole in the rete. They are filled with a black colour. The two circles are connected by a circular path, which is silver-burned. A third circle is called "circle of the sphere of the sun." This circle must be opposite to the circular path and, if it has the same size as the other two circles, it must be tangent to the two circles. The purpose of the circle is not defined. The text indicates that the third circle is close to the suspension of the instrument, and this fixes the orientation of the entire plate (see Figure 1.2).

The function of the first two black circles and the silvery circular path in between them seems clear. The scales on the rim determine that 180° degrees correspond to one lunar month. Therefore, by turning the rete by 180°, the phases of the moon during one month appear in the circular hole.

On the other side of the plate another layout is described in the text. A circle is mentioned that has a scale with twelve units on it by which the eclipsed part of the moon is measured during an eclipse. Obviously, these are the traditional twelve fingers or digits by which the magnitude of an eclipse is measured. Later the text refers to it as the sphere of the moon. It is said that at the right side of the circle the value of the distance of the Moon from the node appears, but no details are given. In the last part of the text where the instruction is given concerning how to determine the size of an eclipse, it becomes clear that there must be a special scale on which the distances of the moon from the node are layed out, and that the rete must be turned until the pointer called "pointer of the node" indicates the correct value (see Figure 1.3).

The text also mentions a sphere of the dragon that is said to be movable relative to the sphere of the moon. It is a part of the rete and has the function of

---

18 Al-Ḥusaynī, *Istīʿāb al-wujūh al-mumkina fī ṣanʿat al-aṣṭurlāb*, 230; cf. Wiedemann, "Ein Instrument, das die Bewegung von Sonne und Mond darstellt, nach al-Bîrûnî," 9.

19 Fabian Käs, *Die Mineralien in der arabischen Pharmakognosie* (Wiesbaden: Harrassowitz 2010), vol. 1, 203.

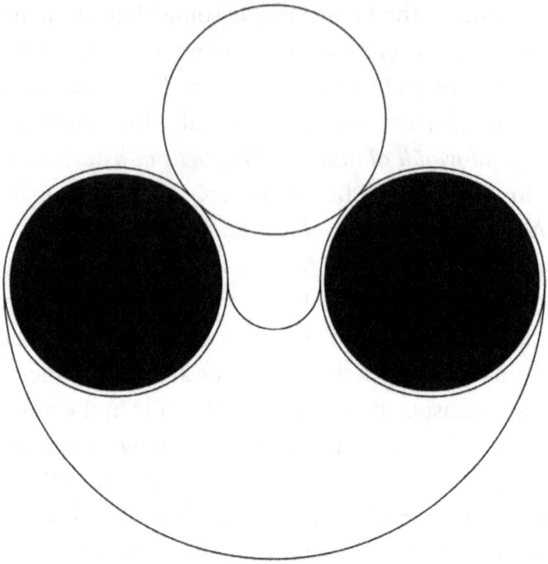

FIGURE 1.2 *The plate of ʿAlī ibn ʿĪsā.*

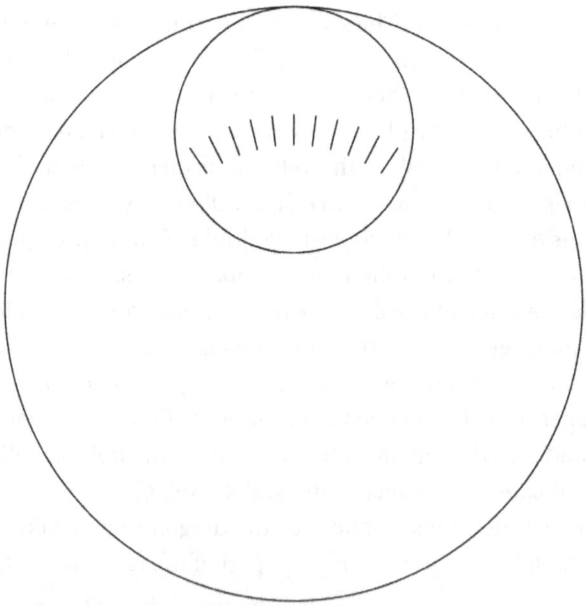

FIGURE 1.3 *Other side of the plate of ʿAlī Ibn ʿĪsā.*

covering the circle of the moon by the desired magnitude. The only other part of the instrument described in the text is a standard diopter, used for altitude measurements, and there is a short chapter on timekeeping. No non-standard parts, gears, or other more complex mechanical devices are mentioned.

There is an explicit statement in the text that the positions of sun, moon, and lunar node should be taken from a *taqwīm*, that is, an astronomical ephemeris. In recent years, a number of fragments of Arabic ephemerides have been discovered in papyrological collections.[20] The oldest known item is an ephemeris for the Persian year 300 (931–932 CE).[21] The daily positions of sun, moon, planets, and the lunar node for a Persian month are listed on a double page. The approximate time of conjunction or opposition of sun and moon can be calculated by simple interpolation. The distance of the moon from the node is found by a simple subtraction of the values. Therefore it affords no higher capabilities to predict the conditions of a lunar eclipse. This is not possible in the case of a solar eclipse, since its conditions vary with the observer's geographical position.

The instrument described by ʿAlī ibn ʿĪsā is mainly a tool for visualization, not for calculation. It provides only a rough idea to the astronomical phenomena, and precision is not intended.

---

20 Johannes Thomann, "An Arabic Ephemeris for the Year 954/955 CE and the Geographical Latitude of al-Bahnasā/Oxyrhynchus (P.Stras. Inv. Ar. 446)," *Chronique d'Égypte* 88.176 (2013): 385–396; Johannes Thomann, "An Arabic Ephemeris for the Year 1026/1027 CE in the Vienna Papyrus Collection," in *Scientific Cosmopolitanism and Local Cultures: Religions, Ideologies, Societies. 5th International Conference of the European Society of the History of Science, Athens 1–3 November 2012. Proceedings*, ed. Gianna Katsiampoura (Athens, Hellenic Research Foundation, 2014), 54–60; Johannes Thomann, "The Arabic Ephemeris for the Year 1149/1150 CE (P.Cambridge UL Inv. Michael. Chartae D 58) and the Arabic Baḫnīṭas, Greek Παχνίτης and Coptic ⲡⲁϣⲟⲛⲥ," *Chronique d'Égypte* 90.179 (2015): 207–224; Johannes Thomann, "Ephemeride für das persische Jahr 363 (994/995 n. Chr.)," in *Orakelspüche, Magie und Horoskope. Wie Ägypten in die Zukunft sah*, ed. A. Zdiarsky (Wien: Phoibos, 2015), 136–137; Johannes Thomann, "Ephemeride für das persische Jahr 371 (1002/1003 n. Chr.)," in *Orakelspüche, Magie und Horoskope. Wie Ägypten in die Zukunft sah*, ed. A. Zdiarsky (Wien: Phoibos, 2015), 137–138; for a general overview see: Johannes Thomann, "Astronomie und Astrologie in arabischen Dokumenten," in *Orakelspüche, Magie und Horoskope. Wie Ägypten in die Zukunft sah*, ed. A. Zdiarsky (Wien: Phoibos, 2015), 63–70.

21 Johannes Thomann, "An Arabic Ephemeris for the Year 931–932 CE," in *From Bawīṭ to Marw. Documents from the Medieval Muslim World*, ed. A. Kaplony, D. Potthast, and C. Römer (Leiden: Brill, 2015), 115–153.

An Anonymous Redaction of ʿAlī ibn ʿĪsā's Epistle (II)

An anonymous text with the title *Bāb fī ʿamal bi-l-ṣafīḥa al-qamariyya* (Chapter on the use of the lunar plate) contains in its second part four sections on an eclipse plate.[22] It was identified with ʿAlī ibn ʿĪsā's epistle.[23] However, a collation of the two texts reveals that the Cairo manuscript contains a redaction of ʿAlī ibn ʿĪsā's epistle in which some passages reappear with only minor variants, while other passages are replaced by a different text.[24] In order to provide full transparency, the Cairo version is edited and translated separately in Appendix B.

Nasṭūlus on the Eclipse Plate (III)

A text on a similar device is preserved in a single manuscript in Mumbai.[25] Its title, "Epistle on the Practice of Using the Plane Lunar Astrolabe," indicates that it describes mainly the use of the instrument, not its construction. It is said that the instrument was invented by Muḥammad ibn Muḥammad, called Nasṭūlus al-Asṭurlābī. Nasṭūlus is known as an important astrolabe maker and he is also credited as an inventor of an eclipse plate by al-Bīrūnī.[26] Three astrolabes and an astrolabe-like instrument with his signature are preserved.[27]

---

22 MS Cairo, Dār al-Kutub, Zakī 706, ff. 20r–22r, seventeenth century CE. Description: David King, *Fihris al-makhṭūṭāt al-ʿilmiyya al-maḥfūẓa bi-Dār al-Kutub al-Miṣriyya* (al-Qāhira: Al-Hayʾa al-Miṣriyya al-ʿĀmma li-l-Kitāb, 1981–1986), 1:666–667, 2:447.

23 Rosenfeld and İhsanoğlu, *Mathematicians, Astronomers, and Other Scholars*, 28.

24 I thank Flora Vafea for providing digital images from a microfilm of MS Zakī 706, also kept in the Dār al-Kutub Library in Cairo.

25 MS Mumbay, Mulla Firuz Library, 86 ff. 44v–45v, twelfth century CE. Descriptions: Edward Rehatsek, *Catalogue raisonné of the Arabic, Hindostani, Persian and Turkish Mss. in the Mulla Firuz Library* (Bombay: Mulla Firuz Library, 1873), 43–44; Sezgin, *Geschichte des arabischen Schrifttums: Band VI*, 288. I thank Prof. Sezgin for providing copies from a microfilm kept in the Institut für Geschichte der arabisch-islamischen Wissenschaft, Frankfurt.

26 Muḥammad Akbar Jawādī al-Ḥusaynī, *Istīʿāb al-wujūh al-mumkina fī ṣanʿat al-asṭurlāb, allafahū Muḥammad Ibn Aḥmad Abū Rayḥān al-Bīrūnī*, 2nd ed. (Mashhad: Majmaʿ al-Buḥūt al-Islāmiyya, 2010), 229.

27 Kuwait, Islamic Archaeological Museum, inv. no. LNS 36M, International Instrument Checklist (IIC) #3501; Cairo, Museum of Islamic Art, inv. no. 15351, IIC #1130 = #4023; see David A. King, *In Synchrony with the Heavens: Studies in Astronomical Timekeeping and Instrumentation in Medieval Islamic Civilization*, 2 vols.. Vol. 1: *The Call of the Muezzin*. Vol. 2: *Instruments of Mass Calculation* (Leiden: Brill, 2004–2005), 2:473–484. Private collection (Switzerland), cf. David A. King, "Two Newly Discovered Astrolabes from ʿAbbasid

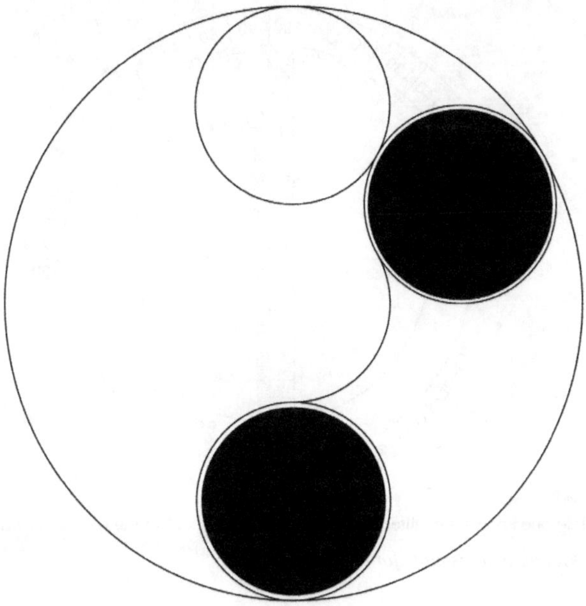

FIGURE 1.4 *The Plate of Nasṭūlus.*

In contrast to the instrument of ʿAlī ibn ʿĪsā, where 180° degrees represented one lunar month, the user is told to divide the lunar elongation by three. This presupposes a different design. As it seems, the astrolabe is supposed to have a standard scale of twelve zodiacal signs with 30° each on its rim. An entire month then is represented by an arc of 120° degree. Instead of a particular scale of days of the month, the user is told to multiply the day of the Arabic month by four, in order to get the desired arc in degrees. Further, the user is told to look up the lunar latitude in a *zīj*.

Again, a circular perforation in which the phases of the moon are displayed is mentioned. The design of the plate consists of two circles of black colour, and a circle called "sphere of the sun," which is adjacent to the second circle. The design is similar to that of ʿAlī ibn ʿĪsā, but the circles are smaller in comparision with the plate, and cover a sector of 120°. The sphere of the Sun is probably placed opposite to the first black circle (see Figure 1.4). This agrees with the design which al-Bīrūnī ascribes to the instrument of Nasṭūlus in his

Baghdad," *Suhayl* 11 (2012): 103–116 (104–108). Doha, Museum of Islamic Art, inv. no. MW278.2006, cf. David A. King, "An Instrument of Mass Calculation Made by Nasṭūlus in Baghdad ca. 900," *Suhayl* 8 (2008): 93–119.

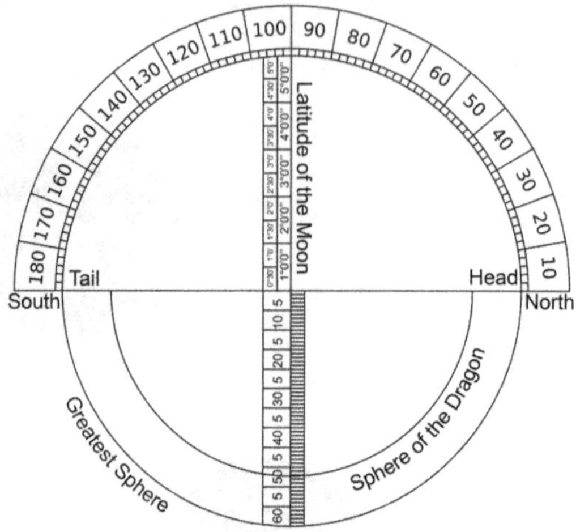

From this circle one knows the latitude of the Moon and the size of the eclipse by way of geometry

FIGURE 1.5 *An anonymous plate for eclipses.*

Istīʿāb.[28] However, al-Bīrūnī's description may contain additions by the other inventors mentioned by al-Bīrūnī.[29]

A Graphical Device for Eclipses (IV)

The third text to be presented here is preserved in an eleventh-century manuscript in Leiden that contains a number of works of al-Sijzī, Ḥabash al-Ḥāsib, and other authors.[30] Therefore, it seems likely that the anonymous text belongs to the tenth century. It is a short text covering one page. Fortunately there is an illustration in the manuscript, which makes the reconstruction easy (see Figure 1.5).

There is no statement concerning the material of the device. It could be made of paper, metal, or other materials, since it has no moveable parts. The operations are executed with a compass, and positions on the scales are marked, probably with wax. The elongation of the Moon from the lunar node

28 Al-Ḥusaynī, *Istīʿāb al-wujūh al-mumkina fī ṣanʿat al-aṣṭurlāb*, 229–236.

29 Al-Ḥasan Ibn Muḥammad al-Ādamī and ʿUṭārid Ibn Muḥammad al-Ḥāsib, see al-Ḥusaynī, *Istīʿāb al-wujūh al-mumkina fī ṣanʿat al-aṣṭurlāb*, 229.

30 MS Leiden, UB Or. 168, ff. 79v–80r, eleventh century CE. Description: Catalogue, available online at https://socrates.leidenuniv.nl (accessed 18 September 2014).

has to be marked on the scale on the rim. Next, the horizontally corresponding point on the upper scale in the middle of the circle has to be marked. This is the lunar latitude. This value has to be found in the lower central scale, which is divided into sixty arc-minutes. The inner half-circle is called "sphere of the dragon" and makes the limit within which a total lunar eclipse occurs. If the lunar latitude is greater, the eclipse will be partial, or, if it is greater than sixty minutes, no eclipse will occur. The instrument is simple and designed for a single limited task. Again, visualization seems to be the main purpose, since for higher precision a numerical table would do better and would not cover more space. Therefore it seems likely that it was used for didactic purposes.

Later Texts on Eclipse Devices

The comprehensive book on the astrolabe of al-Ṣūfī (903–986 CE), which exists in three versions, was the subject of Flora Vafea's PhD thesis.[31] Chapters 149–153 of Version T on lunar and solar eclipses contain detailed instructions on how to find the conditions of an eclipse by means of an astronomical table and a standard astrolabe.[32] There is no description of specialized instruments for eclipses.[33] Version S contains no corresponding section, and in version G chapters on eclipses might have been part of the lost first part.[34]

The book of al-Bīrūnī (973–1048 CE) on the construction of astrolabes contains two chapters on a lunar box and an eclipses plate.[35] For the eclipse plate three creators are mentioned: Nasṭulus al-Asṭurlābī, al-Ḥasan Ibn Muḥammad al-Ādamī and ʿUṭārid Ibn Muḥammad al-Ḥāsib.[36] The design is similar to that which is described in text III, but the description of its constructions is more detailed and precise in al-Bīrūnī's text.[37] In his monumental work on astronomical instruments, al-Marrākushī (fl. thirteenth century CE) included a

31 Flora Vafea, *Les traités d'al-Ṣūfī sur l'astrolabe* (Ph.D. Diss, University Paris 7—Denis Diderot, 2006).

32 Flora Vafea, "ʿAbd al-Raḥman al-Ṣūfī: Study of the Lunar and Solar Eclipses with the Astrolabe," in *Navigating across Mathematical Cultures and Times*, ed. I.M. Vandoulakis and Liu Dun (Singapore: World Scientific, forthcoming).

33 Muḥammad ʿAbd al-Muʿīd Ḫān, *Kitāb al-ʿamal bi l-asṭurlāb li-l-ʿallāma ʿAbd al-Raḥmān Ibn ʿUmar al-Ṣūfī* (Ḥaydarābād : Dāʾirat al-Maʿārif al-ʿUthmāniyya, 1962), 117–127.

34 Vafea, *Les traités d'al-Ṣūfī sur l'astrolabe*, 582–588.

35 Al-Ḥusaynī, Istīʿāb al-wujūh al-mumkina fī ṣanʿat al-aṣṭurlāb, 219–236; see Flora Vafea, "Al-Bīrūnī: The Plate of the Eclipses," *Suhayl* 15 (2016–2017): 297–354.

36 Al-Ḥusaynī, *Istīʿāb al-wujūh al-mumkina fī ṣanʿat al-aṣṭurlāb*, 229.

37 See above and Appendix C.

paraphrase of al-Bīrūnī's chapter on the eclipse plate, but did not add anything new to it.[38]

An ambitious approach to an eclipse computer was made by Ghiyāth al-Dīn al-Kāshī (1380–1429 CE). In his work on the equatorium, which exists in two Arabic versions and a Persian translation, he describes a plate for a geometrical solution for the magnitudes of eclipses and their duration.[39] By means of a ruler, numerical values can be obtained. His aim was to construct a device which made the use of tables superfluous, and he tried to provide models which were designed for precision, not only for a convenient visualization. At the same time, in Europe attempts in the same direction were undertaken by Richard of Wallingford (1292–1336 CE).[40] A sole Plate with eclipse scales of the unique medieval *albion*, constructed after the work of Richard of Wallingford, has survived.[41] Specialised instruments designed exclusively for eclipses were described by Petrus Philomena (fl. 1286–1303 CE). Based on astronomical tables available around 1300 in Paris, he constructed a multi-plate device for eclipses, which he called *eclipsorium*.[42]

Islamic Astrolabes

There is one Islamic astrolabe that can give a realistic impression of the rete in the three first texts: the famous astrolabe with a gear mechanism in Oxford.[43]

---

38 Fuat Sezgin, *Jāmiʿ al-mabādiʾ wa'l-ghāyāt fī ʿilm al-mīqāt. Comprehensive Collection of Principles and Objectives in the Science of Timekeeping by al-Ḥasan b. ʿUmar al-Marrākushī*, 2 vols. (Frankfurt: Institute for the History of Arabic-Islamic Science, 1984), 2: 119–123.

39 Edward Stewart Kennedy, *Planetary Equatorium of Jamshid Ghiyath Al-Din Al-Kashi <d. 1429>: an Ed. of the Anonymous Persian Manuscript 75 <44b> in the Garrett Collection at Princeton University; Being a Description of Two Computing Instruments, the Plate of Heavens and the Plate of Conjunctions* (Princeton, NJ: Princeton University Press, 1960), 222–238.

40 John David North, *Richard of Wallingford. An Edition of his Writings* (Oxford: Clarendon Press, 1976), 1: 334–339, 364–371, 376–379.

41 Rome, Osservatorie astronomico di Monte Mario, cf. North, *Richard of Wallingford*, 3: plate XXIV.

42 Fritz Saaby Pedersen, *Opera Petri Philomenae* (Copenhagen: Societas linguae et litterarum Danicarum, 1983), 473–512; Fritz Saaby Pedersen, *Tractatus instrumenti eclipsium* (Copenhagen: Institut du moyen-âge grec et latin, 1978), 23–48.

43 Oxford, Museum of the History of Science, Inv. No. 48213.

The circular opening is smaller than those described in the texts because the lunar phases are produced by an eccentric wheel.

Concerning the display of eclipses there are only faint echoes of the devices described in the texts. An astrolabe that was acquired by Willy Hartner for the Frankfurt Institute for the History of Science has on the back of its mater a special scale which indicates the lunar latitude for a given lunar elongation.[44] As Silke Ackermann pointed out, this astrolabe was made in the twelfth century and reworked in the seventeenth century.[45] There are two more astrolabes in Oxford with similar designs. One has the design on the back of the mater, as does the Frankfurt astrolabe.[46] The other has it on a separate plate.[47]

Conclusion

The four earliest texts on astrolabe-like eclipse computers initiated a series of attempts to transform the complex Ptolemaic model of eclipses into simple and easy-to-grasp models. None of these attempts seem to reach the ingenuity of Archimedes and his famous globe. For this, we have to wait until the sixteenth century CE when European clockmakers tried to trump each other in complexity. However, even in such an outstanding piece as the small equation clock by Jost Bürgi—the first that produced elliptic movement—we can discover a display of the lunar phases that is almost identical in design with the *ṣafīḥa qamariyya* described seven hundred years earlier.[48] Simplicity and complexity can live in harmony together; neither of them should be blamed by historians of science.

---

44 Frankfurt, Museum für angewandte Kunst (on loan from Frankfurt University), IIC #3922, cf., Silke Ackermann, "Mutabor: Die Umarbeitung eines mittelalterlichen Astrolabs im 17. Jahrhundert," in *Ad Radices. Festband zum fünfzigjährigen Bestehen des Instituts für Geschichte der Naturwissenschaften der Johann Wolfgang Goethe-Universtät Frankfurt am Main*, ed. A. von Gotstedter (Stuttgart: Franz Steiner Verlag, 1994), 193–209, 198; King as in n. 27, 2: 372–374.

45 Ackermann, n. 44.

46 Oxford, Museum of the History of Science, Inv. No. 40744 (Isfahan 1712/1713 CE), IIC #38; King as in n. 25, 2: 373; Ackermann as in n. 41, 198.

47 Oxford, Museum of the History of Science, Inv. No. 40833 (Isfahan early eighteenth century CE), IIC #1087; Ackermann, "*Mutabor*," 199.

48 Fritz Staudacher, *Jost Bürgi, Kepler und der Kaiser. Uhrmacher, Instrumentenbauer, Astronom, Mathematiker, 1552–1632* (Zürich: Verlag Neue Zürcher Zeitung, 2013), 90.

Acknowledgements

My warmest thanks go to Flora Vafea, who helped me a lot in reading the Arabic texts and improving the translations. I also thank James Weaver who made valuable suggestions to make the English translations more readable.

Appendix A: ʿAlī ibn ʿĪsā's *Risāla*

Arabic Version[49]

رسالة علي بن عيسى الحاسب في معرفة العمل بالصفيحة القمرية والحقّ<ـة> الكسوفيّة . رحمه الله ورضى عنه |304v|

1 بسم الله الرحمن الرحيم 2 ثقتي بالله
3 قال علي بن عيسى، رحمه الله 4 بعد حمد الله والثناء عليه لما هو أهله
5 والصلاة على نبيّه وعلى أهله، 6 أمّا معرفة هذه الصفيحة والعمل بما فيها،
فإنّ حجرتها مقسومة بأربعة أقسام، 7 نصفها مقسوم كح قسما وهي
ثمانية وعشرون يوما، 8 وابتداء الأقسام فيها، إذا أخذتها بيدك وجعلت علّاقتها
تلقاء وجهك، يكون ابتداء أحد أرباعها عن يمينك صاعدا إلى العلّاقة. 9 وهو
مقسوم ١٤ قسما. 10 ثمّ من الخطّ الذي نزل من العلّاقة في الربع الأيسر
تمام كح يوما، وهو عدد ما يظهر للناس القمر. 11 وأما النصف الآخر،
وهو أقسام ما يغيب عليه في أول ليلة من كسور الساعات. 12 وابتداء من الربع الذي يلي اليسار 13
المكتوب فيه عدد آ 14 فألف عدد ساعة 15 وه كسر الساعة، 16 وهو خمسة أسباع ساعة، 17 حتّى
يتمّ أقسام هذا الربع من الساعات تمام يبّ ساعة، التي هي ساعات الليل وكسورها، 18 حتّى يتم ما
في حجرة القمر، 19 وهي الحلقة المحيطة بصفيحة العنكبوت، 20 إلى أسفل العنكبوت، فافهم ذلك.
|305r| 21 باب في معرفة العنكبوت وأقسامها

---

49 Brackets are used according to the Leiden convention: {abc} letters in the text considered erroneous and superfluous by the editor; <abc> characters erroneously omitted by the scribe, restored or corrected by the editor. However, since dots below letters should not be used for Arabic texts, «abc» is used instead for characters damaged or otherwise unclear in the text. The superscript numbers correspond to those in the English translation in order to facilitate to compare it to the Arabic text. The Arabic text follows closely the manuscript even in cases where it deviates from standard grammar, or awkward phrases occur. Some minor writing elements like hamza or shadda are normalized according to standard orthography.

[22] فالعنكبوت هي الصفيحة المشبكة، التي في نصفها ثقب، يدور حول الثقب فلك يوري نور القمر. [23] ومن هذا الثقب يعلم، إذا أدير العنكبوت على الصفيحة المرقوم فيها الفضّة والسواد، الذي هو تحت العنكبوت.

[24] وفيها دائرتان فيها فضّة محرقة، [25] والذي فوق هذه الدوائر دائرة فلك الشمس، مما يحاذي العلّاقة.

[26] وبعده، إلى ناحية اليسار، إلى الخطّ الرابع الأيسر، عدد ساعات طلوع القمر بالنهار؛ [27] وفي الربع الذي بعده، عدد ساعات طلوع القمر بالليل.

[28] وفي العنكبوت، من الجانب المحرق دائرة فلك الشمس [29] وعليه مكتوب «فلك الشمس [30] {وعليه مكتوب فلك الشمس}. وعليه أيضا في هذا النصف المحرق موري {موري} الساعات، [31] وهي الشظية التي عليها مرقوم «موري الساعات»؛ [32] وعلى الخطّ الذي يقع العنكبوت بنصفين، [33] ويبتدئ من الخطّ الذي يقع فلك الشمس، [34] ويمرّ بمركز العنكبوت <وب>دائرة فلك القمر، يؤدي نور القمر، [35] حتّى يتمّ الخطّ عليه مرقوم موري القمر. [36] فاعرف علم ما وصفت لك، [37] ليسهل عليك العمل بالصفيحة إن شاء الله. <36.1-36.2>

[38]|305v| وفي الوجه الآخر من الصفيحة، دائرة مقسومة باثنى عشر قسما، [39] يعلم منها ما ينكسف من القمر، إذا كان له كسوف. [40] وفي الجانب الأيمن من هذه الدائرة عدد ما يكون بعد العقدتين والقمر عند الكسوف، [41] فهذه معرفة ما في الوجه الآخر من الصفيحة.

وفي ظهر هذه الصفيحة القمريّة عضادة مثل عضادة الأصطرلاب، [42] وفيها خطوط يعلم منها ما يمضي من النهار وما بقي منه، [43] إذا أقيس على ما أصف لك فيما بعد، إن شاء الله. [44] وفي <ربع> {رفع} من أرباعه أقسام الارتفاع {الارتفاع}، كما يكون في ظهر الأسطرلاب، لأخذ الارتفاع، [45] وفي الربع المقابل له قسم الظلّ. [46] وهو جميع ما يحتاج إلى معرفة ما في هذه الصفيحة، [47] فافهم ذلك وقس عليه إن شاء الله.

[48] **الباب الأول:** في معرفة أخذ الارتفاع.

[49] وهو كما يوخذ بالأسطرلاب سوي، [50] وذلك أن تعلق الصفيحة بيمينك، [51] وتحرّك العضادة بيسارك، وتحاذي بمنكبك الأيسر الشمس. [52] وترفع العضادة أو تنزلها، حتّى يوافق أجزاء الارتفاع، وينفذ نور الشمس من الثقب الأعلى إلى الثقب الأسفل. |306r| [53] ثمّ تنظر رأس العضادة المحدّد ما وافق من الأجزاء، أعني أجزاء الارتفاع، [54] فما كان فهو الارتفاع لذلك الوقت، إن شاء الله.

[55] **الباب الثاني:** في معرفة ما مضى من النهار من ساعات معوّجة، وهي الزمانيّة [56] التي يكون بها النهار يبّ ساعة والليل كذلك، طال أم قصر.

[57] فإذا أردت ذلك، فاعرف ارتفاع نصف نهار يومك ذلك، [58] وضع رأس العضادة المحدّد على غاية الارتفاع لنصف نهار يومك. [59] واستقبل بهذه العضادة عين الشمس، حتّى ترى ظلّ هذه

العضادة على باطن العضادة مستويا، غير منحرف إلى جهة من الجهات. 60 فإذا عرفت الظلّ أين وقع من العضادة، فعدّ الساعات، 61 ونبتدئ بعددها من الهدف الذي قسمت به الظلّ خطوطا قد قطعت باطن العضادة بالعرض، 62 فيما بينهما عدد الساعات بحروف الجمّل آ ب ج د ه و. 63{بـ} الموضوع الذي وقع عليه الظلّ، هذه العضادة هو ما مضى من النهار من ساعة، 64 فإن كان القياس قبل نصف النهار، فعدّ الخطوط التي على العضادة في الجانب المُوَقّع عليه بحروف |306v| الجمّل آ ب ج د ه و، 65 وإن كان بعد نصف النهار، فعدّ الخطوط من العضادة من الجانب المرقوم عليه عدد الساعات بحروف الجمّل هي ز ح ط ي يا يب.

66 **الباب الثالث**: في معرفة الظلّ.

67 إن كان الارتفاع أقلّ من مه، فانظر الرأس المحدّد من العضادة على كم وقع من الظلّ، 68 فما كان من الأجزاء المقسومة الظلّ 69 فاقسم عليه قسم طول د، 70 فما خرج من الظلّ فهو طول الظلّ في الوقت الذي أردت. 71 وإن كان الارتفاع مه، فإنّ الظلّ في الوقت الذي أردت {وإن كان الارتقاع مه} فالظلّ يب سوي. 72 وإن كان الارتفاع أكثر من مه، 73 فالظل ما وقع عليه رأس العضادة المحدّد، فالآن تقسم عليه ق م د، 74 فاعلم ذلك إن شاء الله.

75 **الباب الرابع**: في معرفة ما يظهر من القمر من النور في كلّ ليلة من أيّام الشهر. 76 إذا أردت ذلك، فاردد العنكبوت، حتّى يقـ{ـط}ـع فلك الجوزهر <على> {و}فلك الشمس، 77 ويقع فلك موري |307r| نور القمر على الفضّة المحرقة، 78 ولا تبيّن من الفضّة البيضاء شيء من ثقب موري نور القمر. 79 فإذا فعلت ذلك، فادر العنكبوت جزءا واحدا من الأجزاء المقسومة في العنكبوت في طرف من الأقسام 80 التي بينهما آ <ب> {ح} <ج> و ط يب يه يح، وهو عدد أيّام الشهر القمري 81 إذا كان الشهر ثلـ<ـا>ثين يوما، والجانب الآخر من الحاشية أيضا 82 إذا كان الشهر تسعة وعشرين يوما. فإذ أنت حرّكت الشبكة مقدار جزء واحد من أجزائها، 83 حتّى يكون الجزء مع الخطّ الذي أخرج من العلّاقة إلى وسط الصفيحة، 84 فحينئذ يظهر لك ما في القمر من النور لليلة واحدة. 85 وكذلك تدير كلّ يوم لتنظر ما يزيد من النور في القمر، إلى أن يتمّ نوره في مقابلة الشمس ويكون بدرا تامّا. 86 ثمّ نبتدئ بالنقصان، 87 فلا يزال ينقص من نوره في البياض، ويزيد فيه السواد، حتّى لا يبقى من نوره شيء، 88 ويكون قد دار فلك موري نور القمر حتّى طابق فلك الشمس في الاجتماع. /معها\ 89 فافهم ذلك وقس عليه، إن شاء الله عزّ وجلّ. 90 وإذا شئت أن تعرف ما يكون في الشهر الآخر، فالعمل فيه مثل الأول سواء، وبالله التوفيق.

91 |307v| **الباب الخامس**: في معرفة ما في القمر من النور.

92 فإذا أردت ذلك، فانظر، 93 إذا كان طلوعه على ساعة من النهار، 94 وأكثر من ذلك أو أقلّ، 95 وأردت أن تعمل به، 96 فضع موري الساعات على ساعة من ساعات النهار المقسومة في الربع الأيسر

من الصفيحة، 97 أو على ما أردت من الساعات، فما وجدت قد ظهر من فلك موري نور القمر من الفضّة البيضاء، 98 فهو مقدار ما يرى فيه <من> النور. 99 وكذلك فاعمل إذا أردت علم مقدار ما يرى فيه من النور على ساعات من الليل، 100 فضع موري الساعات على ما أردت من ساعات الليل، 101 ثمّ انظر إلى فلك موري نور القمر، فما يظهر فيه من الفضّة البيضاء، فهو ما في القمر من النور، 102 فاعلم ذلك وقس عليه.

103 **الباب السادس:** في معرفة على كم يغيب القمر أوّل ليلة من الشهر.

104 وذلك بأن تضع موري القمر -وهو الخطّ الذي يبتدئ من طرف الصفيحة، أعني صفيحة العنكبوت، |308r| وتمرّ بفلك الجوزهر، 105 حتّى يقـ{ـطـ}ـع فلك الشمس بنصفين- إلى أن أبتدئ <بـ>ـالعدد من كح إلى عن يمينك في النصف الأعلى من المجرة. 106 فإنّه يوافي عند ذلك موري نور القمر من يسارك في مقابلته سوا. 107 ثمّ حرّك العنكبوت، حتى يقع موري نور القمر على قسم واحد، 108 وهو المكتوب فيه ا، 109 وانظر إلى مقابلة الخطّ من العنكبوت، 110 فإنّه يقع على ما يغيب عليه القمر، 111 وهو و ستّة أسباع ساعة. 112 وكذلك قائما كان تحرك العنكبوت، وتضع موري القمر على عدد الأيّام وأيضا 113 فإن كان يوم تقسم المجرة بنصفين 114 لكي يكون اليوم نصفين. 115 فإذا وضعت موري القمر على ما أردت من الأيام وكسور الأيّام، 116 فانظر إلى مقابلة الخطّ من الجانب الآخر على كم وقع من عدد الساعات وكسورها، 117 فما كان فهو ما يغيب عليه القمر من الليل، 118 وبالله التوفيق.

119 **الباب السابع:** إذا أردت أن تعلم على كم يغيب القمر بالنهار من الساعات وكسورها.

120 فاعلم أنّه لا يكون ذلك إلا بعد انتصاف الشهر. 121 فإذا أردت ذلك، فإذا أنت |308v| وضعت موري القمر على ما مضى من أيّام الشهر، 122 فإنّ الخطّ المقابل له يقع على الساعات وكسورها، 123 فاسقط منه يبّ ساعة، 124 وما بقي فعليه يغيب القمر من النهار، 125 فاعلم ذلك إن شاء الله. 126 فإذا كان تمام كح يوما، يكون القمر قد دخل تحت الشعاع، 127 فلا يحتاج له إلى عمل 128 حتّى يفارق الشمس، 129 ويبتدئ العمل فيه بعد ذلك على معنى ما مضى من القول المذكور فيه المنسوب إليه، 130 فاعلم ذلك.

131 **الباب الثامن:** في معرفة علم ما يكون من كسوف القمر، وكم مقدار ما ينكسف منه.

132 فانظر في التقويم كم بين الشمس والقمر من الدرج في وقت الاستقبال، 133 فإذا عرفت ذلك، وأردت أن تعرف الكسوف، 134 فانظر إذا كانت الشمس والقمر وقت الاستقبال، <ما بين القمر> وبين الرأس والذنب. 135 فإن كان أكثر من يبّ درجة، 136 فإنّ القمر لا ينكسف، 137 وإن كان أقلّ من يبّ درجة، فإنّ القمر ينكسف، 138 وبالله التوفيق.

139 |309r|. الباب التاسع: 140 فإذا علمت أنه ينكـ{ـشـ}<ـسـ>ـف 141 وأردت تعلم كم ينكسف منه، 142 فاقلب الصفيحة التي فيها الفضّة المحرقة، 143 وفيها دائرة القمر مقسومة باثنى عشر قسما. 144 فاطبق العنكبوت على وجهها، 145 ورد القطب والفرس على حاله الأول. 146 وادر العنكبوت حتّى ينطبق دائرة الفلك الشمس على دائرة القمر انطباقا، 147 حتّى لا يبان منه شيأ. 148 ثمّ ادر العنكبوت حتّى يقع طرف الجوزهر على ما يكون بين القمر وبين الجوزهر. 149 فإنه عند ذلك تتنحـ{ـا}<ـى> دائرة الشمس عن دائرة القمر، 150 ويظهر مقدار ما ينكسف من دائرة القمر، 151 ويكون ذلك مقدار ما ينكسف منه، 152 فافهم ذلك. 153 وإن ظهرت دائرة القمر من تحت فلك الجوزهر كلّه، 154 فإنّ القمر ينكسف كلّه؛ 155 وإن بقي منه شيء مغطّا، فهو مقدار ما يبقى لا ينكشف ينكـ{ـشـ}<ـسـ>ـف من القمر، 156 فاعلم ذلك وقس عليه. وإن انكشف انكـ{ـشـ}<ـسـ>ـف كلّه، 157 فما زاد على مقدار دائرة القمر، فهو ما يمكث القمر في الكسوف. 158 وقد عرفت عند ذلك كمّية الكسوف والمكث، من غير حساب ولا تعب، إن شاء الله. 159 وبه الهداية والإرشاد |309v| إلى قصد السبيل، 160 وهو حسبنا ونعم الوكيل .

161 تمّ شرح العمل بالصفيحة القمريّة المنسوبة إلى علي بن عيسى الحاسب، رحمه الله تعالى.

English Translation[50]

Epistle of ʿAlī ibn ʿĪsā al-Ḥāsib on the knowledge of how to use the lunar plate and the eclipse limb—May God have mercy on him and bless him.
1 In the name of God, the Merciful, the Compassionate—2 my trust is in God.
3 ʿAlī ibn ʿĪsā—may God have mercy on him—said, 4 after praising God and extolling Him with what is due to Him, 5 and prayer for his Prophet and for his family: 6 Now, the knowledge of this plate and how to work with what is in it (i.e., its parts), [starts with the information that] its limb is divided into four parts, 7 and its [upper] half is divided into 28 parts, which are twenty-eight days. 8 [As for] the beginning of [numbering] the divisions on it, when you take it in your hand and you suspend it before your face, the beginning [of numbering] of one of its quadrants will be on your right side ascending to the suspension. 9 [This quadrant] is divided into *14* parts. 10 Next, [starting] from the line that descends from the suspension on the left quadrant, the 28 days are completed, being the number [of days] in which the moon is visible. 11 As for the other half [of the plate], which consists of the divisions for the fractions

50 Indian numerals in the Arabic text are transcribed by Arabic numerals in italics, Abjad numerals by regular Arabic numerals. Additions in the translation, which are added for the sake of clarity, are put into brackets: [abc]. Some reconstruction of words and phrases which might have been corrupted in transmission are put into parentheses: (*recte*: abc ?).

of the hours at which [the moon] sets on the first night, [12] the beginning is from the quadrant that is on the left side, [13] the number 1 is written; [14] the *alif* is the number of the hour [15] and 5 the fraction of the hour.[16] This [fraction] is five- [*recte* six-?] sevenths of an hour [for each day], [17] until the divisions of the hours of this quadrant are completed, completing 12 hours. These are the hours of the night and their fractions, [18] so that [the writing] on the limb of the moon is completed down to the lowest [part] of the rete [*recte* limb?]. [19] This [limb] is the ring that surrounds the plate of the rete [20] down to the lowest [part] of the rete. Understand this!

[21] Chapter on the knowledge of the rete and its parts.
[22] The rete is a lattice-like plate, with a hole in one of its halves. Around the pole rotates the circle indicating the light of the moon. [23] Through this hole information is presented, when the rete is rotated on the plate with the silvery and black markings on it, which is below the rete.
[24] On [the plate below the rete] are two circles on which silver is burned, [25] and that which is above the circles is the circle of the sphere of the sun situated near the suspension. [26] After it, to the left, towards the line of the left quadrant, is the numbering of hours at which the moon rises at day, [27] and on the quadrant after it the numbering of the hours at which the moon rises at night. [28] On the perforated side of the rete is the circle of the sphere of the sun. [29] On it is written "sphere of the sun." [30] On this burned [*recte* perforated] half is also the index of the hours; [31] this is a splinter on which "index of the hours" is marked. [32] On the line that stretches (*recte* splits?) the rete along two halves [33] and begins from the line which intersects the sphere of the sun [34] and goes through the center of the rete, is the circle of the sphere of the moon, which displays the light of the moon, [35] until the line on which has been written "the index of the moon" ends. [36] Take note of the knowledge of what I have described for you [37] so that the use of the plate is made easy for you, if God wishes. [38] On the other side of the plate is a circle divided into twelve parts, [39] from which one learns what is eclipsed of the moon when there is a lunar eclipse. [40] On the right side of this circle is a numbering representing the distance between the two nodes and the moon during an eclipse. [41] This is what is on the other side of the plate.

On the back of this lunar plate is an alidade similar to the alidade of an astrolabe. [42] On it are lines from which one learns what will pass of the day and what is left of it, [43] when I measure according to what I will describe to you below, if God wishes. [44] In one of its quadrants are the divisions of the altitude similar to what is on the back of the astrolabe for taking the altitude, [45] and on the opposite quadrant is the division of the shadow. [46] This is all you need to know of what is on this plate. [47] Understand it and measure accordingly, if God wishes.

[48] First chapter: On the knowledge of how to take the altitude
[49] It is done exactly as with the astrolabe: [50] you suspend the plate by your right [hand], [51] and you move the alidade with your left [hand], and you stand with your left shoulder opposite to the sun. [52] You raise or lower the alidade until the degrees of altitude fit and the light of the sun goes from the upper hole through the lower hole. [206r] [53] Next, you look at the pointed tip of the alidade, on which degree it falls among the degrees of altitude; [54] this will be the altitude for that time, if God wishes.

[55] Second chapter
On the knowledge of how many unequal hours have passed of the day, which are the seasonal [hours], [56] by which the day consists of twelve hours, and the night also, [regardless of] it being longer or shorter.
[57] If you want that, you must know the altitude at midday of your day. [58] Put the pointed tip of the alidade on the maximum altitude corresponding to the midday of your day. [59] Let this alidade face the sun, until you see the shadow of the alidade straight on the body of the alidade, without being inclined in any direction. [60] When you know where the shadow falls on the alidade, count the hours. [61] We begin to count them starting from the sighting vane which was used for dividing the shadow with lines that cut the body of the alidade transversely, [62] and between them the number of hours are [written] with the abjad numerals A B J D H W (1, 2, 3, 4, 5, 6). [63] The position of alidade where the shadow falls [indicates] the elapsed time of daytime in hours. [64] When the measurement is before midday, count the lines which are on the alidade on the side which is marked with the abjad numerals A B J D H W (1, 2, 3, 4, 5, 6). [65] When [the measurement] is after midday, count the lines on the alidade from the side on which the numbers of hours are indicated by the abjad numerals Z Ḥ Ṭ Y YA YB (7, 8, 9, 10, 11, 12).

[66] Third chapter: To know the shadow
[67] If the altitude is less than 45 [degrees], look at the pointed tip of the alidade at which division within the shadow [scale] it falls. [68] The division among the divisions of the shadow [scale], [69] by that divide the length of 4 (*recte* Divide the length of 4 by that division—among the divisions—of the shadow [scale]?). [70] What results from the shadow [scale] reveals the length of the shadow at the moment you want. [71] If the altitude is 45, the shadow at the moment you want is exactly 12 [digits]. [72] If the altitude is greater than 45, [73] the shadow is

that on which the pointed tip of the alidade falls; now divide 144 by it. [74] Know that, if God wishes.

[75] Fourth chapter: To know the light of the moon appearing each night of the days of the month
[76] If you want that, rotate the rete until the sphere of the dragon falls on the sphere of the sun, [77] and the circle indicating the light of the moon falls on the burned silver, [78] and nothing of the white silver is shown from the hole indicating the light of the moon. [79] When you have done that, turn the rete by one part of the parts which are allotted on the rete, on its rim, [80] among which are 1, 2, 3, 6, 9, 12, 15, 18, which are the numbers of days of the lunar month [81] when the month has thirty days, and also [on] the other side of the margin [82] when the month has twenty-nine days. When you rotate the rete by the amount of one of its parts, [83] aiming that the part is aligned to the line [84] which goes out from the suspension towards the middle of the plate, then the light on the moon at one [particular] night will be shown to you. [85] Rotate [the rete] likewise to observe how the light on the moon increases each day, until its light is complete in the opposition to the sun, and it will be a complete full moon. [86] After that, we start decreasing and it continues to decrease [87] its light [depicted] as the white [part], and the black [part] increases in it until nothing remains from its light. [88] The circle indicating the light of the moon may turn until it covers the sphere of the sun in the conjunction. [89] Understand that and measure according to it, if God the Glorified and Sublime wishes. [90] When you want to know what will be in another month, the work in it is exactly like that of the first. With God is the success.

[91] Fifth Chapter: To know the illuminated [part] of the moon.
[92] If you want that, look, [93] when it rises at an hour during daylight, [94] or more or less, [95] and [when] you want to work with it, [96] put the index of the hours on one of the daylight hours allotted on the left side of the plate, [97] or [put the index] on what hour you want. The [part] of white silver that you have found to be visible through the circle indicating the light of the moon, [98] will be the extent of the illuminated [part] of [the moon]. [99] Proceed similarly when you want to know the extent of the illuminated [part] of it [when the moon rises] during the hours of night. [100] Put the index of the hours on whichever hour of night you want, [101] next, look at the circle indicating the light of the moon. The [part] of white silver visible through it [depicts] the illuminated [part] of the moon. [102] Know it, and measure according to it!

[103] Sixth Chapter: To know when the moon sets the first night of the month

[104] That [you can find] by putting the index of the moon—and this is the line that begins from the extremity of the plate, i.e. the plate of the rete, and passes through the sphere of the dragon [105]—until it stretches (*recte* splits?) the sphere of the sun along two halves. I will begin with the numbering of 28, which is at your right on the upper half of the limb. [106] Thus, the index of the light of the moon comes to your left straight opposite to it. [107] Next, move the rete until the index of the light of the moon falls upon a part[108] on which 1 is written, [109] and observe the intersection of the line of the rete; [110] it falls upon [the time] at which the moon sets: [111] this is [written on the first compartment at your left as] six-sevenths of an hour. [112] As it is situated, move the rete and put the index of the moon on the number of the days and their halves, [113] when there is a day that [a division of] the limb divides into two halves, [114] so that the day consists of two halves. [115] When you put the index of the moon on the days and the fractions of days you want, [116] observe on which number of the hours and their fractions the intersection of the line on the other side falls; [117] what this shows will be [the time] at which the moon sets at night. [118] The success is with God!

[119] Seventh Chapter: If you want to know when the moon sets during daytime in hours, and fractions thereof
[120] Know that this does not occur, except after the middle of the month. [121] If you want that, you put the index of the moon on [the division corresponding to] what has passed from the days of the month. [122] The opposite line falls on the hours and their fractions. [123] Subtract 12 hours from it [124] and the remainder will be [the time] at which the moon sets during daytime. [125] Know that, if God wishes. [126] When the 28 days are completed, the moon has already gone under the rays [of the sun], [127] and no work is needed for it, [128] until it separates itself from the sun. [129] The work starts after that according to the concept of the elapsed time, as mentioned in the speech attributed to him [i.e. ʿAlī ibn ʿĪsā]. [130] Know that!

[131] Eighth Chapter: How to obtain the knowledge of whether an eclipse of the moon will happen and what the magnitude of the eclipse will be
[132] Look up in the ephemeris how many degrees there are between the sun and the moon at the time of the opposition. [133] When you know that and you want to know the eclipse, [134] look, when the sun and the moon are at the time of the opposition, <how many degrees there are between the moon> and the ascending or descending lunar node. [135] If [the distance] is greater than 12 degrees, [136] the moon will not be eclipsed. [137] If it is less than 12 degrees, the moon will be eclipsed. [138] Success is with God!

[139] Ninth Chapter:
[140] If you know that [the moon] will be eclipsed, [141] and you want to know how much of it will be eclipsed, [142] then turn upside-down the plate on which the burnt silver is, [143] and [use the face] on which the circle of the moon is divided into twelve parts.[144] Next, adjust the rete on [this] face, [145] and bring the pole and the horse to their initial positions. [146] Rotate the rete until the circle of the sphere of the sun is adjusted in congruence to the circle of the moon, [147] so that nothing is visible from it. [148] Next, rotate the rete until the index of the dragon falls on the distance between the moon and the lunar node. [149] In [the process of] that, the circle of the sun (*recte* of the dragon?) is moved away from the circle of the moon, [150] and the magnitude of the eclipse is shown on the circle of the moon; [151] this will be the magnitude of the lunar eclipse. [152] Understand that! [153] If the circle of the moon appears entirely below the circle of the dragon, [154] the moon will be entirely eclipsed. [155] If something of it remains covered, this is the amount of what remains uneclipsed of the moon. [156] Know that, and measure according to it. If it is entirely eclipsed, [157] the magnitude additional to that shown on the circle of the moon, corresponds to the duration of the totality during the eclipse. [158] You have learned by that the magnitude of the eclipse and the [duration of the] totality without calculation and pain, if God wishes. [159] With Him is the guidance and the leadership go the right way. [160] He is sufficient for us; a most excellent guardian.

[161] The commentary on the use of the lunar plate, attributed to ʿAlī Ibn ʿĪsā al-Ḥāsib—may the Exalted God have mercy upon him—has been completed.

Appendix B: Anonymous Redaction of ʿAlī ibn ʿĪsā's *Risāla*

Arabic Text[51]

1 بسم الله الرحمن الرحيم
[2] باب العمل بالصفيحة القمرية
[6] اعلم أنّ حجرتها مقسومة [8]بأربعة أقسام، [7] نصفها مقسوم كح قسما، وهي أيّام، وابتداؤها إذا أخذتها بيدك وجعلت علاقتها تلقاء وجهك بابتداء أحد أرباعها عن يمينك، صاعدا إلى العلّاقة؛ [9] فهو مقسوم ١٤ خطّا. [10] ثمّ من الخطّ الذي نزل من العلّاقة في الربع الأيسر تمام كح يوما،

51 The superscript numbers correspond to those in Appendix A. Passages which do not occur in the former text are indicated by a second number separated by a dot.

وهو عدد ما يظهر للناس القمر.11 وأمّا النصف الآخر، فهو أقسام ما يغيب عليه القمر من أول ليلة من كسور الساعات والساعات. 12 وابتداؤها من الربع الذي يلي يسارك، 13 وهو المكتوب فيه ﹷ 13.1 وَ . ثمّ البيت الثاني فيه آ ةَ ،14 فالألف ساعة15 و ه كسر من ساعة، 16 وهي خمسة أسباع ساعة، 17 حين تتمّ أقسام هذا الربع من الساعات وكسورها17.1 تمام ١٢ ساعة، 17.2 وهو ما يغيب عليه القمر من أوّل ليلة يظهر، إلى ليلة يبدر17.3 ثمّ بعد ذلك إلى تمام كدَ، ما يغيب عليه بالنهار، بعد أن يلقي منه ١٢ ساعات الليل وكسورها، 18 حتّى يتمّ ما في حجرة القمر،19 وهي الحلقة المحيطة بصفيحة العنكبوت .

21معرفة العنكبوت وأقسامها.

22هي الصفيحة التي في نصفها ثقب، تدور حول القطب، عليه مكتوب "فلك مري نور القمر".

23 ومن هذا الثقب يعلم أنّ الذي على الصفيحة المدفونة فيها الفضّة السوداء، التي تحت العنكبوت.

24 والصفحة التي تحت العنكبوت فيها دائرتان فيهما {قصة مخرقة}<فضّة محرقة>، 25 والذي يلي هذه الدائرة [..]<هي>دائرة فلك الشمس مما يحاذي العلّاقة، 26 وبعدها إلى ناحية اليسار إلى خطّ الربع الأيسر عدد ساعات طلوع القمر بالنهار، 27 وفي الربع الذي بعده عدد ساعات طلوع القمر |20v| بالليل.

28 وفى العنكبوت في الجانب المخرق دائرة فلك الجوزهر، 29 وعليه كتابة "فلك الجوزهر". 30 و فيها أيضا في هذا النصف المخرق مري الساعات. 32 وعلى الخطّ الّذي يقطع العنكبوت بنصفين 33 ويبتدئ من الخطّ الذي يقطع فلك نور القمر 34 ويمرّ بمركز العنكبوت وبفلك الجوزهر 34.1 وينتهي إلى آخرها 35 حين يتمّ الخطّ عليه مكتوب مري القمر، 36 فاعرف «كأنّما» وصف لك. 36.1 وفي العنكبوت أيضا، في طرفها، أقسام على عدد أيّام الشهر القمري، في ناحية ٣٠ يوما، 36.2 وفي الناحية الأخرى ٢٩ يوما، لما الشهر القمري كذلك يكون أبدا. 38 وفي الوجه الآخر من الصفيحة دائرة مقسومة ١٢ قسما، 39 يعرف منها ما ينكـ{شـ}<سـ>ف ينكسف من القمر. 40 في الجانب الأيمن من هذه الدائرة عدد ما يكون بين العقدتين والقمر عند الكسوف، 41 فهذا ما في الوجه الآخر من الصفيحة. وفي ظهر هذه الصفيحة القمرية عضادة 42 فيها خطوط للساعات. 44 وفي ربع من أرباعه أقسام الارتفاع، كما في ظهر الأسطرلاب، لأخذ الارتفاع، 45 وفي الربع المقابل له قسمة الظلّ. 46 فهذا جميع ما يحتاج إليه.

46.1 فصل في معرفة الساعات

46.2 ضع مري العضادة على ارتفاع نصف نهارك، 59 واستقبل بالعضادة عين الشمس، حتّى يستوي ظلّ العضادة على باطنها غير منحرف، 59.1 فما بين وقع الظلّ، بعدّ الساعات، 61 وابتداؤها

من الهدف الأعلى التي أخذت منها الظلّ، من الخطوط التي على باطن العضادة، 63 فما كان فهو الماضي من الساعات، 64 إن كان العمل قبل نصف النهار، والباقي، إن كان بعده.

66 فصل في معرفة الظل

/ 67إذا كان الارتفاع أقلّ من ٤٥، فانظر محدد العضادة على كم وقع من الظل، صح \73 فما كان فاقسم عليه ١٤٤، 73.1 فالخارج هو الظلّ. 73.2 وإن كان ٤٥

|21r|

فالظلّ ١٢. 73.3 وإن كان أكثر، فالظلّ ما وقع عليه العضادة، من غير حاجة إلى قسمة.

75 فصل في معرفة ما يظهر من نور القمر في كلّ ليلة من الشهر

76 <إذا أردت ذلك،> ادر العنكبوت حتّى يقع فلك الجوزهر على فلك الشمس، 77 ويقع مري نور القمر على الفضّة المحرقة، 78 ولا يتبيّن من الفضّة البيضاء شيء من ثقب مري القمر. 79 فإذا فعلت ذلك، فادر العنكبوت جزءا واحدا من الأجزاء المقسومة في العنكبوت، في طرفها، من الأقسام 80 التي يليها عدد أيّام الشهر القمري، 81 إذا كان الشهر ٣٠ <يوما>، والجانب الآخر من الحاشية، 82 إذا كان الشهر ٢٩ <يوما>. فإذا حرّكت العنكبوت جزءا من أجزائها، 83 حتّى يكون الجزء مع الخطّ 84 الذي قد أخرج من العلّاقة ويبعد إلى وسط الصفيحة، يظهر لك ما في القمر من النور ليلة واحدة. 85 وكذلك تدير لكلّ يوم واحد، لينظر ما فيه من النور إلى أن يبدر، 86 ثمّ يبتدئ في النقص إلى المحاق، 88 ويكون قد دار فلك <مري> نور القمر حتّى لازق فلك الشمس.

91 فصل في معرفة ما في القمر من النور في الساعات نهارا أو ليلا93 إذا كان طلوعه على ساعة من النهار، 96 فضع مري الساعات على ساعة من ساعات النهار المقسومة في الربع الأيسر، 97 أو على ما أردت من الساعات، فما وجدته قد ظهر في فلك مري نور القمر من الفضّة البيضاء، 98 فهو مقدار ما يرى من النور فيه. 99 وعلى هذا، إذا أردت ما فيه من النور في ساعات الليل، 100 ضع مري الساعات على ساعات الليل، 101 وانظر إلى فلك مري نور القمر، فما ظهر من الفضّة البيضاء، فهو ما في القمر من النور.

|21v|

103 فصل في معرفة القمر على كم يغيب.

إذا أردت أوّل ليلة من الشهر، 104 فضع مري القمر -أيْ الخطّ المبتدئ من طرفي صفيحة العنكبوت، المارّ بفلك الجوزهر، 105 حتّى يقطع فلك الشمس القمر بنصفين- على ابتداء العدد من كح، الذي عن يمينك في النصف الأعلى من المجرة. 106 فإنّه يوافي عند ذلك مري نور القمر

من يسارك في مقابلته سواء. 107 ثمّ حرّك العنكبوت حتّى يقع مري القمر على قسم واحد، 108 وهو المكتوب عليه مري نور القمر ، 109 وانظر إلى مقابلة الخطّ من العنكبوت، 110 فإنّه يقع على ما يغيب عليه القمر، 110.1 وهو مكتوب في أحد يسارك: 111 و ستّة أسباع. 112 وكذلك يعمل بأن يحرّك العنكبوت ويضع مري القمر على عدد الأيّام وأنصافها، 113 فإنّ كلّ يوم قسّم حرك المجرة بقسمين، 114 لكي يكون اليوم نصفين. 115 فإذا وضعت مري القمر على ما أردت من الأيّام وكسورها، 116 فانظر إلى مقابلة الخطّ من الجانب الآخر، على كم وقع من الساعات وكسورها، 117فما كان فهو ما يغيب عليه القمر من الليل.

119فصل في معرفة على كم يغيب القمر من النهار

120وذلك بعد نصف الشهر. 121 اجعل مري القمر على ما مضى من أيّام الشهر، 122 فالخطّ المقابل له يقع على الساعات وكسورها، 123 فاسقط منه ١٢ ساعة أبدا. 126 تعمل كذا إلى تمام كح يوما، ثمّ تدخل القمر تحت الشعاع، 127 ولا يحمل له عملا، 128 حتّى يفارق الشمس، 129 والعمل فيه بعد ذلك مضى.

131فصل في معرفة هيئة الخسوف وكميته

132اعرف أوّلا من التقويم كم بين الجوزهر، والشمس والقمر من الدرج وقت الاستقبال، |22r| 137 فإن كان ١٢ درجة وأقلّ، فينخسف. 141 ولمعرفة كمّيته، 142 اقلب الصفيحة التي فيها {القصة المخرقة}<الفضّة المحرقة> إلى الوجه الآخر، 143 وفيه دائرة القمر مقسومة ١٢ قسما. 144 ويطبق العنكبوت على وجهها، 145 وترد القطب والفرس على حالة الأوّل. 146 وتدير العنكبوت، حتّى ينطبق دائرة فلك الجوزهر من العنكبوت على دائرة القمر 147 انطباقا لا يبيّن منه شيء. 148 ثمّ ادر العنكبوت، حتّى يقع طرف الجوزهر على ما يكون بين القمر والجوزهر، 149 فإنّه ينتحا دائرة الجوزهر عن دائرة القمر، 150 ويظهر مقدار ما ينكـ{شـ}<سـ>ف ينكشف من <دائرة> القمر، 151 فيكون مقدار ما ينكـ{شـ}<سـ>ف منه. 153 فإن ظهر دائرة القمر تحت فلك الجوزهر كلّه، 154 فإنّ القمرينكـ{شـ}<سـ>ف كلّه، 155 وإن بقي شيء مغطّى، مقدار ما لا ينكـ{شـ}<سـ>ف من القمر. 157 {وإن} <وما> زاد على /[..]\ <مقدار> دائرة القمر، فهو مكث القمر في الخسوف.

157.1فصل في معرفة ما يخسف من الأصابع

157.2انظر كم قطع فلك الجوزهر من أقسام دائرة القمر، وهي ١٢، 157.3 فما كان فهو قدر الأصابع المنكسفة منه .

161انتهى والله أعلم .

English Translation[52]

[1] In the name of God, the Merciful, the Compassionate

[2] Chapter: On the use of the lunar plate
[6] Know that its limb is divided into four parts, [7] and its [upper] half is divided into 28 divisions which [indicate] days and [correspond to] their beginning. [8] When you take it in your hand and you suspend it before your face with the beginning of [numbering on] one of its quadrants on your right side, [the numbering] goes up to the suspension, [9] and [the quadrant] is divided by *14* lines. [10] Next, [starting] from the line that descends from the suspension on the left quadrant, the 28 days are completed. This is the number [of days] at which the moon appears to the people. [11] As for the other half [of the plate], it consists of the divisions of fractions of hours and hours at which the moon sets, since the first night [of the month]. [12] Its beginning is from the quadrant that is to your left. [13] On it is written "0," [13.1] "6." *Next on the second compartment* "1," "5" *is written*; [14] the Alif [as the sign for "1"] corresponds to one hour, [15] and "5" to the fraction of an hour, [16] which is five sevenths of an hour. [17] [You continue] until the divisions of this quadrant in hours and their fractions are completed [17.1] *as 12 complete hours* [17.2] *and this [will be the time] at which the moon sets, from the first night it appears up to a night when the moon becomes full.* [17.3] *Next after that [you continue] up to 24 [hours], at which the moon sets during daytime after having encountered the 12 hours of the night and its fractions,* [18] until [the divisions] on the limb of the lunar [plate] are completed; [19][the limb is] the ring which surrounds the plate of the rete.

[21] [Section:] To know the rete and its divisions.
[22] It is the plate on which a hole is in one of its halves. It rotates around the pole, [and] over there is written, "circle indicating the light of the moon". [23] Through this hole information is presented, [when the rete is rotated] on the plates with black silver plate that is below the rete. [24] On the plate beneath the rete are two circles on which silver is burned. [25] Close to this circle is the circle of the sphere of the sun situated near the suspension. [26] After it, on the left side, towards the line of the left quadrant, is the numbering of the hours at which the moon rises at day, [27] and on the quadrant that follows is the numbering of the hours at which the moon rises at night. [28] On the perforated side of the rete is the circle of the sphere of the dragon. [29] On it is written, "sphere of the dragon." [30] On this perforated half is also the index of the hours. [32] On

52 Passages which are not found in the text in Appendix A are in italics.

the line which splits the rete into to halves [33] and begins from the line which intersects the circle of the light of the moon, [34] goes through the center of the rete and the sphere of the dragon, [34.1] *and stops at its end,* [35] where the line is completed, it is written "index of the moon." [36] Understand all that is described to you! [36.1] *On the rete, on its rim, there are also divisions according to the number of days in a lunar month, on one side 30 days,* [36.2] *and on the other side 29 days, because the lunar month is always like that.* [38] On the other side of the plate is a circle divided into *12* parts, [39] from which becomes known what is eclipsed of the moon [40] On the right side of this circle is a numbering representing what is between the two nodes and the moon during an eclipse. [41] That is what is on the other side of the plate. On the back of this lunar plate, there is an alidade, [42] on which are lines for the hours. [44] On one of the quadrants are the divisions of altitude, as on the back of the astrolabe, for taking the altitude, [45] and on the opposite quadrant is the division of the shadow.[46] This is all that is needed.

[46.1] Section: On the knowledge of the hours
[46.2] *Place the pointer of the alidade according to the midday altitude* [*on the current day*]. [59] Let the alidade face the sun, until the shadow of the alidade casts straight on the body of the alidade, without deviation. [59.1] *What is between the cast of the shadow* [61] and the start point of *the hours, counting them* from the upper sighting vane whose the shadow you have taken, on the lines which are on the inner part of the alidade; [63] this will be the past hours, [64] if your action is before midday, and the remaining [hours], if it is after it.

[66] Section: To know the shadow
[67] [in the margin:] If the altitude is less than *45* [*degrees*], look at the pointed [tip] of the alidade and at which division within shadow [scale] it falls, [73] [main text:] and divide *144* by that. [73.1] *The result is the shadow.* [73.2] *If* [*the altitude*] *is 45*[*degrees*]*, the shadow is 12.* [73.3] *If* [*the altitude*] *is greater* [*than 45*]*, the shadow is equal to the indication of the alidade, with no need of division.*
[75] Section: To know the light of the moon appearing each night of the month
[76] <If you want that,> rotate the rete until the sphere of the dragon falls on the sphere of the sun, [77] and the index of the light of the moon on the burned silver, [78] and nothing of the white silver is shown from the hole of the index of the moon. [79] When you have done that, turn the rete by one of the parts allotted on the rete, on its rim, among the divisions [80] that correspond to the numbers of the days of the lunar month, [81] when the month has *30* [days], and on the other side of the rim, [82] when the month has *29* days. When you move the rete by one of its parts [83] aiming that the part is aligned to the line [84] that goes out from the suspension and goes towards the middle of the plate, the light of the moon

at one [particular] night will be shown to you. [85] Rotate [the rete] likewise to observe the light on the moon each day, until it becomes full moon. [86] After that it starts waning until moonlessness. [88] The circle [indicating] the light of the moon may turn, until it sticks to the sphere of the sun.

[91] Section: To know the illuminated [part] of the moon in the hours at day or at night
[93] If its rising is at an hour during daylight, [96] put the index of the hours on one of the daylight hours allotted on the left quadrant, [97] or [put the index] on what hour you want. The [part] of white silver that you have found to be visible on the circle indicating the light of the moon, [98] will be the extent of the illuminated [part] of [the moon]. [99] In the same way, if you want to know the illuminated [part] of it [when the moon rises] during the hours of the night, [100] put the index of the hours on the hours of the night [101] and look at the circle indicating the light of the moon. The visible [part] of white silver [depicts] the illuminated [part] of the moon.

[103] Section: To know at what [time] the moon sets
If you want the first night of the month, [104] put the index of the moon—that is, the line which starts from the extremities of the plate of the rete and passes through the sphere of the dragon [105] until it cuts the circle of the sun divides the moon into two halves (*recte* the circle of the light of the moon?)—at the beginning of the numbering of 28 which is at your right on the upper half of the limb. [106] Thus, the index of the light of the moon comes to your left straight opposite to it. [107] Next move the rete until the index of the moon falls upon the division, [108] on which is written "index of the light of the moon" (*recte* "1"?)[109] and observe the intersection of the line of the rete; [110] it falls upon [the time] at which the moon sets. [110.1] *It is written on the first [compartment] at your left:* [111] 6 [that is] six-sevenths. [112] One will proceed likewise by moving the rete and placing the index of the moon on the number of the days and their halves, [113] [For] each day, divide the motion of the limb into two parts, [114] so that the day consists of two halves. [115] When you put the index of the moon on the days and their fractions you want, [116] observe on which among the hours and their fractions the intersection of the line on the other side falls; [117] what this shows, will be [the time] when the moon sets at night.

[119] Section: To know at what [time] the moon sets during daytime
[120] This happens after the middle of the month. [121] Put the index of the moon on [the division corresponding to] what passed from the days of the month. [122] The line opposite to it falls on the hours and their fractions. [123] Always subtract

twelve hours from that.[126] You work similarly up to the completion of *28* days, then the moon goes under the rays [of the sun], [127] and no work remains for it, [128] until it detaches itself from the sun; [129] after that the work proceeds with that [similarly].

[131] Section: On the knowledge of the form and size of the lunar eclipse.
[132] Learn first from the astronomical ephemeris how many degrees are between the dragon [on one side] and the sun and the moon [on the other side] at the time of the opposition. [137] If there are *12* degrees or less, [the moon] will be eclipsed.[141] For finding the magnitude of [the eclipse] [142] turn the plate on which the burned silver is on the other side, [143] and [use the face] on which the circle of the moon is divided into *12* parts. [144] You adjust the rete on [this] face, [145] and you bring the pole and the horse to the initial position. [146] You rotate the rete until the circle of the sphere of the dragon on the rete covers the circle of the moon [147] completely, so that nothing is visible from it. [148] Next, rotate the rete until the rim/extremity of the dragon falls on the distance between the moon and the dragon. [149] The circle of the dragon moves away from the circle of the moon, [150] and the magnitude of the eclipse is shown on the circle of the moon; [151] this will be the magnitude of the lunar eclipse. [153] If the circle of the moon appears entirely below the sphere of the dragon, [154] the moon will be entirely eclipsed. [155] If something remains covered, [it shows] the amount of what will not be eclipsed of the moon. [157] The magnitude additional to that shown on the circle of the moon, corresponds to the duration of the totality during the lunar eclipse.
[157.1] *Section: On the knowledge of how many digits will be eclipsed*
[157.2] *Look how much the sphere of the dragon cuts off on the divisions of the circle of the moon, which are* 12. [157.3] *What it is, will be the number of the eclipsed fingers of* [*the moon*].
[161] [The chapter] ends—God knows best.

Appendix C: Epistle on the Practice of Using the Plane Lunar Astrolabe

Arabic Text

رسالة في «استخراج ...».

[1] أفادك الله قول الحقّ وأنار بك الفطن [2] سألت وهب الله لك المطلوب عن كيفيةالعملبالأسطرلابالقمري [3] المسطوح في سنة ثمانين ومائتي الهجرة [4] المنخرع فيه عمل الكسوف القمري[5] المدرك به علم ذلك بأسر عمل وأدرّ

مؤنة [6] ولست أترك إيراد مرادك من ذلك عليك بقدر الطاقة وحسب
ما توجبه الأكوان العاليه [7] فكن به سعيدًا.
[8] إذا أردت أن تعرف ما تكسفه كرة القمر من نور الشمس، [9] وهو معنى ما في جرمه من الضوء
الظاهر للعيان، في كلّ وقت أردت ذلك فيه، [10] فانظر بعد ما بين الشمس
والقمر من درج السواء واحفظ ذلك. [11] ثم خذ من المجتمع ثلثه فأعرضه.
[12] ثم ادلّ رأس مري القمر، [13] وهي الشظيّة التي في وسط دائرة الجوزهر
المُوَقّع عليها "مري القمر" ، عن رأس الجدي [14] إلى ما يتلوه من الدرج على
توالي الهجرة بقدر الثلث الذي حفظت. [15] فإذا فعلت ذلك، نظرت إلى
الدائرة المُوَقّع عليها "فلك مظهر زيادة القمر ونقصانه" ، [16] وهي انحر<ا>ف
المستدير، [17] فما كان على دائرة السواد فبذلك المقدار يكون المكتسي
من جرم القمر نورا. [18] وهو إلى أن تبلغ الدائرة الملاصقة لدائرة فلك
الشمس متزائدة في نوره. [19] ثمّ يبتدئ به النقصان من هذه الدائرة
الملاصقة لدائرة الشمس، إلى أن تصير تحت الشمس، فلا يظهر للبصر؛
وهو عند استتاره. [20] فاذا عرفت ذلك، وأردت معرفة الساعة
التي تطلع فيها القمر من ليل أو نهار، فانظر المري المُوَقّع عليه "مري
الساعات" على أيّ قسم وقع من أقسام الساعات، [21] فما وافت

[f. 45r]

فهذه الساعة التي يطلع فيها القمر من ليل أو نهار وهو إلى أن
يصير ذويّا يطلع نهاراً؛ [22] ثمّ إلى تمام الشهر يطلع ليلا، [23] وكذلك تجده
في قسم الساعات المعمولة في باطن الصفيحة المُوَقّع عليها [24] وعلى
أعدادها "ساعات النهار والليل". [25]
وفيه باب آخر، وهو أن تنظر من وقت مفارقة الشمس للقمر، [26] بل في يومك الذي تريد
أن تعلم، وذلك ما مضى من رأس الشهر العربي، [27] فتضرب ذلك في أربعة
[28] فما اجتمع من الضرب حفظته. [29] ثمّ أزلت مري القمر عن رأس الجدي إلى
ما يليه من البروج في توالي الهجرة بقدر المحفوظ من العدد، [30] فإنّه يظهر
لك من نور القمر في مقدار جرمه وفي ساعاته مثل الأوصاف التي تقدمت
بأذن الله. [31]
فأمّا معرفة الكسوف في أيّ زمان توهمت
فيه كسوف القمر، في كلّ عرض أنت فيه، [32] فإنّه <لا> يختلف في العروض ولا خلاف

الزيجات، فابدأ بتصحيح ساعة حقيقة الاستقبال، [33] فاعرفها، فإنّها
إن كان كسوف، فهي ساعة وسط الكسوف بأمر الله. [34] ثمّ انظر
عرض القمر فاعرفه. [35] فإن كان العرض أكثر من أربع وستين دقيقة،
فإنّ القمر لا ينكسف. [36] ثمّ انظر كم بين درجة الاستقبال وبين الرأس والذنب
من درجه. [37] فإنّ كان أقلّ من اثنى عشر درجة، فإنّ القمر ينكسف؛ [38] وإن كان
أكثر، فإنّه لا ينكسف. [39] وهو مثل معرفتك القمر وأوكد بأذن الله.
[40] وإذا عرفت بعد ما بين درجة الاستقبال وبين النقـ<ـطـ>ـتينالتنين فاحفظه، [41] ثمّ
ضع مري الكسوف في الصفيحة على مثل ما حفظت من العدد
المُوَقّع عليه في باطن الصفيحة "أجزاء البعد للقمر والتنين" . [42] فإذا
فعلت ذلك، فانظر إلى فلك الجوزهر كم ستر من فلك الشمس؛ عمّا ستر
منه [43] فهو الذي يبقى لا ينكسف من جرم القمر، [44] وما كان ظاهرا من
[f.45v]
فلك الشمس فهو الذي ينكسف من جرم القمر. [45] وقد قسمت فلك
الشمس باثني عشر قسما، [46] وهي أصابع، يعلم منها، كلّما وضعته، ما ينكسف
من القمر من إصبع، [47] وما يبقى لا ينكسف. [48] وإذا علمت ذلك، فإن لم يبق
فلك الجوزهر من فلك الشمس شيئا، فإنّ القمر ينكسف كلّه، [49] ولا يكون
مكث؛ [50] وإن تباعد عنه بعدا ما، فبقدر الأجزاء التي تباعد عنه، وهي
ساعات، تكون المكث إن شا الله. [51] وإذا فعلت ذلك، فانظر إلى
مري ساعات المكث والانجلاء، [52] وهي شظية في فلك الجوزهر، [53] فانظر إلى
أين وافت من أقسام الساعات المعمولة في باطن الصفيحة المُوَقّع
عليها "ساعات المكث والانجلاء" . [54] فإنّي قد اثبتها بدقائقها؛ [55] فما
كانت من ساعة ودقيقة منها فانقصها من ساعات وسط الكسوف،
وهي ساعات الاستقبال، فما بقي فهو ساعات ابتداء الكسوف. [56] ثمّ زدها
على ساعات وسط الكسوف، فما بلغ [57] فهو ساعات الانجلاء. [58] فإذا فعلت
ذلك، علمت كمّية الكسوف وساعات ابتدائه وساعات انجلائه
ومكثه من غير حساب ولا مؤنة. [59] وهو معمول من أوّل الدهر إلى آخره
في الأقاليم كلّها وعلى الرغبات مصحّح. [60] وهي صفيحة محدثة سطّحها
محمد بن محمد المعروف بنسطولس الأسطرلابي، [61] وزاد فيها عمل الكسوف
وعلمه. [62] ولم يسبق إليها في سنة ثمانين وما<ئـ>ـتي الهجرة. [63] تمت الرسالة بحمد الله.

English Translation

Epistle on finding out ... [?]

1 May God teach you the speech of the truth and enlighten your character. 2 You asked—may God give you the desired—about how to use the plane lunar astrolabe, 3 which was spread in the year two hundred eighty of the Hijra (893 CE), 4 in which the work (related to) the lunar eclipse can be derived, 5 and by which was attained knowledge of that with all its use and plenty of supply. 6 I do not neglect your request [which I owe] you, according to the ability and according to what the higher events make necessary. 7 Be happy with it!

8 If you wish to know how much of the lunar sphere is concealed from the light of the sun, 9 that is the light on its body which is visible for the eyes, at any time you wish, 10 look at the distance which is between the sun and moon in equal degrees, and remember it! 11 Take a third of the result and display it. 12 Next, direct the tip of the index of the moon, 13 which is the splinter in the middle of circle of the dragon, and on which "index of the moon" is inscribed, from the first point of Capricorn 14 to the degree along the limb corresponding to the third of the number that you have retained. 15 When you have done that, you look at the circle, on which "circle showing the increase and decrease of the moon" is inscribed; 16 this will be the alteration of the circular [shape of the moon]. 17 The body of the moon will be deprived of light by the measure of the visible part of the black circle.[1] 18 This goes on until the circle tangential to the circle of the solar disc is reached, steadily increasing in its light. 19 Then it starts to decrease, [while moving] from this circle [that is] tangential to the solar disc, until it arrives below the sun, then it does not appear to the eye; this happens during its occultation (conjunction / new moon).

20 If you have found that, and you want to know the hour at which the moon rises by night or day, look at which subdivision of the hours falls the index on which "index of the hours" is inscribed; 21 what shows up will be the hour of the night or day at which the moon rises, this means, until it reaches the wane it rises during daytime; 22 after that, towards the end of the month, it rises during the night, 23 and in this way you find it in the section of the hours which are constructed in the inner part of the plate, 24 where "hours of day and night" is inscribed on it and its numbers.

25 There is another approach to that, i.e., you consider the time since the conjunction of the sun with the moon,[2] 26 on the day which you want to know, that is, what has passed from the beginning of the Arabic month. 27 You multiply that by four 28 and you retain the result of the multiplication. 29

Then you move the index of the moon from the first point of Capricorn to the corresponding point of the zodiacal signs along the limb, according to the retained amount and the number; [30] there will be displayed to you the light of the moon in proportion to its size and (the corresponding) hours according the aforementioned descriptions, God willing.

[31] As for the knowledge of the eclipse, at any time you suspect a lunar eclipse could occur at any latitude where you are [32]—this <does not> vary with the latitude and there is no dispute in the Zījes—start with the adjustment of the time of the true opposition [33] and find it. If there is an eclipse, this will be the time of the middle of the eclipse, on the command of God. [34] Next, consider the latitude of the moon and find it out. [35] If the latitude is greater than sixty-four minutes, the moon will not be eclipsed. [36] Then observe how many degrees are between the degree of opposition and the ascending and descending node. [37] If there are less than twelve degrees, the moon will be eclipsed, [38] and if there are more, it will not be eclipsed. [39] This is equivalent to your information about the moon and is confirmed with the concession of God.

[40] If you have found the distance between the degree of the opposition and the dragon, remember it, [41] then put the index of the eclipse on the plate at the equivalent of what you have remembered among the enumeration inscribed on it “degrees of distance for the moon and the node” on the inner part of the plate.

[42] When you have done that, look, how much the sphere of the dragon has covered of the solar disc;what is covered of it [43] will be (the part) of the body of the moon that will remain not eclipsed, [44] and the visible (part) of the solar disc is that which will be eclipsed of the body of the moon.

[45] You had divided the solar disc into twelve divisions, [46] and these are digits that allow you to know, whenever you set it up, what will be eclipsed of the moon in digits, [47] and what remains will not be eclipsed. [48] When you learn that, if the sphere of the dragon does not emanate/ stand out from the solar disc, the moon will be eclipsed entirely, [49] but there will be no duration for the totality. [50] If there is some distance from it, the totality will be measured in hours by the measure of the degrees that correspond to the distance from it, God willing.

[51] When you have done that, look at the index of the hours of totality and emersion, [52] which is a splinter on the sphere of the dragon, [53] and observe where it falls among the divisions of the hours constructed on the inner part of the plate, on which “hours of the totality and emersion” is inscribed.

[54] I have determined them with their minutes. [55] Subtract the indication in hours and minutes from the time of the middle of the eclipse, that is the time of the opposition; the remainder will be the time of the beginning of the

eclipse. 56 Then add it to the time of the middle of the eclipse; 57 the sum will be the time (of the end) of the emersion.
58 When you have done that, you know the magnitude of the eclipse, the time of the start and end and the duration of its totality without calculation and further action. 59 It is operational from the beginning of the aeon/age until its end at all latitudes, and it is adjusted to the wishes [of its user]. 60
It is an innovative plate that Muḥammad Ibn Muḥammad, who is known as Nasṭūlus al-Asṭurlābī, made as a plain [plate]. 61 He added the operation concerning the eclipses and the respective knowledge to it. 62 It was not made before the year eighty and twohundred of the Hijra (893 CE). 63
The epistle was completed with the praise of God.

Appendix D: A Graphical Device for Eclipses—Arabic Text

f. 80r:

1 بسم الله الرحمن الرحيم. 2 معرفة عرض القمر وكسوفه، 3 وهو أن تنظر كم بين القمر والرأس أو
الذنب، 4 وتطلب مثله في الدائرة. 5 فإذا أصبت، فضع إصبعك عليه 6 وادخل إلى وسط الد{ا}
ئراة المقسومة بأربعة أجزاء ونصف، 7 فما أصبت من العدد فهو عرضه. 8 فإذا أردت معرفة
الكسوف، فخذ بركار، وضع أحد طرفيه في مركز الدائرة {والصفيحة} <وافتحه> بقدر عرض
القمر. 9 ثم انظر، أين يقع طرف البركار من عرض القمر المقسوم من ستّين دقيقة، في الدائرة.
10 فإن كان عرض القمر داخل <فلك> الجو[ز]هر، فإنّ القمر ينكسف كلّه لا محالة؛ 11 وإن
وقع خارج من دائرة فلك الجوزهر ودائرة الظل، 12 وانظر البركار أين يقع منه، فهو ما ينكسف
من القمر نحو هذه الدائرة، يصلح للأقاليم كلّها. 13 وإنّما <ا>علم أنّ القمر لا ينكسف حين يكون
عرضه أقلّ من ستّين دقيقة. 14 وكلّما نقص من ستّين، كان كسوفه أكثر، 15 فاعرف ذلك،
إن شاء الله تعالى.

English Translation

A plate from which one knows the latitude of the Moon and the magnitude of its eclipse in a geometrical way (see figure 1.5)
1 In the name of God, the Merciful, the Compassionate.
2 Knowledge of the latitude of the Moon and its eclipse.
3 [To know] that, you look how much [the distance is] between the Moon and the ascending or descending node, 4 and you seek its equivalent on the circle. 5 When you come to it, put your finger on it. 6 Enter into [the scale that is in] the middle of the circles, which [scale] is divided into four and half degrees.

[7] The number you came to is the [lunar] latitude. [8] If you wish to know the eclipse, take a compass and put one of its points on the center of the circle open it with the quantity of the latitude of the Moon. [9] Next look on the circle, where the [other] point of the compass falls on [the scale of] the latitude of the Moon, which is divided into sixty minutes [which is] inside the circle. [10] If the latitude of the Moon is inside the [circle of] the dragon, the Moon will be necessarily totally eclipsed. [11] If it falls outside the circle of the sphere of the dragon and the circle of the shadow, [12] look where the compass falls on it; this will be the eclipsed part of the Moon corresponding to this circle; it is suitable for all climates. [13] Know that the Moon is not eclipsed unless its latitude is less than sixty minutes. [14] The more it is less than sixty, the more the lunar eclipse is greater. [15] Know that—God the Almighty willing!

CHAPTER 2

The Astrolabe Finger Ring of Bonetus de Latis: Study, Latin text, and English Translation with Commentary

Josefina Rodríguez-Arribas
Institut für Jüdische Studien Westfälische Wilhelms-Universität, Münster
rodriguezarribas@gmail.com

Abstract

The subject of this article is the treatise on the astrolabe ring (1492/1493) by Bonetus de Latis (Jacob ben Immanuel Provenzale). The treatise belongs to a four-centuries-old tradition of Jewish treatises on the astrolabe, written mainly in Hebrew and more rarely in Judaeo-Arabic, Judaeo-Spanish, Spanish, and Latin, and produced mostly in southern Europe and Turkey. Bonetus's text is the second treatise written in Latin by a Jew, following Abraham ibn Ezra's treatise on the planispheric astrolabe (Rouen 1154). My purpose is to compare it with other contemporary treatises on similar instruments and with a little earlier treatise on the astrolabe in Hebrew (by Eliyahu Cohen of Montalto, fifteenth century) in order to understand the contribution of this instrument and why the treatise was so highly regarded among Bonetus's contemporaries. The instrument depicted in Bonetus's booklet can be considered one of the last contributions of Jewish culture to the history of the astrolabe; these contributions stretch back to the first Hebrew writings on the instrument in the twelfth century. The Latin text and the English translation are included at the end of the article together with the Latin text and translation of the longest version of the introduction to the treatise. The contents of the treatise are exactly the same in all editions of Bonetus's text, but there are two versions of the introduction and one is longer and more complete than the other. I have used both versions in my study, the one in the version printed, among others, in 1557 (shorter) and the one in the version of the treatise printed, among others, in 1507 (longer).

Keywords

Bonetus de Latis (Jacob ben Immanuel Provenzale) – astrolabe finger ring – astrolabe – Jewish astrolabes – medieval and Renaissance astronomy – fifteenth- and sixteenth-century Roman Jews – Jewish scientific literature in Latin

 | DOI:10.1163/9789004387867_005

Introduction

Bonetus de Latis or Jacob ben Immanuel Provenzale became well known in the Renaissance for a short treatise on a special type of astrolabe that was also a finger ring. He gave it the ambiguous Latin name of *Anulus astronomicus* and dedicated it to Pope Alexander VI (Rodrigo Borgia) in 1492/93. Bonetus was born, as his name suggests, in Lattes (or in Aix-en-Provence, according to some), close to Montpellier and its famous school of medicine, around the year 1450, into an important family of Sephardic origin. He worked in Provence as a physician and merchant, but in 1492 or 1493 Bonetus moved to Carpentras (not far from Montpellier), possibly due to the increasing pressure that the arrival of many of the Jews expelled from Spain was putting upon the Jewish comunities of southern France. At this time he also published his book *Anulus astronomicus*, on an instrument that he probably invented before leaving France. It seems that before going to Carpentras he also stayed in Pisa for a year, as a lease signed by him in November 1491 indicates.[1] According to Luzzati, the treatise would have been written, or at least finished, in Pisa, a city in which he would also have made the contacts necessary for having the work printed in Rome, among them Giovanni Borgia, son of Pope Alexander VI, and Giovanni de Medici, who would later become Pope Leo X. Later Bonetus left France, moved to Rome, and settled in the Jewish Quarter of the city, on the other side of the Tiber (Trastevere) near the Fabricius Bridge, and became physician to Pope Alexander VI (Borgia), and later to Pope Julius II (della Rovere) and perhaps Pope Leo X (de Medici).[2] Pope Alexander VI granted him the right to practise medicine as "artium et medicinae doctor" and, as he was a religious man respected in his community, he became the rabbi and possibly also the judge (*dayyan*) of the Roman Jewish community. If the astronomical ring introduces us to Bonetus as an ingenious spirit, his astrological predictions (written in Latin and Italian) caught the attention of the Jewish and Christian communities far beyond Rome, especially his forecast of the arrival of the Messiah for the year 1505, printed in 1498 in Rome.[3] Bonetus died, probably in Rome,

---

1 M. Luzzati, "Bonet de Lattes a Pisa nel 1491," *Italia* 13–15 (2001): 117–124.

2 Regarding the Jews in Italy, see Robert Bonfil, *Jewish Life in Renaissance Italy* (Berkeley, CA: University of California Press, 1994), especially 68–72; and Robert Bonfil, *Rabbis and Jewish Communities in Renaissance Italy* (Oxford: Oxford University Press, 1990).

3 As regards Bonetus's life and works, see Moritz Steinschneider, *Mathematik bei den Juden*, 2nd ed. (Hildesheim: Georg Olms, 1964), 206; Cecil Roth, *The History of the Jews of Italy* (Philadelphia, PA: Jewish Publication Society, 1946), 200; Cecil Roth, *The Jews in the Renaissance* (New York, NY: Harper & Row, 1959), 232–34; R. T. Gunther, *The Astrolabes of the World* (Oxford: Oxford University Press, 1932), 2:326–327; A. Milano, *Il Ghetto Di Roma* (Roma: Staderini, 1964), 67–68, 419; Gian Piero Della Capanna, *Archiatri pontifici ebrei nel*

sometime after May 1510, according to Esposito, or in 1515 if the common date of Reuchlin's letter is to be accepted.[4]

There is textual evidence that while in Rome, he was in contact with several important scholars of the time: the French theologian and humanist Jacques Lefèvre d'Étaples (ca. 1455–1536); the German humanist and kabbalist Johann Reuchlin (1455–1522), who addressed a letter to him asking for the protection of Pope Leo X; and Charles de Bovelles (1475–after 1566), the French mathematician and disciple of Lefèvre.[5] De Bovelles had himself been introduced to Bonetus during his stay in Rome in 1507, for he was curious about the astronomical ring invented by Bonetus and wanted to see it. According to his own textual testimony,[6] he did see it, which means that there was once a ring

---

Rinascimento (Montecatini Terme: Pierini, 1968), 85–87; D. Goldschmidt, "Boneto de Latis e i suoi scritti latini e italiani," in *Scritti in memoria di Enzo Sereni*, ed. D. Carpi, A. Milano, and U. Nahon (Milan: Gerusalemme, 1970), 88–94; J. C. Margolin, "Bonet De Lattes, médicin, astrologue et astronome du pape," in *Ecumenismo della cultura, III, L'umanesimo e l'ecumenismo della cultura. Atti del XIV convegno internazionale del Centro Studi umanistici, Montepulciano—Palazzo Tarugi 1977*, ed. G. Tarugi (Firenze: Olschki, 1981), 107–148; Anna Esposito and Micaela Procaccia, "Un astrologo e i suoi prognostici: Bonnet de Lattes a Roma alla fine del Quattrocento (in collaborazione con M. Procaccia)," *Materia Giudaica: Rivista dell'associazione italiana per lo studio del Giudaismo* 7.1 (2002): 97–104; Daniel Carpi, "Bonet Lattes," *Encyclopaedia Judaica*, ed. Fred Skolnik, 2nd ed. (Detroit, MI: Macmillan Reference, 2007), 12:518–519; and Danièle Iancu-Agou, *Régine-Catherine et Bonet de Lattes: biographie croisée* (Paris: Cerf, 2017). I would like to thank Amos Geula, who provided me with a copy of Goldschmidt's article. Bonetus's astrological predictions are the following according to the *Gesamtkatalog der Wiegendrucke*: Prognostikon für das Jahr 1493/94, lat. [Rom: Andreas Fritag, nicht vor 12.II.1493, nicht nach 1.III.1493], 4°; Prognostikon für das Jahr 1496/97, ital. Rom 13.II.1496 [Venedig: Dionysius Bertochus, nach 13.II.1496], 4°; Prognostikon für das Jahr 1498/99, lat. [Rom: Petrus de Turre, vor 1.III.1498], 4°; Prognostikon für das Jahr 1499/1500, lat. [Rom: Petrus de Turre, nicht vor 26.I.1499, nicht nach 1.III.1499], 4°, catalogue available online at http://gesamtkatalogderwiegendrucke.de/docs/BONELAT.htm (accessed 26 February 2017).

4 See this letter in his Hebrew and Latin version in Margolin, "Bonet De Lattes," 141–147.

5 On the scientific activity of French Jews living in Italy and France, see Bernard R. Goldstein, "The Hebrew Astronomical Tradition: New Sources," *Isis* 72.2 (1981): 237–251, especially 238. Though Goldstein's article focuses on those Jews using Hebrew for their texts, there is also a brief mention of Bonetus's astronomical work.

6 *Questionum theologicarum libri septem*, 1513, fol. 53r (*Dialogi de trinitate*): "L'homme [i.e., Bonetus] nous répondit que l'anneau se trouvait dans sa maison. Et comme nous n'en étions pas très éloignés, nous nous y rendîmes hâtivement en sa compagnie pour y voir l'objet. Alors il sortit l'anneau d'un coffret, et il combla nos voeux" (De Bovelles's words in Margolin, "Bonet De Lattes," 127). According to this testimony, there were at least two astrolabe finger rings: one that Bonetus offered to the Pope and one that Bonetus kept for himself (very likely the model) , which De Bovelles saw.

matching the description in Bonetus's treatise, although as far as I know there is no extant ring matching it. On the other hand, regarding Bonetus's astrological activity, the Jewish geographer and polemicist Abraham ben Mordekhai Farissol (Avignon 1451–1521) vigourously criticised his predictions in his *Magen Abraham* (*Vikkuaḥ ha-dat*).[7] Several mentions in contemporary literature and testimonies about Bonetus's astronomical work indicate his great popularity. The popular French poet and playwright Pierre Gringoire (ca. 1475–1538) refers to him in his play *La sottie du prince des sotz* as a physician and astrologer.[8] Simon de Phares also refers to Bonetus's astrological predictions in his *Elucidaire d'astrologie* (ca. 1498),[9] and François Rabelais in the twenty-first chapter of his *Gargantua and Pantagruel* refers to him as the Jewish physician of the pope.[10] Despite the enormous popularity of Bonetus's astrolabe treatise in Europe and the many prints of this work during several decades, the tradition of Bonetus and his astrolabe treatise in Hebrew sources is non-existent as far as I know.

The Treatise on the Astrolabe Finger Ring (1492/1493)

The astronomical booklet on the astrolabe ring is written in an idiosyncratic Latin that Bonetus feels compelled to excuse at the end of his treatise, explaining that his native language is Hebrew: "Parce, precor, rudibus quae sunt errata latino. Lex hebraea mihi est, lingua latina minus." The presence of Semitic structures is constant throughout the treatise with the one exception of the epilogue, which is the formal dedication of the text to Pope Alexander VI. The treatise appeared in several editions and was reprinted several times alone or with other astronomical treatises by different authors (Johannes Sacrobosco, Lefèvre d'Étaples, and Johann Dryander, among others): in 1492/93, 1493, 1494, 1498 and 1499 (Rome), 1496 (Venice), 1500, 1503, 1506, 1507, 1511, 1516, 1521, 1527, 1531 and 1534 (all in Paris), and finally in 1537 (Marburg), 1557 and 1558 (Paris). The number of editions and the calibre of the astronomical works with which Bonetus's text was printed (see Table 2.2) give us an idea of the success and

7 On Farissol, see David B. Ruderman, *The World of a Renaissance Jew: The Life and Thought of Abraham ben Mordecai Farissol* (Cincinnati, OH: Hebrew Union College Press, 1981), especially 62 and 138 (concerning predictions of the arrival of the Messiah in 1505).

8 A. Hindley, *Le jeu Prince des Sotz et de mere sotte* (Paris: Champion, 2000), 133.

9 Jean-Patrice Boudet, *Simon de Phares, Elucidaire d'astrologie, Recueil de plus célebres astrologues* (Paris: Librairie Honoré Champion, 1997), 1:602–603.

10 François Rabelais, *Gargantua and Pantagruel*, trans. Sir Thomas Urquhart and Peter Anthony Motteux, intro. Michael Randall (New York, NY: Barnes & Noble, 2005), 48.

prestige of the work among his contemporaries and even later.[11] For our study, we have used the 1557 edition, printed in Paris by Guillaume Cavellat (active 1550–1573)[12] and similar to the edition of 1537 printed in Marburg, in which Bonetus's treatise is bound with other treatises on *anuli astronomici* by Pierre Beausard (1536–1577), Gemma Frisius (1508–1555), Johann Dryander (1500–1560), Burchard Mithrob (1501–1564), Oronce Finé (1494–1555),[13] and Johannes Regiomontanus (1436–1476).[14] Bonetus's treatise, and his instrument, are the first and the only of their kind as far as I know, and they remain unique in

---

11 Turner concluded from his study on early printed treatises on astrolabes that Bonetus's treatise was the text with the largest number of editions in the two-and-a-half centuries when books about astrolabes were published. Turner counted eighteen editions of the Annulus, with Stöffler's Elucidatio … astrolabii immediately after it with thirteen editions. However, he also states that "the high number of editions of this short text results from it having become customary to include it in collections of tracts on astronomical subjects," see A.J. Turner, "From Brass to Text: The European Astrolabe in Literature and Print," in *Astrolabes at Greenwich: a Catalogue of the Astrolabes in the National Maritime Museum*, ed. Koenraad van Cleempoel (Greenwich: Oxford University Press and National Maritime Museum, 2005), 33. In any case, the ring was famous among Bonetus's contemporaries, as De Bovelles's testimony shows.

12 *Annuli astronomici: instrumenti cum certissimi, tum commodissimi, usus ex variis authoribus, Petro Beausardo, Gemma Frisio, Ioa[n]ne Dryandro, Boneto Hebraeo, Burchardo Mythobio, Orontio Finaeo; una cum Meteoroscopio per Ioa[n]ne[m] Regiomontanum & Annulo non uniuersali M.T. authore, Lutetiae: Apud Gulielmum Cauellat, 1557* (*P. Beausardi Annuli astronomici usus—P. Beausardi Geometricae scalae usus—Usus annuli astronomici per Gemmam Frisium—Annulorum trium diversi generis instrumentorum astronomicoru[m] co[m]pone[n]di ratio usus … Per Ioan. Dryand.—De compositione meteoroscopi, Ioannis de Regio Mo[n]te epistla – Boneti de Latis. Annuli astronomici utilitatu[m] liber—Compositio alterius annuli non universalis, sed ad certam polarem elevationem instructi authore. M.T.—Annuli cum sphaerici tum mathematici usus et structura, opera D. Burchardi Mithobii—Orontii Finaei. Compendiaria tractatio de fabrica & usu annuli astronomici*). I would like to thank Kirsten van der Veen of the Dibner Library of the History of Science and Technology at the Smithsonian Museum of American History (Washington, D.C.), who sent me a digital copy of Bonetus's treatise.

13 See Isabelle Pantin, "The Astronomical Diagrams in Oronce Finé's *Protomathesis* (1532): Founding a French Tradition?" *Journal for the History of Astronomy* 41.3 (2010): 287–310, especially 290.

14 *Annulorum trium diversi generis instrumentorum astronomicorum componendi ratio atque usus* (Marbourg: Cervicornus, 1537; re-published in Paris in 1557 with some small changes: the original dedication of Bonetus to the pope was replaced by a text possibly written by the editor of Dryander). See Margolin, "Bonet De Lattes," 133.

several respects.[15] This makes them an exceptional testimony about astrolabes and astronomical rings during the Middle Ages and Renaissance, and, more notably, the text is a valuable source of information about Jewish design and authorship of astronomical instruments, especially astrolabes.[16]

In the edition of 1557, Bonetus's treatise has a short introduction (much shorter than the one in the 1492/1493 edition) dedicating the text to Alexander VI, plus thirty-three chapters explaining how to use the instrument, and an epilogue in which Bonetus presents his work to the pope and apologises for his bad Latin. This epilogue (and the long introduction in the edition of 1492/1493) must have been written or corrected by another person, for the language and style are very different from the rest of the text, which is distinctive for the Semitic influence on its syntax.[17] The treatise also includes a woodcut of the

15 Charles Burnett informed me that at the British Library there is a treatise on a related instrument (a horary finger ring). It is a little booklet of 12 folios in the Royal collection of the British Library (Royal 12.A.XXV), an elegantly written explanation of the use of an "horary ring." The dedication is as follows: "Serenissimae invictissimi regis filiae Dominae Elizabetae Virgini omnibus naturae fortunaeque dotibus ornatissimae Guilelmus Bukleus perpertuam optat foelicitatem." The incipit of the dedicatory preface is: "Non sine ratione veteres philosophi…" Buckley goes on to explain that all men are interested in future natural events (illnesses, crops etc.) and so have devised astronomical instruments to predict them. "Inter quas annulis horariis postremus locus non erat tribuendus, partim quia commodo semper et sine molestia in digitis portari possint." This *anulus*, then, is clearly meant to be worn as a ring. The dedicatory preface ends by giving the place and date of writing: "Ex Regio Cantabrigiae Collegio (King's College, Cambridge) 16° Calendo Aprilis 1546." The text begins: "Quoniam annuli horarii reliquaque huius generis astronomica instrumenta …" I thank Charles Burnett for providing me the description quoted above. I have used the two possible forms of the Latin term for ring *annulus* and *anulus* throughout the article, since the two forms were used in the Renaissance (and in Classical and Medieval Latin texts). However, when I am not quoting textual sources, I prefer the form *anulus* (which is only in the long introduction of Bonetus's treatise), which is more correct. See Alfred Ernout and Antoine Meillet, *Dictionnaire étymologique de la langue latine. Histoire des mots* (Paris: Klincksieck, 1994), 37–38.

16 However, Gunther classified it among the Italian astrolabes, for it was designed to be used in Italy. See *Astrolabes of the World*, 327. Gunther also stated that it could be said that Bonetus invented the equivalent of a wrist watch with his astrolabe finger ring. *Ibid.*, 326.

17 What follows are some examples with Hebrew syntactical parallels (in italics). (1) The repetitive use of the Latin *postea* (Latin: *postea*, Hebrew *aḥar kak̠*) at the beginning of sentences: "*Postea* sunt almicantarat qui sunt in medietate circuli superioris descripti, quorum quidam integri apparent, quidam imperfecti … *Postea* linea medii coeli, quae est linea descendens a cruce … *Postea* in circuitu superficiei in spissitudine eius …" (104b); Hebrew parallel (*aḥar kak̠*): עוד אחר כך תחלק הש"ס לי"ב חלקים והוא שתתן לכל חלק ל' מעלות. עוד תכתוב עלהם הי"ב מזלות .עוד אחר כך תוך אותה העגולה על זולת מרכזה אלא

original astrolabe ring, which seems to be the only extant image of the invention (Figure 2.1). This image and the description of the instrument prove that the astrolabe ring is quite different from the standard astronomical rings (Figures 2.2–2.4) described in contemporary texts, although most of their applications are practically identical.

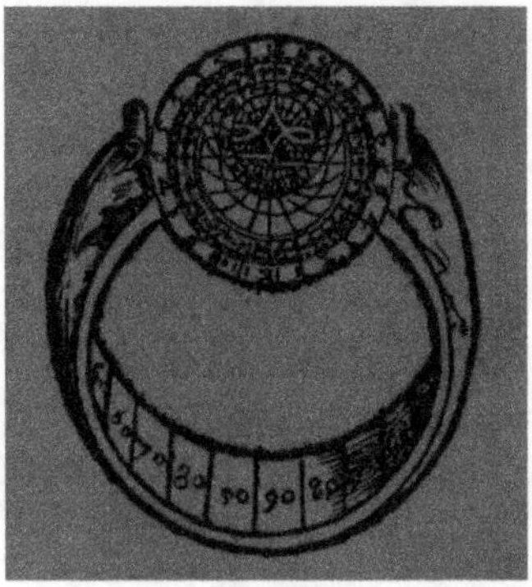

FIGURE 2.1
Bonetus's astrolabe ring (Dryander, Annulorum trium [...] *ratio atque usus, Marburg, 1537).*[18]

במרכז קרוב למרכזה תעגל עגולה אחרת קטנה ממנה שהיא במקום גלגל יוצא המרכז של השמש. אחר כך תעשה מעצר אחד ישר בתכלית היושר (Sibdat on the astrolabe, MS BnF 1030, fol. 46a). (2) The use of the subjunctive (Latin: *si velis*, Hebrew: *'im raṣita*) in the protasis of the conditional sentence: "*Si velis* scire locum solis in quo signo est et in quo gradus eius in illa die, pone filum super diem quem habes, et nota super quem ex gradibus signorum cadat filum ..." (105a); Hebrew parallel (*'im raṣita*): ואם רצית לידע אי זה לוח הוא לוח ארצך ספור המעלות שמהקו השוה עד האפק (Montalto on the astrolabe, MS BnF 1047, fol. 87b). (3) The repetition of the coordinating Latin conjunction *et* (Hebrew *vav*) at the beginning of sentence: "*Et* initium octavae est nadir secundae.... *Et* initium decimae est nadir quartae. *Et* initium undecimae est nadir quintae. *Et* initium duodecimae est nadir sextae (106b); Hebrew parallel (*vav*): שים מקום השמש תחלה באופק המזרחי וַסמן המורה בתא. וַשים מקום השמש שניה באופק המערבי וַסמן המורה בתא. וַספור מה שביניהן מהמעלות וַתן לכל ט"ו מעלה שעה אחת (Montalto on the astrolabe, MS BnF 1047, fol. 85a), and so on. In addition, the punctuation of the Latin text also reveals the influence of the Hebrew language.

18 Woodcut of the *anulus astronomicus* from fol. 476r of the 1537 edition of Bonetus's treatise (*Dryander, Annulorum trium diversi generis instrumentorum astronomicorum componendi ratio atque usus*, Marburg, 1537), currently held at the Bibliotheca Electoralis, Thüringer Universitäts—und Landesbibliothek in Jena. If one compares this woodcut with the woodcut of the *anulus* in the re-print of this edition from 1557, which displays a slightly different image of the ring, the latter is apparently more accurate (Figure 2.5), although the two images present problems of interpretation.

FIGURE 2.2
Standard astronomical ring (attributed to Arsenius, 1567, courtesy of the Museum of the History of Science, Oxford, inv. 48126).[19]

FIGURE 2.3 *Dryander's astronomical ring (Dryander, Annulorum trium* [...] *ratio atque usus, Marburg, 1537).*

19 See http://www.mhs.ox.ac.uk/collections/search/displayrecord/?invnumber=48126 (accessed 4 February 2017). ©2006 Museum of the History of Science.

FIGURE 2.4
Ring dial in the form of a ring with seal (17th/18th century, English, courtesy of the Museum of the History of Science, Oxford, inv. 38676).[20]

The contents of the thirty-three chapters are standard in astrolabe treatises and also in treatises on astronomical rings. Table 2.1 displays the contents in Bonetus's text and in Eliyahu Cohen of Montalto's treatise on the astrolabe, which is a little earlier and was written in Hebrew.[21]

TABLE 2.1 *The contents in Bonetus's text and in Montalto's treatise on the astrolabe*

	Bonetus de Latis's ring	**Montalto's astrolabe**
1	The description of the astrolabe.	
2	The position of the sun any day.	The position of the sun and the planets in the zodiac.
3	The day when one does not know the date.	
4	The altitude of the sun.	The altitude of the sun at noon and the altitude of the planets at midnight.
5	The ascendant degree and the ascendant sign.	The ascendant, the descendant, the culminating sign and the sign in the lower meridian by day and by night.

20 See http://www.mhs.ox.ac.uk/wp-content/themes/mhs-2013-responsive/imu-media.php?irn=21676 (accessed 20 May 2017).

21 For Montalto's texts we have followed MS BnF Hébreu 1047 fols. 84b–91b. Do not confuse Eliyahu Cohen of Montalto with Eliyahu Montalto, the Portuguese *marrano* who was born in Castello Branco and died in Tours (1616). The latter became the physician of Maria de Médici and wrote several treatises on medicine and religion but not a treatise on the astrolabe. I am preparing a critical edition and translation of Eliyahu Cohen of Montalto's text.

TABLE 2.1 *The contents in Bonetus's text and in Montalto's treatise on the astrolabe* (cont.)

	Bonetus de Latis's ring	Montalto's astrolabe
6	The twelve houses of the sky.	The twelve houses by day and by night.
7	Telling the time by the French method.[a]	The equinoctial hours by day and by night.
8	Telling the time in equinoctial hours.	The equinoctial hours by day and by night.
9	Hours elapsed since sunrise.	The diurnal and nocturnal seasonal hours that have elapsed.
10	The duration of the day and of the night.	Morning and evening twilights. The diurnal and the nocturnal arcs.
11	The duration of the day in seasonal hours.	Part of any diurnal or nocturnal seasonal hour.
12	The duration of the equinoctial hour.	On the equinoctial hours of the day that have elapsed.
13	Changing seasonal hours into equinoctial ones.	Conversion of equinoctial hours into seasonal hours.
14	The altitude of the sun in the meridian.	
15	The position of the sun when one does not know the day.	
16	Which day is equal to another [in duration].	
17	The declination of the signs from the equinoctial line (i.e. the equator).	The declination of any northern or southern degree of the zodiac with respect to the equator.
18	The latitude of the place.	The latitude of the place by day and by night. The longitude of the place by day and by night. The azimuthal lines by day. Directions by night. The azimuthal lines by night.
19	The latitude for which the ring has been made.	
20	The ascensions of the signs in the sphaera recta.	The ascensions of the zodiac signs in the sphaera recta, by day and by night.

TABLE 2.1 *The contents in Bonetus's text and in Montalto's treatise on the astrolabe* (cont.)

	Bonetus de Latis's ring	Montalto's astrolabe
21	The ascensions of the signs in any inclined horizon.	The descensions of the zodiac signs in any inclined horizon.
22	Position of the moon in the zodiac without using its altitude.	
23	The ascendant of the natal revolution or of the revolution of the world.	The profections of the years of the world and the nativities.
24	The ruler of the hour by day and by night.	
25	The altitude of the stars.	The altitude of any fixed star by day or by night. The time of the rising of any fixed star by day and by night. The time of the setting of any fixed star by day and by night. The time of the culmination of any [fixed] star by day or by night. The [diurnal] and nocturnal arcs of any fixed star.
26	The degree of the ascendant by night.	The ascendant, the descendant, the culminating sign and the sign in the lower meridian by day and by night.
27	The duration of the night.	
28	The degree of the ascendant without using the latitude, without stars by night and without sun by day.	
29	The degree of the moon without using the conjunction.	The hiding of the moon at the end of the month in the morning.
30	The position of the planets.	The altitude of the planets.
31	Whether a planet is southern or northern.	
32	Whether a planet is retrograde or direct.	

TABLE 2.1 *The contents in Bonetus's text and in Montalto's treatise on the astrolabe* (cont.)

	Bonetus de Latis's ring	Montalto's astrolabe
33	The height of a tower or any other thing.	The height of anything perpendicular to the angle of the sun or other thing. The depth of anything deeper than the surface of the earth.

a In France the diurnal period was divided in 12 hour periods beginning at midnight and noon respectively. In Italy the diurnal period was divided in 24 equal hours reckoned from the moment of sunset, see Silke Ackermann and Robert van Gent, http://www.mhs.ox.ac.uk/epact/article.php?ArticleID=19 (last accessed 2 February 2017).

As the list of contents makes clear, Bonetus's treatise is an instrument-book specifically focused on the use of the astrolabe ring. There is no explanation of its construction, although Bonetus mentions at the beginning of the first chapter that it can be made of gold, silver, or another metal, as is usual for rings. The table of contents presents a combination of astronomical and astrological calculations, which is to be expected from an author like Bonetus, who wrote several prognostications and was one of the astrologers working for Pope Alexander VI. It is also known that Pope Alexander VI and his entourage were most interested in astrological predictions and favoured astrological subjects and practices.[22] If we compare Bonetus's treatise with standard treatises on the astrolabe in Hebrew (such as Montalto's), we can have no doubt that there has been a simplification of the contents that are standard in this genre, but there is a reason for it. Some of the typical components of standard planispheric astrolabes are missing in the astrolabe ring (for instance azimuthal

22 "The courts of Alexander VI, Julius II and Leo X also had their resident astrologers. In 1501, Agostino Vespucci informed Machiavelli that there were 'five or six doctors' in the Borgia court who were fond of discussing poetry and astrology. One of these was Alexander VI's favourite astrologer. When Alexander was still a cardinal, he had correctly predicted that he would one day be pope … Bonatus de Lattis and Gaspar Torrella were Cesare Borgia's astrologers." Letter of Vespucci to Machiavelli, 25 August 1501. See Anthony J. Parel, *The Machiavellian Cosmos* (New Haven, CT: Yale University Press, 1992), 16. For this letter see James B. Atkinson and David Sices (eds.), *Machiavelli and His Friends, Their Personal Correspondence* (DeKalb, IL: Northern Illinois University Press, 1996), letter 25. Gaspar Torrellas (Valencia 1452–Rome 1520) was a Spanish bishop who became the physician and astrologer of Pope Alexander VI and later of Pope Julius II. He wrote several treatises on medicine.

lines, twilight lines, or different plates of latitude), which makes their explanation unnecessary. In addition, Pope Alexander VI presumably would not have been interested in some of the more specific calculations that were standard in astrolabe treatises. Thus, Montalto's text explains how to measure the depth of any object, the culminations, risings, and settings of specific stars, and the longitude of a certain place, all of which are missing in Bonetus's text. Alexander VI was not a practitioner of astronomy but a Renaissance man eager to display his active participation in the culture of his time, which certainly included a basic knowledge of astronomy and astrology. Hence Bonetus refers to the most practical applications of astrolabes in astronomy and astrology and completes most of the thirty-three chapters of his treatise with clear examples using contemporary astronomical and calendrical data. It must be added that Bonetus introduces more explicit astrological information than Montalto does (e.g., the "rulers of the hours," i.e., the planets ruling each hour of every day, or how to find whether a planet is northern or southern and retrograde or direct in its path, positions that have specific interpretations in astrology). Some of the uses of Bonetus's instrument that do not emerge in Montalto's text are common in other astrolabe texts. The description of the astrolabe and the way to ascertain the latitude for which the instrument was made (ch. XIX in Bonetus's treatise) were part of the contents of many astrolabe treatises, and Abraham ibn Ezra referred to the latter in one of his treatises on the astrolabe.[23]

If the astrolabe ring seems to be a standard planispheric astrolabe, what is Bonetus's contribution to the field of astronomical instruments? If we consider the long introduction in the 1492/93 edition of the *Annulus* (Rome), there are some illuminating explanations about the purpose of this specific type of astrolabe and Bonetus's estimation of his invention. In the long introduction to the treatise (see our edition and translation at the end of this article), there are six quotations from the Old Testament, which are very significant coming from a Jewish writer. Furthermore, they place this Renaissance object in the context of the understanding of the science of the stars characteristic of the Middle Ages. The first quotations are introduced to underline two essential ideas: there is no knowledge higher than the knowledge of God (Jeremiah 9:23), and the heavens are the work of God's hands (Psalms 19:1, 8:1, and 8:30), so the study of the heavens places us closest to God and to the knowledge of God. "If we want to know God, we must know the science of the stars," Bonetus states in his introduction. Later he also explains why he decided to make an astrolabe

23 The third Hebrew version of his treatise on the astrolabe, MS Günzburg 937, fol. 8a. See also Josefina Rodríguez-Arribas, "Medieval Jews and Medieval Astrolabes: Where, Why, How, and What For?," in *Time, Astronomy, and Calendars in the Jewish Tradition*, ed. Sacha Stern and Charles Burnett (Leiden: Brill, 2013), 221–271, especially 247–248.

in the shape of a finger ring. He again refers to several biblical quotations to support his choice: the ring is an excellent ornament on the hands of powerful and illustrious men like Pharaoh and Joseph (Genesis 41:42), and it is a reminder of the heavens and the stars, "the work of His fingers." In addition, Bonetus relates his *anulus* to the ring that the high priest Aaron wore on his breast together with the breastplate and the ephod (Exodus 28:28), the nature and purpose of which remain a very intriguing and controversial matter. The most apparent advantages of the instrument are also mentioned, especially at the end of the treatise: it is a small, portable instrument, much easier to make than others; it is a beautiful and visible ornament, always available for use at any moment; and it is new. Bonetus was immersed in the culture of the Renaissance, in which ingenuity was highly prized, and his astrolabe ring was innovative in a certain way and it was presented to the Pope in a persuasive and seductive form.

Did Bonetus de Latis Invent a New Astronomical Instrument?

Bonetus's text was several times included in combined editions with other famous treatises explaining how to make and/or how to use astronomical rings, as if the astrolabe finger ring were one of these. There are two types of rings: the ring dial, which is basically a sundial and a finger ring (see Figure 2.4),[24] and the astronomical ring dial or universal equinoctial ring (see Figures 2.2 and 2.3).[25] According to François Charette, Najm al-Dīn's

24 It is also called the Aquitaine ring dial, for it is said that Eleanor of Aquitaine offered one to her husband, Henry II of England, in about 1152. I have not been able to track down a historical source for this attribution. Another possible name for the large group of portable sun dials is pocket or "poke" dial. See http://www.ukdfd.co.uk/ceejays_site/pages/pocket_sundial.htm (accessed 18 February 2017). To use the ring dial one suspends it from the hand facing the sun. A beam of the sun must then go through the small hole pierced in the ring to fall somewhere among the hour divisions engraved inside the ring. The hole is on a loose ring within the ring so that it can be moved according to the day and month at which the observation is made. The equinoctial ring dial, when it is displayed, resembles an armillary sphere (its components are described in the next footnote). The hour divisions are engraved on the equatorial ring and the meridian ring must be suspended according to the latitude of the place. The solar beam goes through the hole of the inner or declination circle and indicates time.

25 See the astronomical ring in the collections of the British Museum (http://www.britishmuseum.org/research/collection_online/collection_object_details.aspx?objectId=55055&partId=1) (accessed 1 February 2017), © Trustees of the British Museum. The astronomical ring consists of three circles: the outer for the meridian, with a shackle to suspend the ensemble (a further development of the ring); the intermediate for the equator; and the inner circle for the declination. The outer circle contains one or two scales

text on the ring sundial (first half of the fourteenth century) seems to be the earliest extant documentation on this instrument in Arabic.[26] On the other hand, MS 688 (748) of the Bibliothèque Municipale of Arras seems to contain the oldest Latin text on the ring dial (incipit *Ad faciendum anulum*, late fourteenth century).[27] The invention of the astronomical ring was attributed to the Flemish mathematician Gemma Frisius (1508–55) with his *De annuli astronomici usu* (1534),[28] but it seems that astronomical rings and dials were in use well before the sixteenth century, dating back to the Middle Ages and even Roman and Hellenistic times.[29]

Looking at the image of Bonetus's astrolabe ring, we can affirm that it is certainly neither a ring dial (as it can do more than tell the time), nor a universal equinoctial ring (which has several applications besides telling the time and is universal). Bonetus's ring is not universal; it is made for a specific place (like ring dials), and telling the time is just one of its many applications (like universal equinoctial rings). Is Bonetus's astrolabe ring a type of ring sundial? Yes and no; it is a sundial (for it tells time by means of the sun) but it is in addition something else. One can perform many more calculations with Bonetus's ring than with a ring sundial, for instance: telling the time by night, calculating time in equinoctial hours, house division of the sky, diurnal and

---

of 90 degrees. The equatorial circle can contain the months, the zodiac signs, and two twelve-hour scales. The inner circle can turn within the meridian circle by means of two pivots at the twelfth hour. This circle usually has holes so that it can be used as the alidade of an astrolabe and is engraved with the zodiac signs and some of the fixed stars.

26 François Charette, *Mathematical Instrumentation in Fourteenth-Century Egypt and Syria: The Illustrated Treatise of Najm Al-Dīn al-Misrī* (Leiden: Brill, 2003), 142. Any of the more than one hundred astronomical instruments described in the fourteenth-century manuscript that Charette edited shows that constructing intricate instruments was a way to display one's intellect and scientific acumen in the Middle Ages and Renaissance, a tradition in which Bonetus's astrolabe finger ring must be placed.

27 Charette, *Mathematical Instrumentation*, 142, n. 48, and David King, *The Ciphers of the Monks: A Forgotten Number Notation of the Middle Ages* (Stuttgart: Steiner, 2001), 396.

28 Claude Sorgeloos, "Un post-incunable retrouvé: l'*Usus annuli astronomici* de Gemma Frisius, Louvain et Anvers, 1534," *Quærendo* 31.4 (2001): 255–264. Sorgeloos suggests that this cannot be the *editio princeps* of this text, because the words "iam denuo avetus" on the title page (264) make one think there was another edition prior to the year 1534. For a description of Arsenius and similar ring dials, see Samuel Gessner, "The Use of Printed Images for Instrument-Making at the Arsenius Workshop," *Early Science and Medicine* 18 (2013): 144–148.

29 See Lewis Evans, "Portable Sundials" in Alfred Gatty, *The Book of Sun Dials*, 4th edn. (London: George Bell and Sons, 1900), 185–486 (the first-century sundial in the shape of a ham found in Herculaneum) and 190–191 (the fourth-century universal dial in the shape of a flat disk).

nocturnal arcs, etc. We can say that it is a modification of a planispheric astrolabe, as we shall see in the final comments and diagrams appended to the translation. The main difference between Bonetus's ring and astrolabes is that the former has no plates to change according to the latitude of the place (Figure 2.5); rather, it must be used in the place for which it was made, like ring dials. In addition, the small size of Bonetus's astrolabe must have made calculation much less precise, which was a frequent concern among medieval astrolabists in relation to standard astrolabes.[30]

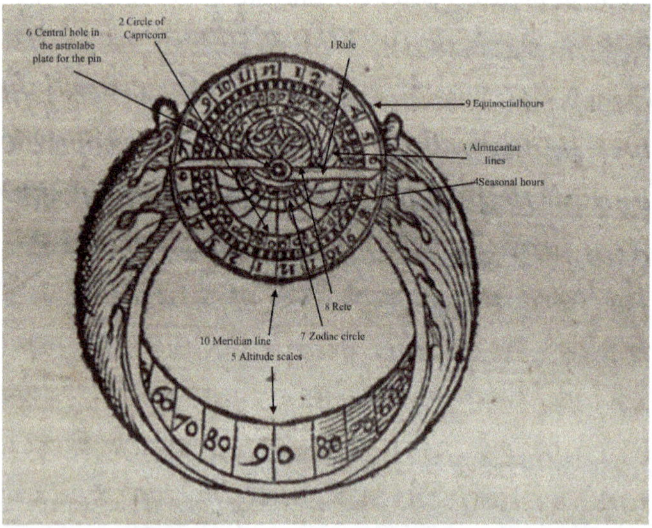

FIGURE 2.5 *Main components of Bonetus's ring according to the image of the ring in the reprint of Bonetus's treatise in 1557.* (*1*) Rule; (*2*) *Circulus Capricorni* (circle of Capricorn); (*3*) *Almicantarat* (almucantar lines); (*4*) *Erraticae horae or horae inaequales* (seasonal hours); (*5*) *Gradus ad inveniendum altitudinem* (altitude scales); (*6*) *Foramina* (central hole in the astrolabe plate for the pin); (*7*) *Rotula zodiaci or circulus signorum* (zodiac circle); (*8*) *Rotula mobilis* (rete); (*9*) *Horae aequales* (equinoctial hours); (*10*) *Linea meridiei, i.e., linea medii coeli et linea mediae noctis* (meridian line consisting of the midday line and the midnight line).
Image reproduced with the permission of the Dibner Library of the History of Science and Technology at the Smithsonian Museum of American History in Washington, D.C.

30 Regarding Ibn Ezra's concern with the accuracy of the calculation with astrolabes, see Rodríguez-Arribas, "Medieval Jews and Medieval Astrolabes," 223, 228–229, 242, and 249.

Bonetus's ring is, properly speaking, an astrolabe, and it should therefore be labelled an "astrolabe finger ring" rather than an "astronomical ring." We have tried, without success, to find images or extant rings that could match Bonetus's description.[31] Examples of astronomical rings were made in Germany in the sixteenth and seventeenth centuries, but there is no evidence of an astrolabe ring like Bonetus's. If the instrument existed, the only visual evidence of it is the woodcut in the 1537 edition of Bonetus's treatise and the slightly different one in its re-print from the year 1557 (Figures 2.1 and 2.5, respectively). The use of printed images of instruments was just starting, which explains the repetition of the few available images. The publishers rather than the authors used to provide the woodcuts (or copperplates) and they were re-used in different works.[32]

31 See George Frederick Kunz, *Rings for the Finger: From the Earliest Known Times, to the Present, with Full Descriptions of the Origin, Early Making, Materials, the Archaeology, History, etc.* (Philadelphia, PA: Lippincott, 1917), 87–88 (in 1917 in the collection of the Imperial Kunstgewerbe Museum in Vienna, "a gold astrolabe ring, which when closed looks like an ordinary one; but when the connected circles are opened up, constitutes a veritable astrolabe"); and 353 (image of an astrolabe ring in the Albert Figdor Collection in Vienna). Neither the image nor the description of either of these two rings matches Bonetus's ring; however, according to De Bovelles's report in his *Questionum*, Bonetus's ring existed and he saw it. Henri Michel, in his *Traité de l'astrolabe* (Paris: Brieux, 1976), 187, thought that the initials H. B. engraved on two extant astrolabes referred to "Hebraeus Bonetus" (the astrolabe of 1490 at the Museum of the History of Science in Oxford and the astrolabe of 1491 at the National Museum of Munich). However, it has been proven that these two astrolabes were made by Hans Bamman, a well-known astrolabe maker who worked in central Germany in the fifteenth century.

32 It seems that the woodcuts in the edition of 1537 and the reprint in 1557 are the only images of Bonetus's ring. This is the reason why we have used the 1537 edition, while also preferring the text in what would be one of the two last of all the editions and printings (1557). There is a different image in the *Annulus astronomicus* of Bonet de Lattes printed by Andreas Freitag in Rome in 1492/1493. See http://collections.countway.harvard.edu/onview/files/original/c80ea519af19455664d969126fdd4198.jpg, Solomon M. Hyams Collection of the Boston Medical Library (Ballard 220). However, this cannot be an image of the ring. There were very few printed instruments in texts between 1475 and 1500, see *Margaret Gaida*, "Reading Cosmographia: *Peter* Apian's Book-Instrument Hybrid and the Rise of the Mathematical Amateur in the Sixteenth Century", *Early Science and Medicine* 21,4 (2016): 277–302, here 288.

TABLE 2.2 *Table of contents of contemporary treatises on the astronomical ring (Paris 1558)*[33]

Petrus Bosardus[a]	Description of the components. The position or the degree of the sun in the sign. The latitude of the place. The equinoctial hours by day, the directions of the world, and other things. The nocturnal hours. The position of the pole and to which part the stars incline. The time in another place. The sunrise and sunset and the duration of the day and the night. The time of sunrise and sunset by another method. The seasonal or planetary hours. The time since sunrise or sunset. The altitude of the sun or the fixed stars. The altitude of the sun and the stars at any time when the sky is cloudy. The declination of any degree of the ecliptic. The declinations of some of the fixed stars. The maximum altitude of the sun or the fixed stars at noon and the latitudes of the places. The position of the sun and the day of the month. The same by observing the altitude of the sun at midday, and at any hour of the day. The minimum visible altitude of the sun at any time. Inscribing the meridian line on a plane. The point of the horizon in which the sun or any fixed star rises or sets any day. The height according to the length of the shadow, and the same by proportions. The shadow of a stake or gnomon. Measuring height with one only observation. The height of inaccessible things and the same without using proportions. The distance of something when its height is inaccessible.

33 Orontius Finaeus's treatise on the use of the instrument is not available in the 1558 edition, in which only a brief explanation on the construction of the astronomical ring is printed.

TABLE 2.2 *Table of contents of contemporary treatises on the astronomical ring* (cont.)

Petrus Bosardus	The length of things that are on a high place and the same by an easier method. The measurement of distances. The measurement of depth.
Gemma Frisius[b]	The components. The use of the ring and the sign in which the sun is any day. The height of the pole. Telling the time. Whether it is before or after noon. The nocturnal hours. The nocturnal hours in an easier way. The sunrise and the duration of the day. Seasonal or planetary hours. The time since sunrise or since sunset. The directions of the world. The altitude of the sun and the stars. Using shadows. The height when the ground is visible. The height of objects out of reach and the same by an easier method. The distance of high-located objects and the same by a different and easier method. Measurement of distance. Measurement of depth.
Ioannes Dryandrus[c]	Components of the ring.[d] How the ring is adapted to the height of the pole. The fourth height of the pole [i.e., the fourth climate]. The height of the pole using the altitude of the sun at noon. The height of the equatorial circle above the horizon. Explanation of the height of the equator and of the pole. The altitude of the sun at noon. The hour of the day with the ring when the sun is shining. Whether the sun is before or after noon. The four directions of the world. The meridian line with the ring and the same by a different method.

TABLE 2.2 *Table of contents of contemporary treatises on the astronomical ring* (cont.)

Ioannes Dryandrus	Finding the day with the ring when the date is unknown. The position of the sun when it is unknown using the date. The declination of the sun and its degree in the zodiac. Identifying the climate when one gets lost in the sea or in the desert. The height of the pole. Constructing sundials on any vertical or horizontal surface using the ring.[e] The measurement of things. The height scale of the ring. The height of anything accessible placed on an uneven ground, and the same by a different method. The height of something according to its shadow. The height of anything placed high up using the ring and without changing one's place. Calculating on the ground the height of something to which access for measuring it is not possible. The height of something placed on a mountain, the top and the base of which are visible to the eyes of the person standing in a valley. The measurement of length, i.e., surface. The measurement of depth.
Burchardus Mithobius[f]	The true position of the sun on any day of the year using the table mentioned above.[g] Another rule on the solar motion. The diurnal hour. The elevation of the pole. How much the zenith is separated from the pole. Whether the hour is before or after noon. When the sun rises and sets. The duration of the day and of the night. The directions of the world. Taking altitudes with the ring. The winds.[h]

a My transcription of the *argumenta* of the *Prioris partis capita* (8a–8b) and *Posterioris partis capita* (by Francisco Hymanno Gandensi, 26b–27a) of the edition of 1558: *Partium declaratio, Solis in signifero locum vel gradum invenire; Loci latitudinem quomodo invenias; Horas aequales interdiu, et mundi plagas, cum aliis nonullis, addiscere; Nocturnae horae inventio; Poli situm, et utram in partem stellae declinent, observare; Quota in alio tractu hora sit per diei ac noctis capienda; Ortus et occasus horam alia ratione investigare; De planetarum seu inaequalibus horis; Ab ortu vel occasu Solis, quota hora sit investigare; De Solis et stellarum altitudine; Solis et stellarum altitudines coelo nubilo ac quolibet tempore inquirere; Cuislibet gradus eclipticae declinationem supputare; Aliquot stellarum declinationes invenire; Solis meridiana vel stellarum maxima altitudinis observata, locorum latitudines indagare; Solis locum et mensis diem venari; Idem meridiana Solis observata altitudine etiam indagare; Idem diei quavis hora inquirere; Solis etiam minime conspicui meridianam altitudinem quolibet tempore obtinere; Meridianam lineam plano inscribere; Quo horizontis puncto Sol vel stella quolibet die oriatur et occidat; Altitudinum per umbras dimensio; Idem calculorum ratione addiscere; Stili seu gnomonis umbra idem investigare; Unica facta statione altitudinem dimetiri; De inaccessarum rerum altitudine capienda; Sine calculorum ratione idem obtinere; Quantum altitudine inaccessa absis, dimetiri; Rerum edito loco sitarum longitudines investigare; Idem faciliori modo obtinere; De distantiarum mensuratione; Profunditatum dimensio*, see http://books.google.co.uk/books?id=V_05AAAAcAAJ&printsec=frontcover&source=gbs_ge_summary_r&cad=0#v=onepage&q&f=false (last accessed 19 February 2017).

b My transcription of the *contenta libelli* (table of contents) of the edition of 1558, 41b: *Declaratio partium; De usu annuli et signi inventione quod Sol quovis die occupet; Eleuatio poli quomodo inveniatur; Horae inventio interdiu; An sit ante meridiem an post; Horae nocturnae investigatio; Qua ratione horae nocturnae facilius inveniantur; De ortu Solis et quantitate diei; De horis inaequalibus sive planetarum; Quota sit hora ab ortu Solis vel occasu; Plagas mundi quomodo invenias; De altitudine Solis et stellarum; Altitudinum dimensio per umbras; De altitudine per solum visum; De altitudinibus rerum inaccessibilium; Facilius idem; De longitudine rerum in aedito sitarum; Aliter idem et facillime; Distantiae dimensio; De profunditatis mensuratione*, see "http://books.google.co.uk/books?id=V_05AAAAcAAJ&printsec=frontcover&source=gbs_ge_sumary_r&cad=0" \l "v=onepage&q&f=false" http://books.google.co.uk/books?id=V_05AAAAcAAJ&printsec=frontcover&source=gbs_ge_sumary_r&cad=0#v=onepage&q&f=false (last accessed 19 February 2017).

c My transcription of the *capitula* (59a–99a) of the edition of 1558 (*Annuli astronomici compositio atque usus*): *De inscriptione meridiani atque aequinoctialis circuli; Aequinoctialis circuli divisio; Zodiaci circuli divisio; De calendario; De tertio sive ultimo annulo; Rimularum tam tertii quam quarti annuli formas hic poni vides; Canones, quibus annulus examinatur nunquid in compositione eius erratum sit; Canon pro rectificatione deferentis qui interior est annulus; Pro rectificationes signorum zodiaci, in deferentis concavitate, descriptorum; De rectificatione circuli latentis in concavitate deferentis, quem Solis indicem vocamus; Hactenus de annuli structura, sequitur eiusdem instrumenti usus; Partes ex quibus annulus constat, declarat; Quomodo optandus sit annulus ad certae regionis poli elevationem; Quarta polaris elevationis; Elevationem poli alia ratione ex Solis meridiana elevatione reperire; Aequatoris circuli elevationem supra horizontem, reperire; Theorica duarum elevationum aequatoris atque poli; Elevationem Solis meridianam, quovis tempore reperire; Quota sit hora diei splendente Sole ex annulo addiscere; Sol sit ne ante meridiem, aut post, addiscere; Quatuor plagae mundi, quomodo indagandae; Lineam meridianam investigare ex annulo; Idem aliter reperire; Tempus calendarii ignotum, quomodo ex annulo depraehendas; Si Solis locum nescias, quomodo ex calendarii noto loco deprehendas; Declinationem Solis, et cuiuscunque gradus zodiaci, addiscere; Si quis in mari aut solitudine erruerit, in quo mundi climate versetur, agnoscere; Elevatio poli; Horologia solaria, in quacunque, sive murali, sive horizontali planitie, adiumento annuli, construere; Horae nocturnae investigatio ex Gemmae Frisii libello; Qua ratione horae nocturnae facilius inveniantur* [*ex Gemmae Frisii libello*]*; De ortu Solis et quantitate diei* [*ex Gemmae Frisii libello*]*; De horis inaequalibus, sive planetarum* [*ex Gemmae Frisii libello*]*; Quando oriatur et quando occidat Sol ex D. Burchardi Mythobii traditione; De metiendis rebus; De scala altimetra annuli; Cuiuslibet rei accessibilis inaequali planitiae erectae, altitudo quomodo sit deprehenda; Idem alio modo; Altitudinem mensurandae rei, per umbram suam observare; Non mutato loco, altitudinem elevatae rei per annulum addiscere; In planitie, si negatur accessus ad rem metiendam, quomodo tum vestiganda sit altitudo; Altitudo rei super montem erectae, cuius altitudinis terminus inferior et summitas videntur, oculo existenti in vale, quomodo dimetiatur; De longitudine, id est, planitie mensuranda; De profunditate mensuranda*, see http://books.google.co.uk/books?id=V_05AAAAcAAJ&printsec=frontcover&source=gbs_ge_summary_r&cad=0#v=onepage&q&f=false (last accessed 20 February 2017).

d The treatise starts with the construction of the instrument: *The inscription of the meridian and equinoctial circle; The division of the equinoctial circle; The division of the zodiac circle; The calendar; The third or last circle; The forms of the inscriptions placed on the third and the fourth circles.* It continues with rules for checking the accuracy of the instrument.

e Here the author introduces several calculations according to other sources: *Finding nocturnal hours according to Gemma Frisius' booklet; Finding the nocturnal hours in an easier way* [*according to Gemma Frisius' booklet*]*; Finding sunrise and the duration of the day* [*according to Gemma Frisius' booklet*]*; Seasonal or planetary hours* [*according to Gemma Frisius' booklet*]*; When the sun rises and sets according to Burchardus Mythobius' method.*

f My transcription of the capitula (125a–149b) of the edition of 1558 (Structura et usus annuli sphaerici): *Aequinoctialis circuli divisio; Appendix compositionis pro hora noctis; Solis locum verum quocunque die anni ex tabula suprascripta indagare; De Solis motu alius canon; Pro hora diurna; Pro elevatione polari; Quantum distet zenith a polo; An sit hora ante meridiem, vel post meridiem; Quando oriatur et quando occidat Sol; Longitudo diei et noctis; De plagis mundi investigandis; Quomodo altitudines per annulum dimetiri liceat; De ventis; Ex aeolia Ioannis Camerarii; Carmen Ausonii de imaginibus coeli; Instrumentum horarum inaequalium; Planetam horae temporalis per annexam tabulam invenire; Annuli mathematici compositio* [*i.e., a sundial ring*]*; Usus annuli mathematici; Quo tempore oriatur vel occidat Sol; Longitudo diei et noctis quomodo investiganda*, see http://books.google.co.uk/books?id=V_05AAAAcAAJ&printsec=frontcover&source=gbs_ge_summary_r&cad=0#v=onepage&q&f=false (last accessed 20 February 2017). Mithobius follows with the description of different types of horologia that are not astronomical rings or sundial rings, which I have not included in the list.

g The treatise starts with some instructions to build the instrument: *Division of the equinoctial circle; Appendix on the construction for the nocturnal time.*

h The text introduces here two poems explaining the winds (from Ioannes Camerarius' *Aeolia* and Ausonius' poem on the constellations of the sky), and continues with the explanation of other instruments, among them the only explanation, in these editions, of the sundial ring: *The instrument of the seasonal hours; To know the planet of the seasonal hour with the attached table; Construction of the sundial ring; Use of the sundial ring; When the sun rises and sets* [with the sundial ring]; *How to find the duration of the day and of the night* [with the sundial ring]. He follows his explanation with other astronomical instruments used for telling the time.

Conclusions

What kind of object was Bonetus's ring? Did it really exist as the treatise and the woodcut suggest? Did it work properly? Was it ever used? There is no reason to doubt that Bonetus's ring existed or that it was functional. Charles de Bovelles's written testimony proves that it existed, for Bonetus showed it to him in Rome, as does the booklet with instructions to use it. In addition, the booklet presented to Pope Alexander VI did not make sense without the offereing of the ring at the same time. The booklet is not a theoretical treatise but a brief and concise list of instructions to handle the object in different situations. Furthermore, the real present to Pope Alexander VI was the ring rather than the booklet, but the former would have been useless without the latter for anyone unfamiliar with astrolabes. Bonetus's ring was a planispheric astrolabe with some structural modifications to make it a finger ring; it displayed an astrolabe and its components, clearly recognizable for anyone who saw it. It had the power of an emblematic object: the astronomical jewel of the Middle Ages, the most prestigious astronomical instrument, the icon of astronomers and astrologers. At the same time, this ring could perform most of the calculations that an astrolabe or an astronomical ring performed. Bonetus's invention crowned the craft of instrument makers, joining the shape and the applications of a planispheric astrolabe with the shape and the function of a jewel, in this case a finger ring, to obtain a beautiful and portable, small yet smart, scientific object. The astrolabe finger ring was charged with all the symbolism of an astrolabe (which represented the pinnacle of medieval science) and a ring (the rings of kings and popes had always been one of the visible tokens of their temporal and/or spiritual power). With this astronomical object one no longer held the world in one's hand, but rather on one's finger. Bonetus carried the image a step further than astronomers in medieval times had in order to convey perfectly the feeling of achievement and success that characterised Renaissance men.

Appendix A: Bonetus's Long Introduction to His Treatise According to the 1507 Paris Edition. Latin Text[34]

Boneti de Latis hebrei medici provenzalis Anuli astronomici utilitatum liber ad Alexandrum sextum pontificem maximum.[35]

Non est gloria sicut gloria notitiam habentis dei nec est exaltatio super exaltationem sapientis operationes suas. Nam, ut ait Hieremias propheta capitulo 9: *Non glorietur sapiens in sapientia sua, nec fortis in fortitudine sua nec dives in divitiis suis, sed in hoc solo glorietur qui gloriatur videlicet scire et noscere me*. Eius autem notitia acquiritur per scientiam stellarum, ut inquit Psalmista psalmo 18: *Celi enarrant gloriam dei et opera manuum eius annunciat firmamentum*. Nam, cum homo corpora celestia et motum accessus et recessus octave sphere ac planetarum eorumque argumenta videt, primam ac magnam et admirabilem causam esse cognoscit, ut ait Psalmista psalmo 8: *Domine dominus noster quam admirabile est nomen tuum in universa terra. Quoniam videbo celos tuos, opera digitorum tuorum Lunam et stellas que tu fundasti.*

Ex cuius verbis tria sunt notanda. Primum quod prius dixit: *videbo celosque Lunam et stellas*, quia universalia sunt nobis prius nota quam singularia, ut vult Aristoteles primo physicorum. Secundum, cum dicit *opera digitorum*, quod dictum voluit intelligi de digitis eclypsis luminariorum quando eclypsantur. Tertio quod ideo de Luna tantum, non autem de Sole qui est candella celi et gubernator mundi mentionem fecit, quia secundum quendam doctorem

34 The transcription of Bonetus's texts has been normalized according to standard Latin.

35 Transcription of the beginning of the *Anulus astronomicus*, which includes the long introduction by Bonetus justifying the study of the stars and the use of its ring, according to the edition printed in Paris in 1507 and now at the Bibliotheca Electoralis, Thüringer Universitäts—und Landesbibliothek in Jena (*Textus De sphera Johannis de Sacrobosco. Cum additione (quantum necessarum est) adiecta: Nouo commentario nuper editio ad utilitate[m] studentiu[m] Philosophice Parisien[sis] Academie illustratus cu[m] Co[m]positione anuli astronomici Boni Latensis. Et Geometria Euclidis Megarensis*, fol. 137r). See http://archive.thulb.uni-jena.de/hisbest/receive/HisBest_cbu_00004311#tab1 (accessed 18 February 2017). I have introduced consistency and changed the punctuation, capitalization, and orthography in the Latin introductions and the Latin treatise to make these texts easier to follow for a modern reader.

nostrum David huiusmodi psalmum de nocte cantavit. Sed vera et melior ratio est quia multa corpora celestia videntur de nocte et propterea ad eorum notitiam habendam ipse David eadem corpora de nocte videre voluit, et tunc maxime cum cognovisse dicitur nomen Tetragrammaton magnum esse et admirabile.

Si ergo Dei notitiam habere desideramus, stellarum scientiam habere oportet. Quod maxime viris ecclesiasticis et cunctis Deum timentibus convenit. Quapropter, pater beatissime Alexander sexte pontifex maxime, etsi tuam humanitatem (divino numine inspirante) omnium virtutum refertam divinarumque ac humanarum rerum notitiam habere intelligam, ad tue tamen sanctitatis laudem et nominis exaltationem ac honorem et gloriam singulorumque tue universali ditioni subditorum commodum et utilitatem quoddam parvum instrumentum adinveni per quod astrorum in firmamento fixorum aliorumque dei operum celestium facili quodam viatico notitia haberi poterit. Quod profecto tue etiam non displicere beatitudini michi facile persuasi ipsumque ad formam anuli formandum censui, cum quia anulus huiusmodi magnorum dominorum et altissimorum virorum ornamentum manuum nobile est, ut apparet libro Genesis capite 45: *Tulit rex Pharao anulum de manu sua et misit eum in manu Joseph*, cum etiam quia quicunque anulum gerens manus suas intuebitur, anulus erit illi figura ante oculos cum quo Solem et Lunam et alia corpora celestia considerabit, et tunc habebit de Deo perfectam notitiam. Perinde ac rex David cum ea de nocte videbat et prosperabitur in ecclesia triumphanti cum isto anulo sicut Aaron pontifex in tabernaculo cum rationali et humerali quod portabat super pectus suum cum quattuor anulis, ut dicitur libro Exodi capite 28: *Stringatur rationale anulis suis cum annulis superhumeralis vitta hiacinthina ut maneat* et cetera. Et *erunt in pectore Aaron quando ingredietur coram domino et gestabit iudicium filiorum Israel* [Ex. 28:30].

Videtur enim michi compositio istius anuli multo facilior astrolabii compositione et quadrati Israelis et aliorum instrumentorum astrologie et minus tediosa, maxime quia est ornamentum nobile ac semper visibile in manibus ad sciendum necessaria et est instrumentum novum. Ideo adaperire debeo suos usus et utilitates que plures sunt. Imprimis autem Deo creatori celi gratias agimus.

English Translation

The Book on the Use of the Astronomical Ring of Bonet de Lattes, Provençal Jewish Physician to Pope Alexander VI

There is no glory like the glory of the one who knows God, nor is there pride higher than the pride of the one who knows His works, as the prophet Jeremiah says in chapter 9: "The wise man does not glory in his wisdom, nor the mighty man glory in his might, nor the rich man in his riches: but he who glories in this alone, namely, in knowing and recognising me."[36] However this knowledge is acquired by the science of the stars, as the Psalmist says in Psalm 18: "The heavens declare the glory of God and the sky proclaims the works of His hands."[37] For when man sees the heavenly bodies and the motion of accession and recession of the eighth sphere and [the motion] of the planets as well as their indications, he knows the first great admirable cause, as the Psalmist says in Psalm 8: "O Lord, our Lord, how excellent is Your name on the whole Earth. When I consider Your heavens, the work of Your fingers, the moon and the stars that You founded."[38]

From these words, three things should be noticed. Firstly, that he mentioned, "I will see the sky, the moon and the stars," because universal things are known prior to individual things, as Aristotle says in the first book of his *Physics*. Secondly, when he says "the work of Your fingers," what he wanted us to understand by "fingers" is the eclipse of the luminaries when they are eclipsed. And thirdly, he mentioned the moon rather than the sun, which is the candle of the sky and the ruler of the world, because, according to a certain master of ours, David sang this psalm at night.[39] But the true and the best reason is because most of the heavenly bodies are visible at night. Therefore, David wanted to see these bodies at night to be able to observe them, and moreover because at that time he is said to have known that the name of the Tetragrammaton was great and wonderful.

---

36 Jer. 9:23 in the King James version. All biblical quotations given here follow this version.

37 Ps. 19:1.

38 Ps. 8:1 and 8:3.

39 There is a double tradition merging in Bonetus's statement regarding King David. On the one hand, David has been considered the author of Psalms, especially those Psalms that begin with the formula: "Mizmor le-David," such as Psalm 8. Regarding this tradition, see Ruth Fidler, "A Pen for All Seasons? Towards a Story of David's Rise as Psalmist" [in Hebrew], *Studies in Bible and Exegesis (Iyyunei Miqra' u-Parshanut)* 10 (2011): 403–428, especially 414, n. 35. On the other hand, there is the Talmudic tradition that David sang his Psalms at night (for instance the Babylonian Talmud, Berakhot 3b). For a basic summary of these Rabbinic traditions, see R. Shraga Weiss, "Kinoro shel David ha-Melech: Masechet Mekorot," *Daf le-Tarbut Yehudit* 251 (2001): 16–18; and Louis Ginzberg, *Legends of the Jews*, 6:262–63 n. 80–84 (and esp. n. 81). I thank Admiel Kosman and Israel Sandman for their valuable information regarding this question.

If we want to know God, we must know the science of the stars, and it is particularly appropriate to churchmen and to all those who fear God. For this reason, most blessed father Pope Alexander VI, although the knowledge of divine and human things should [lead us to] understand that your kindness (by the inspiration of the divine spirit) is packed with every virtue, nevertheless I have invented a certain small instrument to praise your holiness and exalt, honour and glorify your name for the convenience and use of each person who is subject to your universal authority, through which knowledge of the fixed [stars] in the firmament and other celestial works of God can be obtained in a portable and easy way. Certainly I am persuaded that your beatitude would not be displeased by my decision of making it in the shape of a ring, both because a ring of this kind is an excellent ornament on the hands of great lords and illustrious men, as is said in the book of Genesis, chapter 45: "Pharaoh took off his ring from his hand and put it on Joseph's hand,"[40] and also because whoever wears the ring on his hand and looks at it, the shape before his eyes will be for him as if he were looking at the sun, the moon, and the other heavenly bodies. Then he will reach a perfect knowledge of God. Just as king David when he watched them by night and the triumphant church prospered with this ring. And the high priest Aaron, who wore it in the Tabernacle with the breastplate and the ephod upon his breast, with four rings, as is said in the book of Exodus chapter 28: "They shall bind the breastplate by the rings thereof unto the rings of the ephod with a lace of blue, that it may stay …"[41] And "they shall be upon Aaron's heart, when he goes in before the Lord; and Aaron shall bear the judgement of the children of Israel".[42]

I think that the construction of this ring is much easier than the construction of the astrolabe, Israel quadrant,[43] and other astrological instruments, and less tedious, especially since it is a splendid ornament, always visible on the hand to know what is necessary, and a new instrument. Therefore, I have to

40 Gen. 41:42.

41 Ex. 28:28.

42 Ex. 28:30.

43 This is a reference to the *quadrans novus* or *azafea* described by Azarquiel (eleventh century). Jacob ben Makhir ibn Tibbon (thirteenth century) translated its description into Hebrew and there are also several versions of it in Latin. See J. M. Millás Vallicrosa, *Don Profeit Tibbon: Tractat de l'assafea d'Azarquiel* (Barcelona: Arts Gràfiques, 1933), which contains the Hebrew, Latin, and Catalan versions.

disclose its uses and applications, for they are many. First of all, we thank God, creator of the heavens.[44]

Appendix B: Annuli Astronomici, Boneto Hebraeo, Lutetiae 1557 (103b–117b). Latin Text

(103b) ***Boneti de Latis Hebraei medici provenzalis annuli astronomici utilitatum liber ad Alexandrum sextum Pontificem maximum.***

Videtur mihi compositio istius annuli multo facilior Astrolabii compositione et quadrati Israelis et aliorum instrumentorum astrologiae et minus tediosa, maxime quia est ornamentum nobile ac semper visibile in manibus ad sciendum necessaria et est instrumentum novum. Ideo adaperire debeo suos usus et utilitates, quia plures sunt. Imprimis autem Deo creatori coeli gratias agimus.

(104a) ***De declaratione annuli. Cap. I.***

Nota quod annulus debet fieri de auro vel argento aut alio genere metalli ad modum sigilli ad portandum in digito et eius declaratio talis est. Super circuitu annuli sunt 12 signa in suis characteribus: primum est Aries, secundum Taurus, et ita deinceps. Quodlibet signum continet 30 gradus, dividendo de quinque in quinque gradibus per quaelibet puncta. Item sunt duodecim menses in literis, Martius in ista litera m. Aprilis vero in ista litera a. et sic de aliis. Et quilibet mensis dividitur de 5 in 5 dies per puncta, et initium Martii est in 20 gradu signi Piscium, et per istos menses et characteres signorum scietur locus Solis omni die.

Intra annulum sunt 90 gradus de 5 in 5 lineis ad inveniendum altitudinem Solis et stellarum. Et initium numeri istorum 90 graduum est in linea medii foraminis, in quod intrat radius Solis ad cognoscendum eius altitudinem. Item in superficie annuli recta et plana (quae est super annulum) sunt tres circuli, quorum primus est minimus et est circulus Cancri, medius aequinoctialis, maximus vero eorum, circulus Capricorni, qui est ultimus in (104b) circunferentia illius superficiei rectae et planae.

44 This last paragraph is the short introduction of the treatise in the editions that do not have the long introduction I have reproduced here.

Postea sunt almicantarat, qui sunt in medietate circuli superioris descripti, quorum quidam integri apparent, quidam imperfecti, quorum primus horizon in duo dividitur hemispheria, et dicti circuli almicantarat continent 10 gradus, et sunt in summa 90 gradus a primo usque ad centrum inferioris eorum, quod zenith capitis nominatur. Post sunt 4 erraticae horae infra almicantarat. Postea linea medii coeli, quae est linea descendens a cruce, cuius pars a centro in quo est foramen dicitur linea meridiei, et alia dicitur angulus terrae et linea mediae noctis. Postea in circuitu superficiei in spissitudine eius sunt 4 horae signatae in lineis. Postea sequitur rotula mobilis posita super illam superficiem, in qua sunt signa 12 characteribus cum zodiaco constituto et dicitur rotula zodiaci.

Item sunt in dicta rotula zodiaci 4 puncta, quae 4 stellas fixas significant. Primus punctus significat Aldebaran et est in secundo gradu signi Geminorum. Secundus significat cor Leonis et est in 23 gradu signi Leonis. Tertius significat cor Scorpionis et est in secundo gradu signi Sagittarii. Quartus significat caudam Capricorni et est in 15 gradu signi Aquarii. Item in rotula zodiaci sunt 28 lineae, quarum septem 3 signa continent et significant 28 mansiones Lunae.

Deinde (105a) sequitur foramen, quod est in medio lineae rotulae zodiaci, quod transit per caput Arietis et Librae, in quo est axis continens rotulam cum annulo et restringens eam. Postea sunt duo foramina, unum est in medio annuli ad suspendendum annulum ante Solem cum filo, alterum foramen est in summitate annuli pro introitu radii Solis ad sciendum altitudinem eius. Postea est baculus perforatus, qui intrat in foramen altitudinis Solis, tum est mobilis et non est semper cum annulo nisi de nocte per altitudinem stellarum. Postea est filum ad suspendendum annulum.

Ad inveniendum locum Solis in omni die. Cap. II.

Si velis scire locum Solis in quo signo est et in quo gradu eius in illa die, pone filum super diem quem habes et nota super quem ex gradibus signorum cadat filum, ille denotat locum Solis in illa die.

Exemplum: hodie est 10 mensis Martii. Pone ergo filum super decimum mensis Martii, et quia super primum gradum Arietis cadit filum, in illo gradu est Sol hodie.

(105b) *Ad sciendum diem ignotum. Cap. III.*

Si diem ignotum invenire desideras, scias in quo gradu est Sol, et in illo gradu in quo est pone filum, et nota super quem ex diebus mensium cadat filum; ille denotat diem mensis ignoti.

Exemplum: hodie Sol est in primo gradu signi Arietis, pone filum super illum gradum, et cadet super diem 10 Martii. Et si ignoraveris locum Solis, fac sicut dicetur in capitulo 15.

Ad inveniendum altitudinem Solis. Cap. IIII.

Si altitudinem Solis scire desideras, suspende annulum de manu tua dextra per filum eius et pone eum ad Solem donec per foramen transeat radius Solis, et in superficie intrinseca annuli vide in quo gradu ex 90 gradibus cadat radius et illa est altitudo Solis.

Ad inveniendum gradum ascendentis et signum ascendentis. Cap. V.

Si certitudinem ascendentis volueris scire, scias primo locum Solis; post scias altitudinem Solis; post quaere (106a) gradum in rotula zodiaci in quo est Sol, et pone super almicantarat altitudinis ex parte Orientis, si fuerit ante medium diem, aut ex parte Occidentis, si post medium diem accepisti altitudinem, et ille gradus qui ceciderit super lineam horizontis (quae est prima linea almicantarat) est gradus ascendentis, et signum illius gradus est signum ascendens, id est, Oriens. Quod vero in occidentali, Occidens. Quod autem ceciderit in linea medii coeli est in medio coeli et eius nadair[45] in angulo terrae.

Exemplum: hodie locus Solis est in primo gradu Arietis, et altitudo Solis ante meridiem est 30 graduum. Pone primum gradum Arietis ex parte orientali super almicantarat 30 graduum, et super primum almicantarat cadet secundus gradus signi Geminorum; ille gradus secundus Geminorum est gradus ascendentis, et signum Geminorum est signum ascendens.

Ad inveniendum duodecim domos coeli. Cap. VI.

Cum duodecim domos coeli volueris adaequare, quaere primo gradum ascendentis, et ille gradus qui ceciderit super lineam horizontis est prima domus. Post signa gradum qui ceciderit super (106b) lineam medii coeli, in qua linea, volvendo rotulam zodiaci, pone gradum nadair illi gradui medii coeli, et postea vide in parte lineae Occidentis gradum correspondentem horae primae erraticae, et ille gradus est initium secundae domus. Deinde immota rotula zodiaci,

---

45 I keep the form *nadair*, for it seems that this form was circulating in this period together with the common *nadir* (also later in the text). I did likewise with the forms *almicantarat* and *almucantarat*, both frequently used in the medieval period and later, and present in Bonetus's text.

vide gradum correspondentem horae erraticae secundae et ille gradus est initium tertiae domus. Postea vide gradum correspondentem lineae medii coeli et ille gradus est quarta domus. Postea vide gradum correspondentem horae erraticae tertiae et ille gradus est initium quintae domus. Postea vide gradum correspondentem horae quartae erraticae et ille gradus erit initium sextae. Initium autem septimae domus est nadir ascendentis. Et initium octavae est nadir secundae. Initium quoque nonae est nadir tertiae. Et initium decimae est nadir quartae. Et initium undecimae est nadir quintae. Et initium duodecimae est nadir sextae.

Exemplum: gradus ascendentis est secundus gradus signi Geminorum sicut dictum est et est prima domus, et gradus qui cecidit super lineam medii coeli est 14 Aquarii. Volve rotulam zodiaci et pone 14 Leonis gradum super lineam medii coeli et gradus correspondens horae primae erraticae est 26 Geminorum et est secunda domus; et gradus correspondens horae secundae erraticae est 20 Cancri et est initium tertiae domus; et gradus correspondens medio coeli est 14 (107a) Leonis et est initium quartae domus; et gradus correspondens tertiae horae erraticae est 13 Virginis et est initium quintae domus; et gradus correspondens quartae horae erraticae est 14 Librae, qui est initium sextae domus; gradus nadair primae domus est 2 Sagittarii et est septima domus; et gradus nadair secundae domus est 26 Sagittarii et est octava domus; et gradus nadair tertiae est 20 Capricorni et est nona domus; et gradus nadair quartae est 14 Aquarii et est decima domus; et gradus nadair quintae est 13 Piscium et est undecima domus; et gradus nadair sextae est 14 Arietis et est duodecima domus.

Ad sciendum quot sunt horae ex horis xii sicut est in Gallia. Cap. VII.

Cum scire volueris quot sunt horae de horis duodecim, scias primo Solis altitudinem, et pone gradum Solis super almicantarat altitudinis ex parte Orientis si fuerit ante meridiem, et pone super illum gradum Solis filum, et vide in circulo horarum super quam horam ceciderit filum, et numera a linea anguli terrae quae est mediae noctis usque ad illam horam in qua filum cadit, et habebis quot (107b) sunt horae de horis 12.

Exemplum: hodie locus Solis est in primo gradu Arietis et altitudo eius est in 30 gradu; pone primum gradum Arietis zodiaci super almicantarat 30 gradu ex parte Orientis, super hunc gradum pone filum et cadet in circulo horarum super horam 9 numerando ex linea mediae noctis. Et si fuerit post meridiem pone gradum Solis ex parte Occidentis et numera a linea medii coeli quae est meridies, et scies quot sunt horae post meridiem sicut est in patria Gallicana, in qua natus sum.

Ad sciendum quot sunt horae ex horis viginti quatuor. Cap. VIII.

Si scire volueris quot sunt horae ex horis 24 pone gradum Solis super lineam horizontis ex parte Occidentis, et fac signum in circulo horarum cum filo; post volve rotulam zodiaci versus angulum terrae et pone illum gradum Solis super almicantarat altitudinis ex parte Orientis, si fuerit ante meridiem, aut ex parte Occidentis, si post meridiem accepisti altitudinem. Deinde fac signum cum filo in circulo horarum, et numera in eodem per viam anguli terrae ab illo signo primo usque ad signum secundum, et habebis quot de 24 sicut est (108a) in Italia.

Exemplum: hodie locus Solis est in primo gradu Arietis et eius altitudo 30 gradus; pone primum gradum Arietis supra lineam horizontis ex parte Occidentis, et fac ibi signum in circulo horarum cum filo et illa hora est prima. Post volve rotulam zodiaci et pone gradum primum Arietis super almicantarat 30 gradus ex parte Orientis, quia nunc est ante meridiem, et fac signum cum filo, et numera a primo signo quod est prima hora per viam anguli terrae usque ad secundum signum I et erunt 15 horae. Et si est post meridiem, pone gradum illum 30 gradus in almicantarat ex parte Occidentis et erunt 21 horae.

Ad sciendum quot sunt horae ab ortu Solis. Cap. IX.

Si scire volueris quot horae aequales transierunt de die, accipe gradum Solis et eius altitudinem, et pone eundem super almicantarat altitudinis, et tunc signa locum in circulo horarum; postea volve retro gradum Solis usque ad primum almicantarat ex parte Orientis, et secundo nota locum eiusdem in dicto circulo; postea numera ab illo signo primo usque ad secundum et habebis quot sunt horae ab ortu Solis. Et si est post meridiem (108b) pone illum gradum Solis ex parte Occidentis super almicantarat altitudinis, et numera ab illo signo primo usque ad secundum per viam medii coeli.

Exemplum: hodie locus Solis est in primo gradu Arietis et altitudo eius est in 30 gradu; pone illum gradum 30 gradus super almicantarat ex parte Orientis, quia est ante meridiem; post pone illum gradum super horizontem, et fac signum et numera a primo signo usque ad secundum, et erunt 3 horae aequales quae transierunt de die. Et si est post meridiem, pone illum gradum ex parte Occidentis, et erunt 9 horae ab ortu Solis quae transierunt de die.

Ad sciendum quantitatem diei et noctis. Cap. X.

Si volueris arcum diei cognoscere, pone gradum Solis super primum almicantarat ex parte Orientis, et fac signum in circulo horarum. Postea volve rotulam zodiaci per viam medii, et pone illum gradum in linea horizontis ex parte Occidentis, et nota eius locum in circulo horarum cum filo, et horae et partes earum quae sunt ab una nota in aliam est arcus diei; reliqua vero pars circuli est arcus noctis, quia totus circulus continet 24 horas et quantitatem diei et noctis.

Exemplum: hodie Sol est in primo gradu Arietis; pone primum gradum Arietis super horizontem (109a) ex parte Orientis; postea pone illum gradum ex parte Occidentis super horizontem, et numera partes horarum per viam medii coeli quae sunt ab una nota in aliam, et sunt 12 horae. Ergo 12 horae est quantitas diei hodie.

Ad sciendum quantitatem horae erraticae diei. Cap. XI.

Si volueris scire arcum horae inaequalis diei, scias primo arcum diei sicut dictum est, et horas illius quantitatis multiplica per 15, et illam summam divide per 12, et habebis numerum graduum horae inaequalis diei. Quam si subtraxeris a 30, remanebit numerus graduum horae nocturnae, quia hora inaequalis diurna cum hora inaequali nocturna facit 30 gradus in omni die, quae sunt duae horae aequales.

Exemplum: hodie quantitas diei est 14 horae; multiplica eas per 15 et erunt 210; quem numerum divide per 12 et erunt 17 gradus et 5 minuta. Ergo 17 gradus et 5 minuta est arcus horae erraticae diei. Deinde subtrahe dictum numerum a 30 et remanebunt 12 gradus et 55 minuta, et est numerus graduum horae erraticae nocturnae. Et hoc caput est apprime necessarium in Astrologia ad sciendum dominum horae, ut dicitur in capite 24.

(109b) *Ad sciendum quantitatem horae aequalis. Cap. XII.*

Si volueris scire arcum horae aequalis, divide 360 per 24 horas quae sunt in circulo horarum, et habebis numerum graduum horae aequalis de die et de nocte, quia illae horae circuli horarum sunt aequales et quaelibet continet 15 gradus.

Ad reducendum horas erraticas ad aequales. Cap. XIII.

Cum volueris reducere horas inaequales ad horas aequales, scito gradus horarum inaequalium quot sunt et divide gradus earum per 15 et habebis horas aequales.

Exemplum: quantitas horae erraticae est 17 gradus et 5 minuta et sunt 12, qui sunt 210 gradus; divide per 15 et erunt 14 horae aequales.

Ad sciendum altitudinem Solis in meridie. Cap. XIIII.

Cum scire volueris altitudinem Solis in media die, quae est initium recessionis, pone gradum (110a) Solis supra lineam medii coeli, et numerus graduum almicantarat a loco Solis in horizonte est altitudo eiusdem mediae diei, dummodo annulus sit factus ad altitudinem climatis in quo quaeris altitudinem.

Exemplum: hodie Sol est in primo gradu Arietis; pone primum gradum Arietis supra lineam medii coeli, et numera ab Oriente usque ad almicantarat in qua applicatur ille gradus super lineam medii coeli, quot sunt gradus, et erunt 48. Ergo altitudo Solis in media die illius diei est 48 gradus. Ante scito quod circulus signorum dividitur in duos semicirculos, quorum unus est a capite Capricorni in caput Cancri et alius a capite Cancri in caput Capricorni. Et caput Capricorni est solstitium hyemale, et caput Cancri aestivale. Scito etiam quod omnes aeque distantes gradus ab aliquo horum solstitiorum sunt unius declinationis versus septentrionem vel meridiem, et dies eorum vel noctes sunt aequales, et umbrae et altitudines aequales in media die.

Ad sciendum locum Solis sine notitia diei. Cap. XV

Cum scire volueris gradum Solis ignoto die, pone notam super altitudinem mediae diei in me- (110b) -dietate quam sumpsisti prius per foramen quod est ad sciendum altitudinem Solis, deinde volve rotulam zodiaci, cadentque duo gradus super ipsam notam, quorum unum scies esse gradum Solis, quo cognito scies diem mensis.

Exemplum: hodie altitudo Solis est 48 in media die; volve rotulam et vide gradum signorum zodiaci cadentem super almicantarat altitudinis graduum, et erit primus gradus Arietis vel primus gradus Librae, et quia est tempus hyemale, ergo Sol est in primo gradu Arietis. Pone filum in circulo mensium super primum gradum Arietis, et cadet super decimum mensis Martii, et per hoc caput habebis locum Solis et diem ignotum.

Ad sciendum quae dies sit aequalis alteri. Cap. XVI.

Cum volueris scire quae dies alteri sit aequalis, scias hoc per gradus aeque distantes a solstitiis, quia eorum dies sunt aequales, sicut dictum est.

Exemplum: dies 30 mensis Novembris est aequalis die 25 mensis Decembris, quia solstitium hyemale est quando Sol est in capite Capricorni, quod est in die duodecima mensis Decembris, et sic de aliis.

(111a) ***Ad sciendum declinationem signorum ab aequinoctiali linea. Cap. XVII.***

Si scire volueris declinationem cuiuscunque gradus signorum, pone eum super lineam medii coeli et scito eius altitudinem ab horizonte. Postea scito altitudinem capitis Arietis vel Librae in eadem linea, et differentia huiusmodi altitudinum est declinatio huius gradus ab aequinoctiali. Si autem fuerit gradus septentrionalis, septentrionalis est declinatio, si meridionalis, meridiana.

Exemplum: pone primum gradum signi Cancri super lineam medii coeli, et numera ab horizonte usque ad almicantarat altitudinis eius, et erunt hic Romae 72 gradus, et altitudo primi gradus Arietis est 48; ergo declinatio primi gradus Cancri ab aequinoctiali linea est 24 gradus, qui sunt differentia utrarumque ipsarum altitudinum.

Ad sciendum latitudinem civitatis. Cap. XVIII.

Si volueris scire latitudinem civitatis quae est distantia zenith capitis civitatis eius ab aequino- (111b) -ctiali linea, considera altitudinem Solis in media die, quam minues de 90 si fuerit Sol in initio Arietis vel Librae, quae est in 10 die Martii vel Septembris, et residuum est latitudo civitatis; tunc enim erit motus Solis in aequinoctiali linea. Si vero in alio gradu fuerit Sol et in alia die, tunc eius gradus considera declinationem per caput praecedens, quam minues de altitudine Solis in media die, si fuerit septentrionalis, vel adde eidem si fuerit meridiana, et habebis altitudinem initii Arietis in illa civitate; quam subtrahas (sicut prius dictum est) a 90, et quod remanserit erit distantia regionis ab aequinoctiali linea.

Exemplum: quando Sol est in primo gradu Arietis vel Librae, altitudo Solis tunc est 48 in media die; quam minues de 90 et remanebunt 42; ergo 42 est latitudo civitatis Romae.

Exemplum: in aliis diebus quando Sol est in alio signo sicut in Cancro, considera declinationem primi gradus eius, quae est 24, et minue illam declinationem ex altitudine eius, quae est 72, et remanebunt 48; vel adde super altitudinem primi gradus Capricorni, quae est 24, et erunt 42[46]; subtrahes 48 a 90 et remanebunt 42. Ille numerus 42 est latitudo huius civitatis Romae.

(112a) ***Ad sciendum ad quam latitudinem annulus factus est. Cap. XIX.***

Si vis scire ad quam latitudinem factus est annulus, vide quot almicantarat sunt in circulo aequinoctiali usque ad zenith, vel ab axe ad horizontem ad

---

46 This number should be 48, though the number 42 seems to be consistently reproduced in all the editions I have consulted.

septentrionem, et super tantam latitudinem factus est annulus, haec enim sunt aequalia; hic factus est ad latitudinem Romae. Facies tamen tibi unum ad latitudinem quam volueris cum mutatione superficiei almicantarat, ut patet in tabulis astrolabii instrumenti.

Ad sciendum ascensiones signorum in circulo directo. Cap. XX.

Si ascensiones signorum in circulo directo scire desideras, initium cuiusvis signi super lineam meridianam pone, et fac signum in circulo horarum cum filo, et move rotulam zodiaci donec finis signi sit super lineam meridiei, et hoc iterum signa in circulo horarum, et horas quae erunt inter duas notas multiplica per 15 et habebis numerum (112b) graduum ascensionum eiusdem signi. Et similiter facies ad quamlibet partitionem circuli, quia quaelibet hora continet 15 gradus, sicut dictum est.

Exemplum: pone initium signi Arietis super lineam meridionalem, et post move rotulam zodiaci, et fac signum in illis duobus locis, et horae quae sunt inter duo signa sunt una hora et quatuor partes unius. Multiplica per 15 et sunt 27; ergo ascensiones signi Arietis in circulo directo sunt 27 gradus.

Ad sciendum ascensiones signorum in circulo obliquo. Cap. XXI.

Si hoc scire volueris, move rotulam zodiaci ab initio signi ad finem eiusdem super primum almicantarat, et gradus horarum in circulo horarum erunt ascensiones signi in illa regione, quia quaelibet hora continet 15 gradus.

(113a) *Ad sciendum in quo signo sit Luna sine altitudine eius. Cap. XXII.*

Cum hoc scire volueris, scias primo locum Solis, postea scito diem in quo fuit coniunctio Lunae, et numerabis a loco Solis super rotulam zodiaci per lineas quae signant 28 mansiones Lunae tot quot sunt dies quae transierunt a die conniunctionis Lunae. Et illud signum quod inuenies per dictum numerum, ostendet lineam in qua erit Luna.

Exemplum: hodie Sol est in primo gradu Arietis, et a die coniunctionis Lunae usque ad praesentem diem sunt dies 23. Numera a loco Solis lineas 23 et pone filum super lineam 23 et invenies lineam in fine signi Capricorni, in qua est locus Lunae hodie.

Ad inveniendum ascendens revolutionis anni natalis vel mundi. Cap. XXIII.

Cum scire volueris hunc gradum ascendentis anni transacti, super horizontem in Oriente pone gradum Solis et in circulo horarum fac notam cum filo. Post hoc illum gradum move ab eo- (113b) -dem loco per horas 6 et quintam partem unius horae, quae sunt 93 gradus, et gradus qui ceciderit super horizontem est gradus ascendentis eiusdem anni. Si autem plures fuerint anni, pro unoquoque anno deduces illum gradum per sex horas et quintam partem, quae sunt nonaginta tria. Et gradus existens in parte orientali est ascendens ipsius.

Exemplum: ascendens revolutionis istius anni est 8 gradus signi Capricorni; pone eum super horizontem in Oriente; deinde move ipsum per 6 horas et quintam partem unius, et primus gradus Geminorum erit ascendens anni futuri.

Ad inveniendum dominum horae de die et de nocte. Cap. XXIIII.

Si hoc scire desideras scias horam erraticam noctis illius diei per caput II, quia quaelibet hora erratica habet suum dominum unum ex 7 planetis secundum ordinem eorum in coelo, videlicet in die Sabbati dominus primae horae est Saturnus, dominus secundae est Iupiter, et sic per ordinem usque ad 12, quia quaecunque dies sit vel nox longa vel brevis, continet 12 horas inaequales et super quantitatem illius horae erraticae inventae, domi- (114a) -natur unus planeta. Et hoc amplius apparet in tabula inferius descripta.

	1	2	3	4	5	6	7	8	9	10	11	12
Nox dominicae	M	L	s	I	m	S	V	M	L	s	I	m
Dies dominicae	S	V	M	L	s	I	m	S	V	M	L	s
Nox Lunae	I	m	S	V	M	L	s	I	m	S	V	M
Dies Lunae	L	s	I	M	S	V	M	L	s	I	m	S
Nox Martis	V	M	L	S	I	m	S	V	M	L	s	I
Dies Martis	m	S	V	M	L	s	I	m	S	V	M	L
Nox Mercurii	s	I	m	S	V	M	L	s	I	m	S	V
Dies Mercurii	M	L	s	I	m	S	V	M	L	s	I	m
Nox Iovis	S	V	M	L	s	I	m	S	V	M	L	s
Dies Iovis	I	m	S	V	M	L	s	I	m	S	V	M
Nox Veneris	L	s	I	M	S	V	M	L	s	I	m	S
Dies Veneris	V	M	L	S	I	m	S	V	M	L	s	I
Nox Sabbati	m	S	V	M	L	s	I	m	S	V	M	L
Dies Sabbati	s	I	m	S	V	M	L	s	I	m	S	V

M, Mercurius. L, Luna. s, Saturnus. I, Iupiter, m. Mars. S, Sol. V, Venus.

Ad sciendum altitudinem stellarum. Cap. XXV.

Cum hoc scire volueris, suspende de manu tua dextra per filum, et per baculum per- (114b) -foratum vide stellam, et in superficie intrinseca annuli vide in quo gradu de 90 gradibus cadat baculus, et ille stellarum est altitudo.

Ad sciendum gradum ascendentis de nocte. Cap. XXVI.

Cum hoc scire desideras, accipe altitudinem unius stellae fixae de illis quatuor quae sunt in punctis in rotula zodiaci, et pone gradum in quo est illa stella fixa in almicantarat suae altitudinis, ex parte Orientis si stella est in Orientis parte, vel ex parte Occidentis si est in Occidente, et ille gradus qui ceciderit super lineam horizontis quae est prima linea almicantarat est gradus ascendentis.

Exemplum: hodie locus cordis Leonis stellae est in gradu 23 signi Leonis, et altitudo eius stellae in parte Orientis est 30 gradus; pone gradum 23 Leonis super almicantarat 30 gradus, et super horizontem cadet 20 gradus signi Virginis. Ille gradus 20 Virginis est gradus ascendentis.

(115a) *Ad sciendum quot sunt horae de nocte. Cap. XXVII.*

Si ad illius notitiam pervenire volueris, accipe altitudinem unius stellae fixae ex illis quatuor quae sunt in rotula zodiaci, et pone gradum in quo est in almicantarat suae altitudinis ex parte Orientis, si est in Oriente, vel ex parte Occidentis, si est in Occidente. Postea pone filum supra gradum Solis, et fac signum in circulo horarum. Post volve rotulam zodiaci et pone dictum gradum Solis super primum almicantarat Occidentis, et fac signum cum filo in circulo horarum, et numera horas quae sunt inter duas notas; tantae erunt horae noctis. Et si volueris scire quando erit media nox, vide quot sunt horae ex primo signo ad lineam mediae noctis. Et si volueris scire quot sunt horae de 12, numera ex linea medii coeli usque ad gradum Solis in circulo horarum per viam lineae Occidentis, si gradus Solis est ante lineam mediae noctis, et si est post, numera a linea mediae noctis per viam Orientis usque ad lineam medii coeli.

Exemplum: gradus cordis Leonis est 23 Leonis et eius altitudo 30 in Occidente, et Sol est hodie in primo gradu Arietis; pone 23 Leonis super almicantarat suae altitudinis, et fac signum super primum gradum (115b) Arietis. Post volve rotulam zodiaci, et pone primum gradum Arietis super almicantarat primum Occidentis, et numera horas quae sunt inter duas notas, et erunt horae 7 et 45 minuta, et post mediam noctem erit una hora et 45 minuta de horis 12.

Ad sciendum gradum ascendentis sine latitudine et stellis de nocte et sine Sole de die. Cap. XXVIII.

Si ad illius notitiam pervenire volueris, scias quot sunt horae; postea pone gradum Solis in illo puncto horae in circulo horarum, et ille gradus signorum zodiaci qui ceciderit super primum almicantarat in parte Orientis est gradus ascendentis.

Exemplum: hodie Sol est in primo gradu Arietis et nunc sunt tres horae post meridiem; pone gradum Solis correspondentem illi horae in circulo horarum, et cadet super primum almicantarat 10 Geminorum, et hic est gradus ascendentis.

(116a) ***Ad sciendum in quo gradu signi sit Luna sine notitia coniunctionis. Cap. XXIX.***

Cum hoc cognoscere cupis, considera altitudinem Lunae, et eam nota in almicantarat in qua parte fuerit; deinde aliquam quatuor stellarum quae sunt in rotula zodiaci tibi notam pone super suam altitudinem in eadem hora cum altitudine Lunae accepta in parte in qua fuerit, et gradus circuli zodiaci qui ceciderit in almicantarat super notam altitudinis Lunae erit gradus eius. Si autem apparuerit in die, idem facies cum altitudine illius et altitudine Solis. Considera igitur cuius signi sit gradus ille.

Exemplum: nunc altitudo Lunae est in Occidente 30 gradus, et altitudo stellae cordis Leonis in Oriente est 40. Pone 23 gradum Leonis in quo est stella super almicantarat in Oriente 40 gradus, et quia sex gradus Piscium cadit super almicantarat altitudinis Lunae, quae posita est 30, in illo igitur gradu est Luna.

(116b) ***Ad inveniendum locum planetarum. Cap. XXX.***

Cum volueris scire locum planetarum, facias sicut dictum est in capite praecedenti ad sciendum locum Lunae; sic etiam altitudines earum in nocte poteris elicere.

Ad sciendum quando planeta sit australis vel septentrionalis. Cap. XXXI.

Si id deprehendere volueris, vide altitudinem planeta quando erit in linea meridiana, et si est aequalis altitudini gradus Solis, tunc est in via Solis. Si autem fuerit maior altitudine gradus Solis, tunc planeta est septentrionalis a via Solis; si minor, australis. Et tantum declinat a via Solis, quantum altitudo fuerit maior vel minor.

Ad sciendum an planeta sit retrogradus vel directus. Cap. XXXII.

Utrum autem planeta sit retrogradus aut directus si cognoscere velis, quaere eius altitudinem (117a) cum altitudine stellae fixae, et ambas commenda memoriae. Deinde post tertiam noctem vel quartam insensibilis est qua stellae motus expecta quousque stella habeat eandem quam prius altitudinem, et tunc rursus altitudinem planetae considera; quae si fuerit minor altitudine sua prima, erit planeta directus, si fuerit in parte orientali; et si fuerit in parte occidentali, erit retrogradus. Si vero secunda altitudo planetae fuerit maior prima, est retrogradus, hora acceptae altitudinis consistente in parte Orientis; et si fuerit in parte Occidentis, erit directus. Et oppositum est de Luna.

Ad sciendum altitudinem turris vel aliarum rerum. Cap. XXXIII.

Pone baculum perforatum super gradum 45 in numero 90, qui est pro altitudine Solis et stellarum, et accipe altitudinem turris non dimovendo baculum ab illo gradu 45. Tanta enim est altitudo turrus cum altitudine tua, quanta est latitudo quae est inter te et radicem rei. Et si est impossibile mensurare illam latitudinem quae est inter te et rem, vide sine mutando locum in terra, et mensura a loco tuo usque ad locum terrae quem vi- (117b) -disti, et tanta est altitudo rei.

Bonetus de Latis Provenzalis Alexandro VI *Pontifici Maximo*

Haec sunt, beatissime pater, annuli astronomici puncta peregregia, una mecum ad sanctos tuos pedes humillime oblata, quae positis superciliis hilari vultu, ut spes fovet, recipias. Nec mirum si grammaticae metas qui hebraeus sum latinitatis expers nonnunquam excesserim, nolens utile per inutile vitiari. Malui tibi rosulas in vili, quam urticas aut lolium in preciosa offerre sportula, ut quae ad salutem tuam totiusque Reipublicae commodum omniumque rerum opificis laudem utilia comperta sunt, ob humilem contextum vocabulorum non omitterentur. Quin potius sub summa tua maiestate atque authoritate ab omnibus amplius cognoscerentur.

Parce, precor, rudibus quae sunt errata latino. Lex hebraea mihi est, lingua latina minus.[47]

BONETI DE LATIS HEBRAEI ANNVLI ASTRONOMICI FINIS.

47 The last paragraph is completely different from the style and language of the rest of the text. It has been clearly written by a more expert Latinist than Bonet.

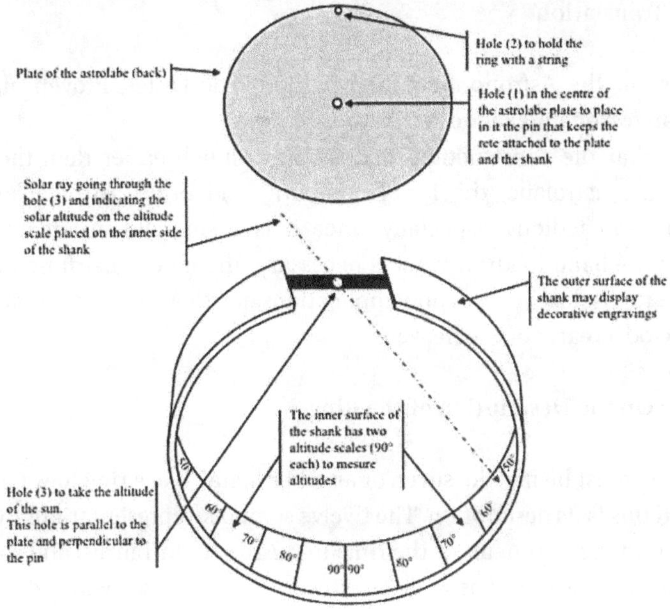

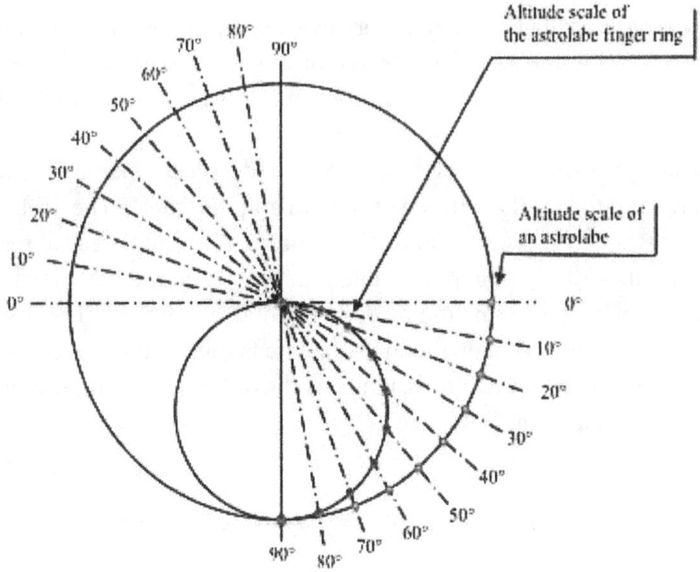

FIGURES 2.6 AND 2.7 *Ring in perspective (above) and divisions of the altitude scale on the shank of the ring (below).*

English Translation

Book On the Use of the Astronomical Ring by Bonet de Lattes, Provençal, Hebrew Physician to Pope Alexander VI[48]

(103b) I think that the construction of this ring is much easier than the construction of the astrolabe, the Israel quadrant, and other astrological instruments, and less tedious, especially since it is a splendid ornament, always visible on the hand to know what is necessary, and a new instrument. Therefore, I have to disclose its uses and applications, for they are many. First of all, we thank God, creator of the heavens.

(104a) Chapter 1: On the Description of the Ring

Know that the ring must be in gold, silver, or another metal, like a ring to wear on the finger, and this is its description. The twelve symbols/representations of the signs are around the perimeter of the ring: first Aries, then Taurus, and so successively.[49] Each of the signs has 30 degrees, every five separated with a dot. The same as regards the twelve months with their [initial] letters: March is represented with the letter M, April with the letter A, and so with the remaining. Each of the months is divided into days, every five [separated] with a dot. The beginning of March is in the twentieth degree of the sign Pisces. One knows the position of the sun every day through these months and the symbols of the signs.

Inside the ring there are ninety degrees, every five [degrees] separated with a line, which are used to find the altitude of the sun and the stars.[50] The beginning of these ninety degrees is on the line of the central hole that the ray of the sun enters to find its altitude (Figure 2.6, hole number 3). On the flat straight surface of the ring, which is on the top of it, there are three circles.[51] The first and smallest is the circle of Cancer, the intermediate is the equinoctial circle, and the largest is the circle of Capricorn, which is the outermost (104b) on the circumference of this flat straight surface.

---

48 Bonet de Lattes (Jacob ben Immanuel Provinciali), Astrolabe finger ring (Paris, 1557), 103b–117b.

49 This is a description of the back of the astrolabe plate with its zodiac and calendar scales rather than of the shank of the ring, which the term *circuitu annuli* seems to suggest (see Figure 2.10).

50 Now Bonetus starts the description of the altitude scale.

51 Bonetus now introduces the description of the front of the astrolabe plate with its distinctive engravings (see Figure 2.9).

After this there are the almucantars, which are on the upper half of the circle described,[52] some of which appear complete and some incomplete. The first of them, the horizon, is divided into two hemispheres.[53] The almucantars of the aforesaid circle are every ten degrees, [making] ninety degrees in total from the first one to the centre of the smallest one, which is called the zenith. Below the almucantars there are four seasonal hours (Figure 2.9).[54] After this is the line of the midheaven, which is the line going down from the cross; the [upper] part from the hole [2] at the centre is called meridian and the other part is called the angle of the earth or the line of midnight. Then on the rim [lit. thickness] of the circumference four hours are engraved with lines.[55] Then a turning wheel [*rotula mobilis*] is placed on that surface on which the twelve symbols of which the zodiac consists are engraved. It is called the wheel of the zodiac [i.e., the rete (*rotula zodiaci*)].[56]

On this wheel of the zodiac there are four points that indicate four fixed stars. The first point indicates Aldebaran, which is in the second degree of Gemini. The second indicates the Heart of Leo, which is on the twenty-third degree of Leo. The third indicates the Heart of Scorpio, which is on the 2nd degree of Sagittarius. The fourth indicates Capricorn's Tail, which is on the fifteenth degree of Aquarius. There are twenty-eight lines on the wheel of the zodiac, of which every seven contain three signs; they indicate the twenty-eight mansions of the moon.[57]

---

52 Literally, "on half the higher circle."

53 More correct: "the horizon divides the world into two hemispheres."

54 There are normally 12 seasonal hours and the woodcut shows them all. However, this puzzling mention of only 4 hours is consistent in the different versions of Bonetus's text that I have consulted. A possible explanation is that Bonetus is using hour divisions that encompass two hours, so that there are four divisions (one for the 1st and 2nd hours, one for the 3rd and 4th, one for the 9th and 10th, and one of the 11th and 12th), which, added to the two hours placed at each side of the lower meridian, makes a total of 6 divisions of two hours each (as in Figure 2.9). This might have been necessary because of the tiny size of the astrolabe finger ring.

55 This number should be 24 (equinoctial hours) and the woodcuts show 24 hours (12 + 12), see Figures 2.1 and 2.5. However, the different editions of Bonetus's text have 4 hours here too. I should also mention that the description here does not include any mention of the circle of 360 degrees just inside the hour circle, which is also visible in the woodcuts.

56 Bonetus starts the description of the tiny rete of this particular astrolabe, which only displays four fixed stars (Aldebaran, the Heart of Leo, the Heart of Scorpio, and Capricorn's Tail).

57 The lunar mansions are usually displayed on the back of astrolabes. Bonetus's description is confusing and implies that the lunar mansions are engraved on the zodiac of the rete, which cannot be.

(105a) There is also a hole (see Figure 2.6, hole number 1), which is in the centre of the line of the zodiac wheel that crosses the beginning of Aries and Libra, in which the axis (i.e., the pin) holding together and binding fast the wheel and the ring is placed. Then there are two holes, one in the middle of the ring to suspend the ring from a thread in front of the sun (see Figure 2.6, hole number 2),[58] the other in the upper part of the ring to make the solar ray enter it to know its altitude (see Figure 2.6, hole number 3). Then there is a hollow stick,[59] which is inserted into the hole of the altitude of the sun [see Figure 2.8 and Figure 2.6, hole number 3].[60] It is movable and is not always with the ring unless by night for taking the altitude of the stars.[61] Then there is a string to suspend the ring.

Chapter 11: To Find the Position of the Sun Any Day

If you want to know the position of the sun, in which sign it is and in which degree of it any day, place the thread on the day you have, and mark on which degree of the sign the thread has fallen.[62] It indicates the position of the sun that day.

Example: Today is March tenth, place the thread on the tenth day of the month of March. As the thread falls on the first degree of Aries, the sun is in that degree today.

---

58 The ring has to be suspended with the disc vertical so that the altitude scale on the inner surface of the shank operates correctly (see Figures 2.10 and 2.11).

59 This is a hollow sighting tube that is a removable accessory. It has to be looked through to spot a star and yet at the same time indicate the star's altitude on the solid ring. For an explanation regarding the use of this hollow stick and the positions in the astrolabe finger ring of the three holes mentioned in the text, see my comments and diagrams at the end of the English translation of the treatise.

60 This might or might not imply that when the hollow stick is in use, the ring is deprived of its suspensory part (hole number three, placed on the astrolabe plate).

61 The roles of the alidade and the ruler as pointers and calculators in standard astrolabes are played here by a thread used to mark the measures on the scales and divisions engraved on the back and the front of the astrolabe plate and on the rete, while the role of the alidade as a visual tool is played by the hollow stick (see Figure 2.11) and the hole that I have numbered 3 (see Figure 2.6).

62 The thread is used here as an alidade.

(105b) Chapter III: To Find the Date

If you want to know the date, determine in which degree the sun is. Place the thread on the degree in which it is and mark on which of the days of the month the thread has fallen. It indicates which is the day of the month that you do not know.

Example: Today the sun is in the first degree of Aries, place the thread on that degree, it will fall on the tenth of March. If you do not know the position of the sun, do as is explained in the fifteenth chapter.

Chapter IV: To Find the Altitude of the Sun

If you want to know the altitude of the sun, suspend the ring from your right hand by its string, and place it aligned with the sun so that the ray of the sun goes through the hole (hole number 3 in Figure 2.6). On the inner surface of the ring [on the altitude scale], look on which of the ninety degrees the ray has fallen, it is the altitude of the sun.

Chapter V: To Find the Ascendant Degree and the Ascendant Sign

If you want to know the ascendant with accuracy, first find the position of the sun. Then find the altitude of the sun and look for (106a) the degree of the zodiac wheel in which the sun is. Place it among the eastern almucantars of altitude if you have taken the altitude before noon, or among the western ones if you have done it after noon. The degree that falls on the horizon line, which is the first almucantar line, is the ascendant degree and the sign of that degree is the ascendant sign, i.e., in the east, while the sign in the descendant [degree] is in the west. The sign cutting the midheaven line is the sign in the midheaven, while its nadir[63] is in the angle of the earth.

Example: Today the place of the sun is in the first degree of Aries and the altitude of the sun before noon is thirty degrees. Place the first degree of Aries on the almucantar of thirty degrees in the east, the second degree of Gemini will fall on the first almucantar. The second degree of Gemini is the ascendant degree and Gemini is the ascendant sign.

Chapter VI: To Find the Twelve Houses of the Sky

If you want to calculate (lit. "equalise") the twelve houses of the sky, find first the rising degree. The degree falling on the horizon line is the first house. Then

63 Nadir is a technical term that means "the diametral opposite."

mark the degree falling on (106b) the [upper] midheaven line. Turning the zodiac wheel/rete, place on this line (i.e., the upper midheaven line) the degree that is the nadir of that degree at the midheaven. Then look at the degree on the western line that falls on the first seasonal hour, this degree is the cusp of the second house. Keeping the zodiac wheel still, look at the degree falling on the second seasonal hour, this degree is the cusp of the third house. Look at the degree falling on the line of the [lower] midheaven, this degree is the fourth house. Then look at the degree falling on the third seasonal hour, this degree is the cusp of the fifth house. Look at the degree falling on the fourth seasonal hour, this degree will be the cusp of the sixth. However, the cusp of the seventh house is the nadir of the ascendant; the cusp of the eighth [house] is the nadir of the second one; and the cusp of the ninth is the nadir of the third. The cusp of the tenth is the nadir of the fourth, the cusp of the eleventh is the nadir of the fifth, and the cusp of the twelfth is the nadir of the sixth.[64]

Example: The ascendant degree is the second degree of Gemini as has been said, and it is the [cusp of the] first house. The degree falling on the line of the [upper] midheaven is the fourteenth degree of Aquarius. Turn the zodiac wheel and place the fourteenth degree of Leo on the line of the [upper] midheaven. The degree falling on the first seasonal hour is the twenty-sixth of Gemini and it is the [cusp of the] second house.[65] The degree falling on the second seasonal hour is the twentieth of Cancer and it is the cusp of the third house. The degree falling on the [lower] midheaven is the fourteenth (107a) of Leo and it is the cusp of the fourth house. The degree falling on the third seasonal hour is the thirteenth of Virgo and it is the cusp of the fifth house. The degree falling on the fourth seasonal hour is the fourteenth of Libra, which is the cusp of the sixth house. The degree that is the nadir of the first house is the second of Sagittarius and it is the [cusp of the] seventh house. The degree that

64 The hour divisions remain a problem in this chapter as well. There seem to be only two seasonal hour lines between the meridian and horizon, making a total of four curved lines below the horizon according to Bonetus's words. The division of the houses using the standard twelve hour lines engraved on astrolabes is feasible but making changes in the standard hour divisions that are not explained anywhere in the treatise. For a possible explanation of the hour divisions on this astrolabe, see note 54. However, Bonetus explanation for dividing the houses seems to work on two conditions: the counting of the hours must be made in reverse (starting from the west rather than the east, as if these divisions were just divisions rather than hours), and, after finding the rising degree, the rete does not move at all to find the cusps of the remaining houses, these are read off on the hour divisions.

65 The description does not work. The 26th degree of Gemini is above the horizon and it is at about 43 degrees altitude to the west. However, it works if the changes explained in the previous note are introduced.

is the nadir of the second house is the twenty-sixth of Sagittarius and it is the [cusp of the] eighth house. The degree that is the nadir of the third is the twentieth of Capricorn and it is the [cusp of the] ninth house. The degree that is the nadir of fourth is the fourteenth of Aquarius and it is the [cusp of the] tenth house. The degree that is the nadir of the fifth is the thirteenth of Pisces and it is the [cusp of the] eleventh house. The degree that is the nadir of the sixth is the fourteenth of Aries and it is the [cusp of the] twelfth house.

Chapter VII: To Find How Many of the Twelve Hours Have Elapsed according to the French Method

When you want to know how many of the twelve hours have elapsed, find first the altitude of the sun. If it is before noon, place the degree of the sun on the [corresponding] almucantar of altitude in the east. Place the thread on the degree of the sun and look on which hour of the horary circle (i.e., the hour scale) the thread has fallen. Count from the line of the angle of the earth, which is midnight, to the hour on which the thread has fallen and you shall have (107b) the time.

Example: Today the position of the sun is in the first degree of Aries and its altitude is thirty degrees. Place the first degree of Aries on the almucantar of thirty degrees in the east. Place the thread on this degree and it will fall on the ninth hour of the horary circle (i.e., the hour scale), counting from the midnight line.[66] If it is after noon, place the degree of the sun in the west and count from the midheaven line, which is the meridian. You shall know what the time is in the afternoon according to the method in France, where I was born.

Chapter VIII: To Find How Many of the Twenty-Four Hours Have Elapsed

If you want to know how many of the twenty-four hours have elapsed, place the degree of the sun on the horizon line in the west. Mark with the thread on the horary circle (i.e., the hour scale) and turn the zodiac wheel/rete towards the angle of the earth. Place the degree of the sun on the [corresponding] almucantar of altitude in the east if it is before noon, or in the west if you took the altitude after noon. Then mark with the thread on the horary circle (i.e., the hour scale) and count on it from the first mark to the second one passing through the angle of the earth. You shall get how many of the twenty-four hours have elapsed (108a) according to the Italian method.

66 The reference here is to a plate with more than four hour divisions, which is clearly in contradiction to Bonetus's apparent statement in chapter I.

Example: Today the position of the sun is in the first degree of Aries and its altitude is thirty degrees. Place the first degree of Aries on the horizon line in the west and mark there with the thread on the horary circle (i.e., the hour scale). This is the first hour. Then turn the zodiac wheel and place the first degree of Aries on the almucantar of thirty degrees in the east, because now it is before noon. Mark with the thread and count from the first mark, which is the first hour, to the second mark passing through the angle of the earth. This is fifteen hours. If it is after noon, place the degree on the thirty-degree almucantar in the west and this will be twenty-one hours.

Chapter IX: To Know How Many Hours Have Elapsed since Sunrise

If you want to know how many equinoctial hours of the day have elapsed, take the degree of the sun and its altitude, and place it on the [corresponding] almucantar of altitude. Mark the position on the horary circle (i.e., the hour scale) and move backward the degree of the sun up to the first almucantar in the east; mark its position on this circle. Then count from the first mark to the second and you shall have the hours that have elapsed since sunrise. But if it is after noon, (108b) place the degree of the sun on the [corresponding] almucantar of altitude in the west and count from the first mark to the second passing through the midheaven.

Example: Today the position of the sun is in the first degree of Aries and its altitude is thirty degrees. Place this degree on the thirty-degree almucantar in the east, for it is before noon. Then place the degree on the horizon, mark [it], and count from the first mark to the second. This will be three equinoctial hours that have elapsed of the day. If it is after noon, place the degree in the west, the hours that have elapsed of the day since sunrise will be nine.

Chapter X: To Know the Duration of the Day and of the Night

If you want to know the diurnal arc, place the degree of the sun on the first almucantar in the east and make a mark on the [corresponding] horary circle (i.e., hour division). Then turn the zodiac wheel/rete passing through the midheaven and place that degree on the horizon line in the west. Mark with the thread this position on the [corresponding] horary circle (i.e., hour division). The hours and their parts, which are between the two marks, are the diurnal arc. What remains of the circle is the nocturnal arc, for the complete circle has twenty-four hours, which is the duration of the day and the night together.

Example: Today the sun is in the first degree of Aries, place the first degree of Aries on the horizon (109a) in the east. Then place that degree on the horizon in the west and count the hour divisions between the two marks passing through the midheaven. They are twelve hours, so twelve hours is the duration of the daylight today.

Chapter XI: To Know the Duration of the Day in Seasonal Hours

If you want to know the diurnal arc in seasonal hours, find first the diurnal arc as has been explained. Multiply the hours of this duration by fifteen and divide the result by twelve. You shall have the number of degrees of a seasonal hour of the day. If you subtract it from thirty, the remainder is the number of degrees of a nocturnal hour, for the addition of any diurnal seasonal hour with [its] nocturnal seasonal hour makes 30 degrees any day, which are two equinoctial hours.

Example: Today the duration of the day is fourteen hours, multiply them by fifteen and the result shall be 210. Divide this number by twelve and you will get seventeen degrees and five minutes. So seventeen degrees and five minutes is the arc of a seasonal hour of the day. Then subtract this number from thirty and the remainder shall be twelve degrees and fifty-five minutes, which is the number of degrees of a seasonal hour of the night. This chapter is necessary mainly in astrology to know the ruler of the hour, as is explained in chapter twenty-four.

(109b) Chapter XII: To Know the Duration of the Equinoctial Hour

If you want to know the arc of the equinoctial hour, divide 360 into the twenty-four hours of the horary circle and you shall have the number of degrees of the equinoctial hour by day and by night, because the [equinoctial] hours of the horary circle are equal and each of them has fifteen degrees.

Chapter XIII: To Change Seasonal Hours into Equinoctial Ones

When you want to change seasonal into equinoctial hours, find how many the degrees of the seasonal hours are and divide their degrees by fifteen, then you shall have the equinoctial hours.

Example: The duration of the seasonal hour is seventeen degrees and five minutes, and there are twelve hours, i.e., 210 degrees. Divide them by fifteen and the equinoctial hours shall be fourteen.

Chapter XIV: To Know the Altitude of the Sun at Noon

When you want to know the altitude of the sun at noon, which is the beginning of its going down, place the degree (110a) of the sun on the midheaven line. The number of degrees of the almucantar from the position of the sun on the horizon is its altitude at noon, as long as the ring has been made for the altitude of the climate in which you are searching the altitude.

Example: Today the sun is in the first degree of Aries. Place the first degree of Aries on the midheaven line and count how many degrees there are between the east and the almucantar on which this degree on the midheaven line is placed. They are 48, so the altitude of the sun at noon this day is forty-eight degrees. Previously, know that the circle of the signs is divided into two semicircles, of which one is from the beginning of Capricorn to the beginning of Cancer and the other is from the beginning of Cancer to the beginning of Capricorn. The beginning of Capricorn is the winter solstice and the beginning of Cancer is the summer one. Know that all the degrees of the solstitial hours equally distant from one solstice have the same declination either northern or southern and their days and nights are equal, as are their shadows and altitudes at noon.

Chapter XV: To Know the Position of the Sun without Knowing the Date

If you want to know the degree of the sun and you do not know the date, make a mark at the altitude of the sun at noon on (110b) the half [of the plate] that you used before with the hole to take the altitude of the sun (see Figure 2.6, hole number 3). Then turn the zodiac wheel/rete and two degrees shall fall on this mark. You shall know that one of them is the degree of the sun. When you know this, you will know the day of the month.

Example: Today the altitude of the sun is forty-eight at noon, turn the wheel, and look the degree of the zodiac falling on the almucantar of altitude. It shall be the first degree of Aries or the first degree of Libra. As it is spring time, the sun is in the first degree of Aries. Place the thread on the first degree of Aries in the circle of the months and it shall fall on the tenth day of March. According to this chapter, you shall have the position of the sun and the date that you did not know.

Chapter XVI: To Know which Day is Equal to Another

If you want to know which day is equal to another, you may know this through the degrees that are equidistant from the solstices, because their days are equal as has been said.

Example: The thirtieth day of November is equal to the twenty-fifth day of December, because the winter solstice takes place when the sun is in the beginning of Capricorn, which happens on the twelfth day of December, and the same as regards the remainder days.

Chapter XVII: To Know the Declination of the Signs from the Equinoctial Line

If you want to know the declination of any of the degrees of the zodiac, place it on the midheaven line and find its altitude from the horizon. Then find the altitude of the beginning of Aries or Libra in this line. The difference between these two altitudes is the declination of this degree with respect to the equator. If the degree is northern, the declination is also northern; if it is southern, southern.

Example: Place the first degree of Cancer on the midheaven line and count from the horizon to the almucantar of its altitude. This, here in Rome, is seventy-two degrees. The altitude of the first degree of Aries is 48, so the declination of the first degree of Cancer with respect to the equinoctial line is twenty-four degrees, which is the difference between these two altitudes.

Chapter XVIII: To Know the Latitude of a Place

If you want to know the latitude of a place, which is the distance of the zenith of this place from the equinoctial (111b) line, take the altitude of the sun at noon. If the sun is at the beginning of Aries or Libra, which is on March tenth or September tenth, respectively, you shall subtract it from ninety. The remainder is the latitude of the place. Then [at the equinoxes] the motion of the sun is on the equinoctial line. But if the sun is in a different degree, on a different day, then take the declination of its degree according to the previous chapter. You shall subtract it from the altitude of the sun at noon if it is northern; or shall add it to the altitude if it is southern, and you shall have the altitude of the beginning of Aries in that place. You shall subtract it (as has been previously explained) from ninety, and the remainder shall be the distance of the region from the equinoctial line.

Example: When the sun is in the first degree of Aries or Libra, the altitude of the sun is forty-eight at noon. You shall subtract it from ninety; the remainder is forty-two, then forty-two is the latitude of the city of Rome. Example: On other days, when the sun is in another sign, such as Cancer, take the declination of its first degree, which is twenty-four. Subtract this declination from its altitude, which is seventy-two. The remainder will be forty-eight. Or you shall add it to the altitude of the first degree of Capricorn, which is twenty-four and the result will be forty-two,[67] or subtract it from ninety and the remainder shall be forty-two. This number forty-two is the latitude of this city, Rome.

(112a) Chapter XIX: To Know the Latitude for which the Ring has Been Made

If you want to know the latitude for which the ring is made, look how many almucantars there are from the equinoctial circle up to the zenith or from the pole up to the northern horizon; the ring has been made for that latitude, for these are equal. This was made for the latitude of Rome, however you shall make for yourself one for the latitude that you want just by changing the almucantars as appears on the latitude plates of the astrolabe.

Chapter XX: To Know the Ascensions of the Signs in the Sphaera Recta

If you want to know the ascensions of the signs in the *sphaera recta*, place the beginning of any sign you like on the meridian line and mark with the thread on the horary circle (i.e., the hour scale). Turn the zodiac wheel so that the end of the sign is on the meridian line. Mark again on the [corresponding] horary circle (i.e., hour division) and multiply by fifteen the hours between the two marks.[68] Then you shall have the number (112b) of the degrees of the ascension of the sign. You shall do the same as regards any part of the circle, for every hour has fifteen degrees, as has been said.

Example: Place the beginning [and the end] of Aries on the meridian line, turn the zodiac wheel, and mark on those two places. The hours between the two marks are one hour and four parts of one (i.e., 4/5). Multiply them by fifteen [degrees per hour], which is twenty-seven [degrees].[69] So the ascension of Aries in the sphaera recta is twenty-seven degrees.

Chapter XXI: To Know the Ascensions of the Signs in any Inclined Horizon

If you want to know this, turn the zodiac wheel from the beginning of the sign to its end on the first almucantar. The degrees of the hours on the horary circle [i.e., the hour scale] will be the ascensions of the sign in that place, because every hour has fifteen degrees.

---

67 Surely this should be forty-eight (declination 24 + altitude 24), but the number forty-two seems to be consistently reproduced in all the editions I have consulted.

68 Presumably the second mark is taken with the beginning of Aries as well.

69 $[1 + 4/5] \cdot 15 = 15 + 60/5 = 15 + 12 = 27$.

(113a) Chapter XXII: To Know in Which Sign the Moon is Without Knowing its Altitude

When you want to know this, first find the position of the sun and then the day on which the moon was in conjunction [with the sun] (i.e., the new moon). You shall count from the position of the sun on the zodiac wheel (i.e., the zodiac scales on the back of the astrolabe plate) how many days have elapsed since the day of the lunar conjunction using the lines indicating the twenty-eight mansions of the moon. The sign upon which you shall come with this counting shall show the line on which the moon is.

Example: Today the sun is in the first degree of Aries. Twenty-three days have elapsed since the day of the lunar conjunction up to the present day. Count twenty-three lines [on the scale of the lunar mansions] from the position of the sun and place the thread on the twenty-third line. You shall come upon the line at the end of Capricorn, which is the position of the moon today.

Chapter XXIII: To Know the Ascendant of the Natal Revolution or the Revolution of the World

When you want to know the degree of the ascendant of the present year, place the degree of the sun on the horizon in the east. Mark on the [corresponding] horary circle (i.e., hour division) with the thread. Afterwards, move this degree from this (113b) position passing through six hours and one fifth of hour, which is ninety-three degrees. The degree falling on the horizon is the degree of the ascendant of this year. If the years are many, you shall apply that degree to every single year with the six hours and one fifth, which are ninety-three [degrees]. The degree in the east is the ascendant of that year.[70]

Example: The ascendant of the revolution of this year is eight degrees of Capricorn. Place it on the horizon in the east. Then move it through six hours and one fifth of hour. The first degree of Gemini shall be the ascendant of the next year.

---

70 When Bonetus says "place the degree of the sun on the horizon in the east" he means "place the [known] degree of the sun [for the previous year's ascendant] on the horizon in the east, and then turn through 93 degrees to get the ascendant of this year." This astrological procedure is standard in astrolabe treatises. See, for instance, Mercè Viladrich i Grau, *El "Kitab al-amal bi-l-asturlab" d'Ibn al-Samh/ Llibre de l'us de l'astrolabi* (Barcelona: Institut d'estudis catalans, 1986), 67–68.

Chapter XXIV: To Find the Ruler of the Hour by Day and by Night

If you want to know this, find the seasonal hour of the night of that day using [the instructions given in] the second chapter, for any seasonal hour has its ruler, one of the seven planets according to their order in the heavens. So on Saturday the ruler of the first hour is Saturn, the ruler of the second hour is Jupiter, and so on according to the order until twelve, because any daytime, whether its night is long or short, has twelve seasonal hours and one planet only rules the duration of that specific seasonal hour. (114a) This is clearly visible in the table below:

	Hour											
	1	2	3	4	5	6	7	8	9	10	11	12
Sunday night	Me	Mo	Sa	Ju	Ma	Su	Ve	Me	Mo	Sa	Ju	Ma
Sunday day	Su	Ve	Me	Mo	Sa	Ju	Ma	Su	Ve	Me	Mo	Sa
Monday night	Ju	Ma	Su	Ve	Me	Mo	Sa	Ju	Ma	Su	Ve	Me
Monday day	Mo	Sa	Ju	Ma	Su	Ve	Me	Mo	Sa	Ju	Ma	Su
Tuesday night	Ve	Me	Mo	Sa	Ju	Ma	Su	Ve	Me	Mo	Sa	Ju
Tuesday day	Ma	Su	Ve	Me	Mo	Sa	Ju	Ma	Su	Ve	Me	Mo
Wednesday night	Sa	Ju	Ma	Su	Ve	Me	Mo	Sa	Ju	Ma	Su	Ve
Wednesday day	Me	Mo	Sa	Ju	Ma	Su	Ve	Me	Mo	Sa	Ju	Ma
Thursday night	Su	Ve	Me	Mo	Sa	Ju	Ma	Su	Ve	Me	Mo	Sa
Thursday day	Ju	Ma	Su	Ve	Me	Mo	Sa	Ju	Ma	Su	Ve	Me
Friday night	Mo	Sa	Ju	Ma	Su	Ve	Me	Mo	Sa	Ju	Ma	Su
Friday day	Ve	Me	Mo	Sa	Ju	Ma	Su	Ve	Me	Mo	Sa	Ju
Saturday night	Ma	Su	Ve	Me	Mo	Sa	Ju	Ma	Sn	Ve	Me	Mo
Saturday day	Sa	Ju	Ma	Su	Ve	Me	Mo	Sa	Ju	Ma	Su	Ve

Me, Mercury; Mo, Moon; Sa, Saturn; Ju, Jupiter; Ma, Mars; Su, the sun; Ve, Venus.[71]

71 To avoid confusion with the reading of the table we have decided to replace the initial letters of the Latin names of the planets in the original with the first two letters of their English translations.

Chapter XXV: To Know the Altitude of the Stars

When you want to know this, suspend it [the ring] from your right hand using the string and look at the star (114b) through the hollow stick.[72] Look on which of the ninety degrees of [the altitude scale of] the interior surface of the ring the stick has fallen. That is the altitude of the star.

Chapter XXVI: To Know the Degree of the Ascendant by Night

When you want to know this, take the altitude of one of the fixed stars among the four that are on the pointers of the zodiac wheel/rete. Place the degree on which that fixed star is on the almucantar of its altitude, in the east if the star is eastern, or in the west if it is western. The degree falling on the horizon line, which is the first almucantar line, is the ascendant degree.

Example: Nowadays the position of the star Heart of Leo is in the twenty-third degree of Leo and its altitude is thirty degrees in the east. Place the twenty-third degree of Leo on the thirty-degree almucantar, then the twentieth degree of Virgo shall fall on the horizon. This twentieth degree of Virgo is the degree of the ascendant.

(115a) Chapter XXVII: To Know How Many Hours the Duration of the Night Is

If you want to arrive at this knowledge, take the altitude of one of the fixed stars among these four on the zodiac wheel/rete. Place the degree in which it is on the almucantar of its altitude, in the east if it is eastern, or in the west if it is western. Then place the thread on the degree of the sun and make a mark on the [corresponding] horary circle (i.e., hour division). Turn the zodiac wheel/rete and place this degree of the sun on the first western almucantar. Mark with the thread on the [corresponding] horary circle (i.e., the hour division) and count the hours between the two marks. The hours of the night are as many as them. If you want to know when it is going to be midnight, look how many hours there are from the first mark to the midnight line. If you want to know which of the twelve hours it is, count from the midheaven line to the degree of the sun on the horary circle passing through the western line if the degree of the sun is before the midnight line. If it is after, count from the midnight line to the midheaven line passing through the east.

---

72 Bonetus is now using the sighting tool of his astrolabe finger ring that is specifically for observing the sky at night.

Example: The degree of the Heart of Leo is the twenty-third of Leo and its altitude is thirty in the west. The sun is today in the first degree of Aries. Place the twenty-third of Leo on the almucantar of its altitude and make a mark on the first degree (115b) of Aries. Then turn the zodiac wheel/rete and place the first degree of Aries on the first western almucantar. Count the hours between the two marks; they shall be seven hours and forty-five minutes, and of the twelve hours [of night-time] it will be one hour and forty-five minutes after midnight.

Chapter XXVIII: To Know the Degree of the Ascendant without the Latitude, without Stars by Night and without the Sun by Day

If you want to arrive at this knowledge, find what is the time. Then place the degree of the sun on the point of the [corresponding] hour in the horary circle (i.e., the hour scale). The degree of the zodiac falling on the first almucantar in the east is the degree of the ascendant.

Example: Today the sun is in the first degree of Aries and the time is now three hours after noon. Place the degree of the sun corresponding to that hour on the horary circle, then the tenth of Gemini shall fall on the first almucantar. The degree of the ascendant is here.

(116a) Chapter XXIX: To Know in Which Degree of the Sign the Moon is without Knowing the Conjunction

When you want to know this, inspect the altitude of the moon and mark it on the almucantar of the part [of the horizon] in which it (i.e., the moon) is. Then place one of the four stars engraved on the zodiac wheel on its altitude at that hour, when the altitude of the moon has been taken in the part in which it is. The degree of the zodiac wheel/rete falling on the almucantar on the mark of the altitude of the moon shall be its degree. But if it is visible by day, do the same with its altitude and the altitude of the sun, and inspect of which sign the degree is.

Example: The altitude of the moon is today thirty degrees in the west and the altitude of the star the Heart of Leo is forty in the east. Place the twenty-third degree of Leo in which the star is on the almucantar of forty degrees in the east. As the sixth degree of Pisces falls on the almucantar of the altitude of the moon, which was on thirty, the moon is in this degree.

(116b) Chapter XXX: To Find the Position of the Planets

When you want to know the position of the planets, do as has been said in the previous chapter to find the position of the moon. You can also find their altitudes by night in this way.

Chapter XXXI: To Know When a Planet is Southern or Northern

If you want to know this, observe the altitude of the planet when it is on the meridian line. If it is equal to the altitude of the sun, then it is on the solar path (i.e., the ecliptic). But if it has an altitude larger than the sun's, then the planet is northern with respect to the solar path. If it has one smaller, it is southern. It departs from the solar path as much as its altitude is larger or smaller.

Chapter XXXII: To Know If the Planet is Retrograde or Direct

If you want to know whether the planet is retrograde or direct, find its altitude (117a) together with the altitude of a fixed star, and keep both in mind. After three or four nights the motion of the star is indiscernible; wait until the star has the same altitude as before. Inspect again the altitude of the planet. If its altitude is less than its previous one, the planet is direct if it is in the east. If it is in the west, it is retrograde. However, if the altitude of the planet is greater than the previous one, it is retrograde as long as the time at which the altitude is taken remains the same. If it is in the west, it is direct. <It works> the opposite as regards the moon.

Chapter XXXIII: To Know the Height of a Tower or Any Other Thing

Place [an extremity of] the hollow stick on the degree forty-five of the ninety degrees used for taking the altitude of the sun and the stars. Take the height of the tower without moving the stick from the degree forty-five, for the height of the tower with respect to your height is like the distance between you and its base. If it is impossible to measure this distance between you and the object, keeping your position on the place look and measure [the distance] between your position [on the ground] and the position on the ground that you (117b) have observed; this is the height of the object.

Bonet de Lattes Provençal to Pope Alexander VI

Blessed father, these are the distinguished points of the astronomical ring that I humbly present at your holy feet together with myself. May you receive them with a cheerful expression of your eyebrows, as one that hope warms. It is not surprising if I, being a Jew and inexperienced in Latin, have sometimes wandered outside the bounds of Latin grammar, refusing to stain the useful with the unuseful. I have preferred to offer you little roses in a cheap [vase] rather than nettles or a darnel in a costly gift basket, so that those things which are found to be useful for your health, for the benefit of the state and for the praise of the Maker of everything should not be omitted on account of the humble weaving of the words. On the contrary, under your high majesty and power everybody shall understand them better.

Pardon, I pray, the rough things which are errors for a Latin. The Jewish Law belongs to me, but not so much the Latin language.

THE END OF THE ASTRONOMICAL RING OF THE JEWISH BONET DE LATTES.

Notes on Chapters I and IV of the Latin Text and the English Translation

Chapters I–IV of Bonetus's treatise on the astrolabe finger ring introduce us to the question of how the instrument works, which in the text is directly related to the right interpretation of some of the terms and expressions used in the description of the instrument and some of its uses. There is mention in chapter I of three holes or perforations in the astrolabe finger ring. The position of the first one is clear, it is the centre of the plate and is placed on the axis of the equinoxes that divides the plate into two equal halves (hole 1 in Figure 2.6).

The position of the second perforation is "in the middle of the ring," and its purpose is "to hold the ring from a thread in front of the sun." If the plate must turn for the altitude of the sun to be read using the third hole mentioned in the text, or the altitude of the stars with the hollow stick fixed in the astrolabe plate, the second perforation should serve to hold the plate (and the ring) and place it vertical to the altitude scale (according to the altitude of the star, the sun, or the moon). Hence this perforation must have been on the limb of the astrolabe (like the throne of standard planispheric astrolabes), i.e., in the upper border of the plate (hole 2 in Figure 2.6). A string placed through it might have been enough to suspend the whole device in a reasonably stable way. Together they (hole and string) constitute the suspensory part of this very particular astrolabe.

According to chapter I, the third hole is "in the upper part of the ring" and serves "to make the solar ray enter it to know its altitude." This is the most difficult perforation of the three to locate (hole 3 in Figure 2.6). Its purpose is double: to take the altitude of the sun and to fix in it the hollow stick described in the text that is essential and specific to taking the altitude of any fixed star. As regards the hollow stick, Bonetus explained that this is a removable piece indispensable for taking the altitude of the stars at night. This tool is barely explained in the treatise and poorly represented on the woodcuts. It must have worked as an alidade and a ruler with their two main functions: a sighting device (with two sighting vanes) and an indicator (together with the thread mentioned here and there in the text). The hollow stick must have had some component (Figure 2.8) that allowed it to be adjusted in the central hole of the ring or hole of the altitude (Figure 2.6, hole 3), as the first chapter of the treatise explains. In this case, the astrolabe plate might or might not have been removed to place the hollow stick in it, but this position would have been inconvenient: without the astrolabe plate the ring would have been deprived of its suspensory part (hole number 2) and must have been held directly with the fingers. The woodcuts in Bonetus's text display a tool on the front that looks like an alidade, but that must be interpreted as a bad representation of the hollow stick or as a ruler not mentioned in the text (see Figures 2.8 and 2.9).

The next question to solve is how the hollow stick was used to take the altitude of the stars. This feature of the instrument is mentioned nowhere in Bonetus's text and must be deduced from other details of the description. A part of the hollow stick is placed through the hole of the axis in the shank (hole 3) so that the altitude can be read on the inner part of the band of the ring. It is apparent that the axis must be able to turn and allow one of the extremes of the hollow stick to point at the star while the other points at the altitude scale (see Figures 2.10 and 2.11).

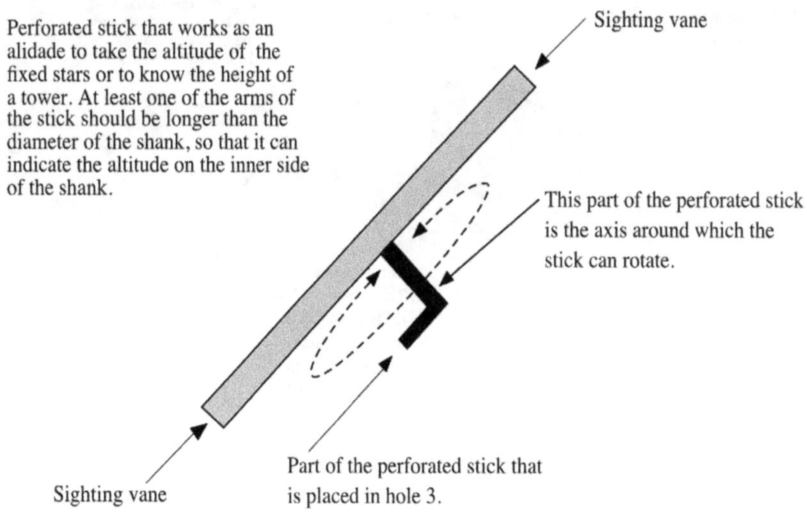

FIGURE 2.8 *Hollow stick.*

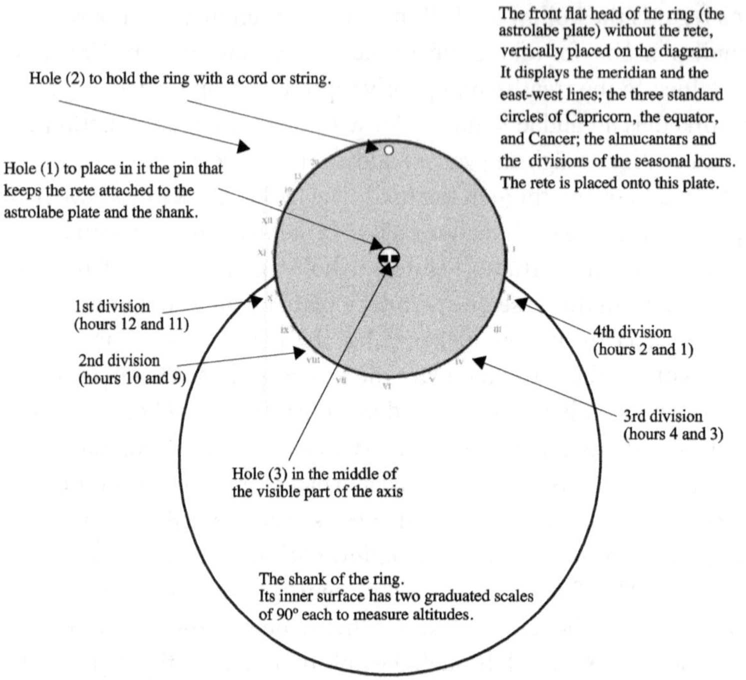

FIGURE 2.9 *Front of the astrolabe plate.*

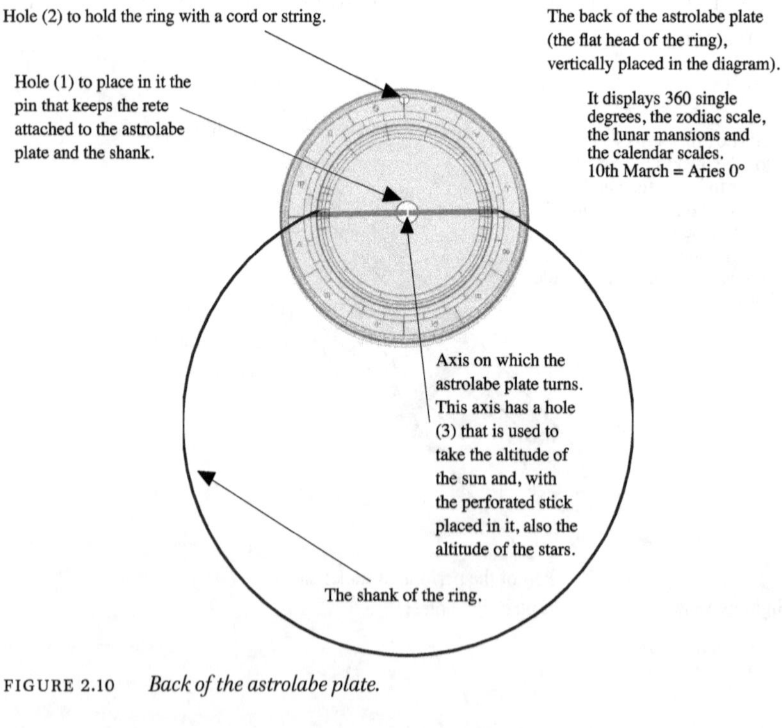

FIGURE 2.10 *Back of the astrolabe plate.*

This axis is also the key for the position of the third hole of the astrolabe finger ring. The third perforation is on the upper part of the shank, which is made in such a way as to make it possible for the astrolabe plate to turn on it to measure altitudes. The third perforation is placed on the middle of this axis and serves to take the altitude of the sun and to place on it the hollow stick to take the altitude of the fixed stars (see Figure 2.11 and 2.12).

We have to conclude that the treatise was written in a straightforward style with many examples, making it easy to follow the instructions for using the instrument. However, without seeing the physical ring, any person not familiar with the components of the instrument would have found it difficult to perform some of the calculations, especially those that require the astrolabe to be suspended, the plate to turn on its axis, and the hollow stick to be positioned in its place to measure altitudes. The woodcuts do not provide any details about these components and the treatise does not much explain them. Those like me

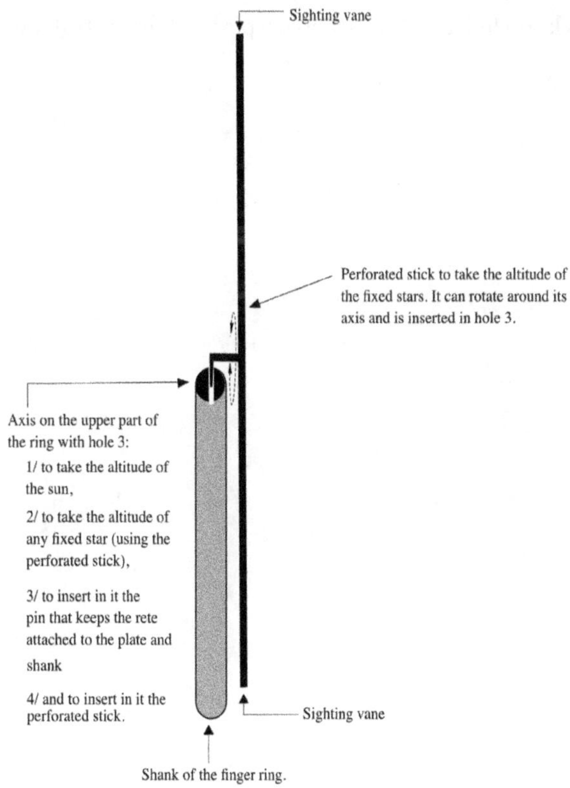

FIGURE 2.11 *The ring side view.*

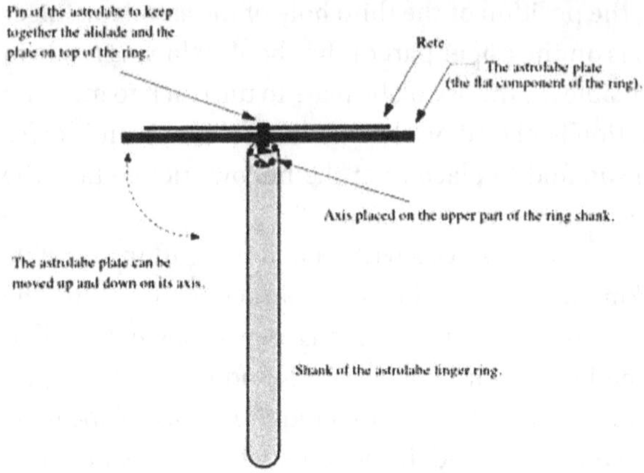

FIGURE 2.12 *Complete ring.*

who never saw Bonetus's ring must have had to deduce this information using the text and their previous knowledge of how standard planispheric astrolabes work.

CHAPTER 3

Some Features of the Old Castilian Alfonsine Translation of ʿAlī Ibn Khalaf's Treatise on the *Lámina Universal*

Emilia Calvo
Department of Semitic Philology, University of Barcelona, Gran Vía 585, 08007 Barcelona, Spain
ecalvo@ub.edu

Abstract

The aim of the paper is to present some features of the treatise on the *lámina universal*, an astronomical instrument devised by ʿAlī ibn Khalaf, an eleventh-century Andalusi mathematician and astronomer who belonged to the scientific circle of Ṣāʿid al-Andalusī. ʿAlī ibn Khalaf was a contemporary of Ibn al-Zarqālluh (al-Zarqālī, Azarquiel), also a mathematician and astronomer working under Ṣāʿid's patronage, and the inventor of the instrument known as *azafea*. Both instruments, the *lámina universal* and the *azafea*, are universal instruments devised to overcome the limitations of the standard astrolabe. The only text describing ʿAlī ibn Khalaf's instrument is the thirteenth-century old-Castilian Alfonsine translation, which has not been studied in detail up to now, although some preliminary studies have been published. The present study deals with some linguistic and technical difficulties of the text. In many passages, it seems to follow literally the grammatical structure of the Arabic language while in others, the lack of technical terms forced the translators to resort either to a literal transcription of the original Arabic terminology or, in some cases, to approximate translations that make the text somewhat difficult to follow. The paper provides additional information related mainly to the astronomical parameters and the technical vocabulary used in the translation.

Keywords

Alfonso x El Sabio – Alfonsine translations – ʿAlī ibn Khalaf – *Andalusi* astronomers – astrolabe – *azafea* – *lámina universal* – Ṣāʿid al-Andalusī – Toledo – universal astronomical instruments

 | DOI:10.1163/9789004387867_006

Introduction

The aim of this paper is to present some features of the treatise on the *lámina universal*. This name refers to an astronomical instrument devised by ʿAlī ibn Khalaf, an eleventh-century Andalusi mathematician and astronomer who belonged to the scientific circle of Ṣāʿid al-Andalusī. The latter was a historian of science and ideas and a mathematical scientist with a special interest in astronomy. In 1068, Ṣāʿid was appointed chief religious judge (*qāḍī*) of Toledo, an office he held until his death. His interest in the history of science, and science itself, especially astronomy, led him to patronize the work of contemporary scientists as well as to compose several scientific works himself. Ṣāʿid's only extant book, *Al-Taʿrīf bi-ṭabaqāt al-umam* (*Exposition of the generations of nations*), from 1068, deals with the "history of science," describing the achievements of the Indians, Persians, Chaldeans, Egyptians, Greeks, al-Rūm ("Byzantines" and other Christians), Arabs, and Jews.[1]

ʿAlī ibn Khalaf was a contemporary of Ibn al-Zarqālī (Azarquiel),[2] also a mathematician and astronomer working under Ṣāʿid's patronage and the inventor of the instrument known as *azafea*. Both instruments, the *lámina*

1 Abū l-Ḥasan ibn Aḥmar al-Ṣaydalānī. See Ṣāʿid Al-Andalusī, *Tabaqāt al-Umām*, ed. Ḥayāt Bū ʿAlwān (Beirut: Dār al-Ṭalīʿa, 1985). French translation by Régis Blachère, *Livre des catégories des nations* (Paris: Larose, 1935). English translation by Semaʿan Salem and Alok Kumar, 2nd ed. (Austin, TX: University of Texas Press, 1997), 180; David A. King, "On the Early History of the Universal Astrolabe in Islamic Astronomy, and the Origin of the Term 'Shakkaziyya' in Medieval Scientific Arabic," *Journal for the History of Arabic Science* 3 (1979): 244–257, reprinted in *Islamic Astronomical Instruments* (London: Aldershot, 1987); Julio Samsó, *Las Ciencias de los Antiguos en al-Andalus* (Madrid: Mapfre, 1992), 180–187; Roser Puig, "Ibn Jalaf al-Ṣaydalānī," in *Biblioteca de al-Andalus* (Almería: Fundación Ibn Tufayl, 2004), 3:28–29; Roser Puig, "ʿAlī Ibn Jalaf," in *The Biographical Encyclopedia of Astronomers* (hereafter *B.E.A.*), ed. Thomas Hockey et al. (New York, NY: Springer, 2007), 34–35. On Ṣāʿid, see Lutz Richter-Bernburg, "Ṣāʿid al-Andalusī," in *B.E.A.*, 1005–1006.

2 Abū Isḥāq Ibrāhīm b. Yaḥyā al-Naqqāsh, known as Ibn al-Zarqālluh or al-Zarqālī. See Ṣāʿid Al-Andalusī, *Tabaqāt al-Umām*, 180; José M. Millàs Vallicrosa, *Estudios sobre Azarquiel* (Madrid-Granada: Consejo Superior de Investigaciones Científicas, 1943–1950; repr., Toledo: Diputación Provincial de Toledo: Instituto Politécnico F.P. Azarquiel, 1993); Juan Vernet, "al-Zarqali," in *Dictionary of Scientific Biography*, ed. Charles C. Gillispie, 16 vols. (New York: Scribner, 1970–1980), 14:592–595; Julio Samsó, *Las Ciencias de los Antiguos*, 71–240; Julio Samsó, "al- Zarqālī," in *The Encyclopaedia of Islam*, 13 vols. (Leiden: Brill, 1960–2012), 11:461–462; Emilia Calvo, "Ibn al-Zarqālluh," in *Encyclopaedia of the History of Science, Technology and Medicine in Non-Western Cultures*, ed. Helaine Selin (Dordrecht: Kluwer, 1997), 415–416; Roser Puig, "Zarqālī," in *B.E.A.*, 1258–1260. On the *azafea*, see José M. Millàs Vallicrosa, "Un ejemplar de azafea árabe de Azarquiel," *Al-Andalus* 9 (1944): 111–119.

universal and the *azafea,* are devised to overcome the limitations of the astrolabe: they are universal—that is to say, useful—for any given latitude. Both are constructed using the same projection system, resulting in a similar design. They are based on a dual stereographic meridian projection of each celestial hemisphere that superimposes the projection of one half of the celestial sphere onto the projection of the other half.[3] However, there are some features that differentiate them so that they must be considered as two completely different instruments. Maybe the most evident difference is that, while Azarquiel's *azafea* does not need a rete to be functional, ʿAlī ibn Khalaf's instrument uses a rete that displays half of a coordinate system together with some star pointers.

The treatise on the use of Azarquiel's *azafea* is preserved in several original Arabic copies in addition to the Alfonsine Castilian translation and, therefore, its characteristics are well known and have been studied thoroughly.[4] Quite the opposite is the case of ʿAlī ibn Khalaf's instrument. The only text describing it is the thirteenth-century Old Castilian Alfonsine translation, since, to our knowledge, there is no extant copy of the original Arabic text of this treatise, and therefore, the Alfonsine translation constitutes its only witness.

This Castilian translation has not been studied in detail up to now, although some preliminary studies have been published. In a previous paper, the main

3 On the projection of the *azafea,* see Roser Puig, *Al-shakkāziyya de Ibn Naqqāsh al-Zarqālluh: Edición, traducción y estudio* (Barcelona: Universidad de Barcelona Instituto "Millás Vallicrosa" de Historia de la Ciencia Árabe, 1986), 45. On the *lámina universal,* see Emilia Calvo, "La lámina universal de ʿAlī ibn Jalaf (S. XI) en la versión alfonsí y su evolución en instrumentos posteriores," in *Ochava Espera y Astrofísica,* ed. Mercè Comes, Honorino Mielgo, and Julio Samsó (Barcelona: Universidad de Barcelona Instituto "Millás Vallicrosa" de Historia de la Ciencia Árabe, 1990), 221–38. For a comparison between them, see Julio Samsó, "Sobre el trazado de la azafea y de la lámina universal: Intervención de los colaboradores alfonsíes," *Al-Qanṭara* 8 (1987): 29–43; David King, "Universal Solutions to Problems of Spherical Astronomy from Mamluk Egypt and Syria" in *A Way Prepared: Essays on Islamic Culture in Honor of Richard Bayly Winder,* ed. Farhad Kazemi (New York & London: New York University Press, 1988), 153–184 at 160–163; and Françoise Charette, *Mathematical Instrumentation in Fourteenth-Century Egypt and Syria: The Illustrated Treatise of Najm al-Dīn al-Miṣrī* (Leiden: Brill, 2003), 103–108.

4 On the *azafea,* see also Roser Puig, "Concerning the ṣafīḥa šakkāziyya," *Zeitschrift für Geschichte der arabisch-islamischen Wissenschaften* 2 (1985): 123–139; Roser Puig, "El Taqbīl ʿalà risālat al-ṣafīḥa al-zarqāliyya de Ibn al-Bannā' de Marrakuš," *Al-Qanṭara* 8 (1987): 45–64; Roser Puig, *Los Tratados de Construcción y Uso de la Azafea de Azarquiel* (Madrid: Instituto Hispano Árabe de Cultura, 1987); and Emilia Calvo, "La *Risālat al-ṣafīḥa al-muštaraka ʿalà al-šakkāziyya* de Ibn al-Bannā de Marrākuš," *Al-Qanṭara* 10 (1989): 21–50.

issues regarding the construction of the instrument were analysed.[5] The present study will provide additional information related mainly to the parameters used and the technical vocabulary of the translation. More specifically, it will deal with some linguistic and technical difficulties of the text. In most passages, it seems to follow literally the grammatical structure of the Arabic language.[6] Besides, the lack of technical terms in the target language at the time of translation forced the translators to resort either to a literal transcription of the original Arabic terminology or, in some cases, to approximate translations that make the text somewhat difficult to follow.

King Alfonso the Learned and His Quest for Enlightenment

Alfonso X was born in 1221 and was king of Castile from 1 June 1252 until 4 April 1284.[7] He was an enlightened monarch who had a great interest in knowledge in general and in the sciences in particular. Under his patronage, works of literary, legal, historical, as well as scientific content were compiled. Some of these texts were original works, but in many cases, they consisted of translations of Arabic scientific works into Castilian. In addition, we should say that, in some cases, they are not mere translations, since we can find some pieces of information added by the translators, as is the case with the two chapters on the construction of ʿAlī ibn Khalaf's universal instrument that were written by one of Alfonso's collaborators, Rabi Çag (Ishāq ibn Sīd), after the Arabic original was lost.

These Alfonsine translations represent some of the first examples of Spanish scientific language. In addition, some of these works were translated later into Latin. Unlike what happens with most of the literary and historical

5 See Emilia Calvo and Roser Puig, "The Universal Plate Revisited," in *New Perspectives on the History of Islamic Science*, ed. Muzaffar Iqbal (London: Ashgate Publishing, 2012), 411–54. Reprinted from *Suhayl* 6 (2006): 113–157.

6 On the literalism of the Alfonsine collaborators, see José M. Millàs Vallicrosa, "El literalismo de los traductores de la Corte de Alfonso el Sabio," in *Estudios sobre Historia de la Ciencia Española* (Madrid: Consejo Superior de Investigaciones Científicas, 1949), 349–358.

7 On Alfonso's life, see H. Salvador Martínez, *Alfonso X the Learned: A Biography* (Leiden: Brill, 2010); Antonio Ballesteros, *Alfonso X el Sabio* (Barcelona: Salvat, 1963; repr., Barcelona: El Albir, 1984); Julio Samsó, "Alfonso X the Wise," in *Medieval Science, Technology, and Medicine: An Encyclopedia*, ed. Thomas Glick et al. (New York: Routledge, 2005), 24–26; Julio Samsó, "Alfonso X," in *B.E.A.*, 29–31; Robert I. Burns, *Emperor of Culture: Alfonso X the Learned of Castile and His Thirteenth-Century Renaissance* (Philadelphia, PA: University of Pennsylvania Press, 1990).

works produced under Alfonso's patronage, which are anonymous, in the case of the scientific works, we have the names of the authors and translators in almost all cases.

It is not clear who, besides the king, chose the works to be translated, even though there is no doubt about his involvement and his keen interest in these matters. Moreover, written scientific works imply that he had a huge library in the Arabic language at his disposal, about which we have little information.

The production of this scientific corpus started around 1250 and coincided, from 1260 until around 1269, with the compilation of the Alfonsine tables. We know that a program of astronomical observations lasting approximately ten years was launched in order to complete this work. Since these observations resulted in the compilation of the tables, we can deduce that they were based, to some extent, on direct observation. Thus, these tables would be the turning point at which Latin astronomy ceases to be based on copying earlier models and starts to develop original works, although their sources are, undoubtedly, Arabic.[8]

The Translation and Elaboration of Arabic Scientific Texts: The Alfonsine Collaborators

The king's collaborators played a prominent role in producing these scientific works. There were Christians, Jews, and Muslims among them. The latter include Bernardo the Arabian, a Muslim convert who collaborated with Don Abraham on the revision of Azarquiel's treatise on the *azafea* in Burgos in 1278. Since he is called a "master," we can deduce that he was an educated man. Another prominent Muslim collaborator was Muḥammad al-Riqutī al-Mursī, who possessed encyclopedic knowledge and who taught in Murcia at a school sponsored by the monarch, although his involvement in the astronomical works is not documented.[9]

8 On the Alfonsine tables, see José Chabás and Bernard R. Goldstein, *The Alfonsine Tables of Toledo* (Dordrecht: Springer, 2003), 15–18; Emmanuel Poulle, *Les tables alphonsines avec les canons de Jean de Saxe* (Paris: C.N.R.S., 1984); Emmanuel Poulle, "The Alfonsine Tables and Alfonso X of Castille," *Journal for the History of Astronomy* 19 (1988): 97–113; John D. North, "Just Whose Were the Alfonsine Tables?" in *From Baghdad to Barcelona: Studies in the Islamic Exact Sciences in Honour of Prof. Juan Vernet*, ed. Josep Casulleras and Julio Samsó, 2 vols. (Barcelona: Instituto "Millás Vallicrosa" de Historia de la Ciencia Árabe, 1996), 1:453–475.

9 For the Muslim collaborators, see Julio Samsó, "Dos colaboradores científicos musulmanes de Alfonso X," *Llull* 4 (1981): 171–79. For the Jewish collaborators, see Norman Roth, "Jewish Collaborators in Alfonso's Scientific Work" in Burns, ed., *Emperor of Culture*, 59–71; David

There were various Christian contributors of whom only one, Garci Pérez, seems to have been competent as an astronomer. We can also find the names of Italian collaborators like John of Cremona or John of Messina.[10] Some of them were charged with translating into Latin works previously translated into Old Castilian, such as the *Tetrabiblos,* Ptolemy's astrological work, or the *Libro Conplido en los Iudizios de las Estrellas,* a translation from Ibn Abī Rijāl's *al-Kitāb al-Bāriʿ fī Aḥkām al-Nujūm*. The presence of these collaborators in Alfonso's court coincides with the period in which he aspired to be crowned Holy Roman Emperor.

The group of Jewish collaborators was certainly the greatest in number. They collaborated both on the translations and on the creation of original works. Among them, we can mention a translator, Yehuda ibn Moshe, as well as the already-mentioned Ishāq ibn Sīd, known as Rabiçag, an astronomer, observer, and constructor of astronomical instruments.[11]

Three large miscellaneous collections were produced with the help of these collaborators. One of them contained works on magic, including the most representative work, the *Picatrix,* a treatise on talismanic magic whose author was from tenth-century al-Andalus. It was very popular in Renaissance Europe, where it circulated in a Latin translation based on a previous Castilian version. The second compilation is on astrology, and the third consists of astronomical works collected under the title *Libros del saber de astronomía.*

Several other translations can be added to the ones already mentioned, some of them on astronomy, such as Ibn al-Haytham's *Fī hay'at al-ʿalam* (*On the Physical Configuration of the Universe*)[12] and al-Battānī's *al-Zīj al-Ṣābi'* (*The Sabian Astronomical Handbook*),[13] others on astrology, such as Ptolemy's *Tetrabiblos*, the *Libro Conplido*, and the *Libro de las Cruces* (*Book of the Crosses*). There are also some original works: the *Alfonsine Tables* and the *Libro del Cuadrante sennero* (*Book on the Sine Quadrant*), which is not included among

Romano, "Le opere scientifiche di Alfonso X e l'intervento degli ebrei," in *Oriente e Occidente nel Medioevo: Filosofia e Scienze. Convegno Internazionale 9–15 aprile 1969* (Rome: Accademia Nazionale dei Lincei, 1971), 677–711, reprinted in *De historia judía hispánica*, 147–181 (Barcelona, 1991).

10 See, for instance, "John of Messina," in *B.E.A.*, 599.

11 On Ishāq ibn Sīd, Rabiçag, see Y. Tzvi Langermann, "Ibn Sid, Isaac ibn Sid," in *B.E.A.*, 570.

12 Abū ʿAlī al-Ḥasan ibn al-Ḥasan (960–ca. 1040). On Ibn al-Haytham, see Y. Tzvi Langermann, "Ibn al-Haytham," in *B.E.A.*, 556–557. The translation of this work into Spanish, which the king asked Don Abraham to do, is lost. An anonymous Latin translation is preserved. This translation shows the interest in physical astronomy at Alfonso's court.

13 Abū ʿAbd Allāh Muḥammad ibn Jābir ibn Sinān al-Ḥarrānī al-Ṣābi' (858–929). On al-Battānī, see Benno Van Dalen, "Battānī," in *B.E.A.*, 101–103.

the *Libros del Saber*.[14] Although written by contributors under his patronage, they were probably inspired by King Alfonso's own personal interests.

The Libros del Saber de Astronomía

The most important astronomical collection compiled under Alfonso is the one known as the *Libros del saber,* which consists of sixteen treatises. The first of these is the *Libro de la ochava espera* (*Book on the Fixed Stars,* lit. "Book on the Eighth Sphere"), based on the *Kitāb Ṣuwar al-kawākib* (*Book of Constellations*) by al-Ṣūfī.[15] The remaining treatises describe the construction and use of different instruments, such as analogue calculators, the purpose of which is the graphic resolution of problems in spherical astronomy.[16] Among the instruments described in the *Libros del saber* we find the celestial globe, the plane and spherical astrolabe, the *azafea*, the *lámina universal,* the instrument of *ataçir*,[17] the *equatorium*, and the sundial.

One of the main objectives of the king in compiling all these materials is stated clearly in the general prologue to the *Libros del saber*. The author wanted to present this learning in a clear framework in order to make it more accessible for the reader. This is the reason why, in many cases, the works were reorganized and illustrated in the translations. Thus, although the content is directly dependent on Arabic sources, in its structure and presentation (for example, in its illustrations) we find certain original features that appear in a systematic way in the Alfonsine works. One of their characteristics is the search for perfection, which led to successive reworking of most of the translations included in this compendium. Some authors describe it as two "drafting stages." The first would have been made up of early versions of the books on the fixed stars (1254), on the *espera redonda* or *dell alcora* (the celestial globe, 1259), and the *azafea* (undated). The text of this first stage has not survived. The second stage is represented by the text in the codex compiled in the Alfonsine scriptorium, composed in 1276–1277. There, in addition to reviewing pre-existing

14 On Alfonso's scientific endeavour, see also Evelyn S. Procter, *The Scientific Works of the Court of Alfonso X of Castile: Patron of Literature and Learning* (Oxford: Oxford University Press, 1951).

15 Abū al-Ḥusayn ʿAbd al-Raḥmān ibn ʿUmar (903–986). On al-Ṣūfī, see Paul Kunitzsch, "Ṣūfī," in *B.E.A.*, 1110.

16 It is edited by Rico y Sinobas. See Manuel Rico y Sinobas, *Libros del saber de astronomía del rey D. Alfonso X de Castilla*, 5 vols. (Madrid: tip. De Eusebio Aguado, 1863–1867), 3:1–132.

17 This term derives from the Arabic *tasyīr*, "astrological progression."

translations and creating new versions of them, some new books were added. Most of them were composed *ex novo* by the Jew Rabiçag, who actually produced them while compiling the *Alfonsine Tables* (1262–1272), as stated in the foreword to these *Tables.*[18]

The *Libro de la lámina universal*: The Old Castilian Alfonsine Translation of ʿAlī ibn Khalaf's Text on His Universal Instrument

The *Book of the lámina universal* is one of the undated books, but, since it is included in the *Libros del* saber, whose compilation was carried out towards the end of Alfonso's reign, we can assume that it belongs to the last period of the king's translation activity. It is contained in folios 83v–109r of Codex Villaamil 156 of the Library of the Universidad Complutense in Madrid. It is included just after the treatise on the astrolabe and is followed by the treatise on Ibn al-Zarqālī's *Azafea zarqaliyya.*

The book consists of two sections: one of them is devoted to the construction of the instrument and occupies folios 83v–87r of the manuscript. It is followed by the second part, on the use of the instrument, on folios 87r–109 r. This is a common division in the treatises included in the *Libros del saber*, as well as in their Arabic originals. The first of these two sections is considered to be an original draft of one of the collaborators of the king, Rabi Çag, since the author of the Arabic treatise on the use of the astrolabe had not written a treatise on its construction. The treatise on use is attributed to ʿAlī ibn Khalaf, described in the text as a geometer and astronomer.

The book on the construction of the *lámina universal* has been previously studied in depth,[19] and therefore I will focus here on some features of the treatise on the use of this universal instrument.

---

18 See Inés Fernández-Ordóñez, "El taller historiográfico alfonsí: La *Estoria de España* y la *General Estoria* en el marco de las obras promovidas por Alfonso el Sabio," in *El Scriptorium alfonsí: De los Libros de Astrología a las "Cantigas de Santa María,"* ed. Jesús Montoya and Ana Rodríguez (Madrid: Fundación Universidad Complutense, 1999), 4.

19 See Emilia Calvo, "La labor de difusión de la cultura árabe por parte de Alfonso X y su contribución a la formación del lenguaje científico: *Los Libros del* saber *de astronomía,*" in *La Civilización Islámica en Al-Andalus y los aspectos de tolerancia* (Casablanca: Centro de Estudios Al-Andalus y de Diálogo entre Civilizaciones, 2003), 27–42; and Calvo and Puig, "The Universal Plate."

Structure and Contents of the Treatise on the Use of Ibn Khalaf's *Lámina Universal* as Presented in the Old Castilian Translation

The calculations that can be performed with this *lámina universal* are described in one section of the treatise, which is divided into five parts. The first part consists of twelve chapters. The first of them describes the knowledge necessary for handling the plate:

- Calendrical: days and months of the Christian and Arabic calendars
- Arithmetical: numbers, how to add, subtract, multiply, and divide, and how to find the square root
- Astronomical-astrological: zodiacal signs, lunar mansions, stars and planets
- Interpolation: in order to obtain the coordinates of a point located between two meridians or between two parallels
- The section of the plate where it is more difficult to get a good result from that procedure: on the polar section and the centre of the plate the projection becomes more difficult to handle because the space between lines is narrower (Figure 3.1).

The second chapter describes the different parts that make up the instrument and their respective names, which coincide with the ones given in the chapters on the instrument's construction.[20] The rest of chapters give instructions of a practical kind, mainly how to put marks on the lines engraved on the plate according to the instructions given in the following chapters.

The second part consists of sixty-three chapters and is devoted to the calculations related to any degree of the ecliptic. In the first place, how to locate it on the plate and, then, how to determine its coordinates:

- Equatorial: declination, right, and oblique ascension
- Related to the ecliptic: longitude and latitude
- Related to the horizon: altitude and azimuth, rising amplitude
- Geographical calculations: latitude and co-latitude
- Timekeeping: daytime and nocturnal arcs, equal and unequal hours
- Astrological calculations: the four cardines of any horoscope, the division of the twelve houses according to Ptolemy and Hermes

20 See Calvo and Puig, "*The Universal Plate*," 418–430.

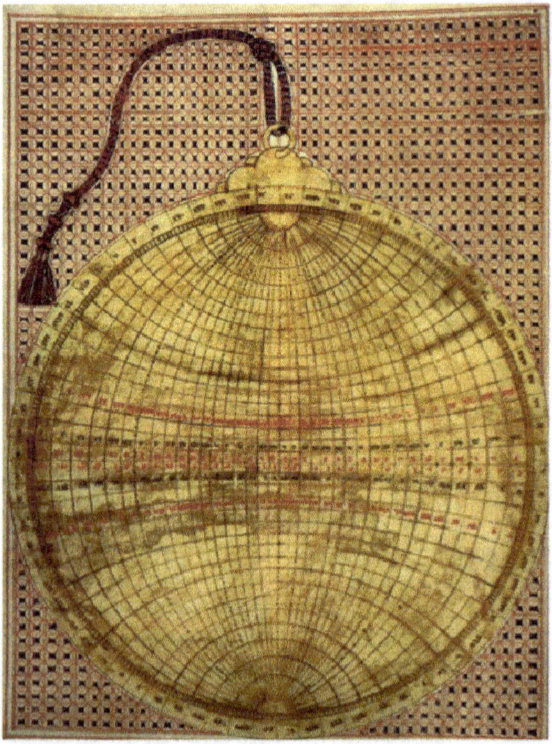

FIGURE 3.1 *Front of the Lámina Universal.* MS *Villaamil 156. Library of the Universidad Complutense, Madrid, fol. 84 v.*

The third part consists of fifty-eight chapters devoted to calculations related to the sun. They can be understood as the application of the general rules described in the previous part for any degree of the ecliptic, to the concrete case of the degree of the sun, and are, therefore, quite repetitive. The few novelties are related specifically to the sun: shadows, calculation and conversion, meridian altitude of the sun and related calculations, solstices and equinoxes, and observation and determination of the four cardinal points. The redundancy of these chapters, as well as the existence of some mistakes similar to those found in other treatises included in the *Libros del saber*, leads one to consider the possibility that all or part of this part is original to the Alfonsine circle and not a mere translation.

The fourth part consists of sixty-four chapters, devoted mainly to calculations related to the stars. The calculations dealt with there are the following:

(i) Geographical:
- Difference of longitudes and latitudes between two places
- The four cardinal points and the azimuth from one place with respect to another

(ii) Timekeeping:
- Calculation of the morning and evening twilight

(iii) Astrological:
- Nativities and projection of rays

The last chapter is devoted to knowing whether a planet is direct or retrograde.

Finally, the fifth part is devoted almost exclusively to the moon. It consists of twelve chapters devoted to the following calculations:

- Lunar longitude and latitude
- Lunar distance to the nodes
- Lunar rising and setting
- The degree with which the moon rises or sets
- The equal hours elapsed in function of the lunar altitude
- The hour of conjunction and opposition

Some Features of the Back of the Lámina Universal (Figure 3.2)

This part of the instrument is called "faz de la yguacion del Sol" (side of the Sun's equation). It is divided by the two perpendicular diameters into four equal parts: the quadrants; the lines and the numbering coincide with the disposition of the basic elements on the back of an astrolabe.

The Zodiacal Calendar

The zodiacal calendar is a set of two circles that allows one to find the equivalence between the day of the month and the degree of the zodiacal sign:

(1) The circle of the zodiacal signs and their divisions is a concentric circle divided into 360°. These divisions are grouped in sections of 30° corresponding to each sign. The name of the signs is written below.
(2) The circle of the months and the days is an eccentric circle divided into 365 parts corresponding to the days of the Roman calendar. Below it, there is the number of the days that correspond to every Roman month.

FIGURE 3.2 *Back of the Lámina Universal. Ms. Villaamil 156. Library of the Universidad Complutense. Madrid, fol. 85 v.*

When one compares the values given in the text of the treatise and the ones found in the drawing that accompanies the text in the manuscript, some discrepancies between them can be noted (Table 3.1).

These inconsistencies between the text and the drawing are possibly due to the fact that the author of the drawings was different from the author of the translation.

The Shadow Square

The shadow square is also found on the back of the instrument. It allows one to convert angles measured in degrees into trigonometric units expressed as digits (a Castilian term?) in such a way that 45° equal twelve digits, which is the maximum value. It is composed of two sides: the converse shadow for angles

TABLE 3.1 *Comparison of the the Values Given in the Text of the Treatise and the Ones Found in the Drawing that Accompanies the Text in the Manuscript*

1st day of the month	Text	Degree of the sign	Drawing
January	19°	Capricorn	21°
February	22°	Aquarius	22°
March	19°	Pisces	19°
April	19°	Aries	20°
May	18°	Taurus	20°
June	18°	Gemini	21°
July	16°	Cancer	21°
August	16°	Leo	21°
September	16°	Virgo	22°
October	16°	Libra	21°
November	17°	Scorpio	21°
December	17°	Sagittarius	21°

between 0 and 45° and the extended shadow for angles between 45 and 90°. The values corresponding to the angle of altitude given for each of the twelve digits are shown in Table 3.2.

Some Features of the Rete

The rete of the *lámina universal* is divided into two halves by a horizontal diameter called *andamio* (a literal translation from the Arabic *mamarr*), i.e., meridian. One half shows a half set of markings corresponding to the meridians and parallels of declination of equatorial coordinates, and the other half shows some star pointers for stars from the northern and southern hemispheres (Figure 3.3).

In Chapter 1.1, the treatise gives a list of stars to be engraved on the rete (Table 3.3).[21]

21 The text adds "et las otras estrellas que son puestas en la red" (and other stars that are engraved on the rete).

TABLE 3.2 *Values Corresponding to the Angle of Altitude Given for Each of the Twelve Digits*

Digit	Solar altitude
1	4;50°
2	9;30°
3	(missing)
4	18;26°
5	22;34°
6	26;33°
7	30;18°
8	30;43°
9	36;54°
10	39;49°
11	42;30°
12	45°

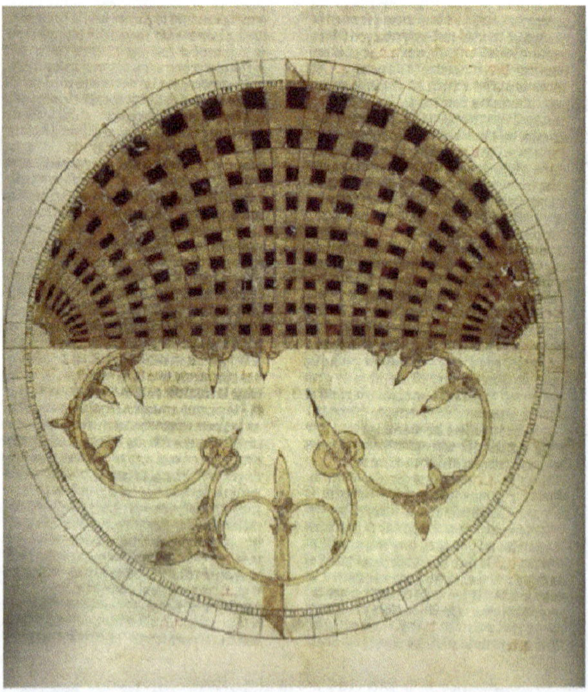

FIGURE 3.3 *Rete of the Lámina Universal. Ms. Villaamil 156. Library of the Universidad Complutense. Madrid, fol. 86 v.*

TABLE 3.3 *A List of Stars to Be Engraved on the Rete of the Lámina Universal*

Old Castilian	Arabic name	Modern star name
Buytre Cayente	*al-nasr al-wāqiʿ*	Vega (α Lira)
Alahabor	*al-shiʿrā al-ʿabūr*	Sirius (α Canis Major)
Algumeyça	*al-ghumayṣāʾ*	Procyon (α Canis Minor)
Alayot	*al-ʿayyūq*	Capella (α Auriga)
Ombro de Gemini	*mankib al-jawzāʾ*	Betelgeuse (α Orion)
Pie de Gemini	*rijl al-jawzāʾ*	Rigel (β Orion)
Alrramech	*al-simak al-rāmiḥ*	Arcturus (α Bootis)
Cor de Leon	*qalb al-asad*	Regulus (α Leo)
Boca del Peç Meridional	*fam al-ḥūt*	Fomalhaut (α Piscis austrinus)
Cabeça Dalgol	*rāʾs al-ghūl*	Algol (β Perseus)
Arridf	*al-ridf*	Deneb (α Cignus)

Some Parameters Provided in the Treatise on the Lámina Universal

Most of the parameters used in this treatise are documented in other astronomical treatises compiled in al-Andalus. This is the case for the value adopted for the obliquity of the ecliptic, 23;30° (chapter 2.2). This is clearly a rounded value, convenient for making calculations with the instrument.[22]

Morning and evening twilights are calculated using the parallel to the horizon of value 18° together with the *nadir* of the degree of the Sun (chapter 4.58).[23] The obliquity of the lunar orbit on the ecliptic: 5° (chapter 5.4) is a Ptolemaic value. Finally, among the calculations described in the text, an absence can be detected, since there is no chapter devoted to determining the azimuth of the *qibla*. This is a calculation usually found in this kind of treatise in Arabic

22 On the obliquity of the ecliptic in al-Andaus, see, for instance, the treatise on the universal astrolabe plate devised by Ibn Bāṣo in Emilia Calvo, *Abū ʿAlī al-Ḥusayn ibn Bāṣo: Risālat al-ṣafīḥa al-jāmiʿa li-jamīʿ al-ʿurūḍ (Tratado sobre la lámina general para todas las latitudes): Edición crítica, traducción y estudio* (Madrid: Consejo Superior de Investigaciones Científicas, 1993), 63, 179, and 132.

23 This is equivalent to the use of the value -18° for the actual degree of the sun. This is a parameter usually found in Andalusi treatises on instruments. See, for instance, Ramón Martí and Mercè Viladrich, "En torno a los tratados de uso del astrolabio hasta el siglo XIII en al-Andalus, la Marca Hispánica y Castilla," in *Nuevos estudios sobre astronomía española en el siglo de Alfonso X* (Barcelona: Consejo Superior de Investigaciones Científicas, 1983), 9–74 at 25.

sources. This absence can also be found in Alfonsine translations of Arabic astronomical texts, such as the *Treatise on the Spherical Astrolabe*, based on Ibn al-Samḥ's treatise on the planispheric astrolabe, which that also omits this chapter, although the original Arabic treatise by Ibn al-Samḥ includes this calculation in chapters 68, 69, 70 and 101.[24]

Technical Vocabulary

The literalism that we find in the translation of the treatise on the *lámina universal* is extended to the other texts in the *Libros*. In this sense, this treatise is a good example of this phenomenon. We find words that are transcriptions of Arabic words, with or without the article, like *axataba, ataçir, alhogera, madahara, zont*. There are some technical words that are literal translations from the Arabic original like *empontizo, andamio, mudamiento, opposito, piértega.*

In Table 3.4 some of these technical terms, present in the Castilian text, together with the English and Spanish translations, are given. In some instances, the likely Arabic source for the Castilian term is given.

TABLE 3.4 *Technical Terms in the Alfonsine Castilian Translation of ʿAlī ibn Khalaf's treatise on the Lámina Universal*

Old Castilian	Arabic	English translation	Modern Spanish translation
alhogera, cabeçon	*hujra*	limb	limbo
anchura delorientamiento	*saʿat al-mashriq*	rising amplitude	amplitud ortiva
anchura del ponimiento	*saʿat al-maghrib*	setting amplitude	amplitud occídua
andamio [mamarrat]	*mamarr*	equatorial meridian	meridiano ecuatorial
archo del dia/de la noche	*qaws al-nahār/al-layl*	diurnal/nocturnal arc	arco diurno/nocturno
ataçir	*tasyīr*	astrological progression	progresión astrológica

24 See Mercè Viladrich, *El "Kitāb al-ʿamal bi-l-asṭurlāb" (Llibre de l'ús de l'astrolabi) d'Ibn al-Samḥ* (Barcelona: Universitat de Barcelona, 1986), 62–66, 131–133, and 141.

TABLE 3.4 *Technical Terms in the Alfonsine Castilian Translation* (cont.)

Old Castilian	Arabic	English translation	Modern Spanish translation
axataba	*shaṭaba*	sighting device/ pinnule	pínula
cerco del mediodía	*dā'irat wasat al-samā'*	meridian circle	círculo meridiano
cercos de la altura/ de los zontes	*dawā'ir al-sumūt*	azimuthal circles	círculos azimutales
cercos de las ladezas	*dawā'ir al-mayl*	latitude circles	círculos de latitud
cercos de las longuras	*dawā'ir al-ṭūl*	longitude circles	círculos de longitud
cinculares		quinary divisions (days, degrees)	divisiones quinarias
circulario/circular	*madār*	equatorial parallel	paralelo ecuatorial
comedio de septentrión et mediodía		local meridian circle	meridiano del lugar
demostrador	*murī*	index	índice
empontizo [almuqantarath]	*muqanṭar*	almucantar	almucantar
estaciones		stations of the planets	estaciones de los planetas
enderezamiento (planeta endereçada)	*istiqāma* (*kawkab mustaqīm*)	the situation in which a planet is direct in its course	(cuando un planeta es directo en su curso)
faz de la yguación del sol	*wajh i'tidāl al-shams*	side of the zodiacal calendar (on the back of the plate)	cara del calendario zodiacal (en el dorso de la lámina)
ladeza	*'arḍ*	latitude	latitud
linna del medio çielo – de comedio de oriente et occidente		East West line	línea Este-Oeste
de comedio de oriente et occidente		East West line	línea Este-Oeste
longura – al yguador dell dia	*ṭūl* – *ila mu'ādil al-nahār*	longitude/distance – to the equator (declination)	longitud/distancia – al ecuador (declinación

TABLE 3.4 *Technical Terms in the Alfonsine Castilian Translation* (cont.)

Old Castilian	Arabic	English translation	Modern Spanish translation
mayor altura del yguador del día	*al- irtifāʿ al-kullī li-muʿādil al-nahār*	meridian altitude of the equator (co-latitude)	altura meridiana del ecuador (colatitud)
mayor declinacion	*al-mayl al-kullī*	maximum declination: obliquity of the ecliptic	declinación máxima: oblicuidad de la eclíptica
meridiano mayor	*manāra*	prime meridian	primer meridiano
mudamiento yuernal/ uerenal	*inqilāb*	winter/summer solstice	solsticio de invierno/ de verano
opposito	*naẓir*	nadir	nadir
piertegas		equatorial coordinates on the rete	coordenadas ecuatoriales de la red
ponimiento	*ghurūb*	setting	ocaso
replectiones	*kusūr*	fractions	fracciones (de hora)
rectificar	*raṣada*	observe	observar
retragtamiento	*rujūʿ*	retrogadation	retrogradación
(planeta retrograda)	(*kawkab rājiʿ*)	retrograde planet	planeta retrógrado
sobimiento	*ṭulūʿ*	rising	orto
sombra espandida	*al-ẓill al-mabsūṭ*	extended shadow	sombra extensa
sombra retornada	*al-ẓill al-mankūs*	reversed shadow	sombra conversa
yguador del día	*muʿādil al-nahār*	equator	ecuador
ygualamiento	*iʿtidāl*	equinox	equinoccio
zodiacho	*minṭaqat al-burūj*	zodiac	zodíaco
zont, zonte	*samt*	azimuth	azimut
zont de la cabeza	*samt al-ra's*	zenith	cenit

CHAPTER 4

From the Celestial Globe to the Astrolabe: Transferring Celestial Motion onto the Plane of the Astrolabe

Flora Vafea
Faculty of Languages and Translation, Al-Azhar University, Cairo, Egypt
fkvafea@gmail.com

Abstract

The celestial globe is an early astronomical instrument that gives a realistic view of the heavens and the stars and can simulate the celestial motion, but has some disadvantages such as difficulties in construction, transportation and in carrying out measurements. The astrolabe is an evolution of the celestial globe; it represents the celestial sphere projected onto the plane, retaining the celestial ratios. It provides accurate measurements and helps in solving a great variety of astronomical problems. The aim of this paper is to show how various processes have been transferred from the celestial globe to the astrolabe, determining the form of each part of the astrolabe.

Keywords

astrolabe – celestial globe – Ptolemy – Geminus – Planisphaerium – stereographic projection

Introduction

Heavens and stars, Sun and Moon, and the celestial motion have been among the wonders that humanity has experienced since its early times. The succession of the seasons dictated specific agricultural works at the appropriate time. Hesiod (early seventh century BC) preserved traces of this knowledge in his poem *Days and Works* ("Εργα και ἡμέραι)[1] In lines 562–625 of the poem, he refers to the positions of Arcturus, Pleiades, Sirius, Orion and Hyades in the sky

1 David Tandy and Walter Neale, *Hesiod's Works and Days: A Translation and Commentary for the Social Sciences* (Berkeley, CA: University of California Press, 1996); and Hesiod, *Works and*

 | DOI:10.1163/9789004387867_007

that must be taken into consideration in order to prune the vines, to harvest and winnow the wheat, to harvest the grapes, to plough the earth, even to haul up the ships from the sea at the correct time.

The positions of the decanal stars in the night sky were used by the ancient Egyptians since the third millennium BC for determining the time.[2] Since the stars played an important role in everyday life, their relative positions had to be recorded for the transmission of this knowledge. Thus, groups of nearby stars were depicted in the form of constellations, according to the understanding of each civilization.

The most realistic depiction of the stars and the constellations in their relative positions is that on a sphere. One of the most ancient descriptions of the relative positions of stars and constellations, in a qualitative but not quantitative description, is preserved in Aratus's poem *Phaenomena*,[3] based on the homonymous treatise of Eudoxus of Cnidus (fourth century BC). Hipparchus (second century BC) in his *Commentary on the Phaenomena of Eudoxus and Aratus* criticizes them both for the errors appearing in their *Phaenomena*;[4] the correct positions of the stars can be expressed only by numbers, so he used astronomical instruments to measure their coordinates. Hipparchus's star catalogue included the name and position for each star and was the principal source for Ptolemy's catalogue.[5] With this, the celestial globe, a realistic model of the celestial sphere, could be constructed (Figure 4.1).

Geminus (first century BC), in his *Introduction to Phaenomena*,[6] describes the celestial globe on a stand as "καταστηριζομένη σφαίρα" (section 5.65),[7] while

---

Days, in *The Homeric Hymns and Homerica* with an English Translation by Hugh G. Evelyn-White (Cambridge, MA, Harvard University Press; London: William Heinemann, 1914).

2 Marshall Clagett, *Ancient Egyptian Science*, 3 vols. (Philadelphia, PA: American Philosophical Society, 1995), 2:50–65.

3 Aratus, *Arati Phaenomena*, ed. J. Martin (Firenze: La Nuova Italia Editrice, 1956); and Aaron Poochigian, *Aratus: Phaenomena*, translated, with an Introduction and Notes (Baltimore, MD: The Johns Hopkins University Press, 2010).

4 Hipparchus, *Hipparchi in Arati et Eudoxi Phaenomena Commentariorum Libri Tres*, ed. Karl Manitius (Leipzig: Teubner, 1894).

5 Gerard J. Toomer, *Ptolemy's Almagest* (Princeton, NJ: Princeton University Press, 1998), 321.

6 Germaine Aujac, *Géminos, Introduction aux Phénomènes* (Paris: Les Belles Lettres, 1975); also James Evans and J. Lennart Berggren, *Geminos's Introduction to the Phenomena: A Translation and Study of a Hellenistic Survey of Astronomy* (Princeton, NJ: Princeton University Press, 2006) and Karl Manitius, *Gemini: Elementa Astronomiae* (Leipzig: Teubner, 1898); in chapter five "The circles on the sphere" are described.

7 Manitius, in *Gemini: Elementa Astronomiae*, 66, uses the term "καταστεριζομέναις" from MS Fragmentum Vat. 318, mentioning the term "καταστηριζομέναις" coming from four

FIGURE 4.1
Celestial globe on a stand (Inv. no. 98115) from the Museum of the History of Science in Oxford.
COURTESY OF THE MUSEUM OF THE HISTORY OF SCIENCE IN OXFORD.

the stand is called "σφαιροθήκη" (section 5.63). These terms are encountered in medieval Arabic treatises as *dhāt al-kursī* and *kursī*, respectively. The celestial globe on a stand, according to the early description by Geminus, contains the following elements (Figure 4.2):

MSS (Vindobonensis 89, Vaticanus 318, Vaticanus 381 and Parisinus 2385) and the term "κατηστερισμέναις" as *editionum consensus*. Aujac, in *Géminos, Introduction aux Phénomènes*, accepts the term "καταστηριζομέναις", while Evans and Berggren in *Geminos's Introduction to the Phenomena*, 159, translate the term in question as "figured with stars", which corresponds to the term "κατηστερισμέναις". The term "καταστηριζομένη" is feminine present participle passive of the verb "καταστηρίζω", which in passive form means, "to be propped up", and denotes a globe propped on a stand. The terms "καταστεριζομένη" and "κατηστερισμένη" are respectively the feminine present participle passive and the feminine perfect participle passive of the verb "καταστερίζω", meaning "to figure with stars". Since the stand "σφαιροθήκη" is mentioned, the term "καταστηριζομένη" is preferable.

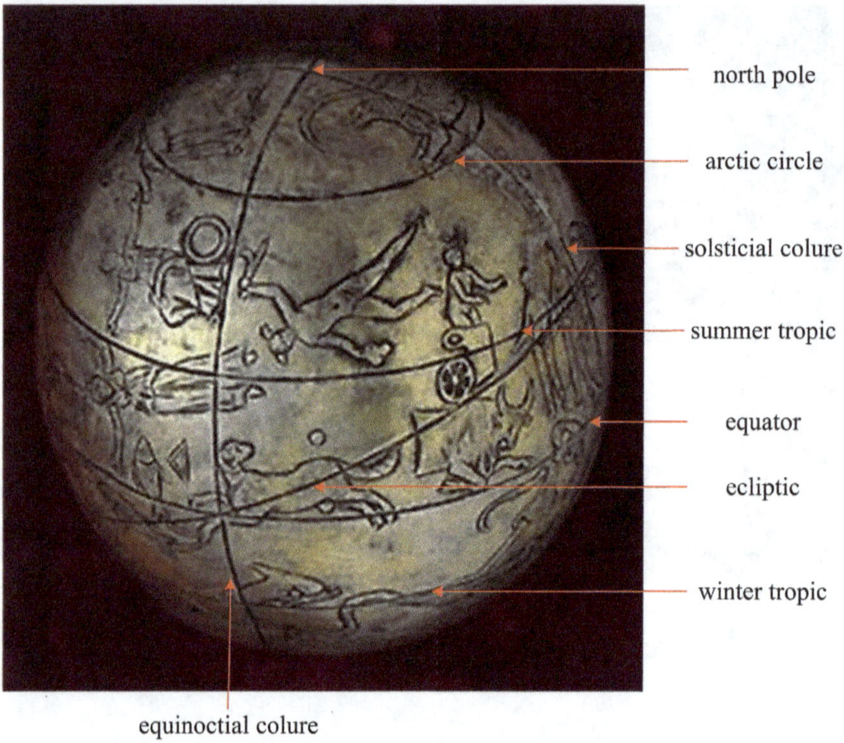

FIGURE 4.2 *The circles drawn on the Kugel Globe according to Geminus.*
COURTESY OF NICOLAS AND ALEXIS KUGEL.

1. Five parallel circles are drawn on the globe (sections 5.12, 5.15–16, 5.45–5.48):

   i. The arctic, namely the greatest of the circles of perpetual apparition;
   ii. the summer tropic;
   iii. the equator;
   iv. the winter tropic;
   v. the antarctic, or the greatest of the circles of perpetual occultation, namely the limit of visibility of the sky throughout the year.[8]

8 Arctic and antarctic circles are considered as the greatest circles of perpetual apparition and occultation, respectively, for observers situated to the north of the terrestrial equator. Summer and winter tropics are also considered for observers situated to the north of the terrestrial equator. In the following we shall always consider the observer's locality to be north of, or on the terrestrial equator, namely we presume $\varphi \geq 0°$.

The positions of the arctic and antarctic circles depend on the latitude of the locality (16.12). The distances between these circles measured in sixtieths of the meridian circle and degrees are presented in Table 4.1.

The distance of 36° between the arctic circle and the north pole corresponds to the latitude of Rhodes (similarly for the distance between the antarctic circle and the south pole). Geminus had probably lived and worked in Rhodes.[9]

TABLE 4.1 *Distances between the Poles and the Parallel Circles on the Celestial Globe, according to Geminus (Sections 5.46 and 16.10–12)*

Distance	Sixtieths of a circle	Degrees
From north pole to arctic circle	6	36
From arctic circle to summer tropic	5	30
From equator to summer tropic	4	24
From equator to winter tropic	4	24
From winter tropic to antarctic circle	5	30
From south pole to antarctic circle	6	36

2. The two colures—through the celestial poles and the equinoctial and tropical points—are depicted on the globe, dividing the ecliptic into four equal quarters (5.49–50).
3. The stars are depicted on the celestial globe (5.13–14).
4. The horizon is not depicted on the globe, because it must not move; the position of the horizon is defined by the stand: "Ὑπὸ μέντοι γε τῆς σφαιροθήκης ἡ τοῦ ὁρίζοντος θέσις κατανοεῖται" (5.62–63).
5. The meridian, namely the circle through the celestial poles and the zenith of a locality, is also stable for this locality and not depicted on the globe (5.64–65). Since the division of the meridian into sixtieths was used for drawing the five parallel circles on the globe (5.45–46), it could be implied that the meridian was a ring surrounding the globe.

9 D.R. Dicks, "Geminus", in In *Dictionary of Scientific Biography*, ed. Charles Coulston Gillispie et al., 25 vols. (Detroit, MI: Charles Scribner's Sons, 2008), 5:344–347.

6. The great galactic circle is not depicted on the majority of the globes, because its width varies (5.69).

Ptolemy, taking into consideration the precession of the equinoxes, used ecliptic coordinates for his star catalogue, so that the latitude of the stars would remain constant, while their longitude would vary uniformly for all the stars. In book VIII.3 of his *Almagest*, he describes the construction of a celestial globe (Figure 4.3), where the only circles drawn on it are the ecliptic (divided into 360°), a great circle through the poles of the ecliptic and the galactic circle.[10] The equator is not depicted on the globe, in order not to define the equinoctial points on the ecliptic. A rotatable ring is mounted around the globe through the poles of the ecliptic; half of it—from one pole to the other—is divided into 180°; this represents the solstitial colure and the poles of the equator are marked on it in a distance of ε=23;51 degrees from the poles of the ecliptic. A similar ring, slightly larger and representing the meridian, is mounted to the previous one on the two poles of the equator. Sirius is positioned on the globe, on the great circle drawn through the poles of the ecliptic, according to its ecliptic latitude. The point of intersection of the ecliptic with the semicircle through the poles of the ecliptic and Sirius is considered as the beginning of the graduation on the ecliptic, for positioning the rest of the stars. The system of the globe with the two rings is placed on a stand (βάσις), whose upper horizontal ring defines the horizon. The north celestial pole can be elevated according to the latitude of the place. This sphere can be adjusted by rotating the solstitial ring, so that it passes through the current positions of the solstices and can eternally be functional.

Among the treatises on the celestial globe written in the first millennium AD we have the following:

1. The partially preserved treatise "On the construction of the Aratean sphere" (Περὶ κατασκευῆς Ἀρατείας σφαίρας) written by Leontius (seventh century) in Greek, where the construction of a celestial globe is

10 Claudii Ptolemaei, *Opera quae extant omnia*, ed. J.L. Heiberg (Leipzig: Teubner, 1903), VIII.3, vol. 1.2: 179–185; and Toomer, *Ptolemy's Almagest*, 404–407.

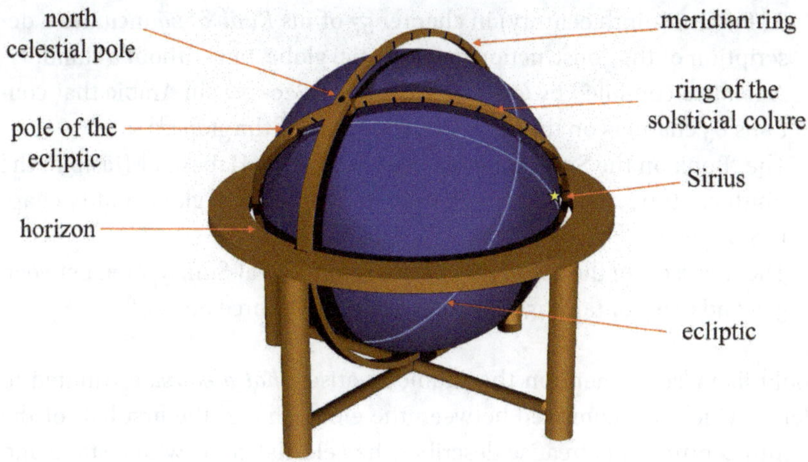

FIGURE 4.3 *The globe described in* VIII.3 *chapter of Ptolemy's Almagest. Based on the design of the precession globe by Hadravová and Hadrava.*[11]

described.[12] The six circles to be drawn on the globe are the following: arctic, summer tropic, equator, winter tropic, antarctic, zodiac. The distance of the equator from each one of the tropics is 24° and the distance of each one of the arctic and antarctic circles is 41° from the north and south pole, respectively. This distance corresponds to a latitude φ=41° (Constantinople). The globe is surrounded by two equal rings (section five): i. The horizon which is represented by a ring on pillars "ἡ ἐπὶ τοῖς κίοσιν ἴτυς, ἣ δὴ καὶ ὁρίζοντος τάξιν ποιεῖται"; it is not mentioned if the ring is graduated. ii. The meridian that supports the globe "ἕτερος δὲ ἴσος μὲν τῇ βάσει καὶ φέρων τὴν σφαῖραν κρίκος, μεσημβρινοῦ δὲ λεγόμενος τάξιν ἐπέχειν" ("another ring equal to the base and carrying the globe, which is said that it represents the meridian"). The two rings bisect each other at right angles. The half-ring above the horizon is divided into 180 degrees.

11 Alena Hadravová and Petr Hadrava: "Sphaera octava, Myty a veda o hvezdach" [Myths and Science on Stars]: IV, *Katalogy hvězd a přemyslovský nebeský glóbus* [Catalogues of Stars and the Premyslid Celestial Globe] (Prague: Artefactum/Academia, 2013), 334.

12 Leontius, *De sphaerae Arateae constructione*, ed. E. Maass, *Commentariorum in Aratum reliquiae* (Berlin: Weidmann, 1898).

2. Al-Battānī (ninth century) in chapter 57 of his *Kitāb al-zīj* includes a description of the construction of a celestial globe, but without a stand.[13]
3. A treatise compiled by Qusṭā ibn Lūqā (ca. 820–912) in Arabic that contains 65 chapters on the description and use of the globe.[14]
4. The "Book on the Sphere and its Use", written by Ḥabash al-Ḥāsib in the ninth century, contains a description of the celestial globe and 14 chapters on its use.[15]
5. The treatise "On the use of the celestial globe" by al-Ṣūfī, written between 983 and 986, contains 157 chapters arranged in three books.[16]

I would like also to mention the Arabic treatise *Dhāt al-kursī*, attributed to Ptolemy, which was compiled between the eleventh and the first half of the sixteenth century. This treatise describes the celestial globe with a stand and contains 33 chapters on its use; it is mainly a summary of the treatise of Qusṭā ibn Lūqā with some enhancement.[17]

The celestial globe could be used for didactic purposes; it could also help in solving certain astronomical problems. For solving problems, the globe has to be positioned according to the position of the celestial sphere. This means:

---

13 Al-Battānī, *Al-Battānī sive Albatenii, Opus Astronomicum*, ed. Carlo Alfonso Nallino, 3 vols. (Milan: Pubblicazioni del Reale Osservatorio di Brera in Milano, 1899–1907). The text is in Arabic (3:210–214) with a Latin translation (1:139–142) made by Nallino himself. A detailed description of the instrument in English, with instructive figures, is presented in Emilie Savage-Smith, *Islamicate Celestial Globes: Their History, Construction, and Use* (Washington DC: Smithsonian Institution, 1985), 18–21.

14 W. H. Worrell, "Qusta Ibn Luqa on the Use of the Celestial Globe", *Isis* 35 (1944): 285–293. This article contains an English translation of chapters 1–14 and only the titles of the rest of the chapters. A Latin translation (thirteenth century) of this treatise is presented in Richard Lorch and José Martínez Gázquez, "Qusta ben Luca: De sphera uolubili", *Suhayl* 5 (2005): 9–62. A comparison between the treatise of Qusṭā ibn Lūqā and the Alfonsine Treatise "Libro de la fayçon dell espera" is presented in Julio Samsó, "Qustā ibn Lūqā and Alfonso X on the Celestial Globe", *Suhayl* 5 (2005): 63–79. The treatise of Qusṭā ibn Lūqā is studied as the source of the treatise *Dhāt al-kursī*, in Flora Vafea, "The treatise *Dhāt al-kursī* attributed to Ptolemy" in *Ptolemy's Science of the Stars in the Middle Ages* (forthcoming within the project Ptolemaeus Arabus et Latinus).

15 Richard Lorch and Paul Kunitzsch: "Ḥabash al-Ḥāsib's Book on the Sphere and its Use" in *Zeitschrift für Geschichte der Arabisch-Islamischen Wissenschaften*, 2 (1985), 68–98.

16 MS Istanbul, Topkapi Saray, Ahmet III, 3505, 1 ff. 1–62, also E.S. Kennedy, "Al-Ṣūfī on the Celestial Globe" in *Zeitschrift für Geschichte der Arabisch-Islamischen Wissenschaften* 5 (1989), 48–93. In the appendix of this article, Kennedy presents the titles of the 157 chapters in English translation and a short summary of the content for most of the chapters.

17 See Flora Vafea, *The Treatise Dhāt al-kursī Attributed to Ptolemy*.

i. The elevation of the north pole of the globe must be equal to the latitude φ
ii. The point of the equator rising on the celestial sphere must be the same on the globe
iii. In some cases, the meridian ring must be aligned to the local meridian.

The celestial globe gives a realistic view of the stars and the diurnal motion, but has some major disadvantages, such as the difficulties:

i. in constructing a perfect sphere
ii. in making measurements on the sphere
iii. in transporting a sphere, in particular when it is bulky and on a stand.

Some sharp mind conceived the idea of "unfolding" the sphere and depicting the stars and the celestial motion on a plane. A "flattened" sphere could solve the problems mentioned above. But the transition from the sphere to the plane should definitely retain the celestial ratios, the circles of the celestial sphere and the circular motion of the heaven.

The solution to the above problems is preserved in the work of Ptolemy *Planisphaerium* or "Ἅπλωσις ἐπιφανείας σφαίρας".[18] Ptolemy begins with mentioning that "it is possible and useful in many matters that we construct on a flat surface the circles which lie on the solid sphere, as if it were spread out"[19] and continues with the description of what later was called stereographic projection from the south celestial pole onto the plane of the equator.

Ptolemy, in his *Planisphaerium*, describes the construction of the equator, the tropics and the ecliptic as circles, while the horary circles are depicted as straight lines (section one). He proceeds with the construction of the horizon, taking as an example the climate of *Rhodes* (φ=36°) and calculating the radius of the horizon and the distance of its center from the center of the equator (section seven).

---

18 Christopher Anagnostakis, "The Arabic Version of Ptolemy's Planisphaerium" (PhD Diss., Yale University, New Haven, CT, 1984); and Nathan Sidoli and J. L. Berggren, "The Arabic version of Ptolemy's Planisphere or Flattening the Surface of the Sphere: Text, Translation, Commentary", SCIAMVS 8 (2007): 37–139.

19 Anagnostakis, "The Arabic Version of Ptolemy's *Planisphaerium*", 69. The Arabic translation of the apparently lost Greek text is: إنّه لما كان من الممكن، وممّا ينتفع به في ابواب كثيرة أن يوجد في بسيط مسطح الدواير التي تقع في الكرة المجسمة كأنّها مبسوطة.
(Sidoli and Berggren, "The Arabic Version of Ptolemy's *Planisphere*", 55).

He always gives proofs to show the conformity of the flat figure to what appears on the solid sphere; so there are proofs for bisection of the great circles of the sphere in their projection onto the plane:

1. The ecliptic bisects the equator (section one).
2. The great circles through the celestial poles intersect the ecliptic in diametrically opposite points of the sphere (section two). [All the circles through the celestial poles are projected as straight lines].
3. The horizon bisects the equator and intersects the ecliptic in diametrically opposite points of the sphere (section three).
4. The circles through the poles of the ecliptic, which intersect the ecliptic in diametrically opposite points of the sphere, bisect the equator (section fifteen).

Ptolemy calculates the sizes of the circles parallel to the equator and shows relations between their radii. He also calculates the rising times of the zodiacal signs for the right sphere (*sphaera recta*), namely for the terrestrial equator (sections eight and nine), and for the oblique sphere that corresponds to the climate of Rhodes (sections twelve and thirteen). At the end, the construction of the circles parallel to the ecliptic (sections sixteen to ninteen) and the positioning of the stars using equatorial or ecliptic coordinates are described (section twenty). The construction of the circles parallel to the horizon, the so-called almucantars, is not discussed, but it is similar to the construction of the circles parallel to the ecliptic.

We notice that wherever the latitude of the locality is requested, it is taken as $\varphi=36°$, which is the latitude of Rhodes, not the latitude of Alexandria ($\varphi=31°$) where Ptolemy was living.[20] This is strong evidence that the transition from the sphere to the plane was conducted in Rhodes. Combining this evidence with the direct testimonies of Synesius that Hipparchus was the inventor of the projection of a sphere onto a plane and he had placed sixteen stars on the

20 The locality is requested for calculating: (i) the radius of the horizon and the distance of its center from the center of the equator (section 7); (ii) the difference between the longest (or the shortest) daytime and the equinoctial daytime, in degrees and equinoctial hours (section 11); and (iii) the rising times of the zodiacal signs in the oblique sphere (sections 12–13). Ptolemy mentions that he uses the latitude of Rhodes, as he had used it on the solid sphere (Sidoli and Berggren, "The Arabic Version of Ptolemy's Planisphere", section 7, p. 90) and as he had used it in the *Almagest* (Sidoli and Berggren, "The Arabic Version of Ptolemy's Planisphere", section 10, p. 95).

instrument,[21] and some indirect ones by Severus Sebokht,[22] we may conclude that the person who initiated the transition from the sphere to the plane was Hipparchus. The study of the circles parallel to the ecliptic (sections sixteen to nineteen) and the great circles through the poles of the ecliptic (section fifteen) must be new ideas initiated by Ptolemy, since these circles are the reference circles for the ecliptic coordinates he presents for the fixed stars in the *Almagest*.[23]

It is not sufficient to have all the circles and the stars on a plane; the celestial motion, as modelled by the celestial globe, should be retained on the plane. Rotating a celestial globe such as the one described by Geminus, we observe that the five circles (the equator and the parallel circles) remain in their position during the rotation. The circles of the horizon and the meridian, also the poles of the equator and those of the horizon (zenith—nadir) remain unmoved. The stars, the Sun, the ecliptic, the solstitial and equinoctial colures,

21 "Σφαιρικῆς ἐπιφανείας ἐξάπλωσιν, ταυτότητα λόγων ἐν ἑτερότητι τῶν σχημάτων τηροῦσαν, ᾐνίξατο μὲν Ἵππαρχος ὁ παμπάλαιος, καὶ ἐπέθετό γε πρῶτος τῷ σκέμματι", "οἱ ἑκκαίδεκα ἀστέρες..., οὓς μόνους Ἵππαρχος μετατιθεὶς ἐγκατέταξε τῷ ὀργάνῳ" ("Hipparchus, the ancient, had darkly discussed the unfolding of a spherical surface, preserving the proportions in the difference of the appearances, and he was the first to contribute to the solution of the problem", "the sixteen stars ... which are the only [stars] Hipparchus transposed and placed in the instrument"). Section 5 in *Synesius, Synesii Cyrenensis opuscula: Ad Paeonium de dono astrolabii*, ed. N. Terzaghi (Rome: Polygraphica, 1944), 132–142.

22 M.F. Nau, "Le Traité sur l'astrolabe plan de Sévère Sabokt", *Journal Asiatique*, série 9e, 13 (1899): 56–101, 238–303. Sebokht refers nine times to the Philosopher who had constructed the astrolabe, making clear that this Philosopher is not Ptolemy (in chapter XXIV); he also mentions that Ptolemy was using the astrolabe (chapter VII). In Joannes Philoponus, "Joannis Alexandrini, cognomine Philoponi, de usu astrolabii eiusque constructione libellus", ed. Heinrich Hase, *Rheinisches Museum für Philologie* 6 (1839): 127–171 at 139, it is mentioned that Ptolemy placed the lines of the unequal hours below the horizon on the astrolabe; the same text appears in Joannes Philoponus, *De usu astrolabii eiusque constructione, Über die Anwendung des Astrolabs und seine Anfertigung*, ed. Alfred Stückelberger (Berlin, München, Boston: De Gruyter, 2015), 22–25. These testimonies, combined with the fact that Ptolemy had written the *Planisphaerium*, show that the Philosopher who had constructed the astrolabe could not be after Ptolemy and that the astrolabe was already present in Ptolemy's time.

23 Sidoli and Berggren mention that there is a clear division of subject matter that allows us to separate the treatise into two parts: The first part is concerned with the equator and the circles related to it, the ecliptic and the horizon, and includes a computational treatment of rising-time phenomena. The second part is concerned with circles related to the ecliptic and addresses special topics in the practical implementation of the planisphere (Sidoli and Berggren, "The Arabic Version of Ptolemy's Planisphere", 51–54).

and the poles of the ecliptic continuously change their position during a complete rotation of the globe. Thus these circles and points must be drawn in two separate plates. One of them should be stable; this is called *τύμπανον*, plate or disc (Figure 4.8). The other should rotate on the common point, the center of the motion of the heavens: the north celestial pole; the movable plate is called *ἀράχνη*, spider or *rete* (Figure 4.9, right).

In the following, I use the English term *spider* instead of the Latin term *rete*; the Greek term *ἀράχνη* and the Arabic "عنكبوت" mean "spider", not "net" (*rete* in Latin). Additionally, the shape of the six great circles through the poles of the ecliptic that divide the ecliptic into twelve zodiacal signs resembles a spider, when these circles are projected through a stereographic projection from one of the celestial poles onto the plane of the celestial equator (Figure 4.4). It is probable that the name *ἀράχνη* (spider) was given to this part of the astrolabe, because those six circles always drawn on the celestial globes were transferred onto this turnable part of the instrument, where also the ecliptic and the fixed stars are depicted.

Another problem of the chosen transition from the solid globe to the plane is that the south celestial pole and a circle around it cannot be depicted on the plane. Thus, the antarctic circle, the greatest of the always invisible circles, could be the limit, the outer circle of the representation; in any case, the rest of the heavens is never visible from the place of observation: The first version of the planispheric astrolabe is ready to be constructed!

According to Geminus (and Leontius later in the 7th century), the globe was constructed for a certain latitude (36° for Geminus, 41° for Leontius),[24]

---

24 Leontius, *De sphaerae Arateae constructione*, 561–567, section 4, lines 73–75: "(δεῖ δὲ) … ἑκάτερον δὲ τῶν ἄλλων (λέγω δὴ τόν τε ἀρκτικὸν καὶ τὸν ἀνταρκτικόν) ἀπέχειν τοῦ καθ' ἑαυτὸν πόλου τῆς σφαίρας μοίρας μα'" ("Each one of the others [I mean the arctic and antarctic (circles)] must have a distance of forty-one degrees from the closest to it pole of the globe."); and Geminus, in Germaine Aujac, *Géminos, Introduction aux Phénomènes*, sections 5.45–5.48: "οἱ δὲ τροπικοὶ κύκλοι ἀπὸ τῶν ἀρκτικῶν οὐ τὴν αὐτὴν ἔχουσι διάστασιν κατὰ πάντας τοὺς ὁρίζοντας, ἀλλ' οἱ μὲν ἔλασσον, οἱ δὲ πλέον διεστήκασιν. Ὁμοίως δὲ οὐδ' οἱ ἀρκτικοὶ ἀπὸ τῶν πόλων τὴν ἴσην ἀπόστασιν ἔχουσι κατὰ πᾶν ἔγκλιμα, ἀλλ' οἱ μὲν ἐλάσσονα, οἱ δὲ πλείονα. Καταγράφονται μέντοι γε πᾶσαι αἱ σφαῖραι πρὸς τὸν ἐν τῇ Ἑλλάδι ὁρίζοντα." ("The tropic circles have not the same distance from the arctic ones in all of the horizons, but some have lesser and some have greater distance. Similarly, the arctic circles have not the same distance from the poles in every latitude, but some [have] lesser and some greater. Indeed, all the globes are inscribed according to the horizon of Greece"); and 16.10–16.12: "ἀπὸ τοῦ πόλου ὁ ἀρκτικὸς διίσταται μοίρας λς … ὁ δὲ ἀρκτικὸς ἀπὸ τοῦ θερινοῦ τροπικοῦ διίσταται μοίρας λ"—"Πρὸς γὰρ τοῦτο τὸ ἓν κλίμα καὶ αἱ κρικωταὶ σφαῖραι κατασκευάζονται

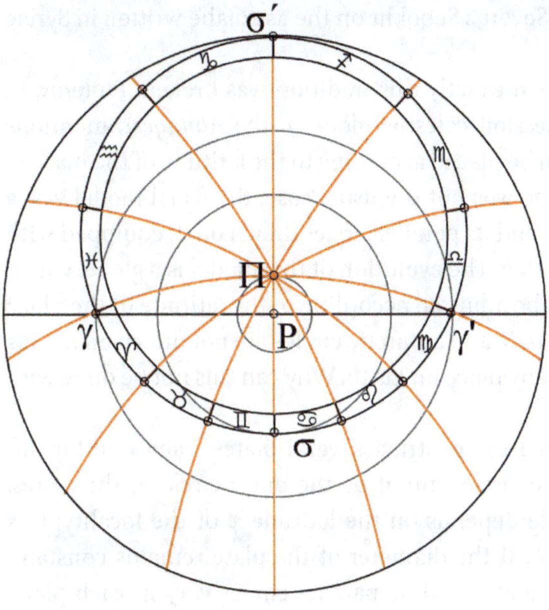

FIGURE 4.4 *The stereographic projection (from a celestial pole onto the plane of the celestial equator) of the six great circles through the poles of the ecliptic which divide the ecliptic into 12 zodiacal signs.*

since the arctic and antarctic circles were drawn on the globes. Thus, the astrolabe, either identical or relative to "the instrument of the hours",[25] as Ptolemy calls it in the *Planisphaerium* (section fourteen), was designed for a unique latitude. Descriptions of this initial form of the astrolabe are preserved in a letter of Synesius to Paionius (fourth century AD); there are also scattered

καὶ αἱ στερεαί." ("The arctic [circle] has a distance of thirty-six degrees from the pole … the arctic [circle] has a distance of thirty degrees from the summer tropic. Both the armillary spheres and the solid [globes] are constructed according to this sole latitude.").

25 The term is preserved in Arabic as "آلات الساعات", see: Sidoli and Berggren, "The Arabic Version of Ptolemy's Planisphere" p. 73, line 402. Ptolemy also refers to observation "δι' ἀστρολάβων ὡροσκοπείων" (through instruments determining the time using the stars) in Tetrabiblos—Apotelesmatica: III.3, lines 123–124; see: *Claudii Ptolemaei opera quae exstant omnia*, vol. 3.1, ed. W. Hübner, post E. Boer and F. Boll (Stuttgart-Leipzig: Teubner, 1998), 172–173.

references in the treatise of Severus Sebokht on the astrolabe written in Syriac (seventh century AD).[26]

We cannot determine when exactly this tradition was broken. Ptolemy, in the description of the precession celestial globe in the *Almagest*, mentions that the pole of the globe can be placed according to the latitude of the place.[27] The precession celestial globe was not the usual case; the usual model was a solid globe, with the equator and its parallel circles drawn on it, equipped with a meridian and a horizontal ring. The evolution of this model is a globe, where the poles of the equator can be adjusted according to the latitude of the place of observation,[28] while the arctic and antarctic circles are not drawn on it. This type of globe can be used at any place on Earth. Why can this not be done with an astrolabe?

For the astrolabe, we need to construct several plates, each one for different latitude. The antarctic circle cannot be the outer circle of the plates. The declination of this circle depends on the latitude φ of the locality; it is $-(90°-\varphi)=\varphi-90°$, where $\varphi \geq 0°$. If the diameter of the plate remains constant, then the diameters of the equator and its parallel circles vary in each plate; therefore, we cannot use the same spider above these plates. If we keep the diameter of the equator constant, then the diameter of the plate varies with the latitude, thus the spider will be too large for some plates and too small for others. A solution is to restrict the plates to the tropic of Capricorn. This way the same spider can be used above the plates of all latitudes. These plates can be kept in a receptacle, *δοχεῖον* or mater.[29]

---

26 See M.F. Nau, "Sévère Sabokt: Traité sur l'astrolabe plan". In the Introduction and chapters I and XXIV of Sebokht's treatise, there is a "hidden" description of an astrolabe with five zones / circles and having as bounding circle the "antarctic", which corresponds to a latitude $\varphi=36°$. This provides evidence for the initial form of the astrolabe, designed for a unique latitude ($\varphi=36°$), probably by Hipparchus.

27 "τὸν δὲ μεσημβρινὸν ὀρθὸν προσαρμόσομεν τῷ κατὰ τὴν βάσιν ὁρίζοντι διχοτομούμενον μὲν ὑπὸ τῆς φαινομένης ἐπιφανείας αὐτοῦ, δυνάμενον δὲ περιάγεσθαι περὶ τὸ ἴδιον ἐπίπεδον, ὅπως ἐξαίρειν ἑκάστοτε δυνώμεθα τὸν βόρειον πόλον ἀπὸ τοῦ ὁρίζοντος διὰ τῆς τοῦ μεσημβρινοῦ διαιρέσεως ταῖς οἰκείαις τῶν ὑποκειμένων κλιμάτων περιφερείαις" ("Then we fix the meridian ring in position perpendicular to the horizon defined by the stand [of the globe], in such a way that it is bisected by the visible surface of the latter, but can be moved round in its own plane: this is in order that we may, for any particular application, raise the north pole from the horizon by the appropriate arc for the latitude in question, using the graduation of the meridian [to place the ring correctly]"), Claudii Ptolemaei, *Opera quae extant omnia*, VIII.3, vol. 1.2:184.5–12, and Toomer, *Ptolemy's Almagest*, 406.

28 There exist some examples of celestial globes whose meridian ring has holes where the poles of the globe can be mounted.

29 The term "δοχεῖον" appears in Joannes Philoponus, *De usu astrolabii*, 146; and Joannes Philoponus, *Über die Anwendung des Astrolabs und seine Anfertigung*, 40.

An astrolabe of the latter form is described in detail by Joannes Philoponus (sixth century AD).[30] He also mentions that there are instruments without a receptacle, consisting of only one plate.[31] On the back side of the astrolabe, each of the upper two quadrants of the circle (or only one of them) is divided into 90°; the altitude of any celestial body can be measured (Figure 4.5) with a diopter pivotable about the center of the mater on the graduation of the quadrants.[32]

In chapter I.3 of his treatise *On the Use of the Celestial Globe*, al-Ṣūfī describes the construction of a concave external quadrant (Figure 4.7) for measuring the altitude of the Sun using the shadow of a pin-gnomon and mentions that the altitude of the stars cannot be directly measured by observation using the globe (since the stars cast no shadow). This quadrant was incorporated on the back of the astrolabe.

For measuring the altitude of a celestial body with the astrolabe (Figure 4.5), the celestial body must be targeted through the two sights of the diopter (alidade); when the celestial body is visible through both sights, the index of the diopter provides the altitude.

This method for measuring the altitude is simple and accurate and replaces the laborious and less accurate method of measuring the altitude with a gnomon mounted on the celestial globe, as it is preserved in the treatise *Dhāt al-kursī* (chapter nine), and the treatises of Qusṭā ibn Lūqā (chapter fifty-three)[33] and al-Ṣūfī (chapters I.4 and I.8)[34] *On the Use of the Celestial Globe*. The general idea for measuring the altitude of the Sun with a globe is to install a gnomon in

30 Joannes Philoponus, *De usu astrolabii*, 129–156; Joannes Philoponus, *Über die Anwendung des Astrolabs und seine Anfertigung*, 2–62.

31 Joannes Philoponus, *De usu astrolabii*, 146–147.

32 Joannes Philoponus, *De usu astrolabii*, 130.

33 See chapter P9: "On the method of obtaining the altitude of the Sun on the globe" (in the Commentary), in Vafea, "The treatise *Dhāt al-kursī* attributed to Ptolemy". W.H. Worrell, in *Qusta Ibn Luqa on the Use of the Celestial Globe*, 292, gives as the title "On the division-parts of the elevation of the Sun on the globe at any time you wish" for chapter fifty-three, while the title according to the MS Cairo, Dār al-kutub, K 3824 is "في أخذ ارتفاع الشمس في الكرة في أي وقت شئت" ("To take the altitude of the Sun on the globe at any time you wish"). The title in the Latin translation is "In accepcione solis altitudinis in spera quocumque tempore et quacumque hora uolueris". Lorch and Martínez Gázquez, "Qusta ben Luca: De sphera uolubili", 55.

34 A summary of the methods presented in chapters I.4 and I.8 is presented in E.S. Kennedy, *Al-Ṣūfī on the Celestial Globe*, 55–56.

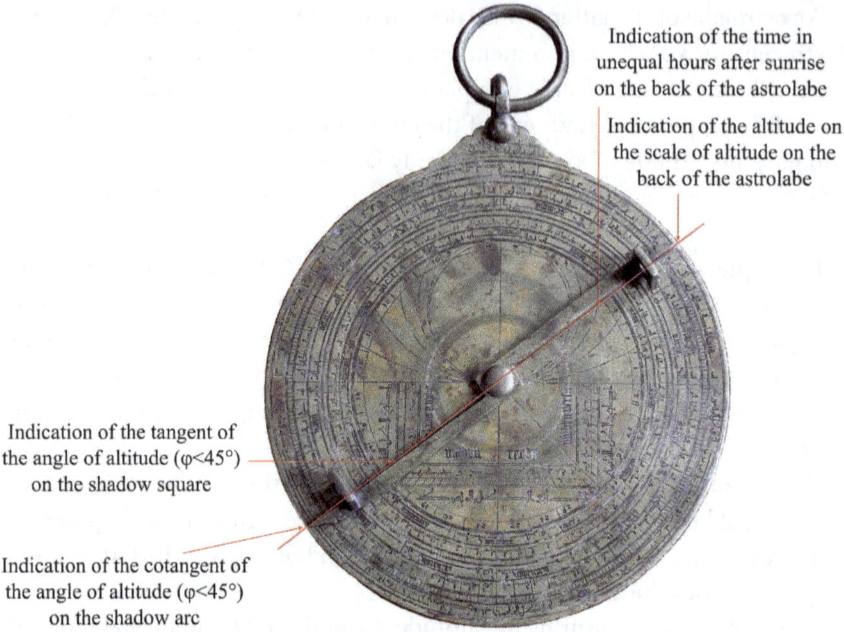

FIGURE 4.5 *Measuring the altitude with the astrolabe.*

the form of a pin on the degree of the Sun, perpendicular to the surface of the globe (Figure 4.6). The north pole of the globe is placed according to the latitude of the locality and the meridian ring is aligned to the local meridian. Then we rotate the globe to reach the position where the gnomon casts no shadow on the globe; this happens when the direction of the gnomon is the same as that of the rays of the Sun. We keep the globe fixed and we fit a quadrant (Figure 4.7) divided into 90° on the globe, in such a way that the tip corresponding to 0° is on the horizon, the tip corresponding to 90° at the zenith and the quadrant passes through the position of the gnomon. The indication of the quadrant that reaches the position of the gnomon provides the altitude of the Sun. The rising degree of the equator defines the position of the celestial sphere at that moment.

The concave quadrant could measure the altitude of any star on the celestial globe that corresponds to its altitude in the sky, provided that the globe has been positioned accordingly to the celestial sphere. On the plates of the astrolabe, the altitude is measured on the circles parallel to the horizon. They are drawn every one, two, three, or six degrees, depending on the size of the astrolabe. These circles, called the almucantars, are constructed on the plate of each latitude (Figure 4.8), in a similar way to that which Ptolemy

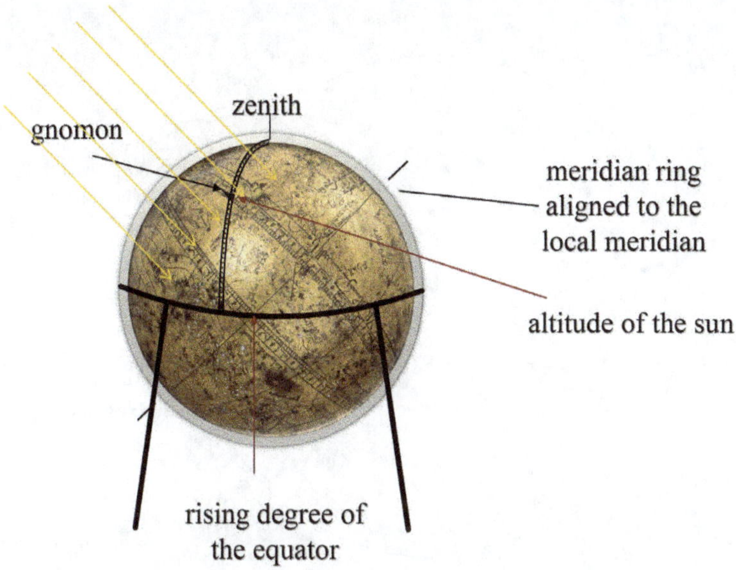

FIGURE 4.6 *Measuring the altitude with a globe and an external quadrant.*

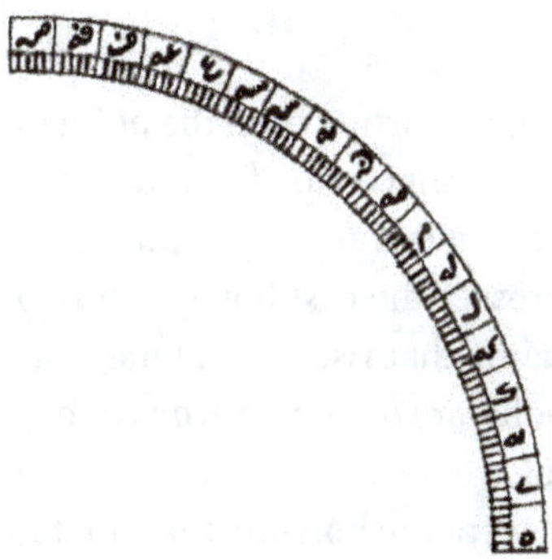

FIGURE 4.7
The quadrant for measuring the altitude on a globe, from the MS of al-Ṣūfī.

describes for the circles parallel to the ecliptic. The east-west line that corresponds to the horizon of the right sphere, namely to the latitude of the terrestrial equator, is drawn in each plate.

The celestial equator was always depicted on the celestial globe (except for the precession celestial globe). According to the description in the treatise *Dhāt al-kursī* and those of Qusṭā ibn Lūqā and al-Ṣūfī,[35] the equator is divided

35 *Dhāt al-kursī*: MS Cairo, Dār al-kutub, Mīqāt Ṭalʿat 189, Arabic, Introduction; Qusṭā ibn Lūqā: Worrell, "Qusta Ibn Luqa on the Use of the Celestial Globe", chapter 1, 287; al-Ṣūfī, MS Istanbul, Topkapi Saray, Ahmet III, 3505,1 chapters I.1–2 (the celestial equator is not

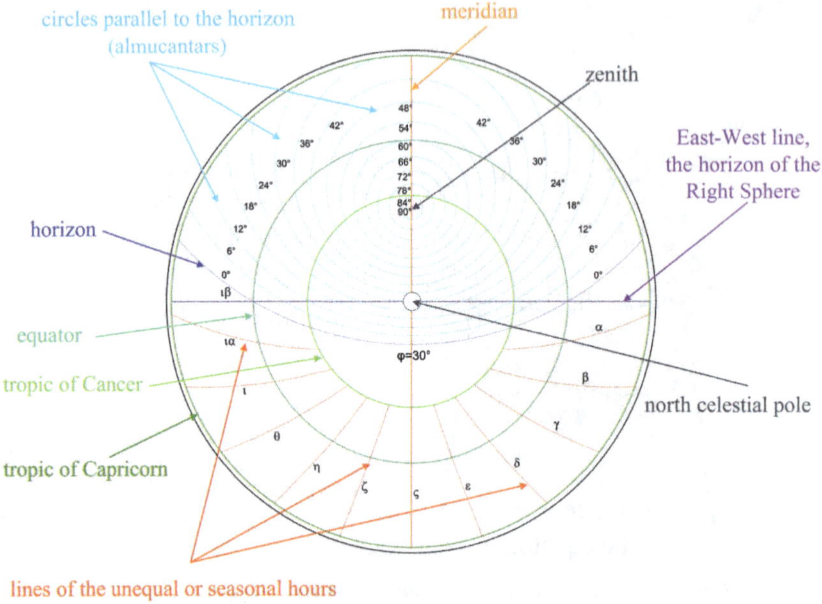

FIGURE 4.8 *Plate of latitude 30°, according to the description of Joannes Philoponus.*

into 360 degrees and the graduation is depicted on it, starting from the point of the vernal equinox γ. The graduation of the equator is necessary for finding the right and oblique ascension of the Sun, the stars and generally any point on the globe: when a point of the globe rises, namely crosses the east horizon during the rotation of the globe, the degree of the equator that rises simultaneously with it is the ascension of this point. When the poles are on the horizon ($\varphi=0°$), the equator is perpendicular to the horizon, thus we have the right ascension of this point. When the poles are inclined in respect to the horizon, the equator is oblique to the horizon, thus we obtain the oblique ascension of this point; the oblique ascension depends on the latitude of the locality.

The rising degree of the equator determines the position of the globe in connection with the time for a given latitude. Thus, it is very important to transfer the graduation of the equator on the astrolabe. The problem now is that the equator is drawn on the stable plates of the astrolabe, while the point γ corresponding to the vernal equinox, which is the beginning of the graduation, is rotating. On the celestial sphere (Figure 4.9, left) we notice that when the vernal equinoctial point γ rises, the point σ′ corresponding to the winter solstice and the point Π corresponding to the north pole of the ecliptic reach the meridian

---

mentioned in the summary of chapters I.1–2 in E.S. Kennedy, *Al-Ṣūfī on the Celestial Globe*).

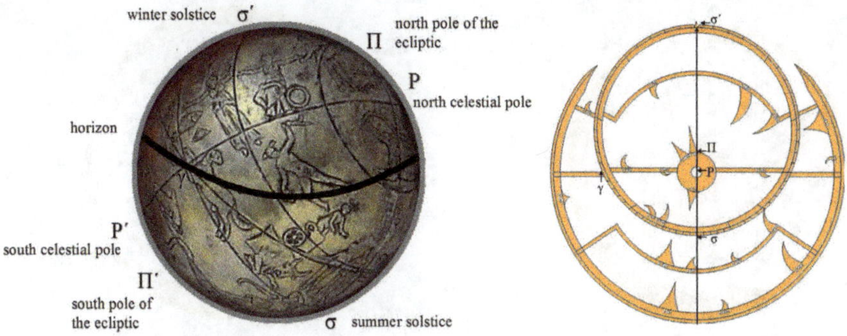

FIGURE 4.9 *When the point γ corresponding to vernal equinox rises, the point σ′ of the winter solstice and the pole Π of the ecliptic reach the meridian ring. The Kugel globe (left) is courtesy of Nicolas and Alexis Kugel.*

ring, that is, the solstitial colure σ′ΠPσΠ′P′ coincides with the meridian.[36] The solstitial colure, according to Geminus, is drawn on the globes.[37] Since this circle passes through the poles, its projection is a straight line on the astrolabe, and because it rotates with the diurnal motion, it is depicted on the spider of the astrolabe. This line has been chosen as an index for the ascension, and a protrusion at the point σ′ is designed (Figure 4.9, right). This index on the spider is called "τὸ τῆς ἀράχνης μοιρογνωμόνιον" in Greek,[38] simplified as *murī* in Arabic. To keep the astrolabe consistent with the celestial globe, at the moment of rising of the point γ of the vernal equinox, the index must point to 0° (Figure 4.10, top left panel); at the moment of rising of the point σ of the summer solstice in the right sphere, the index must point to 90° (Figure 4.10, top right panel); at the moment of rising of the point γ′ of the autumnal equinox, the index must point to 180° (Figure 4.10, bottom left panel); and at the moment of rising of the winter solstice σ′ in the right sphere, the index must point to 270° (Figure 4.10, bottom right panel). These numbers are written on the

36 P, P′ are respectively the north and south poles of the equator, Π, Π′ are respectively the north and south poles of the ecliptic, and σ, σ′ are respectively the points corresponding to the summer and winter solstices.

37 Aujac, *Géminos, Introduction aux Phénomènes*, 5.49–5.50: "Διὰ τῶν πόλων δὲ εἰσι κύκλοι οἱ ὑπό τινων κόλουροι προσαγορευόμενοι, … Γράφονται δὲ οὗτοι οἱ κύκλοι διὰ τῶν τροπικῶν καὶ ἰσημερινῶν σημείων καὶ εἰς δ μέρη ἴσα διαιροῦσι τὸν διὰ μέσων τῶν ζῳδίων κύκλον." Translation from Evans-Berggren, *Geminos's Introduction to the Phenomena*, 157: "[Passing] through the poles are the circles that some call the colures … These circles are drawn through the tropic and equinoctial points and divide into 4 equal parts the circle through the middles of the signs."

38 Joannes Philoponus, *De usu astrolabii*, 141; and Joannes Philoponus, *Über die Anwendung des Astrolabs und seine Anfertigung*, 26.

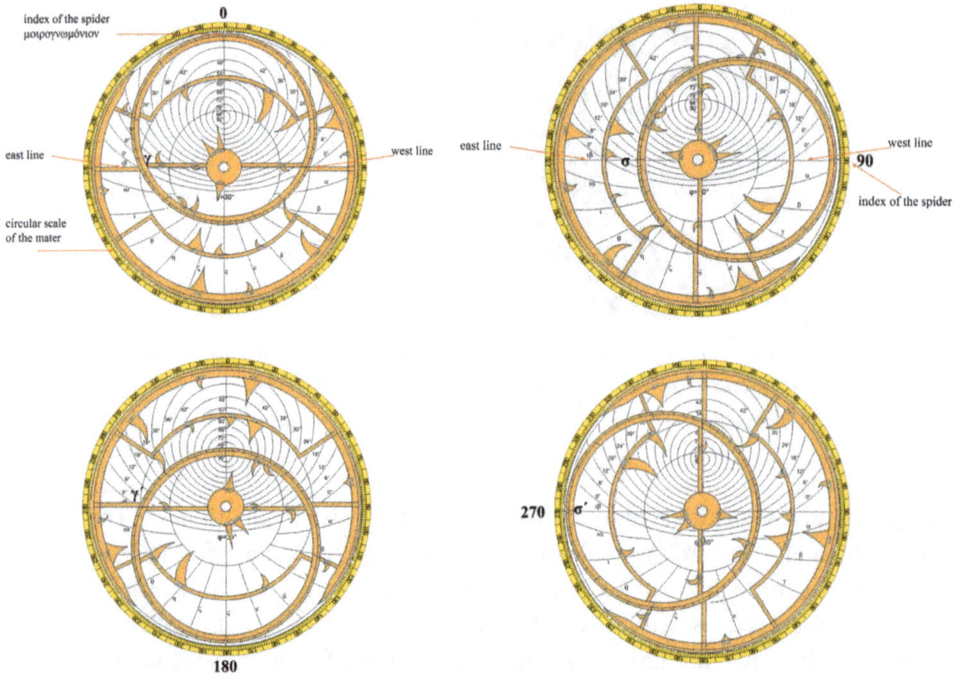

FIGURE 4.10 *From top left: The point γ corresponding to vernal equinox rises and the index of the spider points to 0°; top right: the point σ of the summer solstice rises in the right sphere and the index of the spider points to 90°; bottom left: the point γ′ corresponding to autumnal equinox rises and the index of the spider points to 180°; bottom right: the point σ′ of the winter solstice rises in the right sphere and the index of the spider points to 270°.*

stable part of the astrolabe, on the rim of the mater, which is graduated from 1–360°. On the plates for the various latitudes both the horizon of this latitude, as an arc, and the horizon of the right sphere, as the east-west straight line, are drawn. Thus, when we place any star or degree of the ecliptic on the east horizon, we obtain its oblique ascension, and when we place it on the east line we obtain its right ascension, as indication of the index. A complete rotation of the spider represents the diurnal motion of the celestial sphere. In this way, the celestial motion has been transferred onto the plane.

The descriptions of the celestial globe coming from Greek sources do not mention any graduation of the horizon. Similarly, the descriptions of the astrolabe from the same sources do not mention azimuth lines. The azimuth lines are the projections of the great circles through the zenith and nadir and determine the directions on the horizon. The directions on the horizon were introduced later by the Arabs. The treatises *On the Use of the Celestial Globe* by Qusṭā ibn Lūqā and *Dhāt al-kursī* mention the four cardinal points on the horizon on the globe, the risings and settings during summer and winter and the division of the horizon ring into 360°.[39] Al-Ṣūfī describes the graduation of the horizon from 0° to 90° on each one of the quadrants, starting from the east and the west points (rising and setting at the equinoxes) towards the north and the south.[40] The graduation of the horizon is used for determining the ortive amplitude of the stars and the degrees of the ecliptic, and the azimuth of the *qibla*.[41]

The azimuth on the astrolabe appears in the treatise of ʿAlī ibn ʿĪsā (fl. 844 AD) on the astrolabe, where in chapter 17, the azimuth of the Sun is determined[42] and more detailed in the treatises of al-Khwārizmī (eighth to ninth century AD) on the astrolabe.[43] Azimuth lines are not drawn on the plates of the preserved astrolabes coming from the ninth and tenth centuries.

The astrolabe, containing all the elements of the celestial globe projected on the plane of the front side, has still plenty of free space. On this space additional elements appear on the astrolabe, such as:

i. The lines of the unequal hours drawn on the plates below the horizon (Figure 4.8). Knowing the position of the Sun or a star in the sky, we can read directly the time in unequal hours on the astrolabe.

39 *Dhāt al-kursī*: MS Cairo, Dār al-kutub, Mīqāt Ṭalʿat 189, Arabic, Introduction; Qusṭā ibn Lūqā: Worrell, "Qusta Ibn Luqa on the Use of the Celestial Globe", chapter 1, 287. See also the introduction of treatise P and the commentary on it in Vafea, *The Treatise Dhāt al-kursī Attributed to Ptolemy*.

40 Al-Ṣūfī, MS Istanbul, Topkapi Saray, Ahmet III, 3505,1 chapter I.2 (the description of the horizon is not mentioned in the summary of chapter I.2 in E.S. Kennedy, *Al-Ṣūfī on the Celestial Globe*).

41 *Dhāt al-kursī*, chapter 13: ortive amplitude of the stars and the degrees of the ecliptic; chapter 28: the *qibla*. Qusṭā ibn Lūqā, chapters 25, 27, 31: ortive amplitude of the stars and the degrees of the ecliptic; chapter 57: the *qibla*. Al-Ṣūfī, chapters I.16–19 azimuths of the Sun, stars and planets, any degree of the ecliptic; chapters II.38–39: ortive amplitude of the degrees of the ecliptic and the stars; for a summary of the procedures by al-Ṣūfī see E.S. Kennedy, *Al-Ṣūfī on the Celestial Globe*, 58–59, 72–73.

42 See ʿAlī Ibn ʿIsā, "L'astrolabe et la manière de s'en servir" "كتاب العمل بالأصطرلاب", ed. L. Cheikho, *Al-Mashriq* 16 (1913): 29–46.

43 Josef Frank, "Die Verwendung des Astrolabs nach al-Chwārizmī", Abhandlungen zur Geschichte der Naturwissenschaften und der Medizin, Heft III (Erlangen: 1922), 1–32; and François Charette and Petra Schmidl, "Al-Khwārizmī and Practical Astronomy in Ninth-Century Baghdad. The Earliest Extant Corpus of Texts in Arabic on the Astrolabe and Other Portable Instruments", *SCIAMVS* 5 (2004): 101–198.

FIGURE 4.11 *The quadrant of the sine on the astrolabe.*

ii. The shadow square and shadow arc on the back side (Figure 4.5): When the diopter, or alidade, measures the altitude of a celestial body, we can obtain directly the tangent or the cotangent of the angle of altitude.[44]
iii. The quadrant of sine on the back side (Figure 4.11): knowing the angle, we may obtain directly the sine of the angle on the alidade, when its edge, between the center and one of the tips is divided into 60 or 150 parts.
iv. The lines of the unequal hours (Figure 4.5) on the back side: These lines allow us to find out the time in unequal hours, while measuring the altitude of the Sun with the alidade/diopter. There are two different types of these lines: for all latitudes or for a given latitude; the lines of the midday and afternoon Muslim prayers (for a given latitude) can be depicted on the same quadrant.

The procedures followed for solving astronomical problems with the celestial globe and the astrolabe are similar. As an example, we compare the procedures for determining the time during daytime with the globe and the astrolabe in unequal and equal hours:

44 For more details on the way the tangent and cotangent of an angle can be obtained using the shadow square or the shadow arc, see Flora Vafea, "Les traités d'al-Ṣūfī sur l'astrolabe" (PhD Diss., University Paris 7 – Denis Diderot, 2006), 596–597.

	Celestial globe	Astrolabe
1.	We place the north pole of the globe at an elevation equal to the latitude φ.	We use the plate for latitude φ on the astrolabe.
2.	We determine the position of the sun on the ecliptic for the current day.	
	We place a small pin-gnomon on this position, perpendicular to the surface of the globe.	We mark the position of the sun on the ecliptic on the spider.
3.	Measuring the altitude of the sun	
	We orientate the meridian of the globe to the local meridian and then we rotate the globe to reach the position where the gnomon on the degree of the sun casts no shadow. We measure the altitude of the sun with a graduated quadrant.	We let a sunray pass through the two sights of the diopter and we read the indication of the tip of the diopter on the quadrant of altitude on the back side of the astrolabe.
4.	The position of the globe corresponds to the position of the celestial sphere at this moment.	We rotate the spider on the front side in order to place the degree of the sun on the almucantars according to its altitude, east or west of the meridian depending on whether the measurement is before or after noon. The front side of the astrolabe is representing the position of the celestial sphere at this moment.
5.	We observe and retain the degree $d1$ of the equator that crosses the east horizon (the rising degree of the equator).	We read the indication $d1$ of the index of the spider.
6.	Determining the time in equal hours	
	We rotate the globe in the opposite sense of the zodiacal signs, so that the degree of the sun comes to the east horizon; we observe and retain the degree $d2$ of the equator that crosses the east horizon.	We rotate the spider of the astrolabe in the opposite sense of the zodiacal signs, so that the degree of the sun comes to the east horizon; we read the indication $d2$ of the index.

(cont.)

	Celestial globe	Astrolabe
	The time t_e in equal hours since the sunrise is $t_e = \begin{cases} \frac{d_2 - d_1}{15}, \text{if } d_1 \leq d_2 \\ \frac{360° + d_2 - d_1}{15}, \text{ if } d_1 > d_2 \end{cases}$ (1 hour corresponds to a rotation of 15° of the celestial sphere).	
7.	Determining the time in unequal hours	
	The time t_u in unequal hours since the sunrise is $t_u = \begin{cases} \frac{d_2 - d_1}{n}, \text{if } d_1 \leq d_2 \\ \frac{360° + d_2 - d_1}{n}, \text{ if } d_1 > d_2 \end{cases}$ where n is the number of the degrees of revolution corresponding to one unequal hour	We place the degree of the sun on the almucantars according to its altitude, east or west of the meridian depending on whether the measurement is before or after noon; the position of the degree diametrically opposite to the degree of the sun, among the lines of the unequal hours, gives directly the time in unequal hours since sunrise.

Conclusions

The construction of the planispheric astrolabe presupposes the stereographic projection of the globe onto the plane of its equator; in this way, all the functions of the globe can be transferred to the astrolabe.

Although the celestial globe is suitable for didactic purposes, it is more difficult to construct, transport and make measurements with. The representation of the heaven on the astrolabe is more abstract, but suitable for more accurate measurements.

In medieval times, the astrolabe became a popular instrument enriched with many applications, which could not only replace the celestial globe in solving astronomical problems, but also offer faster and more accurate results and broaden the range of astronomical problems that could be solved.

CHAPTER 5

Knowledge in Motion: An Early European Astrolabe and Its Possible Medieval Itinerary

Petra G. Schmidl
Exzellenzcluster "Normative Ordnungen," Frankfurt University, Frankfurt am Main, Germany
schmidl@em.uni-frankfurt.de

Abstract

This article discusses the place and date of origin, the earlier whereabouts, and subsequent travels of an unsigned and undated early European astrolabe (International Instrument Checklist IIC #0191) now preserved in Oxford, in the Museum of the History of Science. After taking into consideration earlier descriptions, it investigates the evidence given by the astrolabe concerning its localisation and dating, and suggests a possible itinerary of this instrument. In the Appendix a full description of the astrolabe is provided.

Keywords

astrolabes – astrolabe dating – circulation of scientific instruments – International Instrument Checklist number #0191 – Museum of the History of Science (Oxford)

Introduction

The Museum of the History of Science in Oxford is the home of an early European astrolabe (International Instrument Checklist IIC #0191). Although unsigned and undated, the instrument itself provides information that enables us to speculate about its place and date of origin, its earlier whereabouts and its subsequent travels. En passant it tells a story about knowledge in motion in the Middle Ages.[1]

1 For a full description of the astrolabe see "Appendix: Description of IIC #0191," *Knowledge in Motion* is also the title of volume 4 of Muzaffar Iqbal (ed.), *Studies in the Making of Islamic Science* (Farnham: Ashgate, 2012).

 | DOI:10.1163/9789004387867_008

Earlier Descriptions

Before trying to reconstruct a possible medieval itinerary of this astrolabe, it is appropriate to present and to discuss first the available descriptions of it. In the 1930s, Robert Gunther states in his *Astrolabes of the World* that IIC #0191 is an "Early French astrolabe … c. AD 1350–1400 … purchased by L. Evans … at the Roussel sale, Paris, 13 March 1911."[2] Obviously, he based this statement on information provided by Lewis Evans,[3] his friend, a businessman and aficionado of scientific instruments whose collection constitutes the basis of the Museum of the History of Science in Oxford.[4] Evans's card catalogue entry, most probably written by him a day after purchasing the astrolabe in Paris, reports "France" and "ca. 1350–1400."[5] Interestingly, Evans adds, besides other information, that "the form of the star pointers and the rete is Islamic,"[6] a piece of information reappearing in David King's caption related to IIC #0191 (see below) and established in this article, while Gunther writes that "the rete is of a 'Gothic' type."[7]

Evans's and Gunther's conclusion concerning localisation and dating of this astrolabe differ from all later descriptions. Besides Evans and Gunther, all favour a Spanish origin and an earlier date, beginning with Derek de Solla Price and his colleagues. In their *Computerized Checklist of Astrolabes*, compiled in the early 1970s, they list not a French but a "Hispano-Moorish" origin, and give a date around a century earlier, namely "ca. 1260."[8] The online descriptions follow these specifications. The record details on the homepage of the Oxford Museum of the History of Science mention a "creation date: c. 1260" and a

2 Robert T. Gunther, *The Astrolabes of the World*, 2 vols. (Oxford: Oxford University Press, 1932), 340, no. 191, entry in the Roussel sale catalogue does not provide any substantial information and dates the instrument to the sixteenth century (photo kindly provided by Stephen Johnston, Museum of the History of Science, see the Roussel catalogue, available online at http://www.mhs.ox.ac.uk/wp-content/themes/mhs-2013-responsive/emuphp5/media.php?irn=53440&image=yes) (accessed 23 June 2018).

3 An idea kindly provided by Stephen Johnston during the conference.

4 "Sphaera no. 11," available online at http://www.mhs.ox.ac.uk/sphaera/index.htm?issue11/articl4 (accessed 23 June 2018).

5 Photo kindly provided by Stephen Johnston and now also online ("Evans' card," available online at http://www.mhs.ox.ac.uk/wp-content/themes/mhs-2013-responsive/imu-media.php?irn=53439, accessed 23 June 2018).

6 "Evans' card."

7 Gunther, *Astrolabes*, 340, no. 191.

8 Derek de Solla Price, Sharon L. Price, Janice A. Henderson, *A Computerized Checklist of Astrolabes* (New Haven, CT: Yale University Press, 1973), 14 and passim.

"creation place: Spain, Europe."[9] The accompanying description provided by Ilaria Meliconi gives the same date and again a "Hispano-Moorish" origin.[10] In the table of contents based on the research made by David King and his team in Frankfurt in the 1990s, the "Spanish" origin of the astrolabe is followed by a question mark; the date is missing.[11]

Finally, a caption to a figure showing one of the plates of IIC #0191 in David King's book *The Ciphers of the Monks* describes this instrument very briefly as "a composite astrolabe with a rete either modified from or inspired by an Islamic one, French or Italian mater, and northern Spain and French plates (#0191)."[12] He avoids further localisation of its origin or any dating.

Localisation

To some extent, the information given by these descriptions can be checked by having a closer look at the astrolabe itself. Concerning Evans's and Gunther's argument of an origin in France, their main evidence are plates 1a and 2a, for Paris is mentioned on both, so that they link IIC #0191 to France. By counting their altitude circles, it turns out that their grids match the latitude of this city.

This clear finding is clouded by the fact that both A-sides, 1a and 2a both for Paris, are of a design completely different from their respective B-sides, whose engravings are similar to all other plates (henceforth and in the appendix: hand C). Because of the different design of 1a and 1b as well as of 2a and 2b, the simplest explanation—a later replacement of the plates—is out of question. Additionally, all plates fit properly. There is no visible evidence of removing an older layer of engravings or visible differences in shape, size, colour and texture concerning the brass of the plates. Metallurgical investigation supports this observation. Only the ruler, pin and horse of IIC #0191 are of a slightly different alloy.[13] Therefore, it is more plausible that in the beginning the astrolabe

9 "Record details IIC #0191," available online at http://www.mhs.ox.ac.uk/collections/search/displayrecord/?mode=displaymixed&module=ecatalogue&irn=2002 (accessed March 12, 2014).

10 "Oxford narrative IIC #0191," available online at http://www.mhs.ox.ac.uk/object/inv/37878 (accessed 23 June 2018).

11 "Frankfurt catalogue: Table of contents, part 6," available online at http://www.davidaking.org/instrument-catalogue-TOC.htm#part6 (accessed 23 June 2018).

12 David A. King, *The Ciphers of the Monks. A Forgotten Number-Notation of the Middle Ages*, Boethius 44 (Stuttgart: Franz Steiner, 2001), 285.

13 John Davis, Flowton Dials, Flowton (Ipswich, Suffolk), kindly provided me with the results of his metallurgical investigation he undertook in Oxford in July 2014 by "x-ray

comprised fully-engraved plates whose opposite sides were only engraved with the basic grid of the east-west and north-south lines as well as the circles for the equator and the tropics of Cancer and Capricorn. Then, these plates were completed for a specific latitude in a second step when the astrolabe changed place or owner.[14]

From a practical point of view, two plates made for the same city are superfluous in a single astrolabe. As shown above, replacement is no answer. Because plate 1a and 2a display slightly different engravings, it appears conceivable that they were made by different hands. While comparing the scripts is hampered by the small amount of letters available, different distances of the azimuth curves (every 6° on plate 1a and every 5° on plate 2a) and different ways of writing the numerals of the hours (number words on plate 1a and Hindu-Arabic numerals on plate 2a) point to two hands at work rather than one (henceforth and in the appendix: hand D and hand A). Therefore, one of the two plates for Paris could be later than the other. Nevertheless, they must have been Evans's and Gunther's main evidence for associating this astrolabe with France. If they had recognized that plate 2a was most probably engraved in the same hand as the rete (hand A), this may have even strengthened their

---

fluorescence (XRF) using commercial, portable equipment," "Innov-X (now Olympus Instruments) Alpha 2000 XRF analyzer" (John Davis, *Metallurgy of Medieval Scientific Instruments* (*MoMSI*). Project Proposal, unpublished, kindly provided by Lucy Blaxland, Museum of the History of Science, Oxford). In his preliminary conclusion, he writes: "Most components are lattens with medium Zn and Sn. Quite a number of slightly different alloys. The thick band which is sandwiched between the (sheet) mater and limb is a pure leaded brass. The plates (7 in total) have a range of Zn levels from ca. 14% to ca. 18%, all with very low Pb and Fe. The rete gave different Zn levels front to back (unexplained) but is not clearly from a different source to the other components. The rule and horse are most probably modern but the alidade is likely to be original (though with a strange alloy for the sights)."

In a private conversation (26 September 2014) he specifies: "There is a noticeable difference between the mater (with quite low Zn and relatively medium-to-high Fe) and the rest of the major components. Alloys from northern Europe generally have this 'dirty' (high Fe, >0.5%) level whereas those from southern areas are 'cleaner' (<0.2% Fe). (Only in relatively recent times have northern founders been able to produce this 'clean' material.) I am not very confident that this difference is significant but it is something to be aware of—is it conceivable that a replacement mater backing plate could have been added to an existing limb of an otherwise complete astrolabe?"

14 For an example of such a blank plate with the opposite one engraved properly, see IIC #0168, an unsigned and undated European astrolabe in the Oxford Museum of the History of Science (Gunther, *Astrolabes*, 319, no. 168; "Epact entry IIC #0168," available online at http://www.mhs.ox.ac.uk/object/inv/41468, accessed 23 June 2018).

opinion. Additionally, the latitudes of most of the plates are more suitable to an owner north of the Pyrenees than to one on the Iberian Peninsula.

Concerning the other descriptions that assume an origin in Spain, the plates are probably their main evidence, too. Besides the two plates for Paris, the astrolabe contains a third plate, plate 3a, that provides a place name, Cesaraugusta, today Zaragoza, and its geographical latitude ($\varphi = 41°$).

Hence, those who favour France as the astrolabe's origin and those who favour Spain give different answers to the question of which plates came first: one of the two Paris plates or the Zaragoza plate together with the eleven other plates that are of the same design, though with less details than the Zaragoza plate (henceforth and in the appendix: the "standard plates" see Table 5A7). Several features of the standard plates suggest that they were earlier than plates 1a and 2a for Paris. In general, they are less elaborated. The altitude circles are engraved in greater steps (every 6°), except for plates 3a and 3b (every 3°), while plates 1a and 2a display altitude circles every 2°. The azimuth curves are difficult to engrave and are not always displayed on earlier instruments;[15] IIC #0191 presents them only on the Zaragoza plate. Because of the sheer number of plates it appears more plausible that two plates, rather than the twelve, contained only the basic lines and circles. Plates 1b, 3a, and 3b with their latitudes of $\varphi = 38°$, $\varphi = 41°$, and $\varphi = 42°$, respectively, seem to be useful for an owner living and working in the Iberian Peninsula, two of them show a typically Spanish way of writing "40" (the Roman numeral ten "X" with an upside-down breve "˘" between its two brachia). This Visigothic ligature of the Roman numeral "XL" appears in manuscripts until the thirteenth century.[16]

---

15 Only a few plates in the earliest Eastern astrolabes from the ninth and tenth century display azimuth curves (e.g., in the astrolabes of Aḥmad b. Khalaf (IIC #0099) and of al-Khujandī (IIC #0111); David A. King, *In Synchrony with the Heavens. Studies in Astronomical Timekeeping and Instrumentation in Medieval Islamic Civilization. Volume Two: Instruments of Mass Calculation (Studies X–XVIII)*, Islamic Philosophy, Theology and Science 55 (Leiden: Brill, 2005), 467 and 512. The earliest Andalusi astrolabes from the tenth and eleventh century, approximately one hundred years later, all display azimuth curves, while early European astrolabes may include them or not—e.g., IIC #3042, the Destombes astrolabe, comprises plates with and without azimuth curves. See Koenraad van Cleempoel (ed.), *Astrolabes at Greenwich. A Catalogue of the Astrolabes in the National Maritime Museum* (Oxford, Greenwich: Oxford University Press and National Maritime Museum, 2006), 130.

16 Charles Burnett, "Ten or Forty? A Confusing Numerical Symbol in the Middle Ages," in *Mathematics Celestial and Terrestrial. Festschrift für Menso Folkerts zum 65. Geburtstag*, Acta Historica Leopoldina 54, ed. Joseph W. Dauben, Stefan Kirschner, Andreas Kühne, Paul Kunitzsch, Richard P. Lorch (Stuttgart: Wissenschaftliche Verlagsgesellschaft, 2008),

The use of Roman numerals on the standard plates and Hindu-Arabic numerals on plate 2a (plate 1a provides only Latin number words) may also constitute another indication that the Paris plates are later. Hindu-Arabic numerals replace Roman numerals at least in astronomical, astrological and mathematical texts in the twelfth century.[17]

These findings fit with the information provided in the appendix of King's *The Ciphers of the Monks*, in which he does not deem it proper to establish a place of origin for this astrolabe, but detects "Northern Spanish and French plates." Furthermore, he states that the astrolabe comprises a "French or Italian mater," and that the rete is "either modified from or inspired by an Islamic one," an opinion also held by Evans (see above).[18]

Concerning the mater, a link to France or Italy appears plausible, though difficult to establish. This part of the astrolabe is engraved in two different hands. In particular, on the back it becomes evident that two different sets of Latin letters are used. A light minuscule writes twice "vmbra recta" and "vmbra versa" alongside the shadow boxes (hand A). A bold majuscule, specified by Meliconi as "Lombardic script,"[19] abbreviates the month names and the zodiacal signs (henceforth and in the appendix: hand B). Two sets of Hindu-Arabic numerals related to these scripts seem similar to those on the front. There, apparently hand A labelled the hour scale and hand B the degree scale.

The Oxford Museum of the History of Science preserves two astrolabes, IIC #0168 and #0410, whose bold majuscule is like the majuscule on the back of the mater of IIC #0191 (hand B).[20] However, this similarity does not result in further information concerning the localisation and dating of IIC #0191, because both instruments are undated and unsigned. Although the scripts are rather similar, their descriptions vary. Gunther describes IIC #0168 as an "Italian astrolabe c. 1400."[21] Jim Bennett specifies in the epact entry "Italian Brass" and dates it to the "mid-16th century,"[22] while De Solla Price and his

---

81–89; Marie-Thérèse d'Alverny, Charles Burnett and Emmanuel Poulle, *Raymond de Marseille. Opera Omnia.* Tome 1: *Traité de l'Astrolabe. Liber Cursuum Planetarum* (Paris: CNRS, 2009), 10.

17 Burnett, *Ten or Forty*, 87.

18 King, *Ciphers*, 285.

19 "Oxford description IIC #0191."

20 Gunther, *Astrolabes*, 317–319, no. 168 and plate LXXI; "Epact entry IIC #0168"; "Oxford narrative IIC #0410," available online at http://www.mhs.ox.ac.uk/object/inv/54330 (accessed 18 November 2014).

21 Gunther, *Astrolabes*, 317, no. 168.

22 "Epact entry IIC #0168."

colleagues only provide a date *ca.* 1380.[23] Meliconi describes IIC #0410 as "Italian?" and dates it to the "15th century." Because of the Paris plates, she proposes "a French provenance or ownership,"[24] while De Solla Price and his colleagues only list a date ca. 1350.[25] According to King's table of contents of the *Frankfurt Catalogue*, both instruments are Franco-Italian (in the second case with a question mark),[26] while in his *In Synchrony with the Heavens* he refers to IIC #0168 as French.[27] The designs of both retes are completely different from the rete of IIC #0191. Anyway, these two astrolabes just support King's statement concerning a French or Italian origin of the mater, and do not provide additional information.[28]

Concerning King's statement that the rete is "either modified from or inspired by an Islamic one,"[29] a comparison with other astrolabes turns out to be fruitful. The retes of the five Catalan astrolabes inscribed in Latin before 1500 do not show any obvious similarities to IIC #0191,[30] but the Kassel Astronomisch-Physikalisches Kabinett hosts an Andalusi astrolabe made in Valencia at the end of the year 478 (April 1086) by Ibrahīm b. al-Sahlī (IIC #0121) whose rete is the nearly exact image of IIC #0191.[31] Comparison of this rete with the rete of

---

23 De Solla Price et al., *Checklist*, 13 and passim.

24 "Oxford description IIC #0410."

25 De Solla Price et al., *Checklist*, 16 and passim.

26 "Frankfurt catalogue: Table of contents, part 6."

27 King, *Instruments of Mass Calculation*, 358.

28 Even when finding objects of comparison, they often raise more questions than give answers due to the lack of an updated and reliable survey with descriptions of a significant number of astrolabes. Although newly published catalogues such as van Cleempoel (ed.), *Astrolabes at Greenwich* and an increasing amount of online sources are available ("epact database," available online at https://www.mhs.ox.ac.uk/epact/, accessed 23 June 2018) a proper tool that facilitates comparing astrolabes is still a desideratum.

29 King, *Ciphers*, 285.

30 IIC #4560, IIC #3042 (the Carolingian Astrolabe), IIC #0162 (the astrolabe of the Society of Antiquaries in London), IIC #0416, IIC #3053 (the astrolabe by Petrus Raimundi made in Barcelona in 1375). See David A. King, "An Astrolabe from 14th-Century Christian Spain. A Unique Testimonial to an Intercultural Encounter," *Suhayl* 3 (2002/03): 9–156; David A. King, "The Earliest Known European Astrolabe in the Light of Other Early Astrolabes," *Physis* 32 (1995): 359–404.

31 Karsten Gaulke and Petra G. Schmidl, "Astrolabien," in *Museumslandschaft Hessen Kassel. Der Ptolemäus von Kassel. Landgraf Wilhelm IV. von Hessen-Kassel und die Astronomie*, ed. Michael Eissenhauer, ed. Karsten Gaulke, Kataloge der Museumslandschaft Hessen Kassel 38 (Kassel: Museumslandschaft Hessen Kassel, 2007).

IIC #0191 shows similarities in design that are intriguing (see Figures 5A14 and A15). All other known early Andalusi astrolabes resemble less IIC #0191.[32]

The shapes of the star pointers of both retes (IIC #0121 and IIC #0191) are more or less identical; for example, the claw-like pointer of Verar (?) ceti or the candle-like pointer of *Aquilla volās*. Their equinoctial bars are counter-changed twice.[33] Frames surround the star pointers of *Canis Tienorus/Vienorus* (?) and *Humer datis* (?).

In addition to the language and the script, the most obvious differences in design concern the star pointers at the tips of the frame of the tropic of Capricorn (*cauda cap* and *cor scorpionis*), which are less elaborated (and definitely not broken) on the European astrolabe. Further, on IIC #0191 the star pointer for *Cor leonis* does not touch the ecliptic ring as it does on IIC #0121. The shape of the index also looks different. While on IIC #0121 it has the shape of a rectangle, IIC #0191 adds to this basic feature a small tip.

Besides the striking resemblance in design, both retes comprise twenty-six star pointers with a similar list of stars.[34] Their positions differ slightly from

---

32 According to the "Frankfurt catalogue: Table of contents, part 1," available online at http://www.davidaking.org/instrument-catalogue-TOC.htm#part1 (accessed 23 June 2018), the earliest complete Andalusi astrolabes are: IIC #4024 (Khalaf b. al-Muʿādh's astrolabe in a Latin manuscript), IIC #0110, IIC #0116 (the Andalusi astrolabe in Berlin), IIC #1079 (the astrolabe by Ibrāhīm b. ʿAbd al-Karīm; no photo available to the author), IIC #0117, IIC #0118, and IIC #0123 (= #1167; the three astrolabes by Ibrāhīm b. Saʿīd b. Aṣbagh al-Anṣār *thumma* al-Sahlī al-Mawāzīnī), IIC #2527 and IIC #1139 (two astrolabes by Muḥammad b. Saʿīd al-Ṣabbān), IIC #2572 (the astrolabe by Muḥammad b. al-Sahlī preserved in Washington), IIC #1099 (the astrolabe by Aḥmad b. Muḥammad al-Naqqāsh preserved in Nuremberg), and IIC #3622 (the Andalusi astrolabe preserved in Cracow).

Although the rete of IIC #0428 (= #0625; the Caird astrolabe in Greenwich) is also clearly inspired by an Andalusi one, the resemblances to IIC #0118 (the Toledo astrolabe in Oxford) are less striking (van Cleempoel, *Astrolabes at Greenwich*, 115–120 and plates 2–3; Gunther, *Astrolabes*, 253–256, no. 118 and plate LX).

33 Two times counter-changed equinoctial bars are typical of eleventh-century Andalusi astrolabes (King, *Astrolabe from 14th century*, 64).

34 Of the early Andalusi astrolabes preserved, only IIC #1139 (Muḥammad b. Sāʿid al-Ṣabbān's astrolabe from 466 (1073–74, preserved in Munich) includes twenty-six star pointers; even the second astrolabe by this maker (IIC #2527) from 474 (1081/82) comprises one star more. The rete of IIC #1139, as well as of IIC #2527, although of typical early Andalusi design, show considerable differences to IIC #0121 and IIC #0191 (Stautz, 53–54 and 60).

According to King, *Earliest Known Astrolabe*, 369: "[...] some very early European astrolabes (for example, #191, #428 and #300) do have star-pointers that are based on al-Majrīṭī's table."

each other. These variances, however, do not show an interpretable pattern that allow us to explain them. Apart from *Cor leonis*, the varying position of *Spica* can be detected even without measuring it.[35] Nevertheless, the rete of IIC #0191 was made or inspired either by IIC #0121 or by an Andalusi rete very similar to it.

TABLE 5.1 *The Star Names on IIC #0191 and IIC #0121*

	IIC #0191			IIC #0121	
No.	**Name**	**Translation and modern designation**	**No.**	**Name**	**Translation and modern designation**
1	*Verar* (?) *ceti*	"the Sea Monster's ..." = ζ Cet	16	بطن قيطوس *baṭn qayṭūs*	"the Sea Monster's belly" = ζ Cet
2	*Vespa*	"the Wasp" = α Tau (Aldebaran)	17	الدبران *al-dabarān*	= α Tau (Aldebaran)
3	Unnamed	= α Aur (Capella)	18	العيّوق *al-ʿayyūq*	= α Aur (Capella)

Although it might be that the rete was made without using a star table, by comparing the star names or their descriptions and considering the co-ordinates determined (see n. 35), "Typ VII," an extract of the star table compiled in Paris 1246 by John of London, shows the closest relationship to the stars on the rete of IIC #0191 (Kunitzsch, *Sternverzeichnisse*, 47–50). However *Serpēs* and *Corona* are missing in the table and the values differ considerably.

35 Flora Vafea kindly provided me with the results of her recalculation of the stellar positions she undertook in Munich in October 2014 by using Voyager (Carina Software & Instruments, Inc.) and re-mastered photos of both retes superimposed by a geographical grid for the latitude of the Arctic circle and for the longitudes. The coordinates determined by this method fit to a wide variance of different years; for a re-computation of the positions of the star pointers on IIC #0121 for the year 1085 also Stautz, *Untersuchungen*, 196, fig. 3.4g.

TABLE 5.1 *The Star Names on IIC #0191 and IIC #0121* (cont.)

IIC #0191			IIC #0121		
No.	**Name**	**Translation and modern designation**	**No.**	**Name**	**Translation and modern designation**
4	*Pides* (*sic*) *Geminorum*	"the feet of the *jawzāʾ*" = β Ori (Rigel)	19	رجل الجوزاء *rijl al-jawzāʾ*	"the foot of the *jawzāʾ*" = β Ori (Rigel)
5	*Humerus geminor*	"the shoulder of the *jawzāʾ*" = α Ori (Betegeuze)	20	منكب الجوزاء *mankib al-jawzāʾ*	"the shoulder of the *jawzāʾ*" = α Ori (Betegeuze)
6	*Canis Tienorus/ Vienorus* (?)	"the … Dog" = α CMa (Sirius)	21	العبور *al-ʿabūr*	"<the *shirʿā*> that crosses (the milky way)" = α CMa (Sirius)
7	*Canis plorans*	"the wailing/ bleary-eyed Dog" = α CMi (Prokyon)	22	الغميساء *al-ghumayṣāʾ*	"the bleary-eyed (*shiʿrā*)" = α CMi (Prokyon)
8	Unnamed	= μ UMa (?)	24	الركبة *al-rukba*	"the <Greater Bear's> knee" = μ UMa (?)
9	*Humer datis* (?)	"the <…> 's shoulder" = α Hya (Alphard)	23	منير الشجاع *munīr al-shujāʾ*	"Hydra's bright <star>" = α Hya (Alphard)
10	*Cor leonis*	"the Lion's heart" = α Leo (Regulus)	25	قلب الأسد *qalb al-asad*	"the Lion's heart" = α Leo (Regulus)
11	*Alacorvy*	"the Raven's wing" = γ Crv	26	جناح الغراب *janāḥ al-ghurāb*	"the Raven's wing" = γ Crv
12	*Eletus/ Eledor* (?)	= η UMa (?)	2	بنات نعش *banāt naʿsh*	"the daughters of the *naʿsh*" = η UMa
13	*Spica*	= α Vir (Spica)	1	الأعزل *al-aʿzal*	"the unarmed <*simāk*>" = α Vir (Spica)

TABLE 5.1 *The Star Names on IIC #0191 and IIC #0121* (cont.)

IIC #0191			IIC #0121		
No.	**Name**	**Translation and modern designation**	**No.**	**Name**	**Translation and modern designation**
14	*Anceator/ Ameator (?)*	= α Boo (Arcturus)	3	الرامح *al-rāmiḥ*	"the lance-bearing <*simāk*>" = α Boo (Arcturus)
15	Unnamed	= α CrB (Gemma)	4	الفكّة *al-fakka*	= α CrB (Gemma)
16	*Serpēs*	"the Serpent" = α or β Ser (?)	5	الحيّة *al-ḥayya*	"the Serpent" = α or β Ser (?)
17	*Cor scorpionis*	"the Scorpion's heart" = α Sco (Antares)	6	قلب العقرب *qalb al-ʿaqrab*	"the Scorpion's heart" = α Sco (Antares)
18	*Arum/Alrum (?) tenēs (?)*	"… 's Bearer (?)" = α Oph	7	الحوّاء *al-ḥawwāʾ*	"the Serpent Bearer" = α Oph
19	Unnamed	= α Lyr (Vega)	8	الواقع *al-wāqiʿ*	"the swooping <Eagle>" = α Lyr (Wega)
20	*Aquilla volās*	"the flying Eagle" = α Aql (Altair)	9	الطائر *al-ṭāʾir*	"the flying <Eagle>" = α Aql (Atair)
21	*Elphinus*	= ε Del	11	الدلفين *al-dulfīn*	"<the tail of> the Dolphin" = ε Del
22	Unnamed	= α Cyg (Deneb)	10	الردف *al-ridf*	"the One Sitting behind the Rider" = α Cyg (Deneb)
23	*cauda cap*	"the Kid's tail" = δ Cap	12	ذنب الجدي *dhanab al-jady*	"the Kid's tail" = δ Cap
24	*Humerus equi*	"the Horse's shoulder" = β Peg	13	منكب الفرس *mankib al-faras*	"the Horse's shoulder" = β Peg
25	*Cauda ceti*	"the Sea Monster's tail" = ι Cet (?)	14	ذنب قيطوس *dhanab qayṭūs*	"the Sea Monster's tail" = ι Cet (?)

TABLE 5.1 *The Star Names on* IIC *#0191 and* IIC *#0121* (cont.)

	IIC #0191			IIC #0121	
No.	**Name**	**Translation and modern designation**	**No.**	**Name**	**Translation and modern designation**
26	*Corona*	"the Crown" = β Cas	15	الخضيب *al-khaḍīb*	"the stained <hand of the *thurayyā*>" = β Cas

Note to no. 8/no. 24: Although Paul Kunitzsch, *Arabische Sternnamen in Europa* (Wiesbaden: Otto Harrassowitz, 1959), 67–68, no. a9 favours θ UMa, the position of the star pointer fits better to μ Uma. This is in accordance with Stautz, *Untersuchungen*, 53 and 60.
Note to no. 16/no. 5: Although Kunitzsch, *Sternnamen*, 217, no. 196 records for α Ser *ʿun(u)q al-ḥayya*, the star position fits better to β Ser. This is in accordance with Stautz, *Untersuchungen*, 54 and 59–60, though contradictory to 196 where α Ser is mentioned, probably a typo.

As mentioned above, the plates of the European astrolabe (IIC #0191) use a specific ligature based on Roman numerals to denote "XL" that is also found on the plates of the Andalusi astrolabe (IIC #0121), in which the latitudes in Arabic *abjad* notation are supplemented by Roman numerals.[36] On IIC #0191 this ligature belongs most probably to the original inscription, while on IIC #0121 it is a later addition made when the astrolabe changed from an owner familiar with Arabic to an owner familiar with Latin.

Dating

To track down the reasons for the dating of this astrolabe in earlier descriptions is much more complicated than finding out about the places where it had been. Evans, whom Gunther most probably followed, suggested 1350–1400, but De Solla Price and others, when they mention a date, prefer ca. 1260.[37] In all cases, no evidence is given.

Because an astrolabe is not only an astronomical instrument, but also a piece of art and a document, two sets of information help in dating any

36 Gaulke and Schmidl, *Astrolabien*, 221.

37 De Solla Price et al., *Checklist*, 14 and passim; "Record details IIC #0191"; "Oxford narrative IIC #0191."

instrument when no inscription displays this information: the astronomical data provided and the characteristics of its design and script. In the first case, the date of the vernal equinox, the longitude of the perihelion (both read off by means of the solar and calendrical scale displayed on the back of the astrolabe), and the coordinates of the stars can be used.[38] Because outdated data may have been used and there might be inaccuracies in engraving and reading off these parameters,[39] recalculating these parameters provides only one of several clues for dating an astrolabe.[40]

IIC #0191 provides March 14 for the vernal equinox. Comparing this information to the astronomical data given on other astrolabes is complicated by two factors: lacking data on the instruments themselves as well as proper means of comparison.[41] The earliest European astrolabes are not dated.[42] An exception is Petrus Raimundi's instrument, which was made in Barcelona in 1375 (IIC #3053) for a vernal equinox of 10 March.[43] Even if one finds early European instruments with a vernal equinox for March 14 such as IIC #0300 or IIC #0420,[44] they provide no certain reference to narrow down the possible years of manufacture, because they are not dated.

---

38 Henri Michel, *Traité de l'astrolabe* (Paris: Gauthier-Villars, 1947), 135–148; see also n. 35.

39 Gerald L'E. Turner, "A Critique of the Use of the First Point of Aries in Dating Astrolabes," in *Sic itur ad astra. Studien zur Geschichte der Mathematik und Naturwissenschaften—Festschrift für den Arabisten Paul Kunitzsch zum 70. Geburtstag*, ed. Menso Folkerts, Richard P. Lorch (Wiesbaden: Otto Harrassowitz, 2000), 548–554.

40 Calculating the vernal equinox by the method mentioned in Michel, *Traité de l'astrolabe*, 139, the vernal equinox in 1100 gives March 13.83, in 1200 March 13.05, in 1260 March 12.582, in 1350 March 11.88 and in 1400 March 11.49; see also Henri Michel, "Un astrolabe du XIIe siècle," *Ciel et Terre*, 64 (1948), 77 providing 1080–1175 for March 14.

41 See n. 28.

42 David A. King and Kurt Maier, "The Medieval Catalan Astrolabe of the Society of Antiquaries, London," in *From Baghdad to Barcelona. Studies in the Islamic Exact Sciences in Honour of Prof. Juan Vernet*, ed. Josep Casulleras, Julio Samsó, *Anuari de Filologia* 19 (Barcelona: Universitat de Barcelona, 1996), 696: "[…] medieval European instruments […] (in fact all but seven from before 1450) are undated (and mostly unsigned)."

43 "Boston description," available online at http://www.mfa.org/collections/object/planispheric-astrolabe-61726 (accessed 23 June 2018).

44 IIC #0300 shares the fate of IIC #0191 concerning its earlier descriptions: either being made in Spain around 1260 or in England (and France) between 1300 and 1350 (Gunther, *Astrolabes*, 477–478, no. 300 and plate CXXXII; "Frankfurt catalogue: Table of Contents, Part 6"; "Oxford record details IIC #0300," available online at http://www.mhs.ox.ac.uk/object/inv/49033 (accessed 23 June 2018); "Epact entry IIC #0300," available online at http://www.mhs.ox.ac.uk/collections/imu-search-page/narratives/?irn=2001&index=0 (accessed 23 June 2018), incorrectly providing March 13 for the vernal equinox. Due to the design of its throne and its rete, as well as the solar and calendar scale on the back IIC

The situation is different concerning Islamic astrolabes because they are mostly dated, although their direct relation to European instruments is more tenuous. IIC #0118, an astrolabe made by Ibrāhīm b. Saʿīd (b. Aṣbagh al-Anṣārī *thumma*) al-Sahlī (al-Mawāzīnī) in Toledo in *Shawwāl* 460 (August 1068) provides 14 March for the vernal equinox and constitutes an exception.[45] One may speculate about family relations between the makers of IIC #0118 and IIC #0121 on the basis of the maker's name and the date. As we have seen, the rete of IIC #0121 resembles strikingly IIC #0191. IIC #0121 was made by Ibrāhīm b. al-Sahlī in Valencia at the end of the year 478 (April 1086) and provides 13 March for the vernal equinox. Two other astrolabes by Ibrāhīm b. Saʿīd (b. Aṣbagh al-Anṣārī *thumma*) al-Sahlī (al-Mawāzīnī) differ in the dates of the vernal equinoxes. IIC #0117 was also made in Toledo in in *Shaʿbān* 459 (June/July 1067) but one year earlier than IIC #0118, and provides 14.5 March for the equinox.[46] However, IIC #0123 (= #1167), which was made in Valencia in *Rajab* 463 (April 1070), slightly later than IIC #0118 and IIC #0117, gives 14 March.[47] To this group of three plus one astrolabes, a fifth one is related: IIC #2572 was made by Muḥammad b. al-Sahlī in Valencia in the year 483 (1090–1091) and provides 15 March for the vernal equinox (Table 5.2).[48] All these astrolabes document one only fact, that a vernal equinox on 14 March was used on eleventh-century Andalusi astrolabes closely related to IIC #0121, and therefore also to IIC #0191. Concerning the star positions, their calculation by modern means shows so great a variance that it does not allow any dating.

#0420 is related clearly to earlier Andalusi astrolabes, although contrary to IIC #0191, it is not related to any extant instrument (van Cleempoel, *Astrolabes at Greenwich*, 126–131 and plate 1; "Frankfurt catalogue: Table of contents, part 6"; "Greenwich object details IIC #0420," available online at http://collections.rmg.co.uk/collections/objects/10720.html accessed 23 June 2018).

45 Gunther, *Astrolabes*, 256, no. 118; "Frankfurt catalogue: Table of Contents, Part 1" gives the maker's name as Ibrāhīm b. Saʿīd b. Aṣbagh al-Anṣārī *thumma* al-Sahlī al-Mawāzīnī.

46 García Franco, *Catálogo crítico de astrolabios existentes en Espanña* (Madrid: Instituto Histórico de la Marina, 1945), 427; and De Solla Price et al., *Computerized Checklist*, 13 and passim.

47 Photo kindly provided by the archive of the Institute for the History of Science, Frankfurt; also "Frankfurt Catalogue: Table of Contents, Part 1"; Gunther, *Astrolabes*, 263, no. 117 dating this astrolabe to 593 (1197); and De Solla Price et al., *Computerized Checklist*, 13 and passim providing 493 (1099) followed by a question mark.

48 Sharon L. Gibbs with George A. Saliba, *Planispheric Astrolabes from the National Museum of American History*, Smithsonian Studies in History ad Technology 45 (Washington, DC: Smithsonian Institution Press, 1984), 174–177.

In the second case, in particular, the rete with its characteristic star pointers, the shape of the equinoctial bar and its specific counter-changings, the equatorial bar if extant, and additional decorative elements may provide further clues, together with the design of the throne, the scripts used and especially the script way of displaying the numbers. Concerning IIC #0191, the design of the rete provides with certainty a *terminus post quem*, April 1086, when Ibrahīm b. al-Sahlī made in Valencia his astrolabe (IIC #0121) with its typical Western Islamic rete that resembles so closely that of IIC #0191.

The scripts by different hands allow us to classify this astrolabe as medieval or early Renaissance, but the way of displaying the numbers points instead to the twelfth century or only slightly later. The Western form of the Hindu-Arabic numerals are first witnessed at the end of the tenth century but only became common among astronomers in the twelfth.[49] A century later, the Visigothic ligature of the Roman numeral "XL" disappeared in the manuscripts.[50] The *terminus post quem* defined by the close resemblance of IIC #0191 to IIC#0121 fits with these circumstances. Concerning a *terminus ante quem*, uttered with the utmost caution, these clues may point at the latest to the fourteenth, if not already to the thirteenth century. Comparison of the numerals written on IIC #0191 with other early European astrolabes is again hindered by the fact that they are not dated and cannot be used as a reference point. Furthermore IIC #3053, the astrolabe made by Petrus Raimundi in Barcelona in 1375, displays Hindu-Arabic numerals with Gothic forms for 4, 5 and 7.[51] Hands A and B do the same on IIC #0191, while IIC #4523, an early Italian astrolabe made in 1420 by Antonius of Pacento, presents the modern form of 4, but keeps 5 and 7 in the Gothic style.[52] The different requirements when writing on metal or on parchment and paper make it difficult to compare the scripts on these dissimilar materials.[53]

49 Paul Kunitzsch, "The Transmission of Hindu-Arabic Numerals Reconsidered," in *The Enterprise of Science in Islam—New Perspectives, Dibner Institute in the History of Science and Technology* (Cambridge, MA: The MIT Press, 2003), 3–21; King, *Ciphers*, Appendix D, 309–317; Burnett, *Ten or Forty*, 87.

50 Burnett, *Ten or Forty*, passim.

51 "Boston description;" see the article by Azucena Hernández in this issue.

52 Burkhard Stautz, "Ein Astrolab aus dem Jahr 1420," *Zeitschrift für Geschichte der Wissenschaften, Technik und Medizin* 5 (1997): 142–159.

53 Of the examples presented by George Francis Hill, *The Development of Arabic Numerals in Europe* (Oxford: Clarendon Press, 1915), no complete set of ten ciphers matches hand A or hand B, though especially 30, III:14 (probably English, end of 13th c.), 32, IV:12 (probably English, 13th c.), V:10 (German, 13th–14th c.), and 46, XII:1 (Italian or French?, ca. 1400–1416) show some similarities.

TABLE 5.2 *Group of al-Sahlī Astrolabes*

IIC	Maker's name	Place of manufacture	Date of manufacture	Vernal equinox
#0118	Ibrāhīm b. Saʿīd al-Sahlī	Toledo	Shawwāl 460 (August 1068)	14 March
#0117	Ibrāhīm b. Saʿīd al-Mawāzīnī al-Sahlī	Toledo	459 (June/July 1067)	14.5 March
#0123 (= #1167)	Ibrāhīm b. Saʿīd b. Aṣbagh al-Anṣārī *thumma* al-Sahlī al-Mawāzīnī (?)	Valencia	Rajab 463 (April 1070)	14 March
#0121	Ibrahīm b. al-Sahlī	Valencia	End of 478 (April 1086)	13 March
#2572	Muḥammad b. al-Sahlī	Valencia	483 (1090–1091)	15 March

Itinerary

After collecting and interpreting this information, what might a possible medieval itinerary of IIC #0191 look like? Although "the following reconstruction is at best hypothetical" and "no end of other hypotheses could be formulated,"[54] taking together all the clues that are found on IIC #0191 justifies such an attempt and promises a rather intriguing story. This theory relies on the condition that all parts of the astrolabe (except for the ruler, pin, and horse) belong together from the beginning, an assumption that is not contradicted by the evidence at hand.[55]

IIC #0191 started its journey most probably in northern Spain (in or near Zaragoza), after April 1086, when IIC #0121 was made. The design of the rete clearly demonstrates that IIC #0121, the astrolabe made by Ibrahīm b. al-Sahlī in Valencia at the end of the year 478 (April 1086), was the template for the rete of IIC #0191. It seems more plausible that an Arabic rete was copied in Spain than in France, but the opposite is not ruled out. IIC #0121 has to find its

54 King, *Astrolabe from 14th-Century*, 110 and 104.

55 See n. 13.

way from eleventh-century al-Andalus to sixteenth-century Kassel where it has been preserved until today. Why not via Paris?

The style of the standard plates also points towards Spain because of the more elaborated plate for Zaragoza. The standard plates look older than the Paris plates. Using Roman numerals and including the Visigothic ligature of the Roman numeral "XL" to denote the latitudes supports the interpretation that hand C made the oldest inscriptions on the astrolabe. The additional markings for the lunar mansions on the back of IIC #0191, if interpreted correctly, represents further Islamic influence and support therefore an origin in al-Andalus.[56]

One can assume that the scales are engraved in two steps, with their numerals added later,[57] an assumption that is supported by the circular frames on the throne giving the impression of an unfinished instrument, for example to be completed by a later owner or because of insufficient skills or external circumstances. The solar and calendar scales with their typical features of early Andalusi astrolabes,[58] also point towards a Spanish origin. The geographical latitudes seem to contradict this assumption, because they are suitable for places too far to the north to be useful for a Spanish owner, but a person travelling frequently in northern Spain and France could have used them.[59] Taken all this information together, al-Andalus (if not Zaragoza) after April 1086 may have been the starting point of IIC #0191's journey.

---

56 IIC #0121 also provides a much more elaborated scale for the lunar mansions on its back (Gaulke and Schmidl, *Astrolabien*, 220 and 223, fig. 2).

57 One example of scales without inscription provides IIC #3042, the earliest known European astrolabe, also called the Carolingian astrolabe, with an unlabelled degree scale on the rim and unlabelled altitude scales on the back, see Marcel Destombes, "Un astrolabe carolingien et l'origine de nos chiffres arabes," *Archives internationales d'histoire des sciences* 58/59 (1962): 1–45, repr. in M. Destombes (1905–1983). *Selected Contributions to the History of Cartography and Scientific Instruments*, ed. G. Schilder, P. van der Krogt, S. de Clercq (Utrecht: HES 1987), 153–200, and repr. in *Astronomical Instruments and Observatories in the Islamic World. Texts and Studies*, ed. Fuat Sezgin, Islamic Mathematics and Astronomy 96 (Frankfurt: Institut für Geschichte der arabisch-islamischen Wisenschaften, 1998), 401–447.

58 King, *Astrolabe from 14th-Century*, 86.

59 Two other undated and unsigned early European astrolabe with a similar wide range of plates, are IIC #0300, now in Oxford, with plates for latitudes from 24° or 25° to 60° and IIC #0558, now in Nuremberg, with plates from 16° to 52°. See Gunther, *Astrolabes*, 478, no. 300 and plate CXXXII; "Oxford record details IIC #0300"; "Epact entry IIC #0300"; David A. King in *Focus Behaim Globus*, ed. Gerhard Bott, Ausstellungskataloge des Germanischen Nationalmuseums, 2 vols. (Nürnberg: Verlag des Germanischen Nationalmuseums, 1992), 574–576.

The order of the following stations and events in the itinerary of IIC #0191 are difficult to sort out chronologically. After its beginnings in northern Spain, the astrolabe continued its journey northwards and came to France, most probably to Paris. The two plates made for this city constitute the evidence of this movement. That knowledge of astronomy, astrology, and the astrolabe moves from south to north, from Spain to France, in the twelfth century is clearly documented by Raymond of Marseille and his oeuvre. He lived in Marseille and wrote three treatises related to these topics: his astrolabe treatise, his *Liber cursuum planetarum capitisque draconis*, written in 1141 (where he converts the Toledan tables for Marseille and the Christian calendar), and his *Liber iudiciorum*.[60]

If one assumes a Spanish origin of IIC #0191, all the features engraved by hand A (the hand who made plate 2a for Paris) must have been inscribed in the French city. Therefore, IIC #0191 reached Paris with an uninscribed rete,[61] as well as unlabelled scales on the rim, the shadow boxes and the hour quadrants on the back.[62] That the scales were also made in Paris seems less plausible, be-

---

60 D'Alverny, Burnett and Poulle, *Raymond de Marseille*, 9–10 and 24; Emmanuel Poulle, "Le traité d'astrolabe de Raymond de Marseille," *Studi medievali, serie terza* 5.2 (1964): 866–900 (866–867).

61 An idea kindly provided by Koenraad van Cleempoel.

There are several astrolabes with retes that do not provide star names and only present the names of the zodiacal signs such as IIC #3042 (the earliest European astrolabe, also called the Carolingian astrolabe; Destombes, *Un astrolabe carolingien*, planche I), IIC #0161 and IIC #0167 (two undated and unsigned early European astrolabes in the British Museum. See King, *Earliest Known Astrolabe*, 369; "British Museum report IIC #0161," available online at http://www.britishmuseum.org/research/collection_online/collection_object_details.aspx?objectId=54858&partId=1&searchText=astrolabe&page=2 (accessed 3 June 2018); "British Museum report IIC #0167," available online at http://www.britishmuseum.org/research/collection_online/collection_object_details.aspx?objectId=54858&partId=1&searchText=astrolabes&page=1 (accessed 23 June 2018); Francis Alan Burnett Ward, *A Catalogue of European Scientific Instruments in the Department of Medieval and Later Antiquities of the British Museum* (London: British Museum Publications, 1981), 110, 112 and plate XLIX, no. 323 and 114, no. 329; Gunther, *Astrolabes*, 306, no. 161 and 317, no. 167.

Even though the similarities concerning the star names of IIC #0191 and their presentation in the "Typ VII" table, an extract of John of London's star table compiled in Paris 1246, are less distinct, these findings fit the general context (Kunitzsch, *Sternverzeichnisse*, 47–50).

62 IIC #0300 (an early European astrolabe with a rete clearly influenced by Andalusi astrolabes and now preserved in Oxford) omits the numerals of the degree scale on the rim, of the solar and calendar scale on the back, of the altitude circles, the azimuth, and the equal

cause several numerals go over the edge of their respective boxes and because, without an engraved rim, the use of the astrolabe is more cumbersome. Since the small metal piece that repairs the rete bears a scale and numerals that fit nicely with the rete, this repairing was probably also done in Paris.[63] The star names are mainly translations, not transcriptions, of the Arabic names and provide another hint that the rete was engraved in a later or second step given that earlier star lists tend in general to show more transcriptions and fewer translations.[64]

Finally, there are two layers left that have to be integrated into the itinerary of IIC #0191. Their interpretation leads to the conclusion that either the astrolabe stayed in Paris or made a journey and came back at the latest in March 1091. One of these two layers is linked with the inscription of the degrees on the rim and on the solar and calendar scales on the back (hand B), the other is related with plate 1a, another Paris plate (hand D). In both cases, they seem to show hands at work that differ from those that inscribed the standard plates (hand C), and from those that engraved the rete and other features (hand A).

Concerning the layers in hand B and assuming that the solar and calendar scales belong to an earlier layer and that the inscriptions followed in a second step, one may wonder if the inscription of the month names, the names of the zodiacal signs and the respective numerals took place before the inscription of the rete and the other features in hand A or afterwards. In the first case, the engraver left the solar and calendar scale unlabelled (either deliberately or involuntarily), while in the second, he left uninscribed the rete and the other features inscribed in hand A, which makes this assumption less plausible. If the solar and calendar scales were labelled after the rete and the other features in hand A, it most probably happened in Paris or somewhere in France or in Italy, as the similarities in the script to those on IIC #0168 und IIC #0410

hour curves on the plates. The shadow boxes are made completely without any inscription, a grid probably for the lunar mansions on the back is blank (Gunther, *Astrolabes*, 477–478, no. 300 and plate CXXXII; "Oxford record details IIC #0300").

63 If one takes into consideration what John Davis suggested, namely that the back plate of the IIC #0191 may be a later addition (see n. 13), then this replacement also has to have taken place in Paris, because the inscriptions of the shadow boxes as well as those of the hour quadrant are in hand A, the hand that engraved plate 2a for Paris.

64 This linkage appears to be more obvious in the astrolabe tradition than in the manuscript tradition van Cleempoel, *Astrolabes at Greenwich*, 115–209; King, *Focus Behaim Globus*, 572–602; Roderick S. Webster and Marjorie Webster, *Western Astrolabes, Historic Scientific Instruments of the Adler Planetarium & Astronomy Museum* 1 (Chicago, IL: Adler Planetarium & Astronomy Museum, 1998), 40–60; for the manuscript tradition Kunitzsch, *Sternverzeichnisse*, passim.

suggest. There is a third explanation supported by the two different forms of the letter "T" in the calendar and solar scales, in "OCTO" (October) and in "TAV" (Taurus). The initial letter of the latter looks like an initial "T," while the last letter looks like the "V" displayed on the rete (for example in "Vespa"). Are the rete and the scales labelled at the same time and place and by the same person? This person used a bold majuscule with distinct serifs to label the solar and calendar scales and inscribed the names of the stars and the zodiacal signs on the rete in a light minuscule, either to emphasize the former[65] or due to lack of space on the rete. However, the different scripts of the numerals make this explanation less convincing.

Hand D, which is represented only on plate 1a, looks very like hand A. Although the differences are small, they cannot be explained by the different amount of space available on the plate and on the rete. Two different hands explain more easily why there are two plates for one city (both definitely not later replacements) in one astrolabe. Similar to the argumentation concerning the inscription in hand B (of the solar and calendar scale, the degrees on the rim), it appears more plausible that plate 1a was made after plate 2a, because in this case fewer parts of the astrolabe were left uninscribed for a longer period of time. One may conclude that plate 1a presents the latest layer, because it includes the smallest amount of additional inscriptions. One can hardly think that plate 2a was meant to be replaced by plate 1a (in spite of the fact that disc 2 has a broken pin), since 2a (with azimuth curves for each fifth degree) is more exact than 1a (with azimuth curves for each sixth degree).

Of course, "various other explanations are conceivable but not particularly convincing given the lack of concrete evidence,"[66] and new information may support, or change (parts of) IIC #0191's itinerary. Nevertheless, it is a beautiful and concrete example of knowledge in motion.

Conclusion

This possible itinerary of IIC #0191 demonstrates that in those times astronomical and astrological knowledge as well as knowledge concerning the astrolabe, represented on instruments and in texts travelled from south to north, from Spain to France, with or without a diversion via Italy. It makes clear that knowledge was in motion between the Arabic and the Latin speaking

65 An idea kindly provided by Silke Ackermann during my stay in Oxford.

66 King, *Astrolabe from 14th-Century*, 112.

worlds. Astrolabes were on the road and with them the knowledge they provided. In particular, the rete of this instrument beautifully testifies these observations and makes the Islamic and the Andalusi influence on early European astrolabes tangible.[67]

Acknowledgements

I would like to thank all the institutions and people that helped me in preparing and finishing this paper, especially Charles Burnett and Josefina Rodríguez-Arribas of the Warburg Institute, London, for organizing the conference and preparing this volume, Silke Ackermann, Stephen Johnston and Lucy Blaxland of the Museum of the History of Science, Oxford, for their kind hospitality, John Davis, Claudia Heimann, Paul Kunitzsch and Flora Vafea for helping me with their specific experience, the Arbeitsgruppe Wissenschaftsgeschichte for their feedback, and last but not least David King, who first kindled my interest in astrolabes.

Appendix: Description of IIC #0191

TABLE 5A1 *General Description of IIC #0191*

Collection	Museum of the History of Science, Oxford, Inventory no. 37878
Maker	Unsigned
Place of creation	Spain (?)
Date of creation	undated (after 1086, probably before thirteenth century)
Material and size	brass, diameter 146mm, thickness 9mm
Condition	complete, though composite
Exhibited	–

67 En passant, the results presented in this article demonstrate that simplifying comparisons of astrolabes could enhance our perception of the history of astronomy and astrology in Eastern and Western civilisations.

TABLE 5A1 *General Description of IIC #0191* (cont.)

Descriptions in print	Gunter, *Astrolabes*, 340–341, no. 191; De Solla Price et al., *Checklist*, 14 and passim; King, *Ciphers*, 285 (a short remark below the picture of plate 3a).
Descriptions online	http://www.mhs.ox.ac.uk/collections/search/display record/?mode=displaymixed&module=ecatalogue& irn=2002 (with a full set of very good photos); http://www.mhs.ox.ac.uk/epact/catalogue.php?E Number=28895; http://www.davidaking.org/instrument-catalogue -TOC.htm#part6.

Mater

The mater of the astrolabe consists of three parts, a plate with a riveted ring and topped by a second, very thin ring that provides the engravings of the scales on the rim. Its inside is blank. Only the rim is engraved with a scale of 360 parts, each denoted by a stroke and grouped in fives. Besides those for fives and tens, these strokes are apparently drawn without support. In several cases they are drawn twice (Figure 5A1). This scale is numbered in two different ways and two different hands: A light script (hand A) numbers the hours in the usual way, clockwise from 1 to 12 starting below the throne. A bold script (hand B) that attaches serifs to most of the numerals, denotes the degrees also in the usual way, clockwise from 1 to 360 in steps of 10 starting below the throne.

The numerals engraved in hand A as well as those in hand B are too individual to be punched. Both hands go past the edge of the respective boxes from time to time. Because the mater is not perfectly circular, the distance between the outermost scale and the outer edge varies. Between 95° and 110°, the rim had been damaged and was then repaired and engraved in a light script, probably in hand A.

At first sight, the back of the mater does not show a specific altitude scale. A closer inspection, though, reveals faint numerals squeezed between the rim and the outermost scale in the upper left hand quadrant.

They start at the east-west line and run in steps of five up to 90 at the north-south-line. These numerals resemble closely those for the hours on the rim, and are, therefore, in all likelihood, also drawn by hand A (see Figure 5A7).

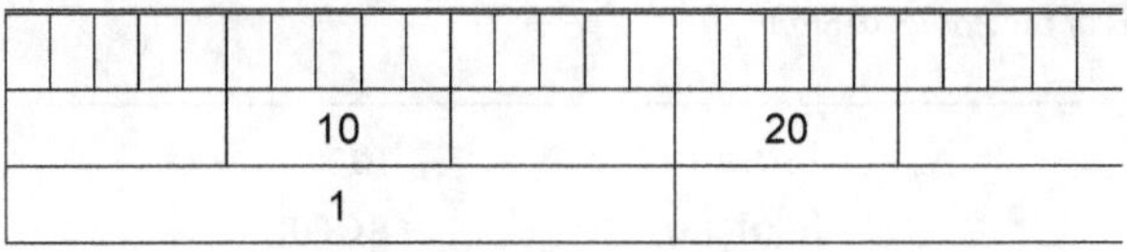

FIGURE 5A1 *Diagram of the beginning of the scale for the hours and the degrees on the rim.*

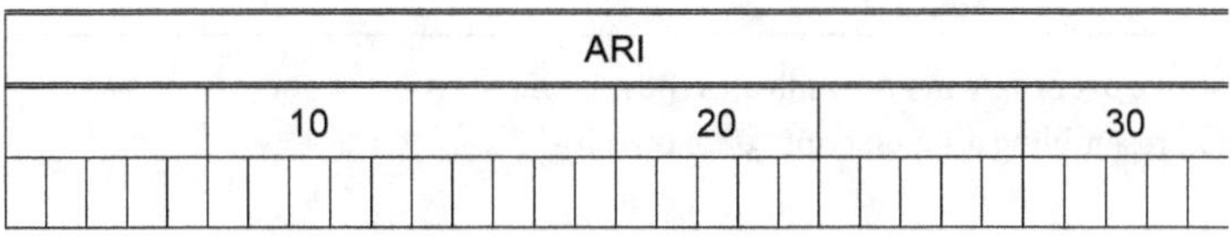

FIGURE 5A2 *Diagram of the beginning of the solar scale on the back of the astrolabe.*

The entries for 85 and 95 spread over the two lower circles engraved on the throne.

The outermost circular scale on the back of the astrolabe provides a solar scale that consists of 360 divisions, each denoted by a stroke, again, besides those for fives and tens, made without support, and grouped in fives (Figure 5A2). It starts at the east-west line and is numbered in the following circle counter-clockwise from ten to 30 for each zodiacal sign, the names of which follow counter-clockwise in an abbreviated form in the following circle.

The numerals in the solar and calendar scales (see also below) resemble closely those used on the rim to denote the degrees, in particular "1," "2," and "3" (Table 5A9).

As stated above, the mater is not perfectly circular, and therefore the distance between this scale and the edge varies too.

Next follows an eccentric calendar scale, also labelled by hand B. It denotes for each day a stroke, grouped in fives, and labelled in the following circle by tens up to 30 or 31 (Figure 5.3). February includes 29 strokes and is labelled up to 28. The month names follow counter-clockwise and abbreviated in the following circle (Table 5A3).

Because both scales are labelled by hand B, they complete our knowledge of this hand by providing a nearly entire set of letters, namely majuscules with distinct serifs. Concerning letters in the shape of a half circle like a waning moon (e.g. "C," "E," "F") these serifs extend so far that their opening seems closed. "n" forms an exemption, it is the only minuscule provided by hand B.

TABLE 5A2 *The Abbreviations of the Zodiacal Signs*

Aries	ARI	Libra	LIB
Taurus	TAV	Scorpio	SCOR
Gemini	GEI	Sagittarius	SAGI
Cancer	CAN	Capricorn	CAP
Leo	LEO	Aquarius	AQ^{3}
Virgo	VGO	Pisces	PIS~

VGO: An oblique stroke over the V in VGO indicates the missing "ir."
AQ: An abbreviation resembling a superscript "3" follows.

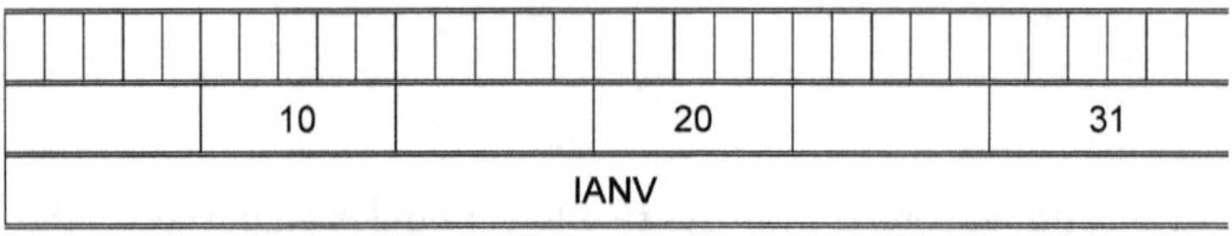

FIGURE 5A3 *Diagram of the beginning of the calendrical scale on the back of the astrolabe.*

TABLE 5A3 *The Abbreviations of the Julian Month Names*

IANV	January	IVLI	July
FEB/	February	AVGV	August
MAR	March	SEP	September
AP'	April	OCTO	October
MAI	May	NOVE	November
IVNI	June	DECE	December

SEP: Close to the "P" is a dent that may be a hint to a correction.

The "T" in "OCTO" and the "T" in "TAV" for Taurus are very different. In the second case, the letter has a curved shank and is very similar to the "T" in Taurus and also to the "V" in Virgo and Vespa (Table 5A10).

The vernal equinox corresponds to 14 March,[68] the autumnal equinox to 15 September, the summer solstice to 14 June and the winter solstice to 15 December.

In the circle between the solar and calendrical scale one can see very fine strokes in more or less regular spaces (Table 5A4). Although their distances vary, these strokes present a rudimentary subdivision of the ecliptic into twenty-eight spaces that may display the lunar mansions.

TABLE 5A4 *The Positions of Very Fine Strokes between the Solar and Calendrical Scale*

Month	Day	Difference	Signum	Degree	Difference
March	26th		Aries	11°	
		13 or 14			14°
April	8th or 9th		Aries	25°	
		13, 14 or 15			13° or 14°
April	22th or 23th		Taurus	8° or 9°	
		13 or 14			12° or 13°
May	6th		Taurus	21°	
		14 and 15			13° and 14°
May	20th and 21st		Gemini	4° and 5°	
		11 and 12			13° and 14°
June	2nd		Gemini	18°	
		23			22°
June	25th		Cancer	10°	
		5			4°
June	30th		Cancer	14°	
		11 or 12			12°
July	11th or 12th		Cancer	26°	
		13 or 14			12°
July	25th		Leo	8°	
		14			14°
August	8th		Leo	22°	
		14			13°
August	22nd		Virgo	5°	
		11 or 12			12°

68 "Oxford description IIC #0191" miscounts and reads: "The first point of Aries corresponds to the 19th of March."

TABLE 5A4 *The Positions of Very Fine Strokes between the Solar and Calendrical Scale* (cont.)

Month	Day	Difference	Signum	Degree	Difference
September	2nd or 3rd		Virgo	17°	
		12, 13 or 14			13°
September	15th or 16th		Virgo	30°	
		12 or 11			12°
September	27th		Libra	12°	
		14			14°
October	11th		Libra	26°	
		19			18°
October	30th		Scorpio	14°	
		12 or 13			13°
November	11th or 12th		Scorpio	27°	
		8 or 9			13°
November	20th		Sagittarius	10°	
		12			8°
December	2nd		Sagittarius	18°	
		12			11°
December	14th		Sagittarius,	29°	
		16 or 17			18°
December	30th or 31st		Capricorn	17°	
		9 or 10			9°
January	9th		Capricorn	26°	
		13			13°
January	22nd		Aquarius	9°	
		13			12°
February	4th		Aquarius	21°	
		13			14°
February	17th		Pisces	5°	
		13			13°
March	1st		Pisces	18°	
		12			12°
March	14th		Pisces	30°	
		12			11°

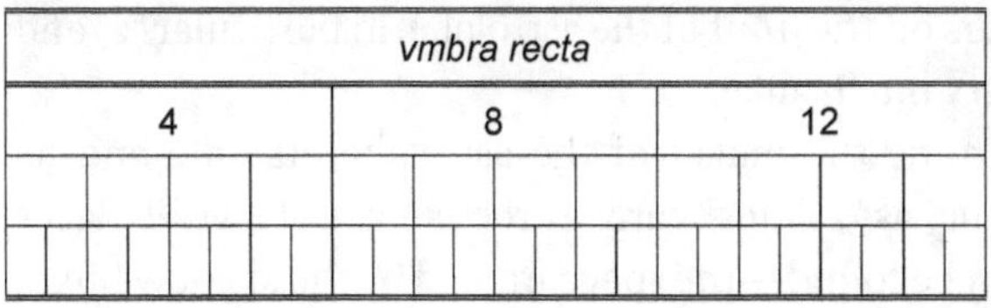

FIGURE 5A3 *Diagram of a part of the scale of the shadow boxes on the back.*

The space in the centre framed by the solar and calendar scale, bears in the lower half-circle two shadow boxes. Both scales, for *vmbra recta* and *vmbra versa*, are for unit 12 (= digits), labelled in steps of four, while the corresponding strokes count up to twenty-four.

Comparing the numerals of the hours on the front of the astrolabe with those used in the shadow boxes, it appears plausible that both are engraved by the same hand (hand A). Assuming that hand A not only numbered but also labelled the shadow boxes, these inscriptions provide more information on hand A, which is obviously different from hand B. First of all, hand A is a minuscule with decorated initial letters, not a majuscule like hand B. Second, while hand A provides distinct ascenders of "b" and the long "ſ" as well as of the initial "r"—descenders are missing—hand B makes use only of the mean line and writes a round "s" as an initial as well as a terminal letter. Third, unlike hand B, hand A writes ligatures. It combines "b" with a rounded "r"—or "r rotunda"[69]—in "vmbra" and superimposes the right vertical stroke of "v" and the left of "e" in "versa." Fourth, altogether, hand A shows a higher degree of fracturing than hand B (Table 5A10 and Figure 5A8). In one case, hand A provides an example that even one hand does not write all the time and in all places all characters in a similar way when it makes use of a double-story lowercase "a," while in all other places it writes a single-story lowercase "ɑ."

In the upper half-circle, two universal hour quadrants are engraved and labelled with Hindu-Arabic numerals not by hand B, for the serifs are missing. With only a set of numerals at hand, it is, though, difficult to tell if hand A numbered the curves of these quadrants or if a third hand made these inscriptions. Comparison of the numerals in the hour quadrant with those in the shadow

69 According to Friedrich Beck and Lorenz Friedrich Beck, *Die lateinische Schrift. Schriftzeugnisse aus dem deutschen Sprachgebiet vom Mittelalter bis zur Gegenwart* (Vienna, Cologne: Böhlau, 2007), 33, the "r *rotunda*" appears in the German-speaking countries first in the tenth century.

boxes and those for the hours on the front of the astrolabe, in particular "2" and "3," shows that they look very much alike.

Any information concerning the maker of the astrolabe, its date and its place of creation are missing, as on most early European astrolabes.[70] Places where such inscriptions can be found—the space framed by the shadow boxes and the circular centre in the middle of the hour quadrants—are blank on IIC #0191.

The three-lobed throne is an integral part of the mater and shows accordingly the same sandwich structure of three metal parts. It is rather small and consists of three circles, the upper one pierced for the shackle and centred upon the two lower ones, which bear two thin circular frames, possibly an incomplete decoration, on the front as well as on the back.[71] The suspensory device consists of a ring to which a stretched shackle of two parts is attached; the lower part of the shackle includes the boss.

The Rete

The ecliptic circle is engraved for each 2° marked by a stroke, grouped in fives, and numbered in steps by tens to 30 for each zodiacal sign (Figure 5A4). The index consists of a lower rectangular part and a tip at its top.

Unlike the zodiacal scale on the back the names of the zodiacal signs on the rete are not abbreviated. They show no vernacular influences and are written in the same hand as the star names, most probably hand A. The right and left spaces between the names are decorated with wavelike lines. Between Virgo and Libra, the ecliptic ring is broken and fixed with a small metal plate. The scales covered by this plate are re-engraved on top of it; the decorative elements appear similar to those on the ecliptic ring.

The equinoctial bar (Figure 5A5) is counter-changed twice on each side. The equatorial bar appears only in the lower part and is broken by two arcs,[72] one

70 See n. 42.

71 The design is not distinct from the three-circle thrones of the astrolabes made in the 16th century by Georg Hartmann (examples in van Cleempoel, *Astrolabes at Greenwich*; 143–146; King, *Focus Behaim Globus*, 593–594; Ward, *Catalogue of the British Museum*, 115, no. 333 and plate LIV; Webster and Webster, *Western Astrolabes*, 53–57).

72 Gunther, *Astrolabes*, 340, no. 191 describes these two arcs as "two cusped Gothic arches," while in King's preliminary description, kindly provided by the author, they are called "*miḥrābs*," i.e. prayer niches in a mosque.

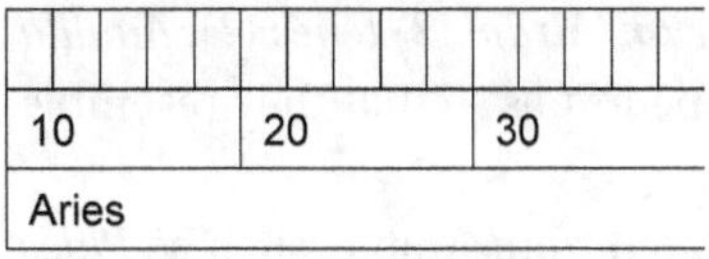

FIGURE 5A4 *Diagram of the beginning of the ecliptic scale.*

TABLE 5A5 *The Zodiacal Signs on the Rete*

Aries	Libra
Taurus	Scorpius
Gemini	Sagittarius
Cancer	Capricornus
Leo	Aquarius
Virgo	Pisces

FIGURE 5A5 *Diagram of the equinoctial bar*

at the right end around the star pointer of *humer datis*, one in its centre framing the star pointer of *Canis Tienorus/Vienorus* (?) (Table 5A6).

The rete bears twenty-six star pointers, most of them flame-like, some with elongated tips reminiscent of claws, and a few worm-like. Most of their bases are shaped like small rings.[73] An exception is the four pointers attached to the central disk around the north-pole at the centre of the astrolabe (for α Aur (Capella), μ UMa (?), α CrB (Gemma) and α Cyg (Deneb). They are less elaborated, their star names omitted, most probably because of lack of space. The two pointers for *Verar* (?) *ceti* and *cauda cap* are formed, though not very distinctly, by the two tips of the circle that presents the southern tropic, while the two pointers for *Vespa* and *Serpēs* are shaped by the circular ring without an elongated tip. Four pointers north of the ecliptic placed above the east-west

73 See also Gunther, *Astrolabes*, 340, no. 191; and "Oxford description IIC #0191" both describing the bases as "perforated bulbars."

line in the upper half of the rete (for *Serpēs, Arum/Alrum* (?) *tenēs* (?), *Aquilla volās* and *Elphinus*) are equipped with square pedestals so that they resemble little candles.

Most of the star names inscribed are translated from Arabic, such as *Pides Geminorum* or provide a name related to the classical myth associated to the star or the constellation such as *Spica*. While *Corona* appears incorrectly attributed to β Cas, *Eletus*/*Eledor* (?) (η UMa) completely defies an explanation. The latter can be considered a corrupt transcription of the Arabic name.

TABLE 5A6 *The Star Names on the Rete Starting at the Vernal Equinox According to Increasing Mediation*

No.	IIC #0191	Translations and modern designation	Remarks
1	*Verar* (?) *ceti*	"the Sea Monster's …" = ζ Cet	Apparently, *verar* (?) is corrupt for *venter*—"belly," the whole name a translation of the Arabic بطن قيطوس—*baṭn qayṭūs*.
2	*Vespa*	"the Wasp" = α Tau (Aldebaran)	One may imagine that الدبران—*al-dabarān* was misread and interpreted either as a form of الدبور—*al-dabbūr*, in the plural الدبابير—*al-dabābīr*, or even of الزنبور—*al-zunbūr*, in the plural الزنابر—*al-zanābir*, both meaning "wasp."
3	Unnamed	= α Aur (Capella)	—
4	*Pides Geminorum*	"the feet of the *jawzā*'" = β Ori (Rigel)	A translation of the Arabic رجل الجوزاء—*rijl al-jawzā*' based on a misinterpretation, because *al-jawzā*' may be related to Gemini as well as to Orion. The incorrect *pides* (for *pedes*) points to little knowledge of Latin, and could be a vernacular influence.

TABLE 5A6 *The Star Names on the Rete Starting at the Vernal Equinox* (cont.)

No.	IIC #0191	Translations and modern designation	Remarks
5	*Humerus geminor*	"the shoulder of the *jawzā'*" = α Ori (Beteigeuze)	A translation of the Arabic منكب الجوزاء—*mankib al-jawzā'* (cf. no. 4).
6	*Canis Tienorus/ Vienorus* (?)	"the ... Dog" = α CMa (Sirius)	Probably a combination of classical and Arabic traditions. The second word may be derived from translating the second part of the Arabic, الشعرى العبور—*al-shiʿrā al-ʿabūr*, "the *shiʿrā* that crosses (the milky way)," into Latin, *transiens*.
7	*Canis plorans*	"the wailing/bleary-eyed Dog" = α CMi (Prokyon)	Again probably a combination of classical and Arabic traditions. The second word may be derived from translating the second part of the Arabic, الشعرى الغميصاء—*al-shiʿrā al-ghumayṣā'*, "the bleary-eyed *shiʿrā*."
8	Unnamed	= μ UMa (?)	—
9	*Humer datis* (?)	"the ... 's shoulder" = α Hya (Alphard)	Probably a confusion of الفرد—*al-fard*, "the solitary one" (α Hya, Alphard), and الفرس—*al-faras*, "the Horse," i.e. Pegasus, as recorded in the sources. Therefore, the second word, *datis*, may be corrupt for الفرس—*al-faras*, while the first word, *humer*, may be added later, possibly because *humerus equi* (= β Peg) is the most favoured astrolabe star of Pegasus.
10	*Cor leonis*	"the Lion's heart" = α Leo (Regulus)	A translation of the Arabic قلب الأسد—*qalb al-asad*.
11	*Alacorvy*	"the Raven's wing" = γ Crv	A translation of the Arabic جناح الغراب—*janāḥ al-ghurāb*.

TABLE 5A6 *The Star Names on the Rete Starting at the Vernal Equinox* (cont.)

No.	IIC #0191	Translations and modern designation	Remarks
12	*Eletus/ Eledor* (?)	= η UMa (?)	A possible link to the Arabic بنات نعش—*banāt naʿsh*, "daughters of the *naʿsh*," or القائد—*al-qāʾid*, "the Leader," both for η UMa, is not evident. A variation of *Alioth*, *aliore* etc. seems more plausible derived either from Arabic الجون—*al-jawn*, "the Black Horse," or الحور—*al-ḥawar*, "the Bull," although it denotes ε UMa.
13	*Spica*	= α Vir (Spica)	According to the classical tradition.
14	*Anceator/ Ameator* (?)	= α Boo (Arcturus)	Probably *Anceator/Ameator* (?) is corrupt for *lanceator* being a translation of the second part of the Arabic السماك الرامح—*al-simāk al-rāmiḥ*, "the lance-bearing *simāk*."
15	Unnamed	(α CrB)	—
16	*Serpēs*	"the Serpent" = α or β Ser (?)	A translation of the Arabic الحيّة—*al-ḥayya*.
17	*Cor scorpionis*	"the Scorpion's heart" = α Sco (Antares)	A translation of the Arabic قلب العقرب—*qalb al-ʿaqrab*.
18	*Arum/Alrum* (?) *tenēs* (?)	= α Oph	The first word may be either a variation of Arabic الراعي—*al-rāʿī*, "the Shepard," or of the first part of راس الحوّاء—*rāʾs al-ḥawwāʾ*, "the head of the Serpent Bearer," or even of its second part, both names for α Oph. The second probably translates the second part of the name with *tenens*, "bearing."
19	Unnamed	= α Lyr (Vega)	—

No.	IIC #0191	Translations and modern designation	Remarks
20	*Aquilla volās*	"the flying Eagle" = α Aql (Altair)	A translation of the Arabic name النسر الطائر—*al-nasr al-ṭā'ir.*
21	*Elphinus*	= ε Del	Probably *Elphinus* is corrupt for *delphinus* (parallel to the possible corruption of no. 14, *Anceator/ Ameator* (?) for *lanceator* where the initial letter may have been also omitted).
22	Unnamed	= α Cyg (Deneb)	—
23	*Cauda cap*	"Capricorn's tail" = δ Cap	An abbreviated translation of the Arabic ذنب الجدي—*dhanab al-jady.*
24	*Humerus equi*	"the Horse's shoulder" = β Peg	A translation of the Arabic منكب الفرس—*mankib al-faras.*
25	*Cauda ceti*	"the Sea Monster's tail " = ι Cet (?)	A translation of the Arabic ذنب قيطوس—*dhanab qayṭūṣ.*
26	*Corona*	"the crown" = β Cas	Although clearly legible, this star pointer belongs to β Cas, in Arabic الكفّ الخضيب—*al-kaff al-khaḍīb*—"the stained hand (of the *thurayyā*)," not to α CrB (Gemma; cf., no. 15).

Note to no. 4: Paul Kunitzsch, *Untersuchungen zur Sternnomenklatur der Araber* (Wiesbaden: Otto Harrassowitz, 1961), 23–24.
Note to no. 6: Paul Kunitzsch, *Typen von Sternverzeichnissen in astronomischen Handschriften des zehnten bis vierzehnten Jahrhunderts* (Wiesbaden: Otto Harrassowitz, 1966), 45, no. 14 (Typ VI) and 48, no. 8 (Typ VII).
Note to no. 9: Kunitzsch, *Sternnamen*, 76, no. 29 and 85, no. 63.
Note to no. 12: Kunitzsch, *Sternnamen*, 78, no. 37; Kunitzsch, *Sternnomenklatur*, 62, no. 106; Paul Kunitzsch, Tim Smart, *Short Guide to Modern Star Names and Their Derivations* (Wiesbaden: Otto Harrassowitz, 1986), 56.
Note to no. 14: Kunitzsch, *Sternverzeichnisse*, 42, no. 26 (Typ VI), 48, no. 17 (Typ VII; with the additional variations *ceator* and *ator*), 55, no. 28 (Typ VIII), 78, no. 9 (Typ XII), 96, no. 10 (Typ XIV), 101, no. 18 (Typ XV), 106, no. 39 (Typ XVI), 113, no. 25 (Typ XVII).
Note to no. 18: Kunitzsch, *Sternnamen*, 81, no. 51 lists *album, albaum, alhaum* etc.; Kunitzsch, *Sternverzeichnisse*, 46, no. 30 (Typ VI) mentions, beside others, *ram alhadde*.

The star names and the zodiacal signs on the rete are engraved in the same hand. It is, however, difficult to decide if it is the same hand that inscribed the small amount of letters included in the shadow boxes. The most obvious commonality between the hand that labelled the shadow boxes and the hand that engraved the rete concerns ligatures of letters with right-facing bowl combined with a rounded "r" (e.g. in "*plorans*"). They look very similar to the ligature of "b" and "r" in "vmbra." Further, in both cases a long "ſ" is used in the middle of a word in addition to a round "s" as initial and terminal letters on the rete, which does not occur in the shadow boxes on the back of the astrolabe. Alas, the most distinct letter of this inscription, the "b," is omitted on the rete and cannot help in deciding if the rete is engraved by hand A. At the same time, the rounded back-sloping "d" on the rete attracts attention with its left-slanted ascender and its left-facing bowl coming close to a circle, but unclosed. However, it is not included in the set of letters provided by the inscriptions of the shadow boxes and cannot help either. Two differences catch one's eye, although they are difficult to assess. First, "a" appears in the shadow boxes double-story as well as single-story lowercases, while only a double-story lowercase "a" is used for the zodiacal signs and the star names (not always very distinct). Further, the initial "V" in "vmbra" on the back of the astrolabe, and in "Virgo" and "Vespa" on the rete look completely different. Possibly, "vmbra" starts with a decorated lowercase "v," while "Virgo" and "Vespa" begin with an uppercase. All things considered, the rete was most probably engraved also by hand A.

The Plates

The mater provides enough space for an extraordinary amount of plates. There are all together seven plates with engravings for fourteen latitudes. The discs do not show obvious differences in shape, size, colour and texture. They fit snugly into the mater, where they can be fixed with their small pins. However, the pin is broken in discs four and six, and repaired in disc two (probably broken again after repairing it a first time).

Twelve of these fourteen plates (Figures 5A8–5A12) (henceforth: the standard plates) are all alike in design, although plates 3a and 3b are more elaborated and provide additional features, while two plates, 1a and 2a, provide a completely different design. Each of the standard plates (1b, 2b, 4a, 4b, 5a, 5b, 6a, 6b, 7a, 7b—for 3a and 3b, see Table 5A7) is engraved with the east-west line perpendicular to the north-south line, the basic circles (the equator, tropics of Cancer and Capricorn), altitude circles for each 6° labelled in steps of 12 up to 90 with Roman numerals (except for plate 1b, in which they are omitted), and seasonal hour lines below the horizon labelled with Roman numerals. All these standard plates bear the word "Latitudo" between the curves of the horizon and the tropic of Cancer, and its value in Roman numerals between the lines of the tropic of Cancer and the equator, but split between the sixth and the seventh hour lines. Only on plate 1b are "Latitudo" as well as the value written in between the crescent. None of these plates includes azimuth curves.

Although similar in design, plate 3a and 3b are more elaborated and therefore slightly different. Outside the circle for the tropic of Cancer they provide altitude circles for each 3° labelled in steps of nine in Roman numerals up to 90, inside for each 6°. It appears that first the altitude circles for each 6° were engraved and afterwards those for each 3° squeezed in the middle between them. Plate 3a displays additionally azimuth curves for each 10° and is supplemented with a place name, Cesaraugusta, today Zaragoza.

TABLE 5A7 *The Plates and Their Latitudes*

No.[a]	Fig.[b]	Place	φ
1a	2-2-1	Parisius	–
1b	1-2-1	–	38°
2a	2-1-4	Parisius	–
2b	1-1-4	–	55°
3a	2-2-3	Cesaraugusta	41°
3b	1-2-3	rhplose (?)[c]	42°
4a	1-1-2	–	45°
4b	2-1-2	–	46°
5a	2-2-2	–	47°
5b	1-2-2	–	48°
6a	1-1-3	–	49°
6b	2-1-3	–	50°
7a	2-1-1	–	57°
7b	1-1-1	–	58°

Note to 4b and 5b: The engravings of both plates are very alike and appear rather suitable for a smaller latitude.

a According to Gunther, *Astrolabes*, 341, no. 191.

b The numbers are related to Figure 5A8 (= photo 1) and Figure 5A9 (= photo 2). In the first place the number of the photo, 1 or 2, is given, in the second place the line, 1 or 2, and in the last place the column, 1 to 4 or 1 to 3 respectively, e.g., "1–1–4" denotes photo 1, line 1, column 4.

c Very lightly scratched inscription that Gunther, *Astrolabes*, 341, no. 191 reads "rhyplose" and suggests "Toulouse"; King, *Ciphers*, 285 interprets "Tolosa," the Latin name of "Toulouse."

Apparently, a third hand (henceforth: hand C) inscribed these plates using a minuscule whose letters are concentrated on the mean line. Only the "L" in "Latitudo," which may be regarded as an uppercase initial letter, has a small ascender as well as in some cases "d" and "t." Different from hand A, hand C uses no round "r"; instead of double-story it uses single-story lowercase "ɑ"; and it connects the horizontal stroke of "t" with following "i" or "u." Both hands write long "ſ," at least in the middle of a word. Both include an uncial "d" with an ascender inclined to the left. It has a left-facing bowl coming close to a circle, but as an unclosed loop. In general, hand C provides a lower degree of fracturing than hand A. In comparison to hand B, a majuscule, hand C does without serifs. To denote the forties, hand C writes "X," the Roman numeral for ten, with an upside-down breve "⌢" between its two brachia, a Visigothic ligature that evolved from a capital "X" combined with a superscript "L."[74] The units follow as usual. Remarkably enough, this ligature is also used for writing "50" with an "X" that follows the ligature (e.g., the latitude on plate 6b).

Plates 1a and 2a are completely different in style and resemble each other. They, however, do not comprise A- and B-sides of the same disc, but bear on their opposite side one of the twelve standard plates. A possible explanation is provided by another astrolabe preserved in Oxford, IIC #0168.[75] Its mater contains a blank plate only engraved with the east-west line perpendicular to the north-south line and the basic circles (equator, tropics of Capricorn and Cancer). When the astrolabe changed its whereabouts, for example, because its owner moved or gave it to a new owner, such a nearly blank plate might have been quite useful and might have served as a basis for engraving additional plates suitable to the new place.

Plates 1a and 2a have in common that they include the east-west line perpendicular to the north-south line, the basic circles (equator, tropics of Capricorn and Cancer), altitude circles for each 2° with every fifth stressed by dots alongside the circle, and seasonal hour lines below the horizon.

They are, however, different in many details (Figures 5A10 and 5A11). While plate 1a omits any numerals related to the altitude circles, plate 2a labels them in Hindu-Arabic numerals in steps of tens up to ninety, although supplementary twenty-four and forty-four are also engraved. Another difference concerns the azimuth curves that plate 1a provides for each 6° and plate 2a for each 5°, both times unlabelled. Further, plate 1a labels the seasonal hour lines with

74 Burnett, *Ten of Forty*, 82; cf. King, *Ciphers*, 284: "Another variant, found already in Carolingian times, is X˘ and the like for 40. This form seems to have been particularly popular in Spain, see also the Spanish plate in a composite astrolabe (#191) where is used."

75 Gunther, *Astrolabes*, 319, no. 168; "Oxford description IIC #0168."

Latin ordinal number words, while plate 2a uses Hindu-Arabic numerals, one time at the tropic of Capricorn and a second time at the tropic of Cancer. Plate 1a includes between the lines of the horizon and the tropic of Cancer the word "Parisius" followed by a floral ornament, while plate 2a writes "Latitudo ad (?) Parisius."

Plate 2a is most probably engraved by hand A. The two sets of Hindu-Arabic numerals related to the unequal hours are both very similar to the numerals engraved in the hour quadrants on the back of the astrolabe, probably by hand A. The "P" in "Parisius," the only inscription that may have replaced an earlier one because it is deeper engraved, looks very alike the "P" in "Pides" on the rete, its "ar" is written as a ligature of "a" with a round "r," its two "s" are written in the middle as long "ſ," in the end as round "s."

Concerning plate 1a, it is doubtful that it was also engraved by hand A. Although showing a similar degree of fracturing and using the long "ſ" as middle, the round "s" as initial and terminal letter, the initial letters on plate 1a are much more elaborated than those on the rete. Although this may be due to lack of space on the rete, further differences give rise to doubts. Plate 2a uses Hindu-Arabic numerals instead of Latin number words, as does plate 1a, to denote the unequal hours. The azimuths are engraved for each 5° on plate 2a, but for each 6° on plate 1a. It becomes more plausible that 1a was engraved by a fourth hand (hence: hand D). Two plates for Paris, from the practical point of view superfluous in one and the same astrolabe, argue also against one hand engraving plate 1a and 2a. Probably, flaws in the first of the Paris plates led to the engraving of the second one.

The Alidade, the Ruler, the Peg, and the Horse

The alidade is counter-changed at the centre where the two brachia form a rectangle of roughly 19 mm × 16 mm. It does not bear any engravings. The holes in its sighting vanes are of four different sizes.

The ruler is also counter-changed at the centre. Both of its brachia are marked with a non-uniform and non-linear scale. On one brachium, it counts for each fifteen units up to ninety and is labelled with Hindu-Arabic numerals. At its tip is shown either an aborted "p," an "h," or a "5." The other brachium comprises nine strokes and a decorative element resembling a tendril. At its tip, a symbol appears that may be a "p" and looks similar to the "p" on the rete.[76]

76 Both nicely reproduced in Gunther, *Astrolabes*, 340, no. 191.

The Hindu-Arabic numerals on the ruler resemble those related to the hours on the front of the astrolabe. Therefore, the ruler may be labelled also by hand A.

TABLE 5A8 *The Elements of the Astrolabe According to the Different Hands at Work*

		Hand A	Hand B	Hand C	Hand D
Mater	Front	Numerals of the hours	Numerals of the degrees		
	Back	Numerals of the altitude scale (?)	Numerals and letters of the zodiacal and the calendrical scale		
		Numerals of the shadow boxes			
		Numerals of the hour quadrants (?)			
Rete		Letters of the zodiacal signs and of the stars			
Plates		2a		1b, 2b, 3a, 3b, 4a, 4b, 5a, 5b, 6a, 6b, 7a, 7b	1a

TABLE 5A9 *Examples of the Different Numerals Engraved On The Astrolabe*

Hand A		Hand B	
Front of the mater	Back of the mater	Front of the mater	Back of the mater

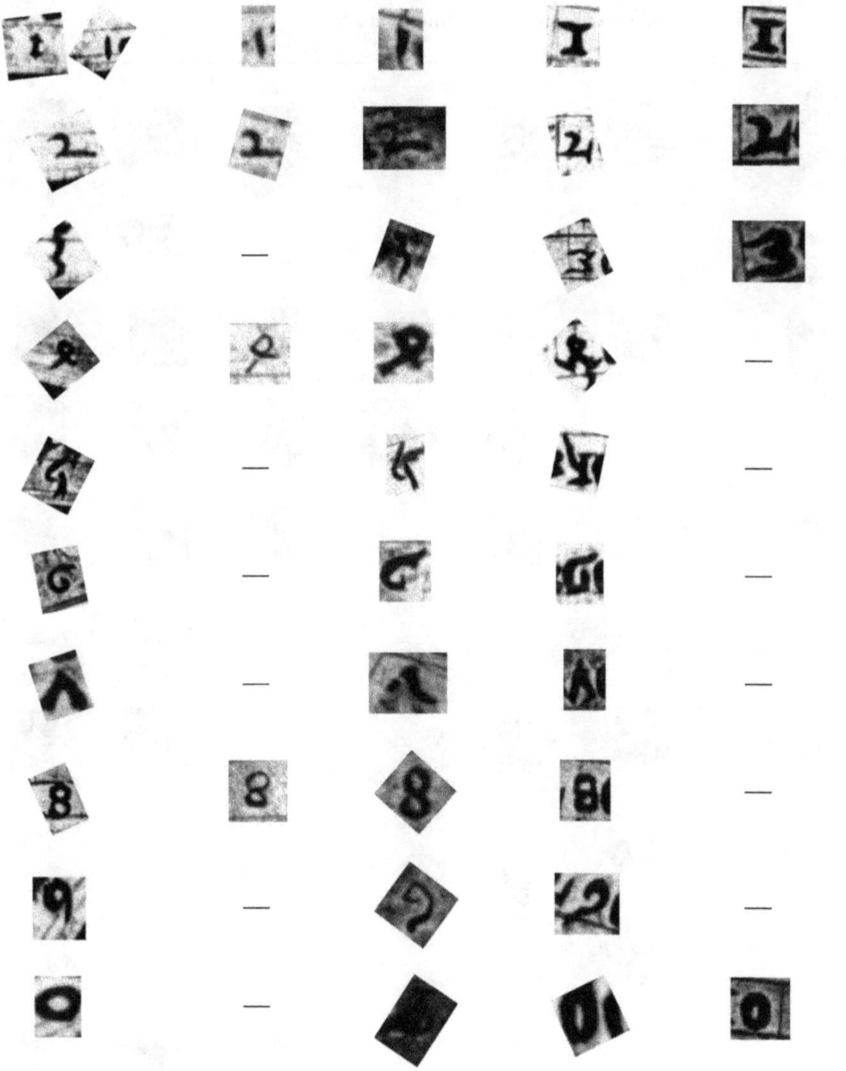

The attribution of the hour quadrants is uncertain (see above). While "2" and "3" resemble the numerals in the shadow boxes and those for the hours, "5," "6," "7," "9," and "0" vary more.

TABLE 5A10 *Examples of the Different Letters Engraved on the Astrolabe*

		Hand A		Hand D		Hand B	Hand C
	Front of the mater	Rete		Paris plate		Back of the mater	Standard plates
		Initial letters		Initial letters			
A	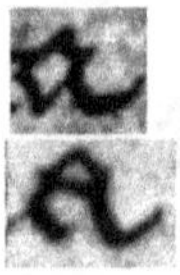			—			
B		—	—	—	—		—
C			—	—			
D	—	—			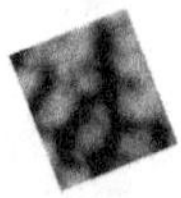		
E				—			
F	—	—	—	—	—		—
G	—			—	—		
H	—	 a  b		—	—	—	—
I/J	—	—		—			
K	—	—	—	—	—	—	—
L	—			—	—		

TABLE 5A10 *Examples of the Different Letters Engraved on the Astrolabe* (cont.)

	Hand A			Hand D		Hand B	Hand C
	Front of the mater	Rete		Paris plate		Back of the mater	Standard plates
		Initial letters		Initial letters			
M							
N							
O	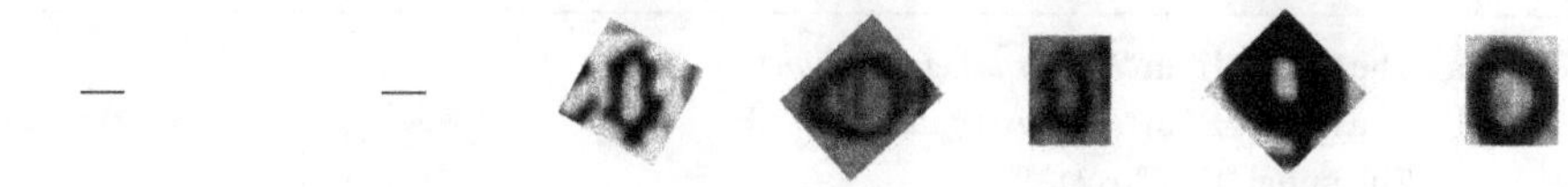						
P							
Q							
R	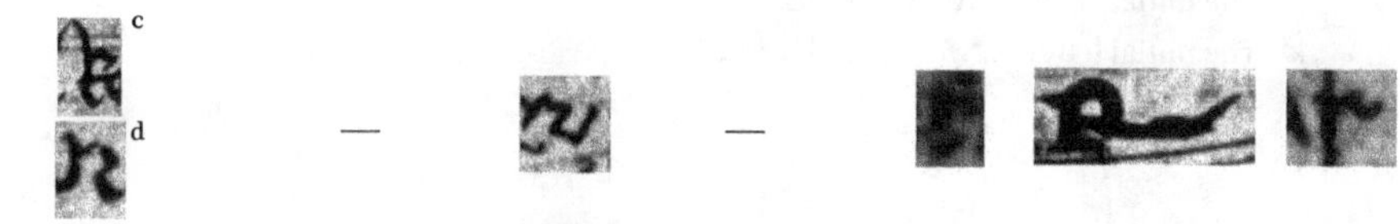						
S	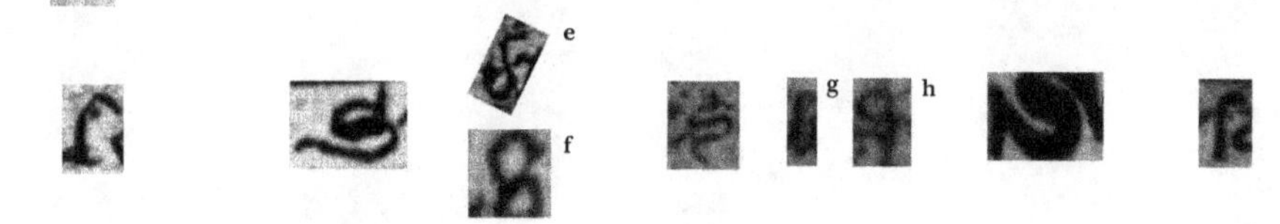						
T	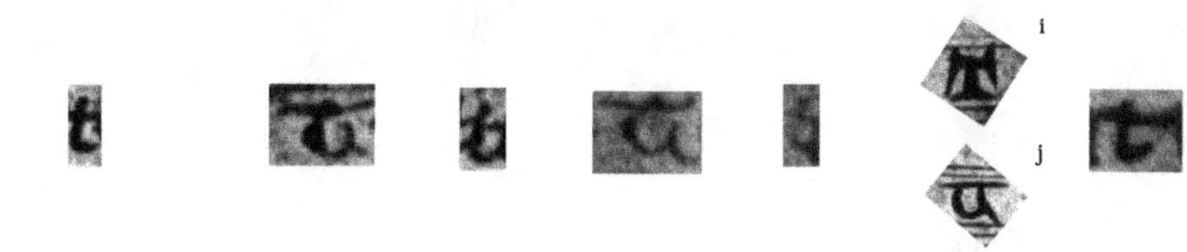						
U/V							

TABLE 5A10 *Examples of the Different Letters Engraved on the Astrolabe* (cont.)

	Hand A			Hand D		Hand B	Hand C
	Front of the mater	Rete		Paris plate		Back of the mater	Standard plates
		Initial letters		Initial letters			
W	—	—	—	—	—	—	—
X	—	—	—	—	—	—	—
Y	—	—		—	—	—	—
Z	—	—	—	—	—	—	—

a The initial "H" in "*Humerus geminor*(*um*)."
b The initial "H" in "*Humerus datis* (?)."
c The initial "r" of "*recta*."
d The middle "r" in "*umbra*."
e The long "ſ" as middle letter.
f The round "s" as terminal letter.
g The long "ſ" as middle letter.
h The round "s" as terminal letter.
i The middle "T" in "OCTO" for October.
j The initial "T" of "TAV" for Taurus.
k The initial letter of "*vmbra*" and "*versa*."

FIGURE 5A6 *Front of IIC #0191.*
ALL PHOTOS OF IIC #191 COURTESY OF THE OXFORD MUSEUM OF THE HISTORY OF SCIENCE.

FIGURE 5A7 *Back of IIC #0191.*

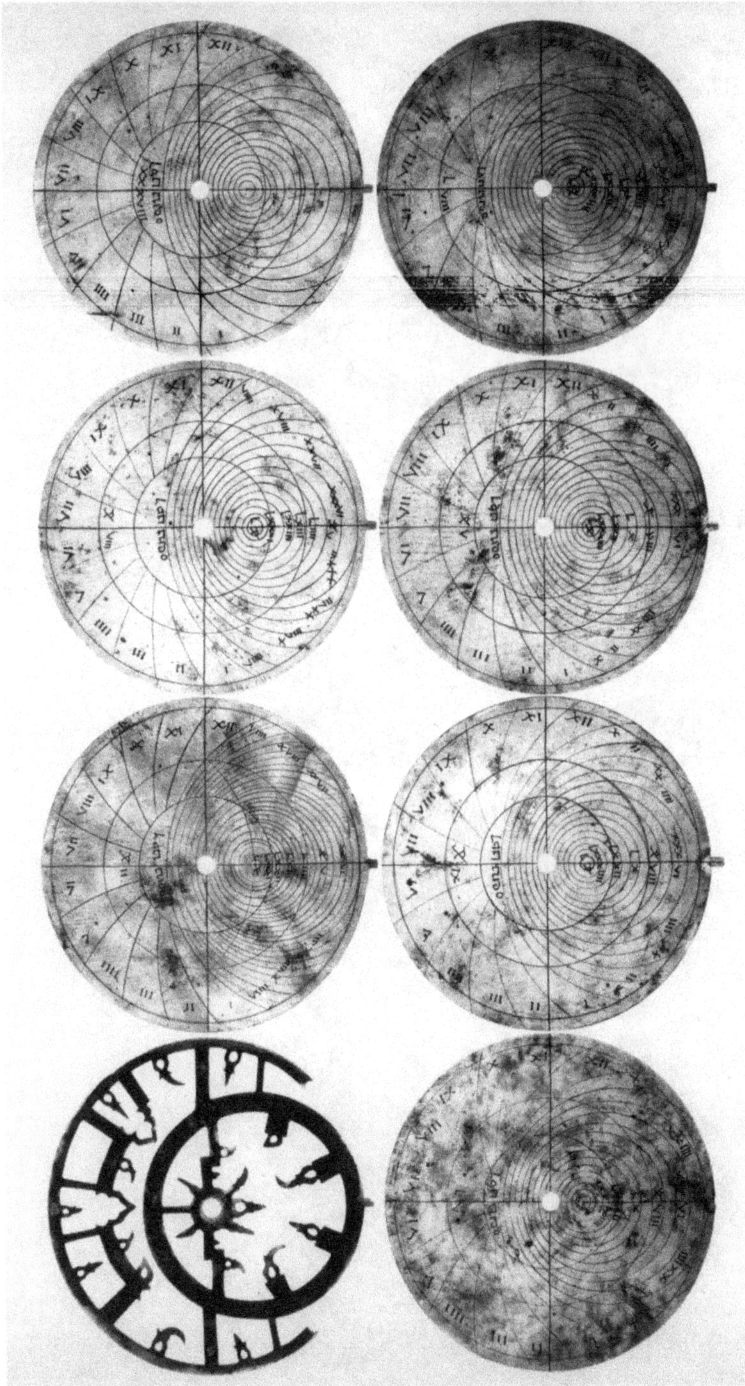

FIGURE 5A8 *The plates of* IIC *#0191 (photo 1).*

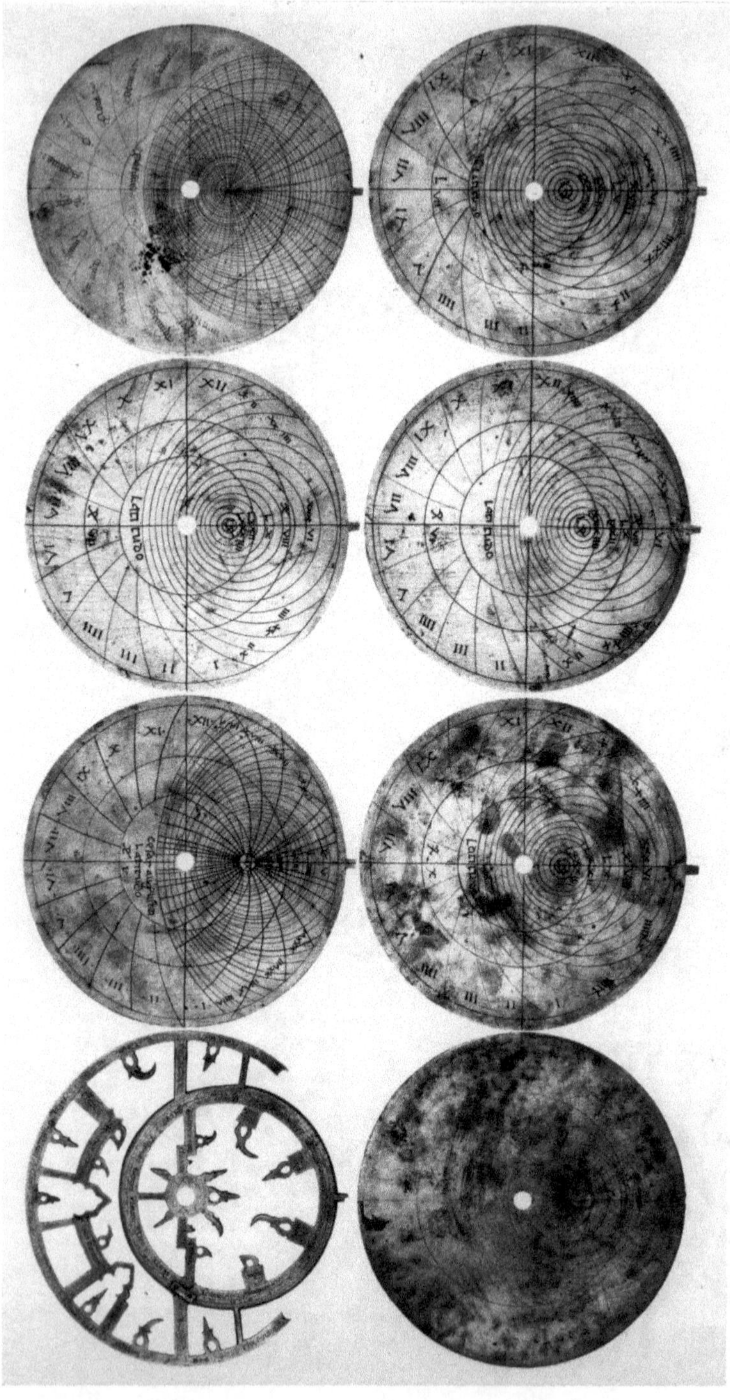

FIGURE 5A9 *The plates of IIC #0191 (photo 2).*

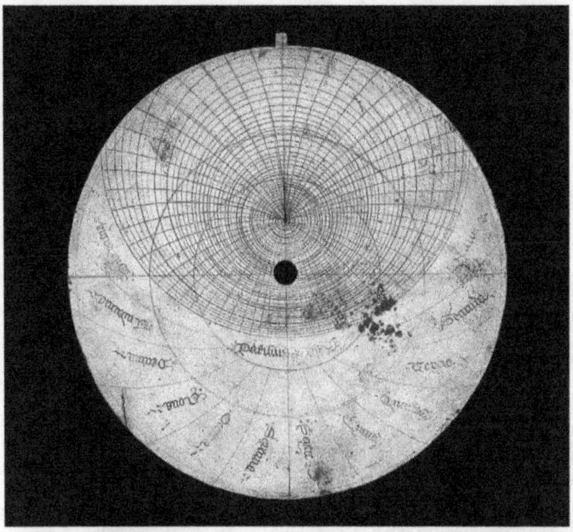

FIGURE 5A10
Plate 1a of IIC #0191 (for Paris).

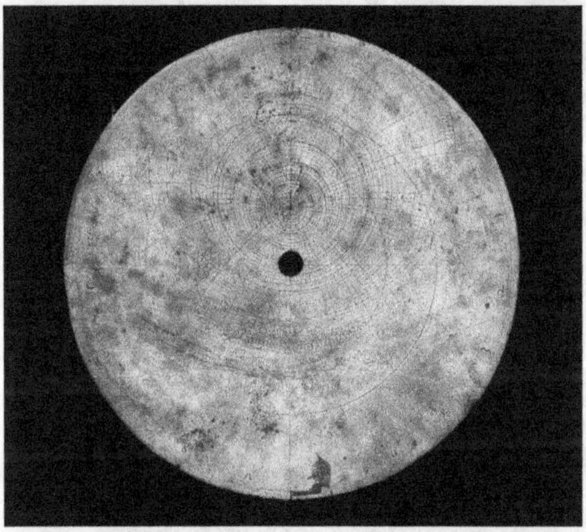

FIGURE 5A11
Plate 2a of IIC #0191 (for Paris).

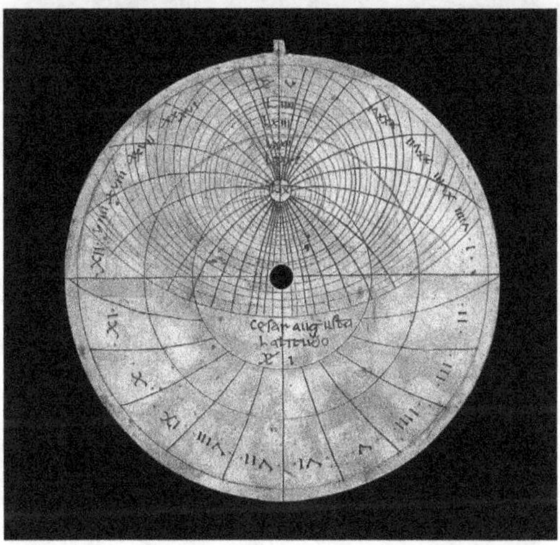

FIGURE 5A12
Plate 3a of IIC #0191 (for Zaragoza).

FIGURE 5A13 *The rete of the undated and unsigned European composite astrolabe (IIC #0191).*

FIGURE 5A14 *The rete of Ibn al-Sahlī's astrolabe made in Valencia at the end of the year 478 Hijra = April 1086 (IIC #0121).*
PHOTO COURTESY OF THE STAATLICHE MUSEEN KASSEL.

CHAPTER 6

A Monumental Astrolabe Made for Shāh Jahān and Later Reworked with Sanskrit Legends

Sreeramula Rajeswara Sarma
Höhenstrasse 28, 40227 Düsseldorf, Germany
SR@Sarma.de

Abstract

When the astrolabe was introduced in India around the eleventh century, it was received with great enthusiasm. While the Muslims continued the Middle Eastern tradition of the study and manufacture of the astrolabe, the Hindus and Jains, who did not read Arabic, composed manuals on the astrolabe in Sanskrit, produced astrolabes with Sanskrit inscriptions, and also occasionally added Sanskrit legends to the Arabic/Persian astrolabes. A very large astrolabe, which is thoroughly reworked in this manner with Sanskrit legends, is the subject of this paper. During the process of reworking the name of the original maker of the astrolabe, the date of its manufacture, and other such details got effaced. But on the basis of the internal evidence, it will be argued that the astrolabe was originally produced between 1648 and 1658 by Ḍiyā' al-Dīn Muḥammad of Lahore for the Mughal Emperor Shāh Jahān. The study continues with a technical description of the components of the astrolabe, in which an attempt will be made to record all the original Arabic inscriptions and the subsequent engravings in Sanskrit.

Keywords

Ḍiyā' al-Dīn Muḥammad – graphs of the solar meridian altitude – Lahore family of astrolabe makers – Nayanānanda – Shāh Jahān –Yantrarāja – Zawraqī horizons

Introduction

When the astrolabe was introduced into India around the eleventh century, it was received with great enthusiasm. Muslim astronomers in India naturally continued the Middle Eastern tradition of the study and manufacture of the astrolabe. The astrolabes produced by them with Arabic and Persian legends

 | DOI:10.1163/9789004387867_009

are generally classified as "Indo-Persian" astrolabes. In the sixteenth and seventeenth centuries, a family of instrument makers from Lahore became renowned for the Indo-Persian astrolabes they produced.[1]

The Hindu and Jain astronomers working with Sanskrit as their scientific language had to overcome the linguistic barriers of Arabic and Persian in order to understand the principles of the astrolabe. Exchanges between Muslim astronomers and Hindu/Jain astronomers had been going on in the northwestern and western parts of India at least since al-Bīrūnī's visit in the eleventh century. These exchanges received an impetus in the latter half of the fourteenth century under Sulṭān Fīrūz Shāh Tughluq (reg. 1351–1388) who initiated a program of translations of Sanskrit scientific texts into Persian. Fīrūz also promoted the manufacture of astrolabes and commissioned manuals on the astrolabe both in Persian and Sanskrit. The Sanskrit manual is extant; it was composed in 1370 by a Jain monk named Mahendra Sūri, who hailed this instrument as the "king of instruments" (*yantra-rāja*).[2] Since then the astrolabe was known as *Yantrarāja* in Sanskrit.

The response of Hindu and Jain astronomers to the astrolabe was three-fold. Between 1370 and 1870, they composed at least fifteen manuals on the astrolabe in Sanskrit.[3] They also caused the production of astrolabes with numerals and legends in the Sanskrit language and in Devanāgarī script. Over a hundred Sanskrit astrolabes are extant today, and more are coming to light.

The third aspect of the response is the "appropriation" of Arabic astrolabes by engraving additional legends and numerals in Sanskrit. Such "appropriation" occurred also in Europe, where additions in Hebrew, Latin or Spanish were made to the Arabic astrolabes.[4] In the course of my survey of pre-modern

1 Sreeramula Rajeswara Sarma, "The Lahore Family of Astrolabists and their Ouvrage," *Studies in History of Medicine and Science* 13.2 (1994): 205–224; reprinted in Sreeramula Rajeswara Sarma, *The Archaic and the Exotic: Studies in the History of Indian Astronomical Instruments* (New Delhi: Manohar, 2008), 199–222. Henceforth page numbers will be cited from this volume of collected papers.

2 Mahendra Sūri, *Yantrarāja*, ed. K. K. Raikva (Bombay: Nirnay Sagar Press, 1936); see also Sreeramula Rajeswara Sarma, "Sulṭān, Sūri and the Astrolabe," *Indian Journal of History of Science* 35.2 (2000): 129–147; reprinted in Sarma, *The Archaic and the Exotic*, 179–198.

3 Sreeramula Rajeswara Sarma, "Yantrarāja, The Astrolabe in Sanskrit," *Indian Journal of History of Science* 34 (1999): 145–158; reprinted in Sarma, *The Archaic and the Exotic*, 240–256.

4 David A. King, "An Astrolabe from 14th-century Christian Spain with Inscriptions in Latin, Hebrew and Arabic: A Unique Testimonial to an Intercultural Encounter," *Suhayl* 3 (2002–2003): 9–157.

astronomical instruments produced in India,[5] I came across some specimens of reworked astrolabes and celestial globes. These constitute an interesting chapter in the dissemination of the Islamic instruments in India.

In India the reworking required a close cooperation between three kinds of specialists: a Muslim astronomer to explain the meaning and significance of the original Arabic inscriptions on the astrolabe, a Hindu astronomer, oftentimes the new owner, who would translate this information into Sanskrit, and a metal worker who engraved the Sanskrit inscriptions upon the Arabic astrolabe.

The additions in Sanskrit were not all uniform. Some Hindu astronomers were content to identify just those latitude plates that were useful to them. Others attempted to duplicate each *abjad* numeral and each Arabic legend with Sanskrit equivalents.

Among this small group of astrolabes reworked in Sanskrit, there is one spectacular specimen, which is the subject of this paper (Figure 6.1). We shall refer to it as the "Khalili Astrolabe" after the name of the collection where it is preserved.

The Khalili Astrolabe, Its Maker and Its Date

This astrolabe (Figure 6.1) is now with the Nasser D. Khalili Collection of Islamic Art at London (accession no. SCI 53) and was briefly described by Francis Maddison and Emilie Savage-Smith in the catalogue of this collection.[6] It is a very large astrolabe with a diameter of 526 mm, a height of 602 mm and a thickness of 14 mm. It is in a good state of preservation, except for the slight patina on some of the plates and on the back. Besides the mater with a ring and a shackle at the top and the rete for 24+29 stars, there are six plates. The alidade, pin, and horse are missing.

It is amazing that, in spite of the huge dimensions of the astrolabe, all the lines, arcs and circles are meticulously and flawlessly engraved and the legends and numerals are written in graceful calligraphy. Thus it is a masterpiece of metal craft and contains also a wealth of technical features. Clearly this

5 Sreeramula Rajeswara Sarma, "Indian Astronomical and Time-measuring Instruments: A Catalogue in Preparation," *Indian Journal of History of Science* 29.4 (1994): 507–528; reprinted in Sarma, *The Archaic and the Exotic*, 294–306.

6 Francis Maddison and Emilie Savage-Smith, *Science, Tools & Magic*, 2 parts (The Nasser D. Khalili Collection of Islamic Art, Vol. XII) (London: Khalili Collections, 1997), 1:230–233; Henceforth "the Khalili catalogue." On this see also David A. King, "Cataloguing Medieval Islamic Astronomical Instruments," *Bibliotheca Orientalis* 57 (2000): 247–258, esp. 254–255.

FIGURE 6.1
The Front of the Khalili astrolabe. The Nasser D. Khalili Collection of Islamic Art, Accession No. SCI 53. PHOTO COURTESY OF THE KHALILI FAMILY TRUST © NOUR FOUNDATION.

monumental astrolabe was not meant for the use of a common astronomer, but was intended primarily as a presentation piece to a high dignitary.

The description in the Khalili Catalogue begins like this: "This remarkably fine, large astrolabe illustrates the symbiosis of Hindu and Muslim astrological and astronomical traditions. The inscriptions are both in Arabic and Sanskrit." This implies that the astrolabe was a bilingual astrolabe on which the Arabic and Sanskrit inscriptions were engraved at the same time.[7] But a close look shows that the Arabic inscriptions are carefully and neatly engraved while the same care and neatness are lacking in the Sanskrit inscriptions; clearly these were incised later by one who was not trained to engrave inscriptions in the Devanāgarī script that is used for Sanskrit. During this process of reworking, the original inscription containing the name of the astrolabe maker, the date of manufacture and other information appears to have been effaced. Fortunately, however, there is enough internal evidence in this astrolabe that allows us to identify the maker of this astrolabe, the time of its manufacture, and also the patron for whom this monumental astrolabe was fabricated.

7 For such a bilingual astrolabe, see Sreeramula Rajeswara Sarma, "A Bilingual Astrolabe from the Court of Jahāngīr," *Indian Historical Review* 38.1 (June 2011): 77–117.

In his essay-review of the Khalili catalogue, David King makes the following remarks about this astrolabe:

> Certainly the piece is not obviously related to the *known* productions of the Lahore school. In fact the piece represents a tradition about which we know virtually nothing ... I strongly suspect that we have here the sole surviving work of Maqṣūd Hirawī, known to have been one of Humāyūn's instrument makers in Lahore, active *ca.* 1550. According to a contemporaneous source, he "manufactured astrolabes and globes in such a manner that the observers of his works were wonderstruck."[8]

King made these remarks on the basis of the Khalili catalogue, which reproduces just the image of the front with the rete on p. 233. What led King to the conclusion that "the piece is not obviously related to the *known* productions of the Lahore school" is this statement in the catalogue: "The mater is engraved as a gazetteer showing the longitude, latitude and *inḥirāf* of a great number of places, which are arranged by region, with India at the centre." *Inḥirāf* is an arc of the horizon, intercepted between the meridian of a particular city and a vertical circle passing through the zenith of Mecca, measured in degrees and minutes. Lahore astrolabes generally do not have *inḥirāf* values in the gazetteers. The very first extant astrolabe made by Allāhdād in 1567 has a column for *inḥirāf*, but it was not filled in. Therefore, King concluded that an astrolabe carrying *inḥirāf* values should be different from the Lahore astrolabes and attributed it to Mawlānā Maqṣūd Hirawī. The fact is that the astrolabe gazetteer does indeed give three values after each place name: these are not longitude, latitude and *inḥirāf* as the Khalili catalogue erroneously states, but longitude, latitude and the maximum duration of the daylight hours (*sāʿāt*), as can be seen from the images of the gazetteer in Figures 6.3 and 6.15 below.

I studied this astrolabe carefully in the summer of 2015. I have also examined, in connection with my project of cataloguing the medieval astronomical instruments produced in India, more than fifty specimens of astrolabes produced by various members of the Lahore family. On the basis of the internal

8 David A. King, "*Cataloguing*," 254–255. This argument is repeated, almost verbatim, in two of his subsequent publications: *World-Maps for Finding the Direction and Distance to Mecca: Innovation and Tradition in Islamic Science* (Leiden: Brill, 1999), xxix, and "A Monumental Astrolabe from 16th-Century India," in *In Synchrony with the Heavens: Studies in Astronomical Time Keeping and Instrumentation in Medieval Islamic Civilization*, 2 vols. (Leiden: Brill, 2004–2005), 2:805–806.

evidence, I shall argue that this astrolabe was indeed a product of the well-known Lahore family of astrolabe makers; in particular, that it was fashioned by the prolific and versatile Ḍiyā' al-Dīn Muḥammad (fl. 1645–1680). The geographical gazetteer engraved on the mater lists Delhi as Shājahānābād; therefore, the astrolabe could have been produced only after 1648 when the Mughal Emperor Shāh Jahān shifted his capital to Delhi which he renamed, after his own name, as Shājahānābād.[9] Finally, it will be shown that the astrolabe was created for no less an exalted personage than Shāh Jahān himself. But first it is necessary to give an introduction to the Lahore family of instrument makers and their products, in order to establish later that the Khalili astrolabe is a product of this Lahore family.

The Lahore Family of Instrument Makers

During the second half of the sixteenth century and in the seventeenth century, the master craftsman (*ustād/Ustādh*) Allāhdād of Lahore and his descendants of four generations dominated the production of astronomical instruments in Mughal India. Neither the names of any members of this Lahore family, nor the instruments produced by them are mentioned in the contemporary documents. The history of this family and their work are reconstructed entirely from the signatures of their dated instruments.

There exist two astrolabes bearing the name of the patriarch Allāhdād (fl. 1567). One of these also carries the year of manufacture as 975 Hijrī (CE 1657/58). This is in fact the earliest extant astrolabe produced in India. Allāhdād's son ʿĪsā (fl. 1600–1604) is known through three astrolabes. ʿĪsā had two sons, Qā'im

9 This fact easily disproves King's attribution of the astrolabe to Maqṣūd Hirawī. If he made astronomical instruments for Humāyūn ca. 1550, he could not have made an astrolabe on which Delhi carries the name of Humāyūn's great-grand-son Shāh Jahān. Incidentally, the only reference in literature to Maqṣūd Hirawī occurs in the *'Āīn-i Akbarī*, a chronicle of Humāyūn's son and successor Akbar by Abū al-Faḍl. Here he is mentioned as one of the seal-engravers at the court of Akbar: "*Maulānā Maqṣūd of Herāt*, one of the servants of Humāyūn, who writes well the *riqā'* and the *nasta'līq* characters. The astrolabe[s], globes, and various *misṭars* that he made, were much admired by people of experience. The patronage of his Majesty perfected his art." Cf. Abū al-Faḍl, *Ā'īn-i Akbarī*, trans. H. Blochmann (vol. 1) and H. S. Jarrett (vols. 2–3), 3 vols. (Calcutta: Asiatic Society of Bengal, 1873–1896; reprint: Institut für Geschichte der Arabisch-Islamischen Wissenschaften, Frankfurt, 1993), 1:52. In the Khalili catalogue (219), this statement is transformed into "He manufactured astrolabes, globes, in such a manner that the observers of his works were wonderstruck."

Muḥammad (fl. 1622–1637) and Muḥammad Muqīm (fl. 1621–1659). Six astrolabes and five celestial globes by Qā'im Muḥammad survive. Qā'im's younger brother Muqīm and Qā'im's son Ḍiyā' al-Dīn Muḥammad (fl. 1645–1680) were very prolific instrument makers. While at least thirty-four astrolabes and one celestial globe crafted by Muḥammad Muqīm are extant in different collections throughout the world, his nephew Ḍiyā' al-Dīn's signature adorns about the same number of extant astrolabes and, in addition, eighteen celestial globes. In comparison, Muqīm's two sons, Ḥāmid (fl. 1655–1691) and Jamal al-Dīn (fl. 1666–1692) have a more modest output. Eleven astrolabes and four celestial globes by Ḥāmid are extant, while Jamāl al-Dīn is known through just four surviving astrolabes.[10] Besides these, there are several unsigned specimens that can be attributed to this family for stylistic reasons. It is also necessary to add that the instruments made by them are not mass products cast in the same mold. In fact, each instrument is a unique piece as regards the size, decorations, ornaments and configuration of the various technical elements. In the medieval period there has been no other family anywhere, comparable to this one in the long continuous family tradition of instrument making, in the large number of instruments produced, in the artistic and technical excellence of production, or in the innovation in design.

The standard northern astrolabes produced by the members of this family distinguish themselves by the high degree of precision in the engraved stereographic projections and by the pleasing floral traceries used to connect the star pointers on the rete, with a matching design on the *kursī*.

The members of this family also show a marked predilection for unusual projections invented in Andalusia in the tenth and eleventh centuries, which they incorporated in some of their standard astrolabes. Anthony Turner aptly remarks that, "the multiplication of complexity and a delight in the unusual seem to be typical of the astrolabe-makers of Lahore."[11] The following are some of their unusual additions.

Allāhdād's undated astrolabe (Oxford, Museum of the History of Science, accession no. 47376) and an unsigned and undated astrolabe that can be attributed to him (Oxford, Museum of the History of Science, accession no. 34611)

---

10 Sreeramula Rajeswara Sarma, "*The Lahore Family of Astrolabists*," 199–222. Since this first comprehensive survey was made in 1994, some more astrolabes and celestial globes by this family have come to light, which are included in the figures mentioned above. Even so, these numbers are still tentative.

11 Anthony John Turner, *Astrolabes, Astrolabe-Related Instruments* (The Time Museum, Catalogue of Collection, Vol. I, Part 1) (Rockford, IL: The Museum, 1985), 83.

contain a plate carrying the universal projection invented by Ibn Bāṣo in the eleventh century.[12]

ʿAlī b. Khalaf al-Ṣaidalānī's universal plate (eleventh century)[13] is imitated in a Lahore astrolabe. The rete came up for auction at Christie's on 4 October 1995,[14] and the mater five and a half years later on 5 April 2001. It is stated in Christie's catalogue of the latter date that "the odds are overwhelming that these two parts once belonged together."[15] But it is yet to be determined precisely which member of the Lahore family may have produced it.

Abū Saʿīd Sijzī's *zawraqī* horizons (late tenth or early eleventh century) are incorporated into the standard retes of some of the Lahore astrolabes, including the present Khalili astrolabe. This will be discussed below in the section on "The Rete."

In 1680 Ḍiyāʾ al-Dīn Muḥammad produced a spectacular specimen measuring 555 mm in diameter of the universal astrolabe *ṣafīḥa zarqāliyya* that was originally invented by al-Zarqālluh's in the eleventh century. This astrolabe is now preserved in Jai Singh's Observatory at Jaipur.[16]

While the standard astrolabes are northern astrolabes (*asṭurlāb shamālī*), there was a certain fascination in India for north-south astrolabes (*asṭurlāb shamālī wa janūbī*). Sulṭān Fīrūz Shāh Tughluq is reported to have commissioned three north-south astrolabes, one in brass, another in silver and the third in brass but with large dimensions which was designated as *Asṭurlāb-i Fīrūz Shāhī*.[17] The first Sanskrit manual on the astrolabe composed at his court by Mahendra Sūri, to which mention has been made earlier, also devotes the third chapter to the construction of northern astrolabes (*saumya-yantra*), southern astrolabes (*yāmya-yantra*) and the composite north-south astrolabes (*saumya-yāmya-yantra*). The first extant north-south astrolabe is produced by Ḍiyāʾ al-Dīn in 1674, which is now with Jai Singh's Observatory at Jaipur. The Victoria and Albert Museum in London holds an unsigned and undated north-south astrolabe, which can be attributed to Ḍiyāʾ al-Dīn, because of the

12 See Emilia Calvo, "Ibn Bāṣo's Astrolabe in the Maghrib and East," in *From Baghdad to Barcelona: Studies in the Islamic Exact Sciences in Honour of Prof. Juan Vernet*, ed. Joseph Casulleras and Julio Samsó (Barcelona: Instituto "Millás Vallicrosa" de Historia de la Ciencia árabe, 1996), 2.755–767.

13 See Emilia Calvo and Roser Puig, "The Universal Plate Revisted," *Suhayl* 6 (2006): 113–157.

14 Christie's Catalogue dated 4 Oct 1995, lot 61, pp. 20–21.

15 Christie's Catalogue dated 5 April 2001, lot 32, p. 3.

16 Sreeramula Rajeswara Sarma, "The *Ṣafīḥa Zarqāliyya* in India," in *From Baghdad to Barcelona*, 2:719–735; reprinted in: Sarma, *The Archaic and the Exotic*, 223–239.

17 Sarma, "Sulṭān, Sūri and the Astrolabe," reprinted in Sarma, *The Archaic and Exotic*, 181–198 (183–189).

close resemblance between this one and the one listed above. A third north-south astrolabe was formerly in the Time Museum, Rockford, IL, USA (Inventory 308).[18]

In 1686 Muqīm's son Ḥāmid produced a large astrolabe together with a massive *qibla* indicator (London, Khalili Collection, accession no. SC15).[19] Ḥāmid's brother Jamāl al-Dīn also produced an astrolabe with a *qibla* indicator (London, Science Museum, accession no. 1985–2077).[20]

With the exception of the north-south astrolabes and *qibla* indicators, there had been no tradition of making the other four types of instruments in the immediately preceding centuries either in India or in the adjacent countries. Therefore the Lahore family members had no actual models to follow and their knowledge must have been derived entirely from Arabic treatises on these instruments.[21]

Royal Patronage

Obviously the Lahore family had a very discerning clientele among the Mughal nobility to appreciate these innovations. For these dignitaries the Lahore astrolabists fashioned highly ornate astrolabes of large dimensions, some inlaid with silver, others with gilded plates,[22] yet others with unusual projections such as those listed above.

In their signatures on the astrolabes and celestial globes made by them, the members of the Lahore family refer to themselves proudly as the descendants of Allāhdād Lāhūrī Asṭurlābī Humāyūnī, "Allāhdād of Lahore, Astrolabe Maker to Humāyūn," the second Mughal Emperor.

The only dated astrolabe by Allāhdād, the patriarch of the family, was produced in 1567, i.e., eleven years after Humāyūn's death, but that does not

18 The Time Museum does not exist anymore. It is not known where this north-south astrolabe is now. For a description, see A. J. Turner, *Astrolabes, Astrolabe-Related Instruments*, 74–83, where the Jaipur astrolabe is also described and illustrated.

19 Francis Maddison & Emilie Savage-Smith, *Science, Tools & Magic*, 1.131.

20 See David A. King, "Weltkarten zur Ermittlung der Richtung nach Mekka," in *Focus Behaim Globus*, ed. Gerhard Brot, 2 vols. (Nürnberg: Germanisches Nationalmuseum, 1992), 1.167–171; David A. King, *World-Maps*, 95; Christie's, New York, Catalogue 31.10.1985, lot 331, pp. 94–95.

21 It should be interesting to search for the manuscript resources that may have been used by the Lahore family in the manuscript collections in Pakistan and India.

22 In the astrolabes by Muqīm and Ḍiyā' al-Dīn which are now preserved in the Salar Jung Museum at Hyderabad and in Jai Sigh's Observatory at Jaipur, the latitude plates look as if they were gilded or rubbed with gold dust.

preclude the possibility of his producing astrolabes in Humāyūn's lifetime for the use of the emperor, who is said to have been interested in astrolabes and celestial globes.[23] The two extant astrolabes by Allāhdād (the astrolabe of 1567 at the Salar Jung Museum, Hyderabad, and the undated astrolabe at the Museum of the History of Science, Oxford) are fairly large instruments with diameters measuring 199 mm and 255 respectively in which the prime vertical, the oblique horizon, and the curves for equal hours as counted from the western and eastern horizons are inlaid with silver on all the latitude plates. Surely these astrolabes were not meant for common astrologers/astronomers but for men of high rank.

Qā'im Muḥammad, the grandson of Allāhdād and the father of Ḍiyā' al-Dīn, is said to have begun the casting of celestial globes in one single piece by *cire perdue* method.[24] He made one such celestial globe in the eighteenth regnal year of Jahāngīr (1622) for Nawāb I'tiqād Khān, brother of Nūr Jahān Begum, the consort of Jahāngīr.[25] The globe is in the Stonyhurst College Library, Clitheroe, Lancashire, UK.[26] He also fashioned a fabulous astrolabe in 1627 for Nawāb Khwajā Abū al-Ḥasan, who held many important positions under Jahāngīr, such as the governor (*sūbahdār*) of Deccan and the general paymaster of the household (*bakhshī-kul*). Only the rete survives in the Oriental Public Library at Patna; in this unique rete the star pointers are joined not by the usual floral tracery but by a calligraphic design that mentions the year of production in Jahāngīr's regnal years and in Hijrī years and the name of the dedicatee.

Ḍiyā' al-Dīn also fabricated unusual instruments for high nobility. As has been mentioned already, he made a huge Zarqālī universal astrolabe for Nawāb Iftikār Khān, the Fauzdār of Jaunpur.[27] A year earlier, in 1679, he made an

23 Emilie Savage-Smith, *Islamicate Celestial Globes: Their History, Construction, and Use* (Washington, DC: Smithsonian Institution Press, 1985), 37, argues "that the name Humāyūnī was used to indicate contemporaneity of Allāhdād with Humāyūn and that even if the founder of the workshop [at Lahore] had enjoyed royal favour, his descendants probably did not." In fact, the descendants also did enjoy royal patronage. Savage-Smith herself describes in the same book a special kind of celestial globe by Ḍiyā' al-Dīn Muḥammad for Aurangzeb in 1679 (pp. 42–43, 232, Figure 6.17) and states that on this globe Ḍiyā' al-Dīn refers to himself as the descendant of Allāhdād Humāyūnī. In this globe made for the Emperor, Ḍiyā' al-Din would not be using the attribute Humāyūnī with the name of his ancestor merely to express Allāhdād's contemporaneity with Humāyūn, but to stress that Allāhdād was the royal astrolabe maker to Humāyūn.

24 On the *cire perdue* method, see Savage-Smith, *Islamicate Celestial Globes*, 91–95.

25 See Thomas William Beale, *An Oriental Biographical Dictionary* (Calcutta: J.W. Thomas, 1881; reprint: Delhi, 1971), 124.

26 See Savage-Smith, *Islamicate Celestial Globes*, Catalogue no. 11, p. 224, Figure 6.12 on p. 37.

27 See Beale, *An Oriental Biographical Dictionary*, 117.

unusual celestial globe for Shāh Jahān's successor Aurangzeb; here the outlines of the constellations figures are cut out in such way that, when the globe is lit from inside, they would appear as silhouettes. Star positions are marked with tiny holes so that they would sparkle when the globe is lit from inside.[28]

Graphs of the Solar Meridian Altitude on the Khalili Astrolabe

After this introduction to the members of the Lahore family and their clientele, we shall now look closely at the vital clues in the Khalili astrolabe that would help in identifying the maker, the date of its manufacture, and the dedicatee for whom the astrolabe was produced.

The astrolabe shows several characteristic features of the astrolabes produced by the members of the Lahore family of astrolabe makers. One of these is the sigmoid graph showing the solar meridian altitude throughout the year (*khaṭṭ niṣf al-nahār*).[29] On the Khalili astrolabe, in the upper right quadrant on the back, three graphs for the solar meridian altitudes were drawn for the latitudes of 27°, 29°, and 32°, which are roughly the latitudes of the three Mughal Imperial cities, Agra, Delhi, and Lahore in this order (Figure 6.2).

Allāhdād's son Mullā ʿĪsā was the first one to introduce this graph in his astrolabes, where he drew a graph for the latitude 32°, roughly the latitude of Lahore.[30] Later his son Muḥammad Muqīm started engraving two graphs for latitudes 27 and 32, which represent the two imperial cities Agra and Lahore. After the Mughal Emperor Shāh Jahān shifted his capital from Agra to Delhi and renamed Delhi as Shāhjahānābād in 1648, a third graph for Delhi at 29° began to be added in these astrolabes by Muqīm's nephew Ḍiyāʾ al-Dīn Muḥammad. Thus the Khalili astrolabe bearing three graphs for the latitudes of 27°, 29°, and 32° was most certainly made by Ḍiyāʾ al-Dīn after 1648.

That it must have been produced after the Mughal capital was shifted in 1648 to Delhi is borne out by other factors as well. An unusual feature of the

28 See Savage-Smith, *Islamicate Celestial Globes*, Catalogue no. 30, pp. 232–33, Figure 6.17 on p. 42.

29 See G. R. Kaye, *The Astronomical Observatories of Jai Singh* (Calcutta: Superintendent, Government Printing, India, 1918; reprint, New Delhi: Archaeological Survey of India, 1982), 22 and Figure 6.37.

30 Three astrolabes by him are extant. An undated astrolabe (in a private collection) does not carry this curve; but in his two other astrolabes, dated 1601 and 1604 respectively, the curve appears in the upper right quadrant on the back. These two are now in the Adler Planetarium & Astronomy Museum, Chicago. For their description, see David Pingree, *Eastern Astrolabes* (Chicago, IL: Adler Planetarium & Astronomy Museum, 2009), 52–63 (53, 59 for the images of the back).

FIGURE 6.2 *The graphs of the solar meridian altitude on the back. The Nasser D. Khalili Collection of Islamic Art, Accession No. SCI 53.*
PHOTO COURTESY OF THE KHALILI FAMILY TRUST
© NOUR FOUNDATION.

astrolabe is that in the rete are incorporated two *zawraqī* horizons for latitudes 29° and 32°. Finally the clinching evidence of its manufacture after 1648 is provided by the geographical gazetteer engraved on the inner surface of the mater where Delhi appears in the changed name as Shāhjahānābād with the attribute *Dār al-Mulk Hind* (the Capital City of India) (Figure 6.3).

The Image on the Back of the Kursī

Another valuable clue is to be found on the back of the suspension bracket (*kursī*) (Figure 6.4). Here instead of a floral pattern as in the front, there is an image in low relief of a finely drawn falcon swooping down and tearing into a bird, while smaller birds, one each on either side, are scuttling away from the fearsome falcon. This is an unusual image; such an image does not occur in any of the one hundred and odd Lahore astrolabes. When I asked Professor Muzaffar Alam for his opinion of the possible significance of this image, he suggested that this image alludes to a particular incident in the life of the Mughal Emperor Shāh Jahān and directed me to Dr. Roy Fischel and to his doctoral dissertation.[31]

31 Roy Fischel, "Society, Space, and the State in the Deccan Sulṭānates, 1565–1636," (PhD dissertation, University of Chicago, Chicago, IL, 2012). I am very grateful to Dr. Fischel for sending me the relevant pages from his dissertation.

FIGURE 6.3 *The name Shāhjahānābād in the Geographical Gazetteer. Here the upper row reads, from right to left, "Akbarābād Dār al-Khilāfat (i.e., Agra), [longitude] 114;0°, [latitude] 27;0°, [the duration of the longest day] 13h 45m." The lower row reads "S[h]āhjahānābād Dār al-Mulk Hind (i.e., Delhi), [longitude] 113;35°, [latitude] 28;39°, [duration of the longest day] 13h 52m." The Nasser D. Khalili Collection of Islamic Art, Accession No.* SCI *53.*
PHOTO COURTESY OF THE KHALILI FAMILY TRUST

FIGURE 6.4 *Falcon attacking Hoopoes on the Back of the Kursī. The Nasser D. Khalili Collection of Islamic Art, Accession No.* SCI *53.*
PHOTO COURTESY OF THE KHALILI FAMILY TRUST

This dissertation discusses and analyses a letter sent by Shāh Jahān around 1634 to the Sulṭān of Bijapur, Muḥammad ʿĀdil Shāh, ordering him to see that *khuṭba* be read in Shāh Jahān's name in the Bijapur territory; otherwise he would pounce on the Sulṭān and tear him to pieces just as a falcon tears into a hoopoe.[32] Hoopoe (*Upupa epops*) is a medium sized bird with extravagant black and white plumage and a distinctive feathery crest on its head. As Fischel explains, the ornithological allusion implies that the Sulṭān of Bijapur and other rulers of the Deccan call themselves kings and wear crowns like hoopoes, but do not have the strength to be independent rulers; Shāh Jahān, on the other hand, has the might to impose his sovereignty over them; if they refuse to accept Shāh Jahān's overlordship, he can subjugate them just as a falcon overpowers hoopoes. The mention of the falcon is also significant because the Mughal rulers were avid falconers and there are miniature paintings portraying them with falcons perching on the gloved hands.

Thus, the image on the *kursī*, of a mighty falcon swooping down on hoopoes is a clear allusion to Shāh Jahān's attempts to subjugate the rulers of the small kingdoms of the Deccan.[33] Indeed, the bird on the right is shown with an erect crest, identifying it as a hoopoe. This would mean that this astrolabe, with its monumental dimensions, lavishly decorated rete, and with a wealth of technical features, was made for none other than the Emperor Shāh Jahān himself.

As has been stated earlier, members of the Lahore family made several astronomical instruments for Mughal nobility and Ḍiyāʾ al-Dīn Muḥammad, the presumptive maker of this astrolabe was to make an unusual celestial globe for Aurangzeb, Shāh Jahān's successor on the Mughal throne. Therefore, there cannot be any doubt that the present astrolabe was made for Shāh Jahān.

---

32 Roy Fischel, "Society, Space, and the State," 253–254:

... [God] made me (lit. us) the emperor of the world (*bādshāh-i jahān*). Therefore, it became my obligation to impose the sermon (*khuṭba*), coinage (*sikka*), and measurement (*dirʿa*) of Shāh Jahān on all parts of the world, especially on the country (*mulk*) of Bijapur, Golkonda, and Hyderabad (*bhāgnagar*) ... You, who live in those districts, are like hoopoes (*hudhud*); each one of you cries "I am the king!" It is most advisable and best [for you] to tie the rope of obedience around your necks and perform the sermon, coinage, and measurement of Shāh Jahān in those cities. Otherwise, I will draw a falcon (*bāz*) [that] will tear the meat off [your] skin [with] its beak ...

33 In his reply, the Sulṭān of Bījapūr agrees to pay tribute as before, but rejects Shāh Jahān's claims to the sovereignty on the Deccan. In a clever repartee, he adds that the hoopoe is not at all a low creature, because it was chosen by King Solomon as his messenger to the Queen of Sheba; See Fischel, "Society, Space, and the State," 255–258.

Ḍiyāʾ al-Dīn may have mentioned this fact in his signature, which he must have engraved on the back of the astrolabe just below the horizontal diameter, as is the general practice in this Lahore family. But the signature was thoughtlessly effaced in the process of reworking.

Shāh Jahān ruled from 1628 to 1658 and he shifted his capital to Delhi in 1648. Therefore this astrolabe which was made for Shāh Jahān and which refers to Delhi as Shāhjahānābād must have been made during the decade between 1648 and 1658.

Reworking in Sanskrit

A Sanskrit inscription engraved on the back shows that a certain Hindu of Delhi by name Nayanānanda acquired this astrolabe in 1701, i.e., in less than half a century of its manufacture, and got it thoroughly reworked with additional labels and numbers in Sanskrit language and in Devanāgarī script.[34] Nayanānanda tried to add Sanskrit equivalents to all legends and inscriptions in Arabic and to duplicate all numbers in *abjad* notation with Devanāgarī numerals.

We do not know whether Nayanānanda translated the Arabic engravings by himself or whether he took the help of a Muslim scholar.[35] But it appears that he was familiar with the functions of the astrolabe and with the Sanskrit terminology related to the astrolabe. However, his zeal for the astrolabe could not be matched by the quality of Sanskrit engravings he caused to be made. These are carelessly incised and are, in parts, illegible. Thre numerals are somewhat neat, but this cannot be said of the letters.

After thus firmly establishing that the Khalili astrolabe was produced by Ḍiyāʾ al-Dīn Muḥammad of the Lahore family, that it was made for the Mughal Emperor Shāh Jahān, and that the time of manufacture lay between 1648 and 1658, we now proceed to describe the individual components of this astrolabe. While doing so, we shall draw attention to the close affinities that exist between some of these components and the corresponding parts in the astrolabes bearing Ḍiyāʾ al-Dīn's signature.

34 The Sanskrit used in the engravings is not always grammatically perfect; it often avoids the case-endings. Sometimes Hindi is used instead of Sanskrit.

35 From the seventeenth century onwards, Hindus in north India began cultivating Persian and even authored books in that language. This is continued until the middle of the twentieth century.

The Suspension Apparatus

The suspension apparatus of the astrolabe consists of a heavy ring and a heavy shackle, both with an octagonal cross section. The terminals of the shackle are simple and circular. Through these two terminals and the tip of the *kursī* passes a pin, attaching the shackle thus to the *kursī* and the main body of the astrolabe. The *kursī* is rather small for the enormous size of the astrolabe. It is not worked *à jour*, as is the general practice in the Lahore astrolabes, but filled with designs in the front and on the back. A heavy rim runs along the lobed profiles in the front and on the back. In the front there is a simple floral design.

The Limb

The *kursī*, the heavy plate at the back, and the limb are cast as a single piece. The limb carries three concentric degree scales: the inner scale is divided in 1/5°; the middle scale is divided in single degrees of arc and numbered in standard Arabic numerals from 1 to 90, separately for each quadrant; in the outer scale are marked groups of 3° and numbered in the *abjad* notation clockwise from 3 to 360°.[36] Devanāgarī numbers are added in the outer scale, wherever space was available, sometimes to the right, sometimes to the left, sometimes above the *abjad* numbers, in groups of 3°, but separately for each quadrant from 3 to 90. All the scales commence from the uppermost point and proceed clockwise.

The Rete

The rete (Figure 6.5), measuring 151 mm in diameter, is the most ornate part of the astrolabe. It is thicker in the middle and tapers towards the rim; the thickness at the centre is about 3 mm and that at the rim is 1.5 mm. It consists of a

36 In Ḍiyā' al-Dīn's Zarqālī astrolabe of 1680, the limb carries three degrees scales: the inner scale is marked for 1/2° and then 1/6°; the middle scale is calibrated for single degrees which are numbered in standard Arabic numerals, separately for each quadrant from one to ninety; the outermost scale is marked for groups of 3°, each of which is labelled in *abjad* notation; See Sarma, "The *Ṣafīḥa Zarqāliyya* in India," 223–239. In his astrolabe of 1651 at Hyderabad, Salar Jung Museum (accession no. 112/2/xxxv) the limb bears three bands of scales, which are divided respectively in 1/3°, 1°, 3° and labelled in 3s from 3° to 360°; See Sreeramula Rajaeswara Sarma, *Astronomical Instruments in the Salar Jung Museum* (Hyderabad: Salar Jung Museum, 1996), 17–20.

FIGURE 6.5
The rete of the astrolabe. The Nasser D. Khalili Collection of Islamic Art, Accession No. SCI *53.* PHOTO BY THE AUTHOR.

Capricorn ring and a much wider ecliptic ring. These two are held together by the equinoctial bar which has counter-changes at its two intersections with the ecliptic ring. The vertical solstitial bar is shown partially. The space between the two rings and inside the ecliptic ring is filled with floral tracery of a creeper with abundant leaves and flowers. The creeper rises from the Capricorn ring at its north point, branches off to the right and left and fills the whole space between the two rings; another branch enters the ecliptic ring, fills the space inside and culminates in a large empty circle with small protuberances issuing from its outer circumference. On the stem of the creeper, a little above the north point, there is a large knob with which to rotate the rete around the centre.

The Ecliptic Ring on the Rete

The ecliptic ring (Figure 6.6) carries four circular scales: (i) the innermost scale is divided into twenty-eight lunar mansions and their names are written in outline against a hatched background; (ii) the next scale is divided into the twelve zodiac signs and their names are embossed in low relief; (iii) in the third scale, groups of 3° are labelled in *abjad* notation, separately for each sign as 3, 6, 9 … 30; and (iv) the outermost scale on the rim of the ecliptic circle is divided into single degrees. The boundary lines between the lunar mansions, those between the zodiac signs, the divisions of 3°, and divisions of 1° are marked by inlaid silver wire.

FIGURE 6.6
The ecliptic ring on the rete (detail showing the zodiac signs Capricorn, Aquarius and Pisces). The Nasser D. Khalili Collection of Islamic Art, Accession No. SCI 53.
PHOTO COURTESY OF THE KHALILI FAMILY TRUST

The names of the zodiac signs engraved on the ecliptic ring are *al-Ḥamal, al-Thawr, al-Jawzā', al-Saraṭān, al-Asad, al-Sunbula, al-Mīzān, al-ʿAqrab, al-Qaws, al-Jadī, al-Dalw, al-Ḥūt*. The names of the twenty-eight lunar mansions will be listed below in Table 6.8.

Next to the Arabic names of the Zodiac signs, Sanskrit names are incised, but unusually with the syllables reversed so that these can be read from right to the left along with the Arabic names: *ṣa-me* instead of *meṣa* (Aries), *ṣa-vṛ* instead of *vṛṣa* (Taurus); *na-thu-mī* (sic!) for *mithuna* (Gemini); *ka-ka* (sic!) for *karka* (Cancer); *ha-sīṃ* (sic!) for *siṃha* (Leo); *nyā-ka* for *kanyā* (Virgo); *la-tu* (sic!) for *tulā* (Libra); *ka-ści-vṛ* for *vṛścika* (Scorpio); *na-dha* (sic!) for *dhanuḥ* (Sagittarius); *ra-ka-ma* for *makara* (Capricorn); *ma-ku* (sic!) for *kumbha* (Aquarius); and *na-mī* for *mina* (Pisces).

The Sanskrit names of the signs are incised once more on the leaves adjacent to the zodiac signs, but this time in the correct sequence from left to right. Owing to the constraints of space, some of the names are written partially: thus *ṣa* for *meṣa, vṛ* for *vṛṣa, rka* for *karka*, and *ka* for *kanyā*. Other names are written in full.

Sanskrit names of the lunar mansions are not added, there being little space to do so.

Stars on the Rete

The pointed tips of the leaves and other projections on them constitute fifty-three star pointers; twenty-nine of these are inside the ecliptic, i.e., to the north of the ecliptic circle and twenty-four outside the ecliptic, i.e. to the south of the ecliptic between the ecliptic and the Tropic of Capricorn. The Arabic star names are inlaid with silver on the leaves of the tracery. In transcribing Arabic star names, the astrolabe makers of Lahore do not always follow Arabic grammar strictly. The definite article *al-* is often omitted either due to the very limited space available on the star pointers or even otherwise. They generally write *rās* instead of *ra's*. Third, the adjectives are not modified according to the gender of the noun. They often use masculine forms of adjectives with feminine names, e.g., *kiffa janūbī, kiffa shamālī, natha sāḥābī.* In the star table given below, the star names will be reproduced exactly as they are engraved on the astrolabe.

Next to the Arabic names are engraved the Sanskrit names of the stars. Some of these are traditional and some are literal translations of the Arabic. The Sanskrit translator, of course, knew the Sanskrit names of certain stars already, but in some cases he appears to have rendered the Arabic names literally. Moreover, sometimes he employs *uttara* (north) for "left" instead of the correct term *vāma.*[37] Six names are illegible; in three cases, there are no Sanskrit names.

The stars are listed in Table 6.1 in the order of increasing right ascension. Each entry contains (i) the serial number, (ii) the Arabic name, (iii) its meaning,[38] (iv) the Sanskrit name as engraved, (v) the correct Sanskrit form, (vi) its meaning, and (vii) the identification of the star.

---

37 The Sanskrit term for "south " is *dakṣiṇa*; it also means "right." But its companion term *uttara* for "north" does not denote "left"; the correct term for "left" is *vāma*.

38 The meanings of the Arabic star names are based on Emilie Savage-Smith, "Celestial Mapping," in *The History of Cartography, Volume Two, Book One: Cartography in the Traditional Islamic and South Asian Societies*, ed. J. B. Herley and David Woodward (Chicago, IL: University of Chicago Press, 1992), 12–70 (23).

The Zawraqī Horizons on the Rete

In this highly ornate, yet standard rete, Ḍiyā' al-Dīn incorporates two movable oblique horizons, obviously to display his virtuosity. The two arcs that intersect at the two points where they cross the equinoctial bar look like the outlines of two boats, with their upright masts represented by the corresponding segments of the vertical meridian. Therefore this projection is called *zawraq* (ship).

The *zawraqī* astrolabe was invented by al-Bīrūnī's contemporary Abu Saʿīd Sijzī.[39] The main source of our knowledge of this variety of astrolabe is Josef Frank's *Zur Geschichte des Astrolabs*,[40] which deals with several non-standard astrolabes described by al-Bīrūnī in his *Kitāb fī istiʿāb al-wujūh fī ṣan'at al-asṭurlāb*. Frank describes the astrolabe as follows:

> Of very special interest is the boat-shaped (*al-zauraqī*) astrolabe of Abū Saʿīd al-Sijzī, which, in its construction, abandons the geocentric model that was generally accepted at his time, in so far as the horizon he designed is movable. The astrolabe consists of a fixed and a movable part. The former is the plate of the northern astrolabe without the azimuth lines. On the plate are drawn, furthermore, the projections of the zodiac and the fixed stars, as they are shown in the rete of the northern astrolabe ... The movable part is an arc cut out from a metal plate which can be rotated ... at the point through which the axis of the astrolabe passes. The convex edge of the arc is the projection of the horizon of the locality for which the plate is calibrated ... Firmly attached to the arc is a ruler with a pointed end—attached in such a way that when the arc coincides with the horizon line drawn on the plate, the ruler rests on the meridian line.[41]

39 Wilfred de Graaf informs me (personal communication dated 4 June 2014): "A treatise by Sijzi on the Zawraqi astrolabe is found in a unique manuscript in Istanbul, Topkapi Palace Library, Ahmet III, 3342. There are no figures in this treatise. As you mentioned, the Zawraqi astrolabe is described in great detail by Biruni. The text by Biruni has recently been edited by Sayyid Muhammad Akbar Jawad Husayni and published in Mashhad in 1422 H (lunar). In this publication however, partly incorrect figures appear. Slightly better figures can be found in an original manuscript, which is now in Leiden (ms. Leiden, Or. 591)."

40 Josef Frank, *Zur Geschichte des Astrolabs* (Erlangen: Universitätsdruckerei von Junge & Sohn, 1920), 18–21.

41 Our translation with slight modifications from Frank's German.

TABLE 6.1 *Star Names on the Rete*

S. No	Star name in Arabic	Meaning	Star name in Sanskrit as engraved
1	*dhanab al-qayṭus [janūbī]*	The [southern] tail of Cetus	*samudra-pakṣī*
2	*baṭn al-ḥūt*	The belly of the fish	*matsyamuṣa*
3	*fam al-qayṭus*	The mouth of Cetus	*Samudrapacchī muṣa*
4	*ṣadr al-qayṭus*	The breast of Cetus	nil
5	*kaff al-jadhmā'*	The cut-off hand	*samudra pacchū taṭala* (?)
6	*rās al-ghūl*	The head of the ghoul	*pretaśirā*
7	*mirfaq al-thurayyā*	The elbow of Thurayya	*ka ta gā*
8	*tālī masāfat al-nahr*	the length of the river	*nadyaṃtaka*
9	*'ayn al-thawr*	The eye of the bull	*rohiṇī*
10	*'ayyūq*	untranslatable	nil[a]
11	*rizl jawzā' yusrā*	The left foot of al-jawzā'	*mithuna pāda uttara*
12	*yad al-jawzā' yusrā*	The left hand of al-jawzā'	*mithuna hasta uttara*
13	*yad al-jawzā' yumnā*	The right hand of al-jawzā'	*mithuna pā … dachina*
14	*rijl jawzā' [yumnā]*	The [right] foot of al-jawzā'	*mithuna pāda dakṣiṇa*
15	*shi'rā yamāniya*	The southern sh'rā	*lubdhaka vaṃsa*, "the hunter's family"
16	*shi'rā shāmīya*	The northern sh'rā	*sārdhā lubdhaka* "one and a half hunter" !
17	*ṭarfat safīna*	The extremity of the ship	*naukāmuṣa*
18	*nathra saḥābī*	The 8th lunar mansion	*karka mi* 3 (?)
19	*fard shujā'*	The isolated one of the serpent	*mahāpuruṣa mu* (?)
20	*rās al-asad*	The head of the lion	*siṃha śī[r]ṣa*
21	*qalb al-asad*	The heart of the lion	*maghā*
22	*ẓahr dubb akbar*	The brighter of the two calves	*ṛḍhapada* (?)
23	*qā'idāt al-bāṭiya*	The base of the bowl	*jāmapakṣī* (?)
24	*ẓahr al-asad*	The back of the lion	*dhāma maṣa*
25	*al-ṣarfa*	The change of weather	*u nāṃ phā*
26	*al-'anāq*	The goat	*a la dha prā*

Correct form in Sanskrit	Meaning	Star ID
	The sea bird	β Ceti
matsya-mukha	The mouth of the fish	β Andromedae
samudrapakṣi-mukha	The mouth of the sea bird	γ Ceti
		π Ceti
samudrapakṣī musa	The sea bird	α Ceti
	The head of the ghost	β Persei
illegible		α Persei
	Belonging to the end of the river	γ Eridani
	The fourth lunar mansion	α Tauri
		α Aurigae
mithuna pāda vāma	The left foot of Gemini	β Orionis
mithuna hasta vāma	The left hand of Gemini	γ Orionis
mithuna hasta dakṣiṇa	The right hand of Gemini	α Orionis
	The left foot of Gemini	κ Orionis
lubdhaka	The hunter[b]	α Canis Majoris
lubdhaka-bandhu[c]	The kinsman/brother of the hunter	
naukā-mukha	The face/beginning of the ship	ρ Puppis
puṣya	The 8th lunar mansion	
mahāpuruṣa[d]	The great man	α Hydrae
	The head of the lion	μ Leonis
	the 10th lunar mansion	α Leonis
illegible		α Ursae Majoris
	The *jāma* (?) bird	α Crateris
illegible		δ Leonis
illegible		β Leonis
illegible		ζ Ursae Maioris

TABLE 6.1 *Star Names on the Rete* (cont.)

S. No	Star name in Arabic	Meaning	Star name in Sanskrit as engraved
27	*janāḥ al-ghurāb*	The [left] wing of the raven	*kākaskandha* "raven's shoulder"
28	*simāk al-aʿzal*	unarmed *simāk*	*citrā*
29	*simāk rāmiḥ*	armed *simāk*	*svāti*
30	*kiffa j[anūbī]*	The southern pan [of the balance]	*tulā kā palaḍā*ᵉ *dachan*
31	*kiffa shamālī*	The northern pan [of the balance]	*tulā kā palaḍā uttar*
32	*nayyīr fakkā*	The luminous one of al-fakka	*viśākhā*
33	*ʿunuq al-ḥayya*	The neck of the serpent	*sarpagrīvā*
34	*qalb al-ʿaqrab*	The heart of the scorpion	*jyeṣṭhā*
35	*rukbat al-ḥawwāʾ al-yusrā*	The left knee of the serpent charmer	*dhanaṣakoṭi jādha vāma*
36	*rās al-jāthī*	The head of the kneeling one	
37	*rās al-ḥawwāʾ*	The head of the serpent charmer	*dhanuṣa koṭi*
38	*ʿayn al-tinnīn*	The eye of the dragon	nil
39	*yad al-ḥawwā al-yumnā al-muqaddam*	The forward right hand of the snake charmer	*dhanuṣakoṭi hasta*
40	*nasr waqīʿ*	The falling eagle	*abhī ji*
41	*minqār al-dajāja*	The beak of the bird	*kruṃja sa ṣṣa ra*
42	*nasr ṭāʾir*	*The flying eagle*	*śravaṇa*
43	*dhanab al-dulfīn*	*The tail of the dolphin*	*kruṃja pachī*
44	*dhanab al-dajāja*	The tail of the bird	*kruṃja pachī pucha*
45	*fam al-faras*	The mouth of the horse	*aśvamuṣa*
46	*dhanab al-jadī*	The tail of the goat	*makarapu[c]cha*
47	*mankib sākib al-māʾ*	The shoulder of the water pourer	*kuṃbhaskaṃdha*
48	*sāq sākib al māʾ j[anūbī]*	The southern leg of the water pourer	*kuṃbhapāda*

Correct form in Sanskrit	Meaning	Star ID
kākapakṣa vāma	The left wing of the raven	γ Corvi
	The 14th lunar mansion	α Virginis
	The 15th lunar mansion	α Boötis
tulāyāḥ dakṣiṇa-śikyam	The southern pan [of the balance]	α1,2 Librae
uttara-śikyam	The northern pan [of the balance]	β Librae
	The 16th lunar mansion	α Coronae Borealis
	The neck of the serpent	α Serpentis
	The 18th lunar mansion	α Scorpii
dhanuṣkoṭi[f] *jaṅghā vāma*	The left shank of the serpent charmer	ζ Ophiuchi
dhanu pa ra jhā	illegible	
dhanuṣkoṭi	extremity of Sagittarius	α Ophiuchi
dhanuṣkoṭi hasta	*The hand of the serpent charmer*	δ Opiuchi
abhijit	The 22nd lunar mansion	α Lyrae
Krauñca-śīrṣa	The head of the curlew	β Cygni
	The 23rd lunar mansion	α Lyrae
krauñca-pakṣī	The curlew bird	ε Delphini
krauñca-pakṣi-puccha	The tail of the curlew bird	α Cygni
aśvamukha	The mouth of the horse	ε Pegasi
	The tail of Capricorn	δ Capricorni
	The shoulder of Aquarius	
	The foot/leg of Aquarius	δ Aquarii

TABLE 6.1 *Star Names on the Rete* (cont.)

S. No	Star name in Arabic	Meaning	Star name in Sanskrit as engraved
49	*dhanab al-qayṭus sh[amālī]*	The northern tail of Cetus	*samudra-pakṣī pu[c]cha uttara*
50	*mankib al-faras*	The shoulder of the horse	*aśvajaṃgha* "shank of the horse"!
51	*surrat al-faras*	The navel of the horse	*aśvanā[bhi]*
52	*janāḥ al-faras*	The wing of the horse	*aśvapuṣa*
53	*dhāt al-kursī*	The one who has a chair, i.e., Cassiopeia	*caukī strī tana (?)*

a In Sanskrit astrolabes, *'ayyūq* is translated as *brahmahṛdaya*, "the heart of Brahmā"; it is mentioned in the very popular astronomical text of about the eighth century, *Sūryasiddhānta*, at 8.11, 20; 9.12, 18; see *The Súrya Siddhánta, or An Ancient System of Hindu Astronomy*, ed. Fitzedward Hall (Calcutta: C.B. Lewis, 1859).

b According to Hindu mythology, the star Sirius is called *lubdhaka*, "hunter," or *mṛga-vyādha*, "the hunter of the deer," because god Śiva in the form of a hunter shot an arrow (represented by the three stars in the belt of Orion) at Brahmā, the creator, who transformed himself into a deer and pursuied his own daughter who metamor-phosed into a doe.

c *Lubdhaka-bandhu* is employed for Procyon in Sanskrit astrolabes because it is a bright star close to *lubdhaka*, Sirius.

d This designation is generally employed in Sanskrit astrolabes, but the reason eludes us.

e The pan of the balance is denoted by the rare Sanskrit term *śikya*. Nayanānanda avoids the Sanskrit term and renders this and the next expression into Hindi.

f In his Sanskrit manual on the astrolabe of 1370, Mahendra Sūri rendered the Arabic star name *rȧs al-ḥawwā'* (head of the serpent charmer) as *dhanuṣkoṭi*, "extremity of the bow" (*dhanus*), i.e., "extremity of Sagittarius" (Arabic *al-qaws*; Sanskrit *dhanus*), because the star belongs to the zodiac sign Sagittarius; see Mahendra Sūri, *Yantrarāja*, 38, 43. This name is often used in Sanskrit astrolabes. But Nayanānanda seems to treat *dhanuṣkoṭi* here and under entry 39 in Table 6.1 as the Sanskrit name of the constella-tion Ophiuchus (Arabic *al-ḥawwā'*).

Frank dwells on the various uses of this astrolabe and concludes with the remark: "From this, one can see that for solving even simple problems two or more operations are necessary, whereas just one operation would often suffice with the common astrolabe."

Astrolabes with this design may have been made and used at the time of its invention in the eleventh century and later, but no such astrolabes are

Correct form in Sanskrit	Meaning	Star ID
	The northern tail of the sea bird	ι Ceti
aśvaskandha	The shoulder of the horse	β Pegasi
	The navel of the horse	δ Pegasi = α Andromedae
aśvapuccha (?)	The tail of the horse (?)	γ Pegasi
	seat woman	β Cassiopeiae

extant today. All that exist now are a few Lahore astrolabes in which one or two *zawraqī* horizons are incorporated in the otherwise standard retes and which contain a special plate to be used along with the *zawraqī* horizons.

Here also the first step seems to have been taken by Ḍiyā' al-Dīn's father Qā'im Muḥammad, who incorporated two arcs in the rete of an astrolabe he made in 1635, in the eighth regnal year of Shāh Jahān.[42] Both the arcs are curved towards north; on the upper arc are engraved the names of six zodiac signs from Aries to Virgo; the function of this arc is not known. The lower arc carries the legends "The Horizon of the Western Solid" and "The Horizon of the Eastern Solid."[43] Obviously this arc is the proper *zawraqī* horizon. This has to be used together with a special type of projection engraved on the reverse side of the plate bearing the tablet of ecliptic coordinates. Gunther describes it thus: "a complicated plate in the centre of which is a miniature tablet for lat. 42°, hours 15. Across the bottom is written 'A table of the projection of rays,' which according to Albiruni is also called 'equalization of houses' (=areola, Nallino, p. 324)."

Following the footsteps of his father, Ḍiyā al-Dīn incorporated two mutually intersecting arcs for the latitudes of 29° (Delhi) and 32° (Lahore) in the rete of

42 Oxford, Museum of the History of Science, accession no. 42730. See also R. T. Gunther, *Astrolabes of the World* (Oxford 1932, 2nd edition, London: The Holland Press, 1976), no. 71, pp. 191–197, Plates XLV–XLVI, figs. 94–97. Gunther refers to the maker as "Muhammad Ibn Isa," omitting the more important part of the name "Qā'im."

43 Rendered thus by Gunther in the *Astrolabes of the World*, 192.

FIGURE 6.7
The oblique horizons for the latitudes of 29° and 32° on the rete. The Nasser D. Khalili Collection of Islamic Art, Accession No. SCI *53.*
PHOTO COURTESY OF THE KHALILI FAMILY TRUST © NOUR FOUNDATION.

his astrolabe of 1663 at the Khuda Bakhsh Oriental Public Library, Patna. For use along with these arcs, he engraved a special plate on the reverse side of the tablet of the ecliptic coordinates.

The Khalili astrolabe, prepared about the same time as the Patna astrolabe mentioned above, follows it exactly in respect of the two *zawraqī* horizons for 29° and 32° and in the design of the plate to be used with the horizons (Figure 6.7).[44] This close resemblance is one more piece of evidence to show that the Khalili astrolabe was made by none other than Ḍiyā᾽ al-Dīn.

44 Besides the three astrolabes mentioned above, I came across the following three more astrolabes containing *zawraqī* horizons with the corresponding plate:
(iv) Not signed, not dated, Oxford, Museum of the History of Science, accession no. 33796; Gunther, *Astrolabes of the World*, pp. 218–220, Plates XLVIII–XLIX, Figure 112–113. The oblique horizons are for 29° and 32° which can be clearly seen on the photo of plate 7b (Plate XLIX). Gunther misreads the values of the two latitudes. On this plate the astrological houses are not marked as in (i) (ii), and (iii). On the back, in the upper right quadrant, there are four sigmoid graphs of solar meridian altitude for the latitudes 24°, 29°, 32°, 36°.
(v) Not signed, nor dated: Chicago, The Adler Planetarium, accession no. M-39; see Gunther, *Astrolabes of the World*, p. 220; David Pingree, *Eastern Astrolabes*, No. 20, pp. 88–93. There are two retes, one standard and another containing just the two horizons for 37° and 38°. The significance of this choice of latitudes eludes us.
(vi) Not signed, nor dated; its present location is not known. In the archives of the Museum of the History of Science, there are three photos of the front, back and gazetteer; a complete set of photos is with Anthony J. Turner. In the rete there is only one oblique horizon for 24°. There are five plates serving the latitudes 18°, 21°, 24°, 27°, 29°, 32°, 34°, 36°, and a tablet of ecliptic coordinates and a tablet of horizons. The plate for 24° has several star positions marked on it; but not the astrological houses as in nos. (i), (ii), (iii); this must be the one to be used in conjunction with the *zawraqī* horizon.

In the present astrolabe, the arc with the convex side to the north is the oblique horizon for latitude 29°. The western and eastern halves of the arc carry the labels *ufq mujassam gharbī bi-ʿarḍ 29*, "western solid horizon for latitude 29" and *ufq mujassam sharqī bi-ʿarḍ 29*, "eastern solid horizon for latitude 29" respectively. The mast corresponding to this arc, i.e., the solstitial bar above the central hole, has the inscription *shaẓiya yatiru zawraq fawq al-ʿard*, "pointer spanning the *zawraqī* horizons above the earth."

The other arc, with the convex side to the south is the oblique horizon for latitude 32°. Two halves of this arc carry the labels *ufq mujassam gharbī bi-ʿarḍ 32*, "western solid horizon for latitude 32" and *ufq mujassam sharqī bi-ʿarḍ 32*, "eastern solid horizon for latitude 32." Its mast, i.e., the solstitial bar below the central hole, bears the label *shaẓiya yatiru zawraq taḥt al-ʿard*, "pointer spanning the *zawraqī* horizons below the latitude." The equinoctial bar carries the label *mastarat tarbīʿ gharbī*, "western perpendicular rule" in the west and the label *mastarat tarbīʿ sharqī*, "eastern perpendicular rule" in the east. All these inscriptions on the two arcs and on the solstitial and equinoctial bars are inlaid in silver. These have not been rendered into Sanskrit, probably because of the lack of space to engrave the additions. The plate to be used along with these *zawraqī* horizons will be described in the following section.

The Plates

There are six plates in this astrolabe (Table 6.2). The first five plates serve different latitudes from 0° to 45°. The obverse side of the sixth plate is engraved with the ecliptic coordinates and the reverse side with the tablet of horizons. The plates have a uniform diameter of 501 mm, but with varying thickness. There is a notch at the north point in all the plates into which the peg affixed at the same place on the mater fits in and holds the plates in position. The following table gives an overview of the specifications of the plates. For convenience,

---

On the back, in the upper right quadrant, there are six sigmoid graphs of solar meridian altitude for the latitudes 21°, 24°, 29°, 32°, 34°, 36°. In his astrolabe of 1651 at the Salar Jung Museum, Hyderabad, Ḍiyāʾ al-Dīn draws as many as eight graphs for latitudes 15°, 19°, 21°, 27°, 29°, 32°, 35°, and 40°. These are the only two astrolabes with such a large number of graphs.

A comprehensive study of the *zawraqī* horizons in these six astrolabes, their function, and their relation to textual sources is beyond the scope of the present paper, but it is a desideratum.

TABLE 6.2 *Specifications of the Plates*

S.No.	Latitude	Longest day (hours;minutes)
1a	*zawraqī* horizons for 29°, 32°	
1b	18°	13;9
2a	0°	12
2b	3°	12;11
3a	27°	13;43[a]
3b	29°	13;52
4a	37°	14;31
4b	39°	14;45
5a	42°	14;0
5b	45°	15;22
6a	66;30°; ecliptic coordinates	
6b	Tablet of horizons	

a The geographical gazetteer, however, has the value 13;45 as can be seen in Figure 6.3.

the two sides or faces of these plates are numbered below as 1a, 1b, 2a, 2b and so on.

On each side of the plate, a horizontal and a vertical diameter intersect each other at right angles. The horizontal diameter represents the straight horizon and the vertical diameter the meridian. The plate faces 1a, 2a, 6a and 6b contain special projections which will be discussed below. The rest of the plates are fashioned in the same style with altitude circles drawn for each 1° and azimuth circles for every 15° above the horizon and for every 3° below the horizon. Each sixth altitude circle is highlighted by a dotted line. The altitude circles are numbered serially from one to ninety, from the horizon up to the zenith, both in the eastern and western halves, the series of numbers in these two parts forming a symmetrical pattern. The azimuth circles are also numbered in a symmetric form. In the lower half, lines for equal and unequal hours are drawn between the tropics of Cancer and Capricorn and numbered.

Below the eastern horizon is written *al-mashriq* (the east) and below the western horizon *al-maghrib* (the west). Below the horizon, in the middle, to the right and left of the meridian line, are mentioned respectively the latitude for which the plate is prepared and the duration of the longest day at this latitude, in hours and minutes. The numerical value of the latitude is preceded by the word *al-ʿarḍ* (the latitude) and the numerical value of the duration of the longest day is preceded by *sāʿātuhu* (its hours).

Nayanānanda caused all the numerical quantities engraved in neat Devanāgarī numerals close to the *abjad* counter parts. Under the eastern and western horizons the Arabic words *al-mashriq* and *al-maghrib* are rendered into Sanskrit as *pūrva* and *paścima* respectively. In the middle of the plates *sāʿātuhu* is rendered *horā* and *al-ʿarḍ* as *akṣāṃśāḥ* (latitudes, plural). However, the artisan engraved the last word incorrectly as *aṣāṃśā*, except on the plate face 1b where the word is correctly reproduced. Amusingly enough, the numerical values of the hours and minutes were reproduced in the Arabic style, i.e., the minutes on the right and hours on the left, because they were so written in Arabic. Thus on plate 5b, the latitude appears as 00 45 (meaning forty-five degrees and zero minutes) and the duration of the longest day as 22 15 (meaning fifteen hours and twenty-two minutes).

Plate 6a for Latitude 66;30°

The obverse side of the sixth plate contains a stereographic projection relative to the latitude of 66;30°, which is a complement of the total obliquity of 23;30° (Figure 6.8). Besides the projections of the equator and the two tropics, it also contains the projection of the ecliptic circle, which is marked with the names of the twelve zodiac signs. The signs commence from the point of the vernal equinox and proceed counterclockwise. The names are engraved with long strokes so that the name occupies the entire stretch of the zodiac sign.

The Sanskrit names are also written from the right to the left in the same manner as the Arabic names, with one syllable at the beginning of the sign and the other syllable/s at the end of the sign: *ṣa-me* (for *meṣa*, Aries), *ṣa-vṛ* (for *vṛṣa*, Taurus), *thana-mi* (sic! for *mithuna*, Gemini), *ha-siṃ* (for *siṃha*, Leo), *nya-ka* (*kanyā*, Virgo), *la-tu* (sic! *tulā*, Libra), *ści-vṛ* (for *vṛścika*, Scorpio), *na-dha* (sic! for *dhanuḥ*, Sagittarius), *kara-ma* (for *makara*, Capricorn), *bha-kuṃ* (*kumbha*, Aquarius) and *na-mī* (for *mīna*, Pisces). The arcs radiating from the centre at intervals of 3° are the lines of the celestial longitudes. They divide each sign into ten units of 3° each. Below the name of the sign, the ten parts are numbered from three to thirty in counterclockwise direction. Every fifth of these arcs is dotted.

The concentric circles of celestial latitudes are drawn for each 1°. Every sixth circle is highlighted with dots. The circles are numbered from the ecliptic up to the North Pole as one to six; thereafter only in even numbers from eight to

FIGURE 6.8
Plate 6a for the ecliptic coordinates. The Nasser D. Khalili Collection of Islamic Art, Accession No. SCI 53.
PHOTO COURTESY OF THE KHALILI FAMILY TRUST © NOUR FOUNDATION.

ninety. Outside the ecliptic, in the lower half, each circle is numbered serially from the ecliptic onwards from one to forty-eight. Circles of longitudes that radiate from the zenith are drawn for every 3°.

By placing the rete on this plate, the longitudes and latitudes of the stars marked on the rete can easily be measured. Hence this projection is technically known as *ṣafīḥa mīzān al-ʿankabūt* (plate for the measures on the rete), which is engraved in the lower half of the plate. The Sanskrit version is engraved above the western horizon; it reads *patrā asāmśa 66 tulā bhāpatra kā* (correct: *patra akṣāṃśa 66 tulā bhapatra kā*, "plate [for the] latitude [of] 66, [which is the] balance of the rete." *Mīzān al-ʿankabūt* is literally translated here.

Plate 6b with the Tablet of Horizons

On the reverse side of the sixth plate are drawn half horizons for every single degree of latitude from two to ninety, in four separate sets in the four quadrants of the plate (Figure 6.9). These are drawn in such a way that they commence outside the circle of equator in one quadrant, pass through the point of intersection of the equator with the radius as a bundle and spread out and end up in the next quadrant.

The four sets of latitudes are as follows:

2, 6, 10, ... 78, 82, 86, 90;
3, 7, 11, ... 79, 83, 87, 90;
4, 8, 12, ... 80, 84, 88, 90;
5, 9, 13, ... 81, 85, 89, 90.

FIGURE 6.9
The tablet of horizons on plate 6b. The Nasser D. Khalili Collection of Islamic Art, Accession No. SCI 53.
PHOTO COURTESY OF THE KHALILI FAMILY TRUST © NOUR FOUNDATION.

In all four quadrants, declination scales are engraved along the radii, once outside the equator, i.e., to the south of the equator and once inside, i.e., to the north of the equator. These scales carry the values 3, 6, 9, 12, 15, 18, 21 and 23;30, the last being the value of maximum obliquity. Each scale is accompanied by a label in florid calligraphy, those to the south of the equator reading *taqsīm mayl kullī janūbī* (divisions of the obliquity, i.e. declination scale, south) and those to the north *taqsīm mayl kullī shamālī* (divisions of the obliquity, i.e. declination scale, north). The Sanskrit labels read *krāṃti aṃśa dakṣiṇa* (more correctly *krānti aṃśa dakṣiṇa*, "degrees of declination south") or *krāṃti aṃśa uttara* (more correctly *krānti aṃśa uttara*, "degrees of declination north") according to the location.

The plate bears the designation *ṣafīha āfāqīya*, "plate/ tablet of horizons" or "universal plate"; this name is engraved in the lower right quadrant. The designation is freely translated into Sanskrit as *patrā sarvadesī* (more correctly: *patra sarvadeśī*, "plate for all localities/latitudes" or "universal plate") and engraved in the upper right quadrant.

Plate 2a for Latitude 0°

On plate 2a, as in the other latitude plates, latitude circles are drawn for each 1°and numbered on the right and on the left; every sixth line is highlighted with dots (Figure 6.10). But unlike in other plates, azimuths are drawn for every 6°; above the horizon they are drawn all over, but below the horizon only inside the circle of equator and labelled.

FIGURE 6.10
Plate 2a for latitude 0°. The Nasser D. Khalili Collection of Islamic Art, Accession No. SCI 53.
PHOTO COURTESY OF THE KHALILI FAMILY TRUST © NOUR FOUNDATION.

In the lower half, an attempt was made to draw lines for *ghaṭīs*. These are traditional units of twenty-four minutes, used throughout India. While the Muslims used the hours of sixty minutes for astronomical, astrological and religious purposes, they also employed the *ghaṭīs* in everyday life. Therefore Lahore astrolabe makers incorporated *ghaṭī* lines in some of the larger astrolabes.[45] Since *ghaṭīs* are smaller units than hours, more space is needed on the astrolabe plates to draw the 30 and more *ghaṭīs* lines.

In this plate are drawn a series of thirty *ghaṭīs* lines and numbered from one to thirty, starting from the western horizon. Because two and half *ghaṭīs* make one hour, certain *ghaṭī* units (i.e., 3, 8, 13, 18, 23, 28) are divided into two parts to synchronize with the equal hours, because 2 ½ gh = 1h; 5 gh = 2 h, 7 ½ gh = 3 h; 10 gh = 4 h; 12 ½ = 5 h, and so on. Thus from these lines time can be measured in both the *ghaṭīs* and hours. The numbers of the *ghaṭīs* are engraved below the circle of the equator and those for the hours above the equator. However, the astrolabe maker erred in drawing the *ghaṭī* lines here as unequal units, for the *ghaṭīs* are always equal and never unequal. In the three other Lahore astrolabes, lines are drawn correctly for equal *ghaṭīs*.

At the middle of the plate, the specifications of the plate are given as *sāʿātuhu 12;0* and *al-ʿarḍ 0;35*. The plate is clearly calibrated for the latitude 0°, i.e., the terrestrial equator; but it is not clear why the latitude is mentioned here as "0;35."

45 Aside from the present astrolabe, three others are known: a zoomorphic astrolabe, unsigned but attributable to Muqīm (Jai Singh's Observatory, Jaipur), Ḍiyā' al-Dīn's astrolabe of 1651 (Salar Jung Museum, Hyderabad) and of 1656 (Jai Singh's Observatory, Jaipur).

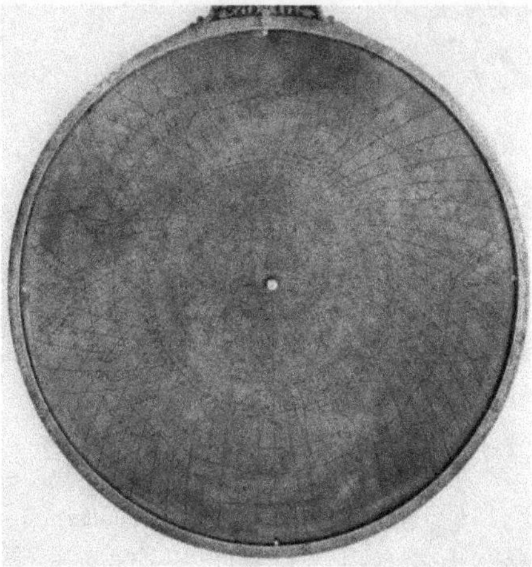

FIGURE 6.11
Plate 1a to be used with the Zawraqī horizons. The Nasser D. Khalili Collection of Islamic Art, Accession No. SCI 53.

Moreover, in 0;35 the zero symbol is unlike the symbol engraved throughout this astrolabe, e.g., in 12;0 just across the line of the meridian. It is written not to the right of "35" but just above it, in a smaller size, as if in an afterthought. The Sanskrit rendering reads more correctly *horā 12* (hours 12) and *patra laṃkā* (plate [for] the terrestrial equator) without any numerical symbol for the latitude.

Plate 1a for Use with the Zawraqī Horizons

This plate is specially designed for use in conjunction with the two *zawraqī* horizons inserted in the rete for the latitudes 29° and 32° (Figure 6.11). Like the other plates, this one is also engraved with the vertical and horizontal diameters, and the three concentric circles of the Tropic of Cancer, the celestial equator and the Tropic of Capricorn.

Around the circle of equator, the twelve astrological houses are marked, once outside the equator and once inside the equator, but with a difference. Outside the equator the enumeration commences from the east point and proceeds counterclockwise, but inside the equator the enumeration commences from the west point and proceeds counterclockwise. The two varying sets of the astrological houses appear to be related to the two oblique horizons engraved on this plate. The names of the astrological houses are engraved in elongated letters so that the names occupy the full stretches of the corresponding houses. Sanskrit names are added just below the Arabic names, the syllables proceeding from the right to the left as in Arabic. But since the Sanskrit

FIGURE 6.12
Plate 1a (detail showing the astrological houses and the miniature tablet inside the Tropic of Cancer). The Nasser D. Khalili Collection of Islamic Art, Accession No. SCI *53.*
PHOTO COURTESY OF THE KHALILI FAMILY TRUST © NOUR FOUNDATION.

letters could not be stretched like the Arabic letters, the first Sanskrit syllable was added just at the beginning of the astrological house and the last syllable/s at the end of the house with a large gap inbetween (Figure 6.12).

In both cases, each house is divided into 5 units of 6° and the numbers in *abjad* notation are engraved above the names of the astrological houses in Arabic. The Devanāgarī numbers are added below the Sanskrit names.

In Arabic, the first house is named *al-ṭāliʿ*, "the ascendant," and the rest are ordinal numbers (Table 6.3). In Sanskrit, the houses (*bhāva*) are named generally according to the area of influence of that particular house in the life of the native. Thus the first house is called *tanu*, "body," because it stands for the body and the physical appearance, behaviour, and character of the person.[46]

Two mutually intersecting oblique horizons are drawn across the plate. As in the rete, here also the arc with the convex side to the north is the oblique horizon for the latitude 29° and that with the convex side to the south is the horizon of the latitude 32°. Where the arcs intersect with the equinoctial colure, there are labels at the end of each horizon. Thus at the two ends of the horizon

46 See Hans-Georg Türstig, *Jyotiṣa: das System der indischen Astrologie* (Wiesbaden: Franz Steiner Verlag, 1980), 6–8.

TABLE 6.3 *Astrological Houses in Arabic and Sanskrit*

	Arabic name	Sanskrit name	Meaning of the Sanskrit name
1	*al-ṭāliʿ*	*tanu*	body
2	*al-thānī*	*dhana*	wealth
3	*al-thālith*	*sahaja*	sibling
4	*al-rābiʿ*	*suhṛd*	friend
5	*al-khāmis*	*suta*	son
6	*al-sādis*	*ripu*	enemy
7	*al-sābiʿ*	*jāyā*	wife
8	*al-thāmin*	*mṛtyu*	death
9	*al-tāsiʿ*	*dharma*	religious life
10	*al-ʿāshir*	*karma*	activity (profession)
11	*al-ḥādī ʿashar*	*āya*	income
12	*al-thānī ʿashar*	*vyaya*	expenditure

for 29° are engraved *ufq sharqī arḍ* 29 (Sanskrit: *kṣitīja pūrva akṣaṃśa*, correct: *kṣitija pūrva akṣāṃśa* 29) on the east, meaning "eastern horizon latitude 29" and *ufq gharbī ʿarḍ* 29 (Sanskrit: *kṣitīja paści akṣaṃśa*, correct: *kṣitja paścima akṣāṃśa* 29) on the west, meaning "western horizon latitude 29." Likewise at the two ends of the horizon for 32° are engraved *ufq sharqī bi-ʿarḍ* 32 (Sanskrit: *kṣitīja pūrva akṣaṃśa*, correct: *kṣitja pūrva akṣāṃśa* 32) on the east and *ufq gharbī ʿarḍ* 32 (Sanskrit: *kṣitīja paści akṣaṃśa*, correct: *kṣitja paścima akṣāṃśa* 29) on the west (Figure 6.13).

The rim of the Tropic of Cancer is divided in 360°; each 1° marked and groups of 5° numbered from five to ninety counterclockwise, separately for each quadrant. Inside this ring of Cancer a miniature tablet is drawn with smaller versions of the two oblique horizons to match those on the rete, and the corresponding altitude circles. Like the larger horizons drawn across the whole the plate, the oblique horizons in the miniature plate also carry labels on both sides. But here no additions were made in Sanskrit as there was not enough space to do so. Some thirty-five star positons are marked within the Tropic of Cancer with dots enclosed in circles along with their names.

The space outside the Tropic of Cancer is filled with two families of arcs. These appear to be lines for unequal *ghaṭīs* (of twenty-four minutes), once counted from the eastern horizon and another time from the western horizon, for the two latitudes of 29° and 32°.

FIGURE 6.13
Plate 1a (detail showing the labels on the oblique horizons in the east). The Nasser D. Khalili Collection of Islamic Art, Accession No. SCI 53.
PHOTO COURTESY OF THE KHALILI FAMILY TRUST © NOUR FOUNDATION.

At the bottom of the plate is a long inscription bearing the designation of this special plate: *ṣafīhat maṭraḥ al-shuʿāʿ keh nazd abī rayhān taswīyat al-buyūt khwānand*, "plate for finding the projection of the astrological rays according to (?) Abū Rayhān [al-Bīrūnī] [and] equalization of the astrological houses" (Figure 6.14).

The Geographical Gazetteer

On the inner face of the mater is engraved an Arabic/Persian gazetteer with the names of 259 localities (*asmāʾ al-buldān*), their longitudes (*al-ṭūl*), latitudes (*al-ʿarḍ*) and the duration of the longest day (*al-sāʿāt*) in five annuli (Figure 6.15).[47]

47 For a comprehensive compilation from several Islamic sources, see E. S. Kennedy and M. H. Kennedy, *Geographical Coordinates of Localities from Islamic Sources* (Frankfurt: Institut für Geschichte der Arabisch-Islamischen Wissenschaften an der Wolfgang Goethe-Universität, 1987). See also David A. King, *World-Maps for Finding the Direction and Distance to Mecca: Innovation and Tradition in Islamic Science* (Leiden: Brill 1999),

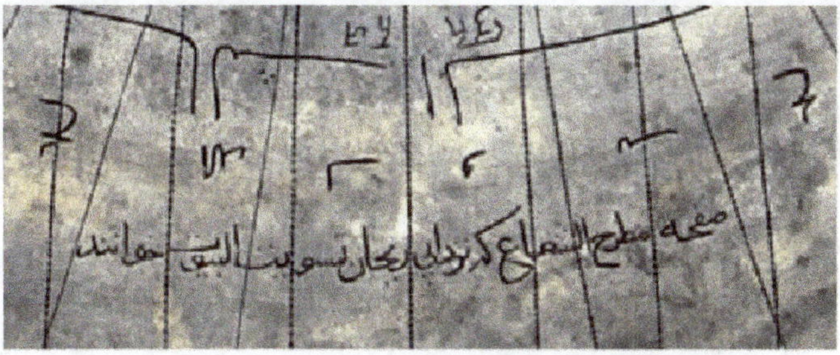

FIGURE 6.14 *The designation of Plate 1a. The Nasser D. Khalili Collection of Islamic Art, Accession No.* SCI *53.* PHOTO COURTESY OF THE KHALILI FAMILY TRUST © NOUR FOUNDATION.

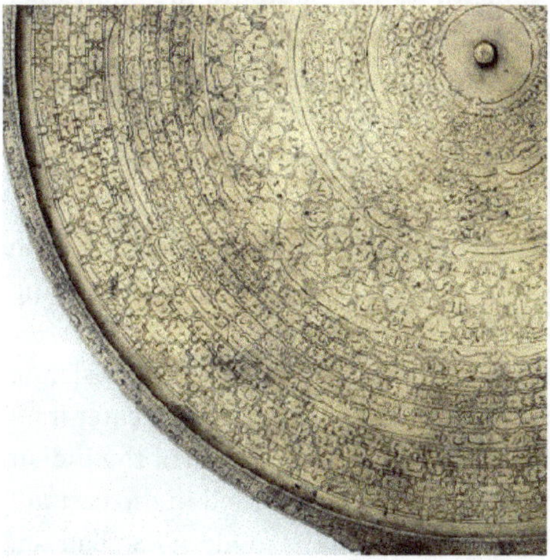

FIGURE 6.15 *The geographical gazetteer on the mater. The Nasser D. Khalili Collection of Islamic Art, Accession No.* SCI *53.* PHOTO COURTESY OF THE KHALILI FAMILY TRUST © NOUR FOUNDATION.

This is the largest gazetteer engraved by Ḍiyā' al-Dīn.[48] In his astrolabe gazetteers, he varies the order of arrangement frequently: sometimes, he begins with

---

455–638, where scores of gazetteers are reproduced; and Sharon Gibbs & George Saliba, *Planispheric Astrolabes from the National Museum of American History* (Washington, DC: Smithsonian Institution Press, 1984), 190–206.

48 Maddison and Savage-Smith erroneously state that the gazetteer contains *inḥirāf*, but that is not the case; cf. Francis Maddison and Emilie Savage-Smith, *Science, Tools & Magic* (The Nasser D. Khalili Collection of Islamic Art, Vol. XII) (London: Khalili Collections, 1997), 1. 230.

the sacred city of Mecca; sometimes he arranges the localities according to the climates (*iqlīm*) of antiquity[49] and some other times, according to regions (also called *iqlīm*). The region-wise arrangement is found in three of his astrolabes, viz., in the astrolabe of 1649 at the Darul Uloom Nadwatul Ulama at Lucknow, in the astrolabe of 1656 at the Jaipur Observatory, and in the present astrolabe. The last two astrolabes, being large in size, contain large gazetteers with 210 and 259 entries respectively. In these two gazetteers, the localities are grouped in about thirty-five regions, commencing from the Maghreb, Andalusia, Sudan, Egypt, Yemen, and concluding with Sind and Hind (i.e., India south of the Indus). The names of the regions are engraved above the names of localities, in narrow bands between the main annuli.[50]

This gazetteer is substantially based on Ulugh Beg's gazetteer, which lists 278 localities.[51] Leaving aside some minor additions and deletions, the sequence of the localities is the same, their longitudes and latitudes are the same. However, The headings mentioning the regions are slightly shifted on the astrolabe with the result that Tajah az Barbar (10) is placed in Andalus and not in Sudan where it rightly belongs to, Qulzum (17) is mentioned under Yemen and not under Egypt and so on. This shift in the headings occurs also in the two other astrolabes mentioned above and probably goes back to the source. While reproducing the geographical gazetteer in Table 6.4 below, the headings will be retained as they are engraved on the astrolabe. The correct positions according to the Ulugh Beg's table will be mentioned in footnotes.

While these astrolabe gazetteers seek to cover the entire area of the Islamic world, with the majority of localities enumerated being from the Greater Iran, the geographical information on the eastern and southern parts of the Indian subcontinent is very meagre. The localities usually mentioned in the east are Benares, Jaunpur and Dhākā Bangālah, i.e., Dhaka in the modern Bangladesh.

49 For a large gazetteer arranged according to the climates, see Abū al-Faḍl, *ʿAin-i-Ākbari of Abul Fazl-i-ʿĀllāmi*, vol. 3, translated into English by H. S. Jarrett, revised and further annotated by Jadu-Nath Sarkar (Calcutta: Royal Asiatic Society of Bengal, 1948; reprint, Frankfurt: Institute for the History of Arabic-Islamic Science at the Wolfgang Goethe University, 1993), 56–116; this gazetteer is reprinted, in alphabetical order of the localities, in Kennedy, *Geographical Coordinates of Localities from Islamic Sources*, 569–580.

50 In the narrow bands above annuli 1, 3 and 4, the enumeration of the regions begins with the argument "*iqlīmhā*."

51 Louis-Pierre-Eugène-Amélie Sédillot, *Prolégominès des tables astronomiques d'Oloug-Beg: Traduction et commentaire* (Paris: Firmin Didot Frères, 1853).

TABLE 6.4 *The Geographical Gazetteer*

S.No.	*asmā' al-buldān*	*al-ṭūl*	*al-arḍ*	*al-sā'āt*	Modern name
Annulus 1					
Iqlīmhā Maghrib (The Region of Maghreb)					
1	Sūs Āqṣā	15;30	32;0	14;08	Sus al-Aksa, Morocco
2	Fars qaṣba Ṭanja	18;0	22;0	13;22	Fez, Morocco
3	Tāhart ʿŪliyā	35;30	29;0	13;53	Upper Tahart
4	Tāhar[t] Suflī az Farīfta	36;0	29;0	13;53	Lower Tahart
5	Sijilmāsa	37;0	31;30	14;6	
6	Qayrawān	41;0	31;30	14;7	Kairouan, Tunisia
7	Mahdiyya	42;0	32;30	14;11	Mahdiya, Tunisia
8	Ṭarābulus Maghrib	45;0	32;0	14;8	Tripoli, Libya
Andalus (Andalusia)					
9	Qurtaba dār al-mulk Andalus	38;26	35;0	14;25	Cordoba, Spain
10	Bajja az Barbar[a]	65;0	14;0	12;54	
Sūdān (Sudan)					
11	Jarmī dār al-mulk Habasha	65;0	9;30	12;10	Jarmī, capital of Ethiopia
12	Danqala shahr Nūba	53;40	14;30	12;56	Dunqula, town in Nubia
Miṣr (Egypt)					
13	Qūṣ az Ṣaʿīd Miṣr	61;30	24;30	13;32	Qus, Egypt
14	Iskandariya	61;54	30;58	14;2	Alexandria, Egypt
15	Miṣr	63;20	30;20	14;1	Egypt = Fustat = Cairo
16	Dumyāṭ	63;30	31;25	14;5	Damietta, Egypt
Yaman (Yemen)					
17	Qulzum az Kanār Darya[b]	64;0	29;30	13;55	Suez on the Red Sea, Egypt
18	ʿAdan	76;0	11;0	12;20	Aden, Yemen
19	Ṣanʿāʾ dār al-mulk Yaman	77;0	14;30	12;56	Sanaʿa, capital of Yemen
20	Zubayd	74;20	14;10	12;55	Zubayd

TABLE 6.4 *The Geographical Gazetteer* (cont.)

S.No.	*asmā' al-buldān*	*al-ṭūl*	*al-arḍ*	*al-sā'āt*	Modern name
'Umān (Oman)					
21	Ṣajā qaṣba 'Ammān	84;0	14;20	12;55	
22	Madīnat Rasūl 'alaihi al-Salām[c]	75;20	25;0	13;35	Medina, Saudi Arabia
Ḥijāz (Hejaz)					
23	Makka Mubārak	77;10	21;40	13;20	Mecca, Saudi Arabia
24	Ṭā'if	77;30	21;20	13;18	Taif, Saudi Arabia
25	Yamāma[d]	82;30	23;30	13;28	Al-Yamama, Saudi Arabia
26	Hajar Baḥrayn	83;0	25;0	13;36	Hajar
27	Hasa Jazīra	83;30	24;0	13;31	
28	Qatīf Baḥrayn	74;0	25;0	13;35	Qatif, Saudi Arabia
Shām (Syria)					
29	Bayt al-Maqdis az Filistīn	66;30	31;50	14;6	Jerusalem
30	Asqalān az Filistīn	66;30	32;15	14;10	Ashkelon
31	Ramla az Filistīn	66;50	32;10	14;9	Ramlah
32	Qaysāriya	66;30	32;30	14;11	Banias
33	Tiberiyyas	68;0	32;0	14;8	Tiberias
34	Ṣawar	68;30	32;40	14;12	Tyre, Lebanon
35	Ṣaydā	68;35	32;40	14;12	Sidon, Lebanon
36	Tarābulus	69;40	34;0	14;19	Tripoli, Syria
37	Dimashq	70;0	33;15	14;16	Damascus, Syria
38	Ba'albak	70;45	33;15	14;16	Baalbek, Lebanon
39	Ḥimṣ	70;45	34;0	14;19	Homs, Syria
40	Anṭākiya az Thaghūr	71;26	35;30	14;27	Antakya, Turkey
41	Sarmīn	71;7	35;15	14;26	Sarmin, Syria
42	Qinnisrīn	72;0	35;30	14;27	Barad
43	Ḥalab	72;10	35;50	14;26	Aleppo, Syria
44	Mankh	72;15	36;15	14;32	
45	Raḥba	74;35	34;10	14;20	
46	Ṭarṭūs	68;40	36;42	14;34	Tartus, Syria

TABLE 6.4 *The Geographical Gazetteer* (cont.)

S.No.	*asmāʾ al-buldān*	*al-ṭūl*	*al-arḍ*	*al-sāʿāt*	Modern name
Thughūr Rūm (The Border Lands of the Byzantines)					
47	Massīsa wa Kagharyānā	69;15	36;45	14;45	Massis
48	Malaṭiya	71;0	37;0	14;36	Malatiya, Armenia
49	Saqalaba Jazīra Barbar Kinat	65;0	37;10	14;37	Saqaliba
50	Rūmiya Kubrā	85;27	41;50	15;6	Rome, Italy
51	Athīna Madīnat al-Ḥikma	65;40	37;20	14;38	Athens, Greece
52	Būr faṭbā (?) wa hua Qasṭanṭīn	59;50	45;0	15;30	Constantinople =
Istanbul, Turkey					
53	Māqadūniyā	60;0	41;0	15;1	Macedonia
54	ʿAmmūriya	64;0	43;0	15;15	
55	Azr al-Rūm	77;0	39;40	14;52	Erzurum, Turkey
56	Arzinkān	76;0	38;0	14;42	Arzingan
57	Sīwās	71;40	39;0	14;48	Siwas
58	Sumaysāt	72;35	40;00	14;54	Samsat, Turkey
59	Qālīqalā	73;25	39;0	14;48	Another name for Erzurum, Turkey
Jazīra (The Region of Ancient Mesopotamia)					
60	Qūniya	66;30	41;0	15;1	Konya, Turkey
61	Harrān az Jazīra	73;0	36;40	14;33	Ancient city, Turkey
62	Raqqa	78;15	36;0	14;30	Raqqa
63	Rās al-ʿAyn	74;0	36;50	14;34	Ras al-Ayn, Syria
64	Mārdīn	74;0	37;55	14;37	Mardin, Turkey
65	Qarqīsiyā	74;40	37;20	14;38	
66	Niṣibīn	75;20	37;0	14;36	
67	Sinjār	76;0	36;0	14;30	
68	ʿĀna	76;30	34;0	14;19	Anah, Iraq
Armīniya (Armenia)					
69	Mawṣil	77;0	34;30	14;21	Mosul, Iraq
70	Irbīl	77;20	35;0	14;25	Erbil, Iraq
71	Takrīt	78;20	34;30	14;21	Takrit, Iraq
72	Amid	73;40	38;0	14;42	
73	Malāzjird	75;00	38;45	14;46	Malazgirt, Turkey

TABLE 6.4 *The Geographical Gazetteer* (cont.)

S.No.	*asmāʾ al-buldān*	*al-ṭūl*	*al-arḍ*	*al-sāʿāt*	Modern name
74	Akhlāṭ	75;50	39;20	14;50	Ahlat, Turkey
75	Arjīsh	77;0	38;30	14;44	Arjish, Turkey
76	Anī	79;0	41;0	15;1	Ancient city, Turkey
77	Salmās	79;15	37;40	14;40	Salmas, Iran
78	Khuway	79;40	37;40	14;40	
79	Urmiya	79;45	37;0	14;36	Urmia, Iran
80	Marand	80;45	37;50	14;51	Marand, Iran
Azarbayjān (Azerbaijan)					
81	Nakhjawān	81;15	38;40	14;46	Nakhchivan, Azerbaijan
82	Marāgha	82;0	37;20	14;38	Maragheh, Iran
83	Tabrīz	82;0	38;0	14;42	Tabriz, Iran
Annulus 2					
Tatimma Azarbayjan (Continuation of Azerbaijan)					
84	Ardabīl	82;30	38;0	14;42	Ardabil, Iran
Irān (Iran)					
85	Mūqān	83;0	38;0	14;42	
86	Bardhaʿa	83;0	40;30	14;54	Barda, Azerbaijan
87	Jankara	83;0	41;20	15;3	
88	Samkūra	83;0	41;50	15;6	
89	Tiflīs dar al-mulk il-Khān	83;0	43;0	15;15	Tbilisi, Georgia
90	Alān	83;0	44;0	15;23	
Karkh (Western Shore of the Tigris River?)					
91	Saflān	83;30	39;50	14; 52	
92	Bākaway	84;30	39;30	14;50	
Alān (?)					
93	Shamākhī qaṣba Shīrwān	84;30	40;50	14;54	
94	Bāb al-Ābwāb	85;0	43;0	15;15	Derbent, Russia
Rūs (Russia)					
95	Balanjar dār al-mulk Khazar	85;20	46;30	15;42	
96	Kūyā Madīnat Rūs	87;0	43;0	15;15	
97	Saqsīn	86;30	43;0	15;15	

TABLE 6.4 *The Geographical Gazetteer* (cont.)

S.No.	*asmā' al-buldān*	*al-ṭūl*	*al-arḍ*	*al-sā'āt*	Modern name
Bulghār (The Bulghars, Region north of the Caspian Sea)					
98	Bulghār	90;00	49;30	16;9	
99	Buzbāl mābīn Rūs Bulghār	83;0	44;0	15;23	
100	Sarmīn Rāyy	79;0	34;0	14;19	
'Arāqayn (The Two Iraqs)					
101	Kūfa	79;30	31;30	14;6	Kufa, Iraq
102	Madā'in	72;0	33;10	14;15	Ctesiphon, Iraq
103	Baghdād	80;0	33;25	14;16	Baghdad, Iraq
104	Wāsit	81;30	32;20	14;10	Wasit, Iraq
105	Baṣra	84;0	30;0	13;58	Basra, Iraq
106	Ubulla	84;0	30;15	14;1	
107	'Abbādān	84;30	30;0	13;18	Abbadan, Iran
108	Jundisābūr	84;5	31;15	14;3	Gundeshapur, Iran
Khūzistān (a region in Iran)					
109	Tastīr	84;30	31;30	14;6	Tustar, Iran
110	'Askar Mukram	84;30	31;15	14;5	Askar Mukram, Iran
111	Ahwāz	85;0	31;0	14;3	Ahvaz, Iran
112	Rāmhurmūz	85;45	31;0	14;3	Ramhormoz, Iran
113	Arrajān	86;30	30;30	14;1	
114	Shāpūr	86;55	30;0	13;18	
Fārs (Persia, a province of Iran)					
115	Kāzirūn	87;00	29;15	13;53	Kazerun, Iran
116	Nawbandjān	87;15	30;10	13;59	Nawbandegan, Iran
117	Fīrū[z]ābād	87;30	28;10	13;50	Firuzabad, Iran
118	Sīrāf	88;00	29;0	13;53	Siraf, Iran
119	Shīrāz	88;0	29;36	13;55	Shiraz, Iran
120	Iṣṭakhr	88;0	30;0	13;18	Ancient city, Iran
'Irāq					
121	Yazd	89;0	32;0	14;8	Yazd, Iran
122	Shahrazūr	82;20	32;30	14;10	Shahrizor, Iraq
123	Hulwān	82;15	34;0	14;19	Ancient city, Iran
124	Qirmāsīn	83;0	34;30	14;21	Kermanshah, Iran
125	Dīnawar mā' al-Kūfa	83;0	35;0	14;25	Ancient city, Iran

TABLE 6.4 *The Geographical Gazetteer* (cont.)

S.No.	*asmā' al-buldān*	*al-ṭūl*	*al-arḍ*	*al-sāʿāt*	Modern name
126	Suhraward	83;20	36;0	14;30	Suhraward, Iran
127	Nahawand mā' al-Ṣara	83;15	36;20	14;21	Nahavand, Iran
128	Zankān	83;40	36;30	14;32	Zanjan, Iran
129	Abhar	84;30	36;45	14;34	Abhar, Iran
130	Hamadān	83;0	35;10	14;26	Hamadan, Iran
131	Karkh	84;45	34;0	14;19	Western half of Baghdad?
132	Sāwa	85;0	35;0	14;25	
133	Qazwīn	85;0	36;0	14;30	Qazvin, Iran
134	Aba	85;0	34;40	14;28	
135	Zarbādqān	85;0	34;15	14;20	
136	Semīrum	85;40	32;15	14;9	
137	Iṣfahan	86;40	32;25	14;10	Isfahan, Iran
138	Kāshān	86;0	34;0	14;19	Kashan, Iran
Dīlmān					
139	Qumm	85;40	34;45	14;28	Qom, Iran
140	Rayy	86;20	35;0	14;19	Ancient city, now part of Tehran, Iran
141	Ḥuwār	87;10	35;40	14;29	
142	Alamūt	85;37	36;21	14;33	The Fortress of Alamut, Iran
Kīlān (Gilan, a province of Iran)					
143	Tāliqān	85;45	36;10	14;31	Taliqan, Afghanistan
144	Hūsum az Jīlān	85;10	37;0	14;36	
145	Kajūr wa Kalār	86;50	36;25	14;32	
146	Dīma qaṣba Damāwand	87;20	37;10	14;37	
Ṭabaristān (or Tapuria, southern coastal region of the Caspian Sea)					
147	Amul qaṣba Ṭabaristān	88;20	36;15	14;31	Amol, Iran
148	Sāriya	88;0	37;0	14;36	
149	Astarābād	89;35	36;50	14;35	Gorgan, Iran

TABLE 6.4 *The Geographical Gazetteer* (cont.)

S.No.	*asmā' al-buldān*	*al-ṭūl*	*al-arḍ*	*al-sā'āt*	Modern name
Jurjān (a region in Iran)					
150	Abaskūn qaṣba Jurjān	89;30	37;15	14;37	Ancient port on the Caspian sea
151	Jurjān	90;0	36;50	14;35	Jurjan, Iran
152	Samnān	88;0	36;0	14;30	Semnan, Iran
Qūmis (a region in Iran)					
153	Dāmghān	88;55	36;20	14;32	Damghan, Iran
154	Bistām	89;30	36;10	14;31	Bastam, Iran
Annulus 3					
Iqlīmhā Tatimma Qūmis (regions of: continuation of Qūmis)					
155	Biyār	89;50	38;45	14;28	
156	Farawā	90;0	39;0	14;48	
157	Marīnān	90;30	37;0	14;36	
Khurāsān (Khorasan, a region in Iran)					
158	Sabzawār	91;30	36;5	14;30	Sabzevar, Iran
159	Nīshābūr	92;30	36;21	14;31	Nishapur, Iran
160	Ṭūs	92;30	36;0	14;30	Tus, Iran
161	Nar[ma]shīr	92;0	35;0	14;25	Narmashir, Iran
162	Ṭūs Gīlakī (?)	92;0	33;0	14;3	
163	Tūn	92;30	34;30	14;21	Ferdows, Iran
164	Qā'īn	93;20	33;40	14;16	Qain, Iran
165	Ṭabas Masinā	94;15	33;15	14;14	
166	Zūzan	93;30	35;20	14;27	
167	Busha'	94;5	34;50	14;28	
168	Harat	94;20	34;30	14;21	Herat, Afghanistan
169	Bādghīs	94;30	35;20	14;27	Badghis, province in Afghanistan
170	Sarakhs	94;30	37;0	14;36	Sarakhs, Iran
171	Marw al-Rūd	97;0	36;30	14;32	
172	Marw	97;0	37;40	14;39	Mary, Turkmenistan
173	Jūzajān	98;0	35;30	14;27	
174	Fāriyāb	99;0	36;45	14;33	Faryab, province in Afghanistan
175	Ashbūrqān	105;0	36;45	14;33	

TABLE 6.4 *The Geographical Gazetteer* (cont.)

S.No.	*asmā' al-buldān*	*al-ṭūl*	*al-arḍ*	*al-sā'āt*	Modern name
176	Balkh	101;0	36;41	14;33	Balkh, Afghanistan
177	Bāmiyān	102;0	34;35	14;20	Bamiyan, Afghanistan
Khatlān (a province of Tajikistan)					
178	Samankān	102;0	36;0	14;30	Samangan, Afghanistan
179	Qayādyān (?)	102;0	37;45	14;39	
180	Walwālij	102;20	36;55	14;34	Afghanistan
181	Tāliqān az Tukhāristān	102;50	37;25	14;37	Taloqan, Afghanistan
Khwārizm					
182	Andarāb	103;45	36;0	14;30	Anadarab, Iran
183	Badakshān	104;24	37;10	14;37	Province in Afghanistan
184	Bulūr	108;0	37;0	14;30	
185	Būkarkānj	93;45	42;35	15;10	
186	Kurkanj Dār al-Mulk Khwārizm	94;30	42;12	15;9	Urgenz, Khorezm Province, Uzbekistan
Mā wrā' al-Nahr (what is beyond the [Oxus] river, Transoxonia)					
187	Hazār [A]sf	95;0	41;0	15;1	Hazorasp, Uzbekistan
188	Kāt	95;0	41;36	15;3	
189	Darqān	94;15	39;30	14;50	
190	Bukhārā	97;30	39;50	14;52	Bukhara, Uzbekistan
191	Samarqand	99;56	39;37	14;38	Samarkand, Uzbekistan
192	Nasaf	98;0	39;0	14;48	Qarshi, Uzbekistan
193	Kash	99;30	39;30	14;50	Kash, Iran
194	Isfījāb az Shāsh	99;50	43;36	15;12	
195	Ṭarāz	99;50	44;31	15;25	
196	Abīr washnah (?)	105;0	40;0	14;54	
197	Ṣafāniyā	105;30	38;14	14;43	
198	Khujand	105;35	41;55	15;6	Sughd, Tajikistan

TABLE 6.4 *The Geographical Gazetteer* (cont.)

S.No.	*asmā' al-buldān*	*al-ṭūl*	*al-arḍ*	*al-sā'āt*	Modern name
199	Nīkab qaṣba Shāsh	101;0	43;0	15;15	
200	Būkah qaṣba Aylān	101;0	43;25	15;27	
201	Akhskab qaṣba Farghāna	101;20	42;25	15;10	
Annulus 4					
Iqlīmhā Tatimma Mā warā' al-Nahr (continuation of what is beyond the river [Oxus], Trans-Oxania)					
202	Ūzkand	102;50	44;0	15;28	
203	Ūsh	102;20	43;20	15;12	
204	Kāshghar	106;30	44;0	15;28	Kashgar, China
Tabaraknān (?)					
205	Jāj wahua Shāsh	109;0	42;30	15;8	
206	Khutan	107;0	42;0	15;8	Khotan, China
207	Almāligh	102;30	44;0	15;28	Almaliq, Xinjiang, China
208	Qabāligh	108;0	44;0	15;28	
209	Ushsh Kalūrān	110;0	46;0	15;38	
210	Pīsh Bāligh	102;0	45;50	15;35	
211	Qarāqūrūm	105;0	46;0	15;38	Karakorum, mountain range
212	Khān Bāligh	124;0	46;0	15;38	Beijing, China
213	Nankyās	130;0	39;0	14;48	
Kirmān (a province of Iran)					
214	Mahrī	140;0	30;0	13;58	
215	Shīrjān	90;20	29;30	13;55	Sirjan, Iran
216	Hurmūz	92;0	25;0	13;35	Hormuz, Iran
217	Bardsīr	92;30	29;50	13;57	Bardsir, Iran
218	Jīruft	93;0	27;30	13;45	Jiroft, Iran
219	Bam	94;0	28;30	13;51	
220	Narmāshīr	94;20	29;0	13;53	Narmashir, Iran
221	Khabiṣ	93;0	31;0	14;3	Khabis, Iran
222	Zaranj Madīnat Sijitān	90;0	32;30	14;10	
Zābalistān (a region in Afghanistan)					
223	Bust	105;0	33;0	14;13	
224	Takinābād	101;55	33;20	14;15	

TABLE 6.4 *The Geographical Gazetteer* (cont.)

S.No.	*asmā' al-buldān*	*al-ṭūl*	*al-arḍ*	*al-sā'āt*	Modern name
225	Mīmand	101;55	33;20	14;15	
226	Ghazna qaṣba Zābalistān	104;20	33;35	14;16	Ghazni, Afghanistan
227	Banjāhīr Kābulistān	104;40	35;0	14;25	
228	Ribāṭ Amīr	105;0	34;0	14;19	
229	Tabar qaṣba Makrān (?)	98;0	26;15	13;41	
230	Kīj az Makrān	99;0	28;0	13;49	
Makrān (Coastal strip in the south of Sindh and Baluchistan, in Pakistan)					
231	Jālq az Makrān (?)	99;0	30;0	13;58	
232	Daybul wa Hūtana	102;30	25;10	13;36	Debal, ancient port near Karachi, Pakistan
233	Mansūra qaṣba Sind[e]	105;0	27;40	13;48	Brahmanabad, Pakistan
234	Tāna az Sāhil Baḥr Hind[f]	102;0	19;20	13;12	
Sind (Sindh)[g]					
235	Nahlāwārā	102;30	22;0	13;22	Anhilwad = Patan, Gujara, India
236	Sawmnāt	106;0	17;0	13;2	Somnath, India
Annulus 5					
Iqlīmhā Tatimma Sind (Continuation of Sindh)					
237	Multān	107;35	29;40	13;46	Multan, Pakistan
238	Qandahār	107;40	33;0	14;13	Qandahar, Afghanistan
239	Lahūr dār al-Salṭanat	109;20	31;50	14;54	Lahore, Pakistan
Hind (India south of Indus)					
240	Qannawj	115;50	26;35	13;42	Kannauj, India
241	Bānārasī	117;20	26;55	13;45	Varanasi, India
242	Sarandīp	130;0	10;0	12;40	Sri Lanka
243	Kashmīr	108;0	35;0	14;25	Kashmir =Srinagar, India

TABLE 6.4 *The Geographical Gazetteer* (cont.)

S.No.	*asmā' al-buldān*	*al-ṭūl*	*al-arḍ*	*al-sā'āt*	Modern name
244	Tubbat	110;0	40;0	14;54	Tibet=Lhasa
245	Akbarābād dar al-Khilāfat	114;0	27;0	13;45	Agra, India
246	S[h]āhjahānābād dār al-mulk Hind	113;35	28;39	13;52	Delhi, India
247	Ajmer	112;0	26;0	13;31	Ajmer, India
248	Kawāliyār	115;0	26;29	13;42	Gwalior, India
249	Burhānpūr	109;0	20;31	13;14	Burhanpur, India
250	Sūrat	110;50	20;15	13;14	Surat, India
251	Dawlatābād	111;0	20;30	13;14	Daulatabad, india
252	Ujjayn	112;0	22;30	13;24	Ujjain, India
253	Bījāpūr	115;30	17;20	13;42	Bijapur, India
254	Kūlkanda	114;39	18;4	13;9	Golconda =Hyderabad, India
255	Dhakā Bankālā	130;0	27;0	13;45	Dhaka, Bangla Desh
256	Patnā	120;45	26;40	13;44	Patna, India
257	Jawnpūr	119;6	26;36	13;44	Jaunpur, India
258	Mankīr	121;45	26;40	13;44	Mungher, India
259	Allāhābād	115;34	26;50	13;44	Allahabad, India

a It belongs to Sudan, the next region.
b Qulzum should be placed in Egypt.
c Medina should be included in Hejaz.
d It should be in Hejaz.
e Manṣūra should be in Sind.
f "Tāna on the shore of the ocean of Hind." It is actually Thane, near modern Mumbai; wrongly placed in Makrān.
g The two places are in Gujarat, i.e., in the region Hind and not in Sindh.

The present astrolabe mentions three more localities, viz., Allahabad, Mungher and Patna. The southern localities generally mentioned are Daulatabad, Golconda (i.e., modern Hyderabad) and Bijapur. This scanty information is particularly surprising in view of the fact that the Mughal Empire extended quite far into the east and south.

Most of the astrolabe gazetteers mention "Sarandīb" which is the Arabic name for Sri Lanka.[52] While the latitude of 10° is a reasonably correct value for the northern tip of the island, the longitude of 130° places the island far in the East on the same meridian as Dhaka.[53] This is inexplicable because the island must have been well known to the Arab sailors. In Sanskrit astronomical texts, *Laṅkā* stands for the imaginary point on the terrestrial equator through which the Prime Meridian of India passes, touching Ujjain and several other localities in the North. But the Sanskrit gazetteer caused to be engraved by Nayanānanda on the present astrolabe repeats the coordinates of the Arabic gazetteer, while rendering "Sarandib" with the traditional Sanskrit name "Siṃhala-dvīpa," from which the word "Ceylon" is derived.[54]

Aside from the variation in the arrangement of the localities, the gazetteers exhibit also varying styles of presentation, because the large area of the inner surface of the mater allows all kinds of stylistic experimentation: while the rows remain circular, the columns are sometimes curved, all slanting in a single direction and sometimes zigzag. The orientation of the letters also varies; in some gazetteers the upper part of the letters is oriented towards the centre and sometimes towards the outer circumference, and sometimes both in different annuli.

But the number of items in each entry remained almost constant with the name of the locality, its longitude and latitude, with one notable exception. Allāhdād, the patriarch of the Lahore family, in his astrolabe of 1567 which is the earliest extant astrolabe produced in this family, allotted a column each for longitudes, latitudes, *inḥirāf* and the duration of the longest day (*al-sāʿāt*) in the gazetteer, but left the last two columns blank without filling in the values. The same is the case with an unsigned and undated astrolabe at Greenwich that is attributed to him.[55] But in his undated astrolabe at the Museum of the History of Science at Oxford, Allāhdād filled the column for the duration of the longest day with appropriate values for each locality. His descendants, ʿĪsā, Qā'im, Muqīm and Jamāl al-Dīn did not have this column in their gazetteers at all. Their gazetteers are limited to just the two coordinates of longitudes and

52 The Arabic name *Sarandīb*, which gave rise to the felicitous word "serendipity" in English, must have been derived from Sanskrit *Śaraṇa-dvīpa*, "the island of shelter," for Sri Lanka offered shelter to the ships tossed about by the fierce monsoon winds. However, the expression is not attested in any Sanskrit source.

53 See item no. 242 in the Geographical Gazetteer in Table 6.4. Even the *'Ain-i-Ākbari of Abul Fazl-i-'Āllāmi*, 3.57, has the same longitude of 130°, but the latitude there is 12°.

54 See item no. 6 in Geographical Gazetteer in Sanskrit in Table 6.5.

55 François Charette, "An Early Moghul Astrolabe," in *Astrolabes at Greenwich: A Catalogue of the Astrolabes in the National Maritime Museum*, ed. Koenraad van Cleempoel et al. (Oxford: Oxford University Press & National Maritime Museum, 2005), 227–231.

latitudes, and eschewed the *inhirāf* and the *qibla* direction which are common elsewhere in the Islamic world.[56] This is the first time after Allāhdād that the column for the duration of the longest day reappears in Ḍiyā' al-Dīn's astrolabe.

In this Khalili astrolabe each annulus contains four rows of cells to accommodate the four elements of the gazetteer. The cartouche-like cells are created by intertwining chains drawn by two parallel lines.[57] Again in each annulus, the design of the cartouches varies. The letters are oriented towards the centre of the astrolabe.

The Sanskrit Gazetteer Engraved on the Edge

The surface of the mater is so completely filled with the Arabic gazetteer that there is hardly any space for the additions in Devanāgarī script. Therefore, Nayanānanda caused a much smaller gazetteer of forty-five localities to be engraved in Devanāgarī script along the narrow surface of edge of the astrolabe measuring 14 mm, starting from the foot of the *kursī* and proceeding downwards (Figure 6.16). The first twenty-three cities in this Sanskrit gazetteer are from India and are copied exactly from the fifth annulus of the Arabic gazetteer. The remaining localities are selected at random from the first and second annuli. Leaving aside a few well-known names like Mecca, Medina, Misr, Maragha, Shiraz, Mosul, Baghdad, Kufah, the rest of the names are so badly written that they cannot be deciphered. Occasionally there are errors in copying the numerical figures. The correct numbers are shown in parentheses in Table 6.5.

Starting from the bottom of the *kursī* on the other side, a table in three columns is engraved on the edge, which eventually meets the geographical gazetteer coming from the opposite direction. This is a Sanskrit version of the table engraved on the back, within the shadow squares, containing the multiples of differences (in minutes and seconds) between the approximately correct length of the tropical year and 365 days. The original Arabic table and the Sanskrit version is reproduced in Table 6.6.

---

56 In some of his astrolabes, Ḥāmid has columns for both *inḥirāf* and *jihat*.

57 The design of the geographical gazetteer in the present astrolabe where the entries are engraved within small cartouches has a distinct resemblance to the design of the geographical gazetteer on another massive astrolabe which is not dated, but bears the signature of Ḍiyā al-Dīn (London, Khalili Collection, accession no. SC283; see Maddison & Savage-Smith, *Science, Tools & Magic*, 1:226–227).

FIGURE 6.16 *The geographical gazetteer in Sanskrit engraved on the edge. The Nasser D. Khalili Collection of Islamic Art, Accession No. SCI 53.*
PHOTO BY THE AUTHOR.

TABLE 6.5 *The Geographical Gazetteer in Sanskrit*

	Place name (*deśanāma*)	Correct form/ modern name	L (*śara*, read *deśāmtara*)	φ (*akṣaṃśa*, read *akṣāṃśa*)	Duration of the longest day (*paramadina*)
1	Mulatāna	Multan	107;35	33;40 (29;40)	14;46 (13;46)
2	Kaṃdhāra	Qandahar	107;40	33;00	14;13
3	Lahaura	Lahore	119;20	31;50	14 19
4	Kanauja	Kannauj	115;50	26;35	13;42
5	Banārasa	Banaras	117;20	26;55	13;45
6	Ma*ladāpa	Siṃhaladvīpa (Sri Lanka)	130;00	10;00	12;40
7	Kasmīra	Kashmir	108;00	35;00	14;25
8	Tidyata	Tibbata (Tibet)	110;00	40;00	14;54
9	Āgarā	Agra	104;00	27;00	13;45
10	Dillī	Delhi	113?;35	28;39	13;52
11	Ajamera	Ajmer	111;50 (112; 05)	26;00	13;31
12	Gvāliyara	Gwalior	115;00	26;29	13;42
13	Burahānapura	Burhanpur	109;00	20;34 (20;31)	13;14
14	Sūrata	Surat	110;50	20;15	13;14
15	Dolatābāda	Daulatabad	110;00	20;30	13;14 (13;24)
16	Ujena	Ujjain	102;0	22;30	13;24
17	Vījā[pura]	Bijapur	15;30 (115; 30)	17;20	13;4
18	Gulakuḍā	Golconda (Hyderabad)	115;40 (114 ;39)	18;4	13;9

TABLE 6.5 *The Geographical Gazetteer in Sanskrit* (cont.)

	Place name (*deśanāma*)	Correct form/ modern name	L (*śara*, read *deśāmtara*)	φ (*akṣaṃśa*, read *akṣāṃśa*)	Duration of the longest day (*paramadina*)
19	Ḍhākā Vaṃgāla	Dhaka	130;0	27;0	13;45
20	Paṭanā	Patna	110;45 (120;45)	26;40	13;49 (13;44)
21	Jonapura	Jaunpur	115;6 (119;6)	26;36	13;44
22	Muṃgera	Mungher	121;45	26;40	13;44
23	Alāhābā[da]	Allahabad	115;24	26;50	13;44
24	Makā	Mecca	77;10	21;40	13;20
25	Madīnā	Medina	74;20 (75;20)	25;0	13;30 (13;35)
26	Tiphaka??	Taif	77;33 (77;30)	21;29 (21;20)	13;20
27	Traphagru??		76;11?	12;20?	3?
28	Misara	Misr = Cairo	63;20	30;20	14;1
29	Phakaparīṣa	Iskandariyā Alexandriya	61;54	30;58	14;35 (14;32)
30	Rahatavava-raphaka	??	65;0	14;0	12;54
31	Mahadāsā	Mahdiyyah	42;0	32;30	14;11
32	Keravāṃ	Qayrawan	41;7 (41;0)	31;40 (31;30)	14;73 (14 7)
33	Tabaja	Tabriz	102;5 (82;5)	38;0	14;40 (14;42)
34	Marāgaha	Maragha	102;0 (82;0)	37;20	14;28 (14;38)
35	Naravajanvā	Nakhichevan	81;15	38;40	14;46
36	Ghoy	Khoy	79;40	37;40	14;40
37	Amada	Amid	33;40 (73;40)	38;0	14;40 (14;42)
38	Mūsa	Mosul	77;00	34;30	14;31 (14;21)
39	Ānaha	Ānah	76;30	34;00	14;19
40	Tamasāva	Sumaysāṭ	72;35	40;00	24;54 (14;54)
41	Yatre	Yazd	89;00	32;00	14;8
42	Sīrāja	Shīrāz	88;00	29;36	13;15 (13;55)
43	Śāhapura	Shābūr	86;55	30;00	13;58(13;18)
44	Namarā	Basra	84;00	30;00	13;58
45	Kūphaha	Kūfah	79;30	31;30	14;6

The Back

The back of the astrolabe (*ẓahr al-aṣṭurlāb*) is also divided into four quadrants by the vertical and horizontal diameters that are marked by dotted lines (Figure 6.17). In the upper half, the two altitude scales on the rim contain three concentric bands as on the limb. The innermost band is divided into 0.25°, the middle one into single degrees and labelled in standard Arabic numerals and in the outermost band groups of 3° are marked in *abjad* notation from three to ninety, starting from the east and west points and reaching the south point at the top. Sanskrit numerals are added in this band, sometimes above the *abjad* numerals, sometimes below and some other times next to them.

On the upper left is the trigonometric quadrant, with sixty equally spaced horizontal parallel lines and the same number of equally spaced vertical parallel lines (Figure 6.18). Every third line is dotted. Two radial lines divide the quadrant into three segments of 30° each. The sixty vertical lines are numbered along the radius at the bottom of the quadrant, first in standard Arabic numerals from one to sixty and then in groups of three in *abjad* notation, both times starting from the apex of the quadrant and then proceeding towards the

FIGURE 6.17
The back of the Khalili astrolabe. The Nasser D. Khalili Collection of Islamic Art, Accession No. SCI *53.*
PHOTO COURTESY OF THE KHALILI FAMILY TRUST © NOUR FOUNDATION.

FIGURE 6.18
The trigonometric quadrant on the back. The Nasser D. Khalili Collection of Islamic Art, Accession No. SCI *53.*
PHOTO COURTESY OF THE KHALILI FAMILY TRUST © NOUR FOUNDATION.

circumference. The horizontal parallels are numbered in the same manner along the vertical radius on the right. The *abjad* numerals in groups of 3° are repeated later in Devanāgarī numerals.

On the upper right is the zodiac quadrant, consisting of ninety equally spaced concentric quarter circles, with each fifth highlighted with dots (Figure 6.19). These are labelled along the horizontal radius from the winter solstice to the summer solstice with the names of the zodiac signs *al-Jadī, al-Dalw, al-Ḥūt, al-Ḥamal, al-Thawr* and *al-Jawzā'*, starting from the circumference and reaching up to the centre of the astrolabe. Above the names of the signs, the quarter circles are numbered in standard Arabic numerals one to thirty, separately for each sign. Sanskrit names of the signs are added above the Arabic names: *makara, kuṃbha, mīna, meṣa, vṛṣa, mithuna.* Instead of serial numbers one to thirty above each sign, Devanāgarī numerals record only the decans (ten, twenty, and thirty).

The quarter circles are labelled along the vertical radius from the summer solstice to the winter solstice with the names of the zodiac signs *al-Saraṭān, al-Asad, al-Sunbula, al-Mīzān, al-ʿAqrab* and *al-Qaws*, starting from the centre of the astrolabe and reaching up to the circumference. To the right of the names of the signs, the quarter circles are numbered in standard Arabic numerals one to thirty, separately for each sign. Sanskrit names of the signs are added next to the Arabic names: *karka, siṃha, kanyā, tulā, vaścika* (sic! read *vṛścika*) and *dhana* (sic! *dhanuḥ*). Instead of serial numbers one to thirty above each sign, Devanāgarī numerals record only the decans (ten, twenty, and thirty).

Upon this grid of quarter circles drawn for each degree of zodiac signs are plotted three sigmoid graphs for the meridian altitude of the Sun, throughout

FIGURE 6.19
The zodiac quadrant on the back. The Nasser D. Khalili Collection of Islamic Art, Accession No. SCI 53.
PHOTO COURTESY OF THE KHALILI FAMILY TRUST © NOUR FOUNDATION.

the year, for the latitudes of 27°, 29° and 32°, one below the other in this order. Upon these graphs are engraved the labels *khaṭṭ nisf al-nahār bi-ʿarḍ 27* (line of midday at latitude 27°), *khaṭṭ nisf al-nahār bi-ʿarḍ 29*, and *khaṭṭ nisf al-nahār bi-ʿarḍ 32*. As mentioned earlier, these graphs are peculiar to the astrolabes produced in the Lahore family of astrolabists and also firmly establish 1648 as the terminus a quo for the time when the astrolabe was produced. Surprisingly the labels of these graphs are not rendered into Sanskrit.[58]

In the lower half, there are two shadow squares to the left and right of the vertical diameter (Figure 6.20). The square on the left is designed for a gnomon of seven feet; the horizontal and vertical scales are divided in seven units and numbered in *abjad* notation from one to seven. The square on the right is designed for a gnomon of twelve feet; the horizontal and vertical scales are divided in twelve units (and each unit is further subdivided into five parts) and numbered in *abjad* notation from one to twelve. These shadow squares carry the following labels:

> Vertical scale left: *ẓill sullam aqdām maʿkūs*, "reverse shadow in feet"
> Horizontal scale left: *ẓill sullam aqdām mustawī*, "direct shadow in feet"

---

58 A Sanskrit astrolabe made for Rāghavajit at Surat in 1669 carries the graph for the meridian altitude of sun at latitude 22° which is roughly the latitude of Surat. There the graph is designated as *madhya rekhā* (middle line); *madhyāhna-rekhā* (midday line) would have been more appropriate. This astrolabe is illustrated in Mahendra Sūri, *Yantrarāja*, ed. K. K. Raikva (Bombay: Nirnaya Sagar Press 1936). The pagination in this book is unusual. There are three sections in this book, each numbered separately starting at 1; the astrolabe in question is illustrated on 1–13 in the middle section; for the back, see p. 1.

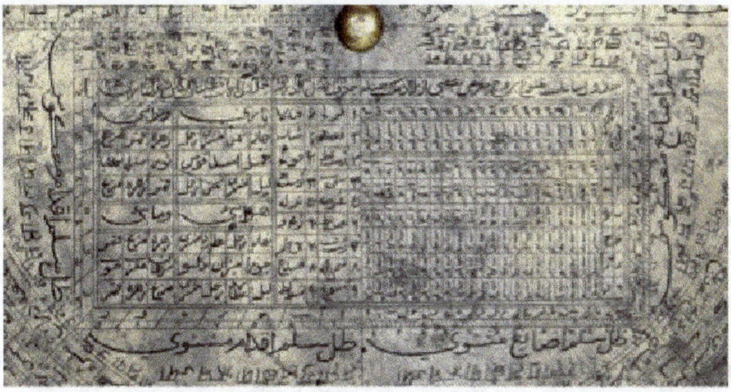

FIGURE 6.20 *The shadow squares on the back. The Nasser D. Khalili Collection of Islamic Art, Accession No. SCI 53.* PHOTO COURTESY OF THE KHALILI FAMILY TRUST © NOUR FOUNDATION.

Horizontal scale right: *ẓill sullam aṣabiʿ mustawī*, "direct shadow in digits"
Vertical scale right: *ẓill sullam aṣabiʿ maʿkūs*, "reverse shadow in digits."

The corresponding Sanskrit labels added later are:

Vertical scale left: *saptāṃgulachāyā utkramajyā*, "shadow of seven digits, *utkramajyā*"[59]
Horizontal scale left: *saptāṃgulachāyā kramajyā*, "shadow of seven digits, *kramajyā* "
Horizontal scale right: *dvādaśāṃgula chāyā kramajyā*, "shadow of twelve digits, *kramajyā*"
Vertical scale right: *dvādaśāṃgula chāyā utkramjyā*, "shadow of twelve digits, *utkramajyā*."

The *abjad* numbers on the scales are repeated later in Devanāgarī script.

59 In Sanskrit astronomy these two terms refer to the trigonometrical functions of an arc of a circle; while the former means the "chord of an arc," the latter is its reverse, i.e., the difference between the radius and the *kramajyā*. It is not correct to use these terms in the present context in the sense of umbra recta or direct shadow and umbra versa or reverse shadow. In fact, Sanskrit writers had difficulty in labelling the shadow squares as they were not familiar with the tangent and cotangent functions involved here. They do not also make the distinction between the "feet" and "digits" in denoting the length of the gnomons.

Astrological Tables on the Back

Inside these shadow squares are engraved three tables. The first table from the right is designated as *Jadwal sāʿāt ṭulūʿ burūj bi-ʿarḍ baʿḍī az bilād mulk Hind* (Table of times of rising of the signs for the latitudes of some localities in India). In eight columns and twenty-three rows, this table gives the rising times of the signs from latitudes 21° to 40°.[60]

The second table, named *Jadwal faḍl al-dawr*[61] (Table of excess of revolution), displays the multiples of differences, in minutes and seconds, between the approximately correct length of the tropical year and 365 days (Table 6.6).[62] The number of days is written in standard Arabic numerals and the difference in *abjad* notation. The Sanskrit version of this table is engraved on the edge of the astrolabe, as mentioned above. In a few cases, the Sanskrit version misreads the Arabic numbers. Both the Arabic and Sanskrit versions are shown in the following table. Where the Sanskrit version differs, the values are shown in italics.

The third table inside the shadow squares, designated as *Jadwal arbāb al-muthallathāt laylī wa nahār* (Table of regents of the triplicities by day and night), shows the signs grouped as triplicities and their planetary regents of the day and of the night (Table 6.7).[63] In the narrow space above this table and the horizontal diameter, the Sanskrit version of this table is engraved, using just the first syllables of the names of the signs and their planetary regents, i.e., the planets that have their diurnal/nocturnal domicile in these signs. Both the Arabic and Sanskrit tables are reproduced below. In the Arabic table, the signs are arranged from the right to the left; the Sanskrit version follows the same order.

The rim of the lower half contains two circular scales of shadows; that on the left is designated as *ẓill aqdām mabsūṭ* (horizontal shadow in feet), with unequal divisions numbered from the vertical radius clockwise from 1 to 24, 26, 28, 30, 33, 36, 42, 48, 54, 64?, 69?, 70, ?, 90, followed by the designation (Figure 6.21). The scale on the right is named *ẓill aṣābiʿ mabsūṭ* (horizontal shadow in digits), with unequal divisions numbered from the vertical radius counterclockwise as

---

60 Ḍiyāʾ al-Dīn's astrolabe of 1656 at Jai Singh's Observatory at Jaipur has a similar table for the latitudes of 29 to 39°; it is reproduced in G. R. Kaye, *The Astronomical Observatories of Jai Singh* (Calcutta: Archaeological Survey of India, 1918; reprint: New Delhi, 1982), 23.

61 On the astrolabe the name is engraved as *Jadwal faḍl al-darū*, obviously by mistake.

62 See Kaye, *The Astronomical Observatories of Jai Singh*, 24; Sreeramula Rajeswara Sarma, *Astronomical Instruments in the Rampur Raza Library* (Rampur: Raza Library, 2003), 50–52.

63 See Kaye, *The Astronomical Observatories of Jai Singh*, 23, where the same table from Ḍiyāʾ al-Dīn's astrolabe of 1656 at Jai Singh's Observatory at Jaipur is reproduced.

TABLE 6.6 *Multiples of Difference between the Length of the Tropical year and 365 Days*

Number of days	Excess in minutes and seconds		Number of days	Excess in minutes and seconds	
	Arabic	Sanskrit		Arabic	Sanskrit
1	87 33	87 33	10	155 31	155 31
2	175 6	175 6	20	61 2	*311 2*
3	262 39	*263 4*	30	106 33	106 33
4	350 12	350 12	12	262 10	262 4
5	77 45	77 45	50	57 35	*57 34*
6	164 19	*165 19*	60	213 6	213 *
7	252 52	252 52	70	08 37	08 37
8	340 25	*340 20*	80	164 8	164 8
9	67 58	*66 58*	90	319 39	*309 39*

TABLE 6.7 *Triplicities of Signs and their Regents of the Day and of the Night*

Arabic					Sanskrit		
	nārī (fiery)				*agni* (fire)		
zuḥal	*mushtarī*	*shams*	*nahā*	*dina*	*śa[ni]*	*bṛ[haspati]*	*sū[rya]*
Saturn	Jupiter	Sun	day	day	Saturn	Jupiter	Sun
qaws	*asad*	*ḥamal*			*dha[nuḥ]*	*si[ṃha]*	*meṣa*
Sagattarius	Leo	Aries			Sagittarius	Leo	Aries
zuḥal	*shams*	*mushtarī*	*layl*	*niśā*	*śa[ni]*	*s[ūrya]*	*bṛ[haspati]*
Saturn	Sun	Jupiter	night	night	Saturn	Sun	Jupiter
	turābī (earthy)				*bhū* (earth)		
mirrīkh	*qamar*	*zuhara*	*nahār*	*dina*	*maṃ[gala]*	*caṃ[dra]*	*śu[kra]*
Mars	Moon	Venus	day	day	Mars	Moon	Venus
jadī	*sunbula*	*thawr*			*ma[kara]*	*ka[nyā]*	*vṛ[ṣa]*
Capricorn	Virgo	Taurus			Capricorn	Virgo	Taurus
mirrīkh	*zuhara*	*qamar*	*layl*	*niśā*	*maṃ[gala]*	*śu[kra]*	*caṃ[dra]*
Mars	Venus	Moon	night	night	Mars	Venus	Moon

TABLE 6.7 *Triplicities of Signs and their Regents of the Day and of the Night* (cont.)

	Arabic					**Sanskrit**	
	hawāʾī (airy)					*vāta* (air)	
mushtarī	*ʿuṭārid*	*zuḥal*	*nahār*	*dina*	*bṛ[haspati]*	*bu[dha]*	*śa[ni]*
Jupiter	Mercury	Saturn	day	day	Jupiter	Mercury	Saturn
dalw	*mīzān*	*jawjāʾ*			*kuṃ[bha]*	*tu[lā]* Libra	*mi[thuna]*
Aquarius	Libra	Gemini			Aquarius		Gemini
mushtarī	*zuḥal*	*ʿuṭārid*	*layl*	*niśā*	*bṛ[haspati]*	*śa[ni]*	*bu[dha]*
Jupiter	Saturn	Mercury	night	night	Jupiter	Saturn	Mercury
	māʾī (watery)					*jala* (water)	
qamar	*mirrikh*	*zuhar*	*nahār*	*dina*	*caṃ[dra]*	*maṃ[gala]*	*śu[kra]*
Moon	Mars	Venus	day	day	Moon	Mars	Venus
ḥūt Pisces	*ʿaqrab*	*saraṭān*			*mīna*	*vṛ[ścika]*	*ka[rka]*
	Scorpio	Cancer			Pisces	Scorpio	Cancer
qamar	*zuhara*	*mirrīkh*	*layl*	*niśā*	*caṃ[dra]*	*śu[kra]*	*maṃ[gala]*
Moon	Venus	Mars	night	night	Moon	Venus	Mars

FIGURE 6.21
Scales in the lower half of the back. The Nasser D. Khalili Collection of Islamic Art, Accession No. SCI 53.
PHOTO COURTESY OF THE KHALILI FAMILY TRUST © NOUR FOUNDATION.

1–30, thereafter 32, 34, 36, 38, 40, 42, 44, 46, 48, 50, 54, 56, 60, 63, 66, 72, 78, 84, 90, followed by the designation. The names of the two circular scales were not rendered into Sanskrit. The numbers on the scales are repeated in Devanāgarī numerals as far as the space permitted.

The space between the rim containing the circular shadow scales and the two shadow squares is filled with seven semicircular scales; each two rows are separated by a narrow band.

The first scale, as counted from the outside, consists of three rows. While the third row is filled with the twelve names of the signs of the zodiac, the first row is divided into single degrees and in the second row groups of 5° are numbered, separately for each sign (as, 5, 10, 15, 20, 25, 30). The names of the signs are repeated in Sanskrit, syllables arranged right to left above the Arabic names, and the *abjad* numbers in the second row are repeated in Devanāgarī.

The remaining scales show various kinds of subdivisions of the signs and their regents or functions: one-twelfth parts (duodenaries, Sanskrit: *dvādaśāṃśa*);[64] one-ninth parts (novenaries, Sanskrit: *navāṃśa*); terms or limits, i.e., one-fifth parts of the signs and the strength of planets which preside over these limits (the name *ḥudūd miṣriyān* is engraved in the first cell of the row);[65] planets and their domiciles in one-thirds of the signs (the name *darjān kawākib* is engraved in the first cell);[66] and finally the regents of the decans of 10° (the designation *wujūh kawākib* is engraved in the first cell).[67]

The very last semi-circular scale carries the names of twenty-eight lunar mansions in Arabic followed by their serial numbers (Table 6.8). Immediately below this, Nayanānanda caused the addition of the Sanskrit names of the lunar mansions. The following table shows the names of the lunar mansions as engraved on the rete in Arabic, as engraved here in Arabic and those added here in Sanskrit.

The Inscription on the Back

The astrolabe makers of Lahore usually engrave their names, together with the name of their ancestors going up to Allāhdād, and the year of manufacture of the astrolabe in the narrow strip of space between the horizontal diameter and the upper edge of the tables engraved inside the shadow squares. In the

64 The *Zarqālī* astrolabe made by Ḍiyā' al-Dīn in 1680 that is now in Jai Singh's Observatory carries a table of duodenaries; it is reproduced in Kaye, *The Astronomical Observatories of Jai Singh*, 125.

65 See Kaye, *The Astronomical Observatories of Jai Singh*, 122–123; Sarma, *Astronomical Instruments in the Rampur Raza Library*, 49.

66 See Kaye, *The Astronomical Observatories of Jai Singh*, 126.

67 See Kaye, *The Astronomical Observatories of Jai Singh*, 123; Sarma, *Astronomical Instruments in the Rampur Raza Library*, 50.

TABLE 6.8 *The Names of the Lunar Mansions in Arabic and Sanskrit*

	Lunar mansions Arabic on the rete	Lunar mansions Arabic on the back	Lunar mansions Sanskrit on the back	Sanskrit corrected
1	*Sharaṭān*	*Sharaṭān 1*	*aśvinī*	
2	*Buṭayn*	*Buṭayn 2*	*bharaṇī*	
3	*Thurayyā*	*Thurayyā 3*	*kattaka*	*kṛttikā*
4	*Dabarān*	*Dabarān 4*	*rohiṇī*	
5	*Haqʿa*	*Haqʿa 5*	*mṛgaśira*	*mṛgaśirā*
6	*Hanʿa*	*Hanʿa 6*	*ādrā*	*ārdrā*
7	*Dhirāʿ*	*Dhirāʿ 7*	*punarvasuha*	*punarvasu*
8	*Nathra*	*Nathra 8*	*puṣya*	
9	*Ṭarf*	*Ṭarf 9*	*aślaṣā*	*aśleṣā*
10	*Jabha*	*Jabha 10*	*maghā*	
11	*Zubra*	*Zubra 11*	*pūrvā*	*pūrvā phālgunī*
12	*Ṣarfa*	*Ṣarfa 12*	*uttarā*	*uttarā phālgunī*
13	*ʿAwwāʾ*	*ʿAwwāʾ 13*	*hasta*	
14	*Simāk*	*Simāk 14*	*citrā*	
15	*Ghafr*	*Ghafr 15*	*svātī*	
16	*Zubānā*	*Zubānā 16*	*viśāṣā*	*viśākhā*
17	*Iklīl*	*Iklīl 17*	*anurādhā*	
18	*Qalb*	*Qalb 18*	*jyeṣṭhā*	
19	*Shawla*	*Shawla 19*	*mūla*	*mūlā*
20	*Naʿāʾim*	*Naʿāʾim 20*	*pūrṣā*	*pūrvāṣādhā*
21	*Balda*	*Balda 21*	*utta*	*uttarāṣāḍhā*
22	*Saʿd Dhābiḥ*	*Dhābiḥ 22*	*abhijita*	*abhijit*
23	*Saʿd Bulaʾ*	*Saʿd Bulaʾ 23*	*śravaṇa*	
24	*Saʿd Suʿūd*	*Saʿd Suʿūd 24*	*śatabhiṣā*	
25	*Saʿd Akhbiya*	*Sʿad Akhbiya 25*	*dhaniṣṭhā*	
26	*Fargh Muqaddam*	*Fargh Muqaddam 26*	*pūrvā*	*pūrvā bhādrapadā*
27	*Fargh Muʾakhkhar*	*Fargh Muʾakkhar 27*	*uttarā*	*uttarā bhādrapadā*
28	*Rishāʾ*	*Rishāʾ 28*	*revatī*	

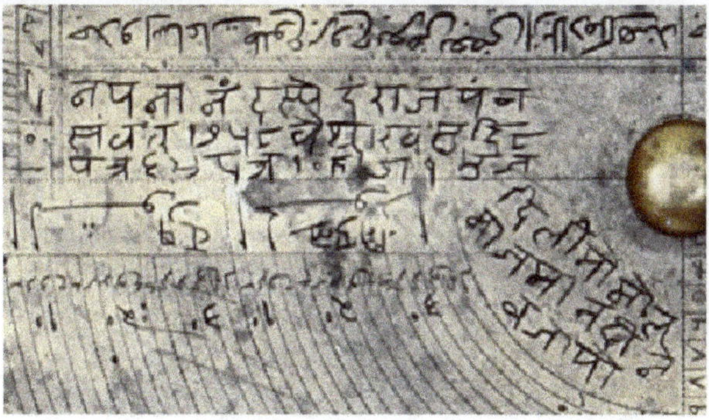

FIGURE 6.22 *Nayanānanda's inscription on the back. The Nasser D. Khalili Collection of Islamic Art, Accession No. SCI 53.* PHOTO COURTESY OF THE KHALILI FAMILY TRUST © NOUR FOUNDATION.

present astrolabe, this inscription appears to have been carefully erased and in its place, new inscriptions are engraved in Sanskrit.

On the left is incised the table of the triplicities and the corresponding regents as mentioned above, and on the right the name of the person who caused the reworking in Sanskrit, the date of reworking and so on in three lines (Figure 6.22):

> *nayanānaṃdasyedaṃ rājayaṃtra* (*sic! yantrarājaḥ*)
> *saṃvat 1758 vaiśākha sudi 8*
> *patra 6 bhapatra 1 o hajā 1 bhuja*
> (This is Nayanānanda's astrolabe (*yantrarāja*).
> Saṃvat 1758 Vaiśākha bright fortnight [day] 8
> [There are] 6 plates (*patra*), 1 rete (*bhapatra*), [and] 1 alidade (*bhuja*)).

This is continued at the apex of the upper right quadrant close to the central hole, again in three lines; this first line appears to be in Hindi:

> *dilī mo mol* (*dillī men mol ?*)
> *mājamānadī ne*
> *bajāyā* (*banāyā ?*)
> (In Delhi purchase (*mol*)
> *Mājamānadī*
> made (has done the engraving?)

The date of reworking, *Saṃvat 1758 vaiśākha sudi 8*, corresponds to Sunday, 15 May 1701 (Gregorian). The person who got the reworking done is Nayanānanda and there is a reference to Delhi. If "Mājamānadī" should be the name of the artisan who made the Sanskrit additions, it is so badly incised that no discernable name can be read out of it.

Nayanānanda must have been a wealthy resident of the Mughal capital—probably an astrologer with important clientele—to have been able to acquire this monumental Mughal astrolabe within fifty years of its manufacture and to fill it so thoroughly with Sanskrit equivalents. His has a beautiful Sanskrit name that means "delight to the eyes." His interest in the astrolabe is no doubt laudable, but regrettably the additions in Sanskrit he caused to be incised on this Lahore masterpiece are not at all *nayanānanda*.

Acknowledgements

Although I had to disagree in the course of this paper with some of the views expressed by Francis Maddison, David A. King, and Emilie Savage-Smith, my admiration for their scholarship remains nonetheless undiminished, so too my gratitude for their many acts of kindness in connection with my studies of Indian astronomical instruments. Moreover, David King went through my paper very meticulously and corrected my transliteration of Arabic terms and suggested several other improvements that are gratefully incorporated. I am highly obliged to Muzaffar Alam and Roy Fischel for the interpretation of the image on the back of the *kursī*, to Michio Yano for his advice on the nomenclature of the astrological houses and to Wilfred de Graaf for his advice on Zawraqi horizons. It is a pleasure to thank Charles Burnett and Josefina Rodriguez-Arribas for giving me an opportunity to present this paper at the conference on "Astrolabes in Medieval Cultures" at the Warburg Institute on 24–25 April 2014. Finally, I am highly indebted to Nahla Nassar, Curator and Registrar, Nasser D. Khalili Collection of Islamic Art, who kindly facilitated my study of the Indian astronomical instruments in 2005 in London, and now overwhelmed me by generously providing me eighteen high resolution images of the present astrolabe, which are reproduced here.

Professor David A. King has declared that a description of this astrolabe is alone "worth a master's dissertation." It is too late for me to aspire for a new university degree and for Professor King to award one. But it is not too late for me to dedicate this piece to the "King of astrolabes" (*yantra-rāja*) as a small token of my admiration for his profound scholarship.

CHAPTER 7

Saphaeae and *Hay'āt*: The Debate between Instrumentalism and Realism in Al-Andalus

Miquel Forcada
Department of Semitic Filology, University of Barcelona, Gran Vía 585, 08007 Barcelona, Spain
mforcada@ub.edu

Abstract

The article explores several texts written in al-Andalus between the twelfth and the early fourteenth centuries that mention Ibn al-Zarqālluh (al-Zarqālī) or the "Toledan astronomers." These texts, contained in the works of physicians and philosophers, express scholarly reactions to the contributions of the circle of astronomers of eleventh-century Toledo, sponsored by the judge Ṣā'id al-Andalusī. The textual evidence reveals attitudes ranging from a somewhat naïve admiration to harsh criticism. The article deals with the idealised image of the Toledan circle, due to a great extent to the significance of their work but also to external factors such as the innovative instruments and devices that these astronomers made. The texts suggest that some of the contributions of the Toledan circle were debated during the twelfth century, and that the echoes of this debate were felt in the following centuries. The discussion centred on the instrumentalism of Toledan astronomy, which was criticised by scholars who discussed the compatibility of Ptolemy's mathematical astronomy and Aristotle's natural philosophy. The "revolt against Ptolemy" that characterises this debate is reconsidered in the light of both the texts under study and the findings of recent research.

Keywords

cosmology – Ibn al-Zarqālī/Ibn al-Zarqālluh – instrumentalism – Islamic astronomy – realism

Introduction

The kingdom of Toledo established itself as the centre of mathematical astronomy in eleventh-century al-Andalus. As far as we know, the main reason for

 | DOI:10.1163/9789004387867_010

this was the presence of a member of the Toledan elite who was passionately committed to astronomy and astrology, the judge and astronomer Ṣāʿid al-Andalusī (1029–1070), who sponsored these disciplines. Our knowledge of the development of astronomy and astrology in Toledo is far from complete. It stems to a large extent from a bio-bibliographical treatise on science and scientists written by Ṣāʿid himself, the famous *Kitāb Ṭabaqāt al-Umam* (The Categories of the Nations) which, despite its thorough documentation, suffers from significant lacunae regarding the Toledan astronomers: the author makes no reference to a well-established team working under his leadership, or to one of the most important works that this team is believed to have produced, the *Toledan Tables*. The existence of this group is nevertheless commonly accepted in the light of several later sources which suggest that Ṣāʿid sponsored a significant number of scholars who worked together over a relatively long period of time and performed astronomical observations.[1] As is well known, these scientists changed the course of the history of astronomy in al-Andalus. Their main contributions were:[2] the compilation of a new *zīj* (the *Toledan Tables* mentioned above), which was translated into Latin at least twice in the twelfth century, the development of universal instruments (namely Ibn al-Zarqālluh's *saphaeae* and ʿAlī ibn Khalaf's "universal plate"); and the reform of Ptolemaic planetary models. Ibn al-Zarqālluh, by far the most important astronomer in this circle and in the history of al-Andalus, made decisive contributions to all three areas.

The works of the Toledan astronomers were widely known and used by the astronomers of the following generations both in al-Andalus and abroad. However, it is difficult to establish how far the activity of these astronomers permeated other sectors of the society of their own time because the information in this regard is merely anecdotal. For instance, the Andalusī astronomer Ibn al-Hā'im says in the twelfth century that Ibn Wāfid, the leading physician of eleventh-century Toledo but not an astronomer, collaborated with ʿAlī b. Khalaf, renowned for his treatise on the universal plate, in the observation of the obliquity of the ecliptic.[3] Since many astronomers were involved in the group and their alleged leader was himself a member of the Toledan elite, it is all the more probable that, already in their own times, their work was known to a wider audience than just the specialised circles of astronomers and astrologers. The twelfth-century sources reveal that scholars other than astronomers—namely physicians and philosophers, or scholars whose activity

1 Cf. Julio Samsó, *Las ciencias de los antiguos en al-Andalus* (Almería: Fundación Ibn Ṭufayl, 2011), 147–159 and 482–483 and the bibliography mentioned there.

2 Cf. Samsó, *Las ciencias*, 144 ff. for an updated synthesis of Toledan astronomy.

3 Samsó, *Las ciencias*, 181.

spanned both disciplines—believed that these contributions merited mention and discussion in their works. The influence and transmission of Toledan astronomy among astronomers between the twelfth century and the Renaissance has already been studied in depth; the purpose of the present paper is to explore the echoes of Toledan astronomy in non-astronomical (though scientific in the broad sense of the word) sources from al-Andalus so as to shed more light on the scientific debate during the eleventh and twelfth centuries and beyond, a decisive period in the history of Andalusī science.

Much as it is today, astronomy was a popular discipline in the Middle Ages, particularly in association with astrology. Furthermore, a grounding in astronomy was necessary for anyone devoted to the growth of high culture and two disciplines were particularly indebted to mathematical astronomy. One was philosophy, not just because of the essential role played by the planets and stars in natural philosophy according to the principle of the interrelation between the celestial and the sublunary worlds, but because philosophers also needed the mathematical astronomers' account of planetary motions to be able to describe the shape and workings of the heavens.[4] The other discipline was medicine, because of the above-mentioned influence of planets on the natural world, and therefore on human body. Hence, it is no surprise that the echoes of the Toledan astronomy we find in non-astronomical sources from al-Andalus are present in philosophical and medical works.

The Echo of the Toledan Astronomers in Andalusī Philosophers: Two Approaches to Astronomy

Ibn al-Zarqālluh is quoted quite often by Ibn Bājja and Ibn Rushd, two of the most important philosophers of twelfth-century al-Andalus. Both these scholars can be regarded to some extent as astronomers,[5] but they were not

4 Cf., Aristotle, *Metaphysics*, XII.8, in which the author uses the astronomical theories of Eudoxus (408–355 BCE) modified by Callippus (ca. 370 BCE–ca. 300 BCE). According to these theories, the planets move as a result of the rotation of earth-centred spheres nested inside each other, which revolve around different axes. These different axes explain one of the main problems of representing the planetary motions: the combination of the clockwise movement of the diurnal motion and the anticlockwise movement of the motions in longitude. The set of spheres attributed to each planet may engender a *hippopede*, a sideways figure of eight, thus named after its similarity with a horse fetter.

5 On Ibn Bājja and Ibn Rushd's contributions to astronomy, see Miquel Forcada, "Ibn Bājja" and "Ibn Rushd," in *Bibliographical Encyclopedia of Astronomers*, ed. Thomas Hockey et al., 2 vols. (New York, NY: Springer, 2007), 1:550 and 564–565, respectively.

professionals in the sense of Ibn al-Zarqālluh, who made his living crafting astronomical instruments, assisting Ṣāʿid al-Andalusī and, possibly, casting horoscopes. Ibn Bājja and Ibn Rushd owed their fairly thorough knowledge of astronomy to the fact that they were philosophers in the etymological sense of the world, and because of the demands that the philosophical life placed on a committed philosopher of the time: the possession of all theoretical knowledge so as to reach intellectual perfection.[6] As philosophers they can be defined as neo-Aristotelians, since the reading and commentary of Aristotle shaped their minds and works. As far as astronomy is concerned, their purpose (especially in the case of Ibn Rushd) was to *Aristotelise* it, and in fact they made decisive contributions to what A.I. Sabra called "the Andalusī revolt against Ptolemy":[7] the attempts to build a planetary system without eccentrics and epicycles, on the grounds that these artifices were incompatible with the Aristotelian theory of the heavens. In this setting, Toledan astronomy (and especially Ibn al-Zarqālluh's contribution) with its two-fold condition of being innovative and, at the same time, orthodoxly Ptolemaic in conception, appears as a sort of epistemological counter figure to the philosophers' conception of astronomy: something that philosophers might disagree with, but were obliged nonetheless to address.

The crux of the problem is the tension between realism and instrumentalism that characterises the history of ancient and medieval astronomy: are planetary models needed to explain the real world, as a philosopher would say, or is their role to accurately describe and predict planetary motions, as a professional astronomer/astrologer would claim?[8] In this connection, it is worth stressing the importance of the term *hay'a*, which will appear recurrently in the following pages and has prompted an interesting debate among specialists.[9]

---

6 This is explained in Ibn Rushd's *Summary of the Almagest* (Julianne Lay, "L'Abrégé de l'Almageste: un inédit d'Averroès en version hebraïque," *Arabic Sciences and Philosophy* 6 (1996): 23–61 (51–59).

7 A.I. Sabra, "The Andalusian Revolt against Ptolemaic Astronomy. Averroes and al-Biṭrūjī," in *Transformation and Tradition in Sciences. Essays in Honour of I. Bernard Cohen*, ed. E. Mendelsohn (Cambridge: Cambridge University Press, 1984, 133–153; reprinted in 2003).

8 For further nuances about this problematic, cf. Gad Freudenthal, "'Instrumentalism' and 'Realism' as Categories in the History of Astronomy: Duhem vs. Popper, Maimonides vs. Gersonides," *Centaurus* 45 (2003): 227–248 (233 and subsequent pages).

9 See, for instance, A.I. Sabra, "Configuring the Universe: Aporetic, Problem Solving, and Kinematic Modeling as Themes of Arabic Astronomy," *Perspectives on Science* 6 (1998): 288–330, George Saliba, "Arabic versus Greek Astronomy: A Debate over the Foundations of Science," *Perspectives on Science* 8 (2000): 328–341 and A.I. Sabra, "Reply to Saliba," *Perspectives on Science* 8 (2000): 342–345.

In its primary sense, *hay'a* means "shape," and it is often used by Ptolemaic astronomers to mean the geometrical models that provide abstract explanations of planetary motions and thereby engender a complex form made up by several elements—a "configuration," but a configuration restricted to the explanation of a single phenomenon. This may be called the "instrumental" meaning of *hay'a*. The concept acquires further nuances when astronomers (or philosophers) aim to give a three-dimensional account of the heavens by representing them in their physicality. In this case, *hay'a* may be rendered as "configuration," in the more general sense of "configuration of the heavens."

Despite the scarce (though real) evidence of the fact that the astronomers of eleventh-century al-Andalus were concerned with the configuration of the heavens,[10] only one work dealing specifically with this issue in the Toledan circles has come down to us: an anonymous *Risālat al-Hay'a* (Epistle on the configuration [of the heavens]) written by an astronomer who knew Ibn al-Zarqālluh personally. This unknown author wrote—or planned to write—a treatise entitled *Kitāb al-Istidrāk* (Book of recapitulation).[11] According to Saliba,[12] the latter work may have contained a recapitulation of the corrections to Ptolemy made by the Islamic astronomers, namely the Andalusīs, such as Ibn al-Zarqālluh's discovery of the motion in the solar apogee. What interests us most about this *Risālat al-Hay'a* is the author's use of the term *hay'a* in connection with the task of the astronomer:[13]

> You should know, that the practitioner of this art (*ṣināʿa*), after having extracted from the observations the motions that are like the principles and the foundations, he should then seek from the art of geometry (*ṣināʿat al-handasa*) the manner in which these motions could be achieved, and which configuration (*hay'a*) would be the one necessitating them. While searching for that, he should not deviate from that which he has accepted from the physical sciences (*al-ʿilm al-ṭabīʿī*) by way of foundations (*mabādi'*) for this art. He should not abandon the spheres, the circles, and

10 Julio Samsó, "On al-Biṭrūjī and the *hay'a* Tradition in al-Andalus," in *Islamic Astronomy and Medieval Spain* (Aldershot: Variorum, 1994), XII. We owe most of these references to Ṣāʿid's *Ṭabaqāt*.

11 This work is unedited and virtually unstudied. Cf. George Saliba, "Critiques on Ptolemaic Astronomy in Islamic Spain," *Al-Qanṭara* 20 (1999): 3–25. The author, who suggests that the title could have been "recapitulation of Ptolemy," contextualizes the book within the cosmological debates held in Islamic astronomy.

12 George Saliba, "Critiques," 12–15.

13 Translation borrowed from George Saliba, "Critiques," 14.

> the circular uniform motions, and pass on to other than that like non-spherical bodies, or non-circular figures. If by virtue of his power he was able to discover many configurations (*hay'āt*) for each planet [text has planets], all of them leading to the same result, and all of them in perfect accord with the observable particular motions, then he should opt for the simplest and most facile and that which resembles the celestial bodies as was done by Ptolemy when he opted for the eccentric in the case of the sun and not the epicycle.

Although the term "*hay'a*" in the title of the book appears to mean "configuration of the heavens,"[14] in this excerpt the author uses it in its instrumental meaning, signifying the complex shape that results from geometrising the planetary motions. His approach to astronomy is purely instrumental in that he only attempts to make an abstract representation of the phenomena. The author mentions the harmonisation between geometry and physics, but he does so in the loose sense that Ptolemy uses in the *Almagest*, and does not explore the issues that a twelfth-century philosopher from al-Andalus would have seen as important: that is, the fact that the homocentricity of planetary motions precludes eccentrics, and that eccentrics and epicycles actually engender non-uniform movements. Moreover, the only argument given to decide whether a configuration is preferable to another is its simplicity, and not the fact that they represent something real. In this connection, it is worth noting that, as we shall see below, the authors who care about natural philosophy and scientific method will criticise the astronomers for their inability to obtain a unique, consistent representation of astronomical phenomena.

There is no evidence that the Toledan astronomers had a philosophical training similar to their twelfth-century successor al-Biṭrūjī, who wrote a treatise on *hay'a* that presented a cosmology without eccentrics and epicycles that tried to explain in physical (i.e. philosophical) terms the principles of planetary motion, as we shall see below. However, this does not mean that the Toledan astronomers were ignorant of philosophy or of the cosmological implications of their work. In *Ṭabaqāt* Ṣā'id shows himself to be an admirer of philosophy; he recommends the reading of al-Fārābī to anyone who wants to pursue a scientific career;[15] and mentions several astronomers of his time who had a thor-

---

14 The reason for the title may be that the book, according to Saliba ("*Critiques*," 9) is an "elementary introduction to astronomy."

15 Miquel Forcada, "Ibn Bājja and the Classification of the Sciences in al-Andalus," *Arabic Sciences and Philosophy* 16 (2006): 287–307 (292–295).

ough knowledge of *hay'a*.[16] One of his disciples, Abū Marwān al-Istijī, wrote a treatise on astrological progressions, which was dedicated to his master Ṣā'id, who in turn mentioned it in very positive terms in *Ṭabaqāt*. This book contains several references to the method used by the philosophers to justify the scientific validity of the astrological procedures he describes, and to harmonise physics and astrology.[17] However, all these references (among which we can include the text of *Risālat al-Hay'a* quoted above) reveal a superficial knowledge of philosophical disciplines that contrasts sharply with the philosophical approach to astronomy of the twelfth-century authors; what is more, the works of Ibn al-Zarqālluh lack similar references. Probably one of the most important aims of the Toledan astronomers was to obtain better tools for the practice of astrology. Al-Istijī, in the treatise on astrological progressions quoted above, complains that the astrologers of his time had no interest in the precision of astrological computations;[18] ultimately, what is the purpose of tables, instruments, and more precise models for predicting the planetary positions?

Therefore, in spite of the hints of evidence quoted above, it seems that the Toledan astronomers did not undertake a serious study of the configuration of the heavens and their physical characteristics. For this reason, the scholars of the twelfth century and beyond saw the Toledan astronomers as specialized professionals whose main concern was the mathematical accuracy of their works. As we shall see immediately below, Ibn al-Zarqālluh's tables and instruments and his reform of certain Ptolemaic models were regarded as exemplifying a kind of astronomy that was based on shaky foundations and was unable to describe reality.

Two Contradictory Opinions Attributed to Ibn Bājja

The scholar who is most critical of Ibn al-Zarqālluh is Ibn Bājja (d. 1139), who starts a letter addressed to a physician who lived in Egypt named Abū Ja'far Yūsuf ibn Ḥasdāy (probably an exiled Jew from Saragossa) in this way:[19]

---

16 Julio Samsó, "On al-Biṭrūjī and the *hay'a* Tradition in al-Andalus," XII, 1–3.

17 Julio Samsó and Hamid Berrani, "The Epistle on Tasyīr and the Projection of Rays by Abū Marwān al-Istijī," *Suhayl* 5 (2005): 163–242 (170 and 177–182).

18 Samsó and Berrani, "The Epistle on Tasyīr," 215–216.

19 *Rasā'il falsafiyya li-Abī Bakr Ibn Bājja. Nuṣūṣ Falsafiyya ghayr manshūra*, ed. Jamal al-Dīn al-'Alawī (Beirut-Casablanca: Dār al-Thaqāfa et al., 1983), 77–78. The text is translated into Spanish and commented in Julio Samsó, "Sobre Ibn Bāŷŷa y la astronomía," *Sharq al-Andalus* 10–11 (1993–4): 669–681 (671–2).

> As for al-Zarqālla, Ibrāhīm ibn Yahyā al-Andalusī, he never treads the path of the art of cosmology (*hay'a*), whose gist he [misses][20] because (he speaks in accordance with the first things that come to his mind).[21] So, his phrases are confused and his discourse is full of nonsense. Later on, he thought that one idea was a principle—and his idea is the mean longitude of a planet—and he believed that it had the light of a [true] principle. It is as if you spoke, for instance, about the beginning of Aries and the point at which the ecliptic meets the line that comes from the centre of the eccentric and goes to the centre of the star. He repeated this idea again and again and he considered it as one of the first, indubitable principles. For this reason, he frequently criticised Ptolemy when referring to him.

The background to this text is Ibn Bājja's criticism of Ibn al-Haytham's and Ibn al-Zarqālluh's attempts to correct Ptolemy's model of Mercury: the former in his most famous *Shukūk ʿalā Baṭlamyūs* (doubts about Ptolemy),[22] the latter in a short tract that has not come down to us which is mentioned for the first and only time in this text by Ibn Bājja, *Maqāla fī ibṭāl al-ṭarīq allatī salakahā Baṭlīmūs fī istikhrāj al-buʿd al-abʿad li-ʿUṭārid* (Chapter on the refutation of the method followed by Ptolemy for determining Mercury's apogee). Ibn Bājja claims that Ibn al-Haytham misunderstands Ptolemy when he criticises the latter's method for determining the eccentricities of Venus and Mercury. Although the part of the text dealing with Ibn al-Zarqālluh is difficult to

20 A word is illegible in the ms. (Bodleian, Pococke 206, 118 r). I follow Samsó's interpretation.

21 A difficult sentence which reads *"fa-huwa yaqūlu bi-ḥasabi sawāniḥi-hi wa-lawāʾiḥi-hi."* The *lectio* seems right according to the ms. (Bodleian, Pococke 206, 118r) and the ed., yet the words *sawāniḥ* and *lawāʾiḥ*, which are followed by a pronoun whose antecedent can only be Ibn al-Zarqālluh, are puzzling here. I translate tentatively according to the literal meaning of the words given in Lane's *Lexicon*, s.v.: *lawāʾiḥ* (sing. *lāʾiḥa*), "the external parts of a thing," from *lāḥa*, "to appear, to shine"; *sawāniḥ* (sing. *sāniḥ*), usually "good omen," also "what appears," and, according to one of the meanings of the verb *sanaḥa*, an "idea or opinion presenting itself." The word *sāniḥ* with the sense of idea is also attested in dictionaries of Modern Standard Arabic such as Wehr-Cowan. Samsó's translation differs from mine but conveys a similar understanding that is still unsatisfactory ("su exposición se basaba en aspectos accidentales y [falsamente] brillantes"). As we shall see below, with slight changes, the sentence may make more sense.

22 Ibn al-Haytham, *al-Shukūk ʿalā Baṭlamyūs*, ed. A.I. Sabra and N. Shehaby (Cairo: The National Library Press, 1971), 29–32.

understand, the gist of Ibn Bājja's criticism is, as Samsó has shown,[23] that neither Ibn al-Zarqālluh nor Ibn al-Haytham properly understood what Ptolemy said, and therefore their amendments to Ptolemy's model are wrong.

I will not go into the astronomical content of the text, as it has already been analysed by Samsó.[24] What is important at this point is Ibn Bājja's attitude towards Ibn al-Zarqālluh, expressed at the beginning and at the end of the text. The philosopher considers the astronomer to be a poor scientist because he builds his theories on mistaken principles. Ibn Bājja expands and generalises this criticism in a short tract entitled *Kalām fī l-Hay'a* (Discourse on cosmology), a work that has come down to us in an incomplete form. The missing part appears to have contained Ibn Bājja's rejection of epicycles.[25] This discrepancy does not address the problem of Aristotle's conception of the universe directly, but it is not less Aristotelian: Ptolemaic astronomers do not meet the requirements that Aristotle lays down in *Posterior Analytics* in order for science to be true, and so what they describe does not correspond to reality.[26] The methodological criticism seems therefore to be a sort of prelude to the main thesis of the "Revolt" mentioned above, but it is not the only criticism contained in the text: Ibn Bājja, with the boldness which (according to some sources) characterised him, explains that Ibn al-Haytham and Ibn al-Zarqālluh were wrong even when judged within the framework of Ptolemaic astronomy (at least in this particular case).

There is still another point in this text that deserves attention. If the obscure phrase mentioned above,[27] "*fa-huwa yaqūlu bi-ḥasab sawāniḥihi wa-lawā'iḥihi,*" is read in a different way, we may gain new insights into Ibn Bājja's opinion of Ibn al-Zarqālluh. Although the manuscript is sufficiently clear on this point, it is possible that the scribe may have misunderstood at least one of the words: he may have read *sawāniḥihi* instead of *ṣafā'iḥihi*, a confusion which is easy to explain in Arabic.[28] In this setting, the term *lā'iḥ(a)*, pl. *lawā'iḥ*, would be understood in the more usual sense of "list" and its cognates, which, although

---

23 Julio Samsó, "Ibn al-Haytham and Jābir b. Aflaḥ's Criticism of Ptolemy's Determination of the Parameters of Mercury," *Suhayl* 2 (2001): 199–225.

24 Julio Samsó, "Sobre Ibn Baŷŷa," 672–674.

25 Miquel Forcada, "Ibn Bājja's Discourse on Cosmology (*Kalām fī l-Hay'a*), and The *Revolt* Against Ptolemy," *Zeitschrift für Geschichte des Arabisch-Islamischen Wissenschaften* 20–21 (2012–2014): 63–167 (142–146).

26 Ibn Bājja, *Kalām fī l-Hay'a*, ed. Forcada, "Ibn Bājja's Discourse," 154–155, paras. 7–8 (commentary, 121 ff).

27 Cf., above n. 21.

28 The initial *sīn* may be easily confused graphically and phonetically with *ṣād; wāw*, graphically, with *fā'*; and *nūn* with *yā'*.

less frequent than in Modern Standard Arabic, is nevertheless attested in the dictionaries of Classical Arabic and Middle Arabic.[29] The sentence would then mean that Ibn al-Zarqālluh "speaks in accordance with his *saphaeae*[30] and lists," where "lists" could be taken as a somewhat pejorative way of referring to Ibn al-Zarqālluh's tabular works.[31] This interpretation is consistent with the beginning of the text and, if correct, the whole may simply mean that Ibn al-Zarqālluh is ignorant of cosmology, the science that deals with the shape (*hay'a*) of the universe; he gets confused with the technical intricacies of instruments and tables and misses what should be the main concern of astronomy. The tension between instrumentalism and realism thus appears because Ibn Bājja seems to be belittling Ibn al-Zarqālluh's work on instruments, suggesting that the universal instruments are mere calculating devices that do not represent the real world. Even Ibn al-Zarqālluh's equatorium, which is supposed to represent the heavenly motions, is built on the basis of Ptolemy's *Almagest* and moreover contains innovations such as the famous representation of Mercury's deferent as an ellipse,[32] which also contradicts the Aristotelian principle of circular motion in the planets.

---

29 R. Dozy and E. Fagnan, in, respectively, *Supplément aux dictionnaires arabes* and *Additions aux dictionnaires arabes* mention *lā'iḥa/lawā'iḥ* as "feuille volante sur laquelle les commis font rapidement leurs comptes"; that is, a sheet on which one writes numbers.

30 Or simply "plates."

31 The following considerations complement the point made in note 21. The text is preserved in a single manuscript, and in the absence of other versions it is difficult to determine the meaning of a sentence which, according to ʿAlawī, should be read as: *amma al-Zarqālla Ibrāhīm ibn Yaḥyā al-Andalusī, fa-lam yaqaʿ qaṭ fī ṭarīq ṣināt al-hay'a* [word missing, probably a verb] *ʿalayhi amruhā fa-huwa yaqūlū bi-ḥasabi sawāniḥī wa-lawā'iḥihi.*" It makes sense to say that Ibn al-Zarqālluh spoke about cosmology according "to the first ideas [that came into his mind]" (*yaqūlū bi-ḥasabi sawāniḥī*), but it is hard to connect this clause with the term *lawā'ih*, since its non-technical meaning would imply that the author was referring to something like the "external parts of Ibn al-Zarqālluh." Indeed, the only meaning of *lawā'iḥ* extant in the dictionaries that fits the context is the technical meaning given by Dozy and Fagnan. So there are two reasonable ways of understanding the sentence: either with *sawāniḥ*, "he speaks in accordance with the first ideas [that come into his mind] and with his sheets full of numbers"; or with *ṣafā'iḥ*, "he speaks in accordance with his *saphaeae* and sheets full of numbers." The first option may well have been the interpretation of the scribe, but for the sake of the internal consistency of the text, I think that the second option is preferable. Besides, the copyist of ms. Pococke 206 was particularly inaccurate, due to his poor understanding of the material he was dealing with (cf. the bibliography given below, n. 34).

32 On this issue, cf. Julio Samsó and Honorino Mielgo, "Ibn al-Zarqālluh on Mercury," *Journal for the History of Astronomy* 25 (1994): 289–296.

A direct, irrefutable reference to Ibn al-Zarqālluh's *saphaeae* appears in another epistle attributed to Ibn Bājja, although its editor considers that it may have written by another author also named Abū Bakr.[33] Despite its dubious authorship, it represents convincing evidence of the reception of Toledan astronomy among the Andalusī philosophers of the twelfth century, inasmuch as it is extant in the above mentioned ms. Pococke 206, which was seemingly copied in that century from a manuscript that had belonged to Ibn al-Imām—Ibn Bājja's favourite disciple, who ended his days in Egypt.[34] Now, although this specific epistle may not belong to Ibn al-Imām's textual collection, it was written in all probability in al-Andalus.[35] Its editor gives it the title *Fī l-ʿilm al-insanī wa-l-ʿuqūl al-thawānī wa-l-ʿilm al-ilāhī* (On human knowledge, the second intellects and the divine knowledge). Its subject is not astronomy but philosophy, the way by which God and human beings perform the process of intellection. Although the context is not related to technical devices, the *saphaea* and the clepsydra (*mīkāna*) made by Ibn al-Zarqālluh are put forward as examples of complexity:[36]

> The First, and there is no other God except Him, intelligizes (*yaʿqul*) His essence only in its perfection, and He comprehends with the intellect of the perfection of His essence all that comes from Him and emanates from His perfection, and so He knows all the things via His knowledge of His essence. [Hence, he is one in any aspect and his union with the creatures that come from Him consists of the fact of His knowing them, according to their rank, with His knowledge of his essence].[37]
>
> He does not start knowing the creatures and then he comprehends finally what he has known, as happens with what is understood by the craftsman who invents some object according to his knowledge of its essence, and [anyone who sees][38] afterwards this object without its inventor may understand it after a while, after having reflected thoroughly enough to attain knowledge of it, and this means that he has understood it (finally).

33 ʿAlawī, *Rasaʾil Falsafiyya*, 37 ff.

34 ʿAlawī, *Muʾallafāt Ibn Bājja* (Beirut-Casablanca: Dār al-Thaqāfa et al., 1983), 43–46; cf. moreover Josep Puig Montada, *Avempace. Libro de la generación y corrupción* (Madrid: Consejo Superior de Investigaciones Científicas, 1995), introd., xxiv–xxv.

35 ʿAlawī, *Rasāʾil Falsafiyya*, 59–60.

36 ʿAlawī, *Rasāʾil Falsafiyya*, 187.

37 The editor omits this passage (MS Pococke 206, 124r: 14–15) without explanation.

38 I follow the editor's amendments to the ms.

> For example, one specimen of the *saphaea* and one of the clepsydra, of the kinds that al-Zarqālluh invented. He [Ibn al-Zarqālluh] understood them on the basis of their essence, not on the basis of [a particular] clepsydra and *saphaea*. Then, who saw them after him, understood them on the basis of these two specimens.

For a philosopher, either Ibn Bājja or another Abū Bakr, the *saphaea* and the famous clepsydra that Ibn al-Zarqālluh is believed to have crafted for the king of Toledo al-Ma'mūn b. dhī l-Nūn[39] were symbols of the complexity of the world, and the understanding of their functioning was a metaphor of the problem of divine knowledge. The second excerpt shows a somewhat naïve admiration that is absent in the first, but the two texts are not necessarily contradictory: a philosopher might consider Ibn al-Zarqālluh a poor scientist while admiring his evident skills as a craftsman. In any case, both texts bear witness to the fact that the Toledan astronomer was well known and that his works were discussed during the first half of the twelfth century in the learned circles of al-Andalus.

Ibn Rushd and al-Biṭrūjī on Ibn al-Zarqālluh's Models of Planetary Motion

One of the authors who shows most clearly the imprint left by the Toledan astronomy and Ibn al-Zarqālluh is the philosopher Ibn Rushd (d. 1198). Like Ibn Bājja, he knew enough astronomy to deal with the bible of mathematical astronomy, Ptolemy's *Almagest*, of which he made a critical summary entitled *Mukhtaṣar al-Majisṭī*.[40] In this book, besides echoing some of the methodological qualms extant in Ibn Bājja's *Kalām*,[41] Ibn Rushd sums up the contributions

39 José M. Millás Vallicrosa. *Estudios sobre Azarquiel* (Madrid-Granada: Consejo Superior de Investigaciones Científicas, 1943–1950), 6–10. The clepsydra is described by the geographer al-Zuhrī (d. 1137), who attributes it to a certain Abu l-Qāsim ibn ʿAbd al-Raḥmān known as al-Zarqāl. Since the first part of this name does not coincide with the name usually given to Ibn al-Zarqālluh in other sources (Abū Isḥāq), it has been suggested that there may have been two people with the same pseudonym crafting complex devices in eleventh-century Toledo. This proposal is tenable, just as it is also possible that they were, for instance, father and son, or that al-Zuhrī may have erred in the first part of the name. The text already translated shows that, at least for some Andalusīs who lived in the first half of the twelfth century, the astronomer and the clockmaker were one and the same person.

40 See on this book Julianne Lay, "*L'Abrégé*" quoted above.

41 Forcada, "Averroes," 75–76.

of Muslim astronomy known to him, among them Ibn al-Zarqālluh's *Maqālat al-kawākib al-thābita* (chapter on the fixed stars) and *Fī sanat al-shams* (on the solar year).[42] The books contain two of Ibn al-Zarqālluh's outstanding contributions to the redefinition of Ptolemaic planetary models: the model of trepidation and the solar model, with his discovery of the motion of the solar apogee.

The trepidation models were the focus of the cosmological debates among the Andalusī philosophers of the twelfth century. This would have been impossible if the Toledan astronomers had not played a major role in the development of the theory.[43] As is well known, the theory of trepidation is an alternative way to explain the precession of the equinoxes. The motion of the equinoxes is explained as a back and forth oscillation of the sphere of the fixed stars over a short arc. The first mention of the theory appears as early as Theon's commentary of the *Handy Tables* (fourth century), attributed to the "ancient astrologers." One of the first treatises on the question was written much later—the *Liber de motu octave sphere*, attributed to the famous Baghdad astronomer Thābit ibn Qurra (d. 901). However, there are solid grounds for maintaining that the book was written by a member of the circle of Ṣāʿid al-Andalusī, if not by Ṣāʿid himself, on the basis of a previous book written by the Baghdad astronomer Ibn al-Ādamī; or that it was in some manner related to the circle of Toledan astronomers.[44] Ibn al-Zarqālluh, who for some reason was particularly concerned with the question of trepidation, went so far as to elaborate three models that reached a wide readership and were discussed by later astronomers. The important point for us is that they were known not only to the astronomers but to the philosophers as well, such as Ibn Rushd. Although Ibn Rushd was reluctant to accept a cosmology that encompassed the theory of trepidation (as we shall see in a moment), the work of the Toledans, and especially that of Ibn al-Zarqālluh, was too important to be simply put aside without further inquiry or discussion. So, in his short commentary (*jāmiʿ*) to

42 Lay, "L'Abrégé," 49–51.

43 Samsó, *Las ciencias*, 219 ff. and 492–495, and the bibliography quoted there, specially F. Jamil Ragep, "Al-Battānī, Cosmology, and the Early History of Trepidation in Islam," in *From Baghdad to Barcelona. Essays on the History of the Islamic Exact Sciences in Honour of Prof. Juan Vernet*, ed. Josep Casulleras and Julio Samsó (Barcelona: Universidad de Barcelona, 1996), 267–298.

44 Samsó, *Las ciencias*, 225; F. Jamil Ragep, *Naṣīr al-Dīn al-Ṭūṣī's* Memoir on Astronomy (*al-Tadhkira fī ʿilm al-hayʾa*) (New York: Springer-Verlag, 1993), 400–408.

Metaphysics, a work that belongs to the first period of his *oeuvre* when he also wrote his *Summary of the Almagest*,[45] Ibn Rushd states:[46]

> We say: There is a consensus that there are thirty-eight [different] motions of celestial bodies, namely five each [performed] by the three supra[-solar] stars, Saturn, Jupiter, and Mars, five by the moon, eight by Mercury, seven by Venus, one by the sun (provided we regard its course as [pertaining] solely to an eccentric sphere, not as [pertaining] to an epicyclic sphere) and one by the all-embracing sphere, i.e. the sphere of the fixed stars.
>
> It is not clear whether there is a ninth sphere. Ptolemy thinks that there is a slow motion of the zodiac different from the diurnal motion [and that] it completes one revolution in thousands of years. Others maintain that it [performs] a motion of accession and recession, namely our fellow countryman from Andalusia called [Ibn] al-Zarqāla and his adherents, who assume for this [theory] a model that accounts for this motion. What motivated them to establish this [kind of] motion was that they observed the [annual] return of the sun to certain points of the zodiac and found that they had changed.
>
> Again others maintained that this change is due to an acceleration of the [sun's] motion or [a combination of several] motions in the sphere of the sun. Still others maintained that this [change] originates from defective instruments or that the instruments as such are insufficient to observe the true nature of this [phenomenon]. On the whole, I consider the existence of such a ninth, starless sphere as far-fetched, for a sphere exists only for the sake of the star, which is the noblest of its parts. This is why the more stars [a sphere] has, the nobler it is, as explained already by Aristotle.

Ibn Rushd is trying to describe the heavens by means of the motions of the planets, since, according to Aristotle, one motion entails a mover and a sphere. The underlying source is neither Aristotle nor Ptolemy, but seemingly Ibn al-Haytham's *Kitāb Hay'at al-ʿālam*;[47] the treatise presents a physical description

45 Lay, "L'Abrégé," 39, says that it was written between 1159 and 1162.

46 Ed. Carlos Quirós, *Averroes. Compendio de Metafísica. Texto árabe con traducción y notas* (Madrid: Imprenta de Estanislao Maestre, 1919), 133–135; English trans. quoted here with minor amendments, Rüdiger Arnzen, *Averroes on Aristotle's* Metaphysics. *An Annotated Translation of the So-called* Epitome (Berlin-New York: De Gruyter, 2010), 146.

47 Miquel Forcada, "La ciencia en Averroes," in Jorge Ayala (coord.), *Averroes y los averroísmos. Actas del III Congreso nacional de filosofía medieval* (Saragossa: Sociedad de Filosofía Medieval, 1999), 49–102 (84–85).

of the configuration of the heavens, and, as Ibn Rushd is aware that Aristotle's astronomical theory has been superseded by the doctrines of Ptolemy and his Islamic followers, he wants to expound what he believes it is the current state-of the-art in astronomy. For this reason, he adds Ibn al-Zarqālluh's models of trepidation to his commentary,[48] even though he is not convinced of their soundness. The theory of trepidation presents him with an essential problem: it entails the existence of a ninth starless sphere beyond the sphere of the fixed stars which, according to Aristotle, should be the outermost body of the heavens.[49] Since he cannot refute the theory, he includes it among an array of other possibilities. In the epitome of *On the Heavens*, written in the same period, he adopts a similar attitude towards trepidation, saying that it is the astronomer's concern to decide whether or not the back and forth motion of the sphere of the fixed stars exists.[50]

As is well known, Ibn Rushd becomes increasingly critical of Ptolemaic astronomy in the Aristotelian commentaries written in the following years. However, even in the books in which he explicitly dismisses epicycles and eccentrics, namely the long commentary (*Tafsīr*) of *Metaphysics* written between 1186 and 1190, he still quotes the findings of the Toledan astronomers. He refers to Ibn al-Zarqālluh's theory of the motion of the solar apogee, without mentioning the author;[51] he says that this motion is doubtful but does not reject it absolutely. What is more, when Ibn Rushd explains his alternative to epicycles and eccentrics in the same source, he says that this other system is quite similar to the trepidation models:[52]

> This spiral motion exists in the heavens out of a combination of the daily motion and the motion of the stars in their inclined orbs. I mean that the astronomers coincided in that the poles of the inclined spheres move in their daily motion around the poles of the universe and this motion

48 Although the motion of the solar apogee is not mentioned, it is possible that Ibn Rushd has taken it into account in his list of planetary motions, see Samsó, *Las ciencias*, 340.

49 James C. Evans, *The History and Practice of Ancient Astronomy* (New York, NY: Oxford University Press, 1998), 276–77.

50 Henri Hugonnard-Roche, "L'Épitomé du *De Caelo* d'Aristote par Averroès: questions de méthode et de doctrine," *Archives d'Histoire Doctrinale el Littéraire de Moyen Age* 52 (1985): 7–39, at 19–22.

51 Ibn Rushd, *Tafsīr Ma baʿda al-ṭabīʿa* ed. Maurice Bouyges, *Averroès. Tafsīr Ma baʿda aṭ-ṭabīʿāt. "Grand commentaire" de la métaphysique*, ed. Maurice Bouyges, 3 vols. (Beirut: Dār al-Machreq, 1983), 3:1680.

52 Ibn Rushd, *Tafsīr Ma baʿda al-ṭabīʿa*, 3, 1675. Cf. a similar passage in Ibn Rushd's long commentary of *On Heavens, Averroes Commentaria Magna in Aristotelem. De Celo et Mundo*, ed. F.J. Carmody and R. Arnzen, 2 vols. (Leuven: Peters, 2003), 2:332 (comm. 35).

is what the mathematicians of this Peninsula, i.e., the Peninsula of al-Andalus, found in the orb of the [fixed] stars and named it motion of access and recess.

It is difficult to discern precisely what Ibn Rushd understood by "spiral motion" (*ḥaraka lawlabiyya*) as an alternative to Ptolemy's eccentrics and epicycles.[53] Apparently, the only conclusion that may be drawn from this passage and others of the same kind is a general intuition of a shape that represents generically the motion of a planet as seen from the Earth. However, if we look at the authors who worked on the theory of trepidation like al-Zarqālluh and al-Biṭrūjī (who we will discuss in a moment), the text makes some sense. The gist of Ibn Rushd's spiral is that the apparent motion of the planets is not described by means of eccentrics and epicycles but by combining the two main motions mentioned in the text, represented by means of earth-centred circles: one, the daily motion around the poles of the equator, and the other, the motion in longitude following the ecliptic which goes in circles whose poles rotate—not around the poles of the equator as Ibn Rushd says, but around a small circle whose centre lies on a circle centred on the pole of the equator.[54]

If finally we want to visualise the spiral, we must turn to al-Biṭrūjī. Al-Biṭrūjī was ultimately the astronomer responsible for conforming planetary models according to the tenets of trepidation in his *Kitāb fī l-Hay'a* (Book on the "shape," i.e. the shape of heavens, therefore, book on cosmology).[55] This book defines more precisely the concept *lawlab* on the same basis as in Ibn Rushd's excerpt, and explicitly mentions the author's indebtedness to Ibn al-Zarqālluh for discovering it in his models of trepidation. Although several passages of *Kitāb fī al-Hay'a* may be quoted to illustrate both points, there are two whose sum is equivalent to Ibn Rushd's excerpt quoted above:[56]

53 For further insights into this issue, cf. Francis J. Carmody "The Planetary Theory of Ibn Rushd," *Osiris* 10 (1952): 556–586, at 570–572; and Forcada, "La ciencia," 90–94.

54 Cf. for instance al-Biṭrūjī's model for Saturn's motion in his *Kitāb fī l-Hay'a*, B.R. Goldstein, *Al-Biṭrūjī: On the Principles of Astronomy*, 2 vols. (New Haven, CT: Yale University Press, 1971), 1:7; and Ibn al-Zarqālluh's third model of trepidation in Bernard R. Goldstein, "On the Theory of Trepidation According to Thābit b. Qurra and al-Zarqāllu and Its Implications for Homocentric Planetary Theory," *Centaurus* 10 (1964): 232–247, at 240, and Julio Samsó, *Las ciencias*, 230 ff.

55 On the author and his work, cf., Samsó, *Las ciencias*, 342 ff., and "On al-Biṭrūjī and the *hay'a* Tradition in al-Andalus."

56 Al-Biṭrūjī, *Kitāb fī l-Hay'a*, Hebrew and Arabic texts in Goldstein, *Al-Biṭrūjī: On the Principles of Astronomy*, 2, 178–179; English trans. I, 91. Translation reproduced here, Mancha, "Al-Biṭrūjī's Theory," 172, paras. 84–85.

(a) Moreover we assert: if we consider a sphere whose poles revolve as we said, and in addition to this motion the sphere is moved with another motion around its own poles (which do not move with this additional motion), and these motions are combined, and a point is marked on the surface of the sphere, and the sphere is simultaneously moved with these two motions, then the marked point will not trace with these two motions a true circle, [all the points of which lie] on the same plane, and it will not return after a complete revolution of the sphere at the place of the circle from which it started, but it will incline to it, and the revolved point will generate something like a gyrating figure, called *leuleb*; that is, the circle will begin at a given point and once completing a revolution it will end at another point in another plane, and when the sphere has made a number of revolutions, the circles generated by that point will be like a gyrating revolution, called *leuleb*. (= *lawlab*)[57]

(b) When Abū Isḥāq al-Zarqāla considered these different motions,[58] he found a way of combining them according to his views, although he did not determine perfectly the truth of the motions of the stars. He constructed for them tables and a theory according to which the poles of this orb are moved on two circles parallel to the celestial equator, so that the motion of the stars follows the motion of these two poles. What he said induced us to discover something never considered by others, namely a motion really doing so[59]—this is indeed the motion of this orb (which is placed immediately below the highest one and follows its motion) about its poles in order to complete the lagging by which it remains behind the motion of the highest one, so that [due to this motion] it can be known and distinguished from it. Thus it is true that the mentioned motion is as al-Zarqāla posited.

The *lawlab* thus described—and according to Mancha's analysis of *Kitāb al-Hay'a*'s models for the motion of fixed stars—is a *hippopede* or, more precisely, "the four parts into which the hippopede is divided by its two axes

57 Al-Biṭrūjī, *Kitāb fī l-Hay'a*, Hebrew and Arabic texts in Goldstein, *Al-Biṭrūjī: On the Principles of Astronomy*, 2, 157–159; English trans. 1, 85–86. I borrow without modification the translation from José Luis Mancha, "Al-Biṭrūjī's Theory of the Motions of the Fixed Stars," *Archive for History of Exact Sciences*, 58 (2004): 143–182, at 167, para. 72. Mancha translates the Latin version written by Qalo Qalonymos (from Hebrew in 1528).

58 The author refers to the daily motion of the fixed stars and their motion in precession.

59 The *lawlab*.

of symmetry."[60] Now, the figure of eight-shaped *hippopede* is precisely the form traced by a planet in its motion, according to Eudoxus's models used by Aristotle in *Metaphysics*.

The narrative of the discovery (or rediscovery) of the *lawlab* is contradictory. One aspect of the issue is its author. On the one hand, the first references are those of Ibn Rushd's long commentaries on *Metaphysics* and *On the Heavens*, if we consider that both books predate al-Biṭrūjī's *Kitāb fī l-Hay'a*, though this dating is far from being conclusively proven; on the other hand, al-Biṭrūjī claims to be the inventor of the *lawlab*, although he states that the philosopher Ibn Ṭufayl (d. 1185) said that he had found a way of representing the planetary motions that was different from Ptolemy's system, although ultimately he was unable to keep his promise to write a book dealing with the matter. A second aspect of the issue is the very origin of the idea. Ibn Rushd and al-Biṭrūjī claim that it is found in what we can call the Toledan models of trepidation, and indeed Ragep has shown that the model of the *Liber de Motu Octaue Spere* engenders a *hippopede*.[61] However, the coincidences between Eudoxus and al-Biṭrūjī are so striking that it is difficult not to think of some kind of transmission from the commentators on Aristotle who explained Eudoxus's astronomy (for instance, Simplicius) to al-Andalus,[62] although no evidence of this has been found to date. Despite all these contradictions, there can be no doubt that the cosmological implications of the astronomy of eleventh-century Toledo were discussed at length in the philosophical circles of the twelfth-century courts of the Almohads—Seville, Marrakech, and other cities in the Maghreb—where Ibn Ṭufayl and Ibn Rushd served; and that in the course of these debates someone—whether cognizant of Eudoxus's astronomy or not—came up with the idea of the *lawlab*, which al-Biṭrūjī developed into a full system of representation of the heavens.

The existence of these debates on trepidation and precision in the Almohad circles is attested by two pieces of evidence that show that scholars who had little or no connection at all with astronomy were interested in, or aware of, these issues. One appears in the *Kitāb al-Istibsār fī 'ajā'ib al-amṣār* (Book on the discernment of the wonders of the cities) by the polymath Ibn 'Abd Rabbihi al-Ḥafīd (d. 1205), who states that he was taught about precession by Ibn

60 Mancha, "Al-Biṭrūjī's Theory," 156–157 and subsequent pages.

61 Ragep, *Naṣīr al-Dīn al-Ṭūṣī's* Memoir on Astronomy, 403–404 and 345, fig. C3. Compare this figure with fig. 1 in Mancha, "Al-Biṭrūjī's Theory," 149: they are the same except for the fact that the hippopede is traced by the poles of the ecliptic in Ragep's figure, and by a fixed star in Mancha's.

62 Mancha, "Al-Biṭrūjī's Theory," 143–144.

Rushd.[63] The second piece of evidence will be described in the following section.

Toledan Astronomy and the Physicians

Ibn Zuhr

One unexpected echo of the Toledan theories of trepidation appears in the *Kitāb al-Taysīr* written by Abū Marwān ibn Zuhr (d. 1159). In spite of his connections with Ibn Rushd, this author cannot be ranked among "the physician-philosophers" of his time. We know nothing about his philosophical training or about his interest in the discipline aside from some fleeting references in his books.[64] Ibn Zuhr was not a theoretician but an experienced physician who was a member of the leading medical dynasties of al-Andalus's history, active from the eleventh to the thirteenth centuries. What characterises his most important books is, on the one hand, their pragmatic character focusing on healing and its methods, and on the other, his recourse to his actual experience, which in some cases was quite innovative.[65] For this reason, Ibn Rushd had a very high opinion of Ibn Zuhr and considered the latter's *Kitāb al-Taysīr fī l-mudāwā wa-l-tadbīr* (Book for facilitating medication and regime) as a natural complement of his *Kitāb al-Kulliyyāt fī l-ṭibb* (Book on the general principles of medicine),[66] which approached medicine from a theoretical point of view. In a section of the *Taysīr* on purging and purgatives, we read:[67]

> The time in which taking a purgative is beneficial for health is the *iʿtidāl* (balanced time). By *iʿtidāl* I do not imply what the astronomers mean (equinox) because they reckon the days, whereas it is not the concern of medicine to do so. We (the physicians) estimate [this time] according

63 Muḥammad ibn Sharīfa, *Ibn ʿAbd Rabbihi al-Ḥafīd. Fuṣūlu-hu min sīra mansīya* (Beirut, 1992), 196.

64 Miquel Forcada, *Ética e ideología de la ciencia. El médico filósofo en al-Andalus (siglos X–XII)* (Almería: Fundación Ibn Ṭufayl, 2011), 303–305.

65 Rosa Kuhne, Cristina Álvarez Millán and Expiración García Sánchez, "Ibn Zuhr, Abū Marwān" in *Biblioteca de al-Andalus*, ed. Jorge Lirola et al. (Almería: Fundación Ibn Ṭufayl, 2009), 6:353–368.

66 See Rosa Kuhne, Cristina Álvarez Millán and Expiración García Sánchez, "Ibn Zuhr, Abū Marwān," 6:365, for a discussion of the relationship between the two authors and works.

67 Ibn Zuhr, *Kitāb al-Taysīr fī l-mudāwā wa-l-tadbīr*, ed. Michel Khūrī (Damascus: Munaẓẓama al-ʿArabiyya li-l-Tarbiya wa-l-Thaqāfa wa-l-ʿUlūm, 1983), 289–290.

> to our perception. Hence, when we find that the time is balanced—not too hot or too cold or too humid or too dry—we say that this is the *iʿtidal* according to our complexion, even if it is not the *iʿtidāl* according to the count of months and days, and despite the divergences among those who are devoted to observe the time of what we actually call the *iʿtidāl* (equinox). The differences between them have persisted to the point that, according to what the Toledans and their followers say, they have observed something different from what their predecessors found. They (the Toledans) claim that they have provided a demonstration with which they emended what was assumed before and they call this [emendation] "the forth and back motion." Probably others will come who will contradict them. The medicine is based on intuition and on professional assessment.

The author plays with the concept of *iʿtidāl*, whose first meaning is "balance." In astronomy it signifies equinox, but for Ibn Zuhr it means the time in which the elements are most balanced, and this according to Hippocrates and Galen is the spring,[68] which is also the best time for purging.[69] The central topic of the excerpt is obviously the difference between two classes of calendars: on the one hand astronomical calendars, and on the other calendars based on natural phenomena, where the beginning of the seasons is not determined by solstices and equinoxes; among these, medical calendars may be considered as a subclass.[70] To my knowledge, there is no other medical treatise that so acutely stresses the differences between the two ways of defining the seasons and pays so little attention to the influence of precise astronomical markers for health and, consequently, the work of astronomers.[71] The models of trepidation

68 Galen, *De Temperamentis*, Kühn I, 525–526.

69 Hippocrates, *Aphorisms*, n. 47.

70 Julio Samsó, "La tradición clásica en los calendarios agrícolas hispanoárabes y norteafricanos," in *Astronomía y astrometeorología medievales* (Barcelona: Universitat de Barcelona, 2008), 4:179–186. It is worth noting that the subject of medical calendars in Islamic medicine has not received the attention it deserves, although Teodor Loinaz is preparing a paper that will broaden our knowledge considerably. Samsó, in the article already quoted, discusses a passage from the *Calendar of Cordova*—written in the late tenth century—which says that there is a method "typical of the computists and the school of Hippocrates and Galen" that considers seasons of unequal length (four months for summer and winter, two months for spring and fall). Samsó shows the similarities of the dates attributed to these seasons with Hippocrates's *On Regime*.

71 In contrast, the passage translated above drew the attention of the Italian physician Maino de Manieri (d. 1368), who included it in a Latin translation in one of his *Regimina*

developed in Toledo are given as an example of the inconsistencies of astronomers; they are used to argue in favour of the use of a natural calendar and, especially, of seeking the advice of an experienced physician to determine the periods of time relevant for human health. Again, and in much the same way as the philosophers, the Toledan astronomers are criticised for their inability to explain reality. It is possible that Ibn Zuhr's statements contain a second, more general criticism of medical astrology. This does not mean that the author denies the influence of the macrocosmos on the human microcosmos, but rather that he downplays the role of medical astrology (iatrophysics or iatromathematics), in tune with the anti-astrological atmosphere that characterised the intellectual life of twelfth-century al-Andalus.[72]

The "Book of Royal Castilian Medicine"

This anonymous *regimen sanitatis* was written in Arabic in the early fourteenth century by a Jewish physician for the King of Castile.[73] Although produced in a Christian society, the work is still ascribed to mainstream Andalusī science and medicine, partly because of the language in which it is written but above all because of its sources and contents. In a chapter on bloodletting, there is a digression about Ibn al-Zarqālluh that follows the trends of the texts discussed above.[74] The problem dealt with is the amount of blood that one must extract in a phlebotomy; the author believes that the quantity prescribed by Galen is too great for a fourteenth-century man. Since the author implicitly assumes

---

Sanitatis. See Pedro Gil-Sotres, in *Arnaldi de Villanova Opera Medica Omnia X.1. Regimen Sanitatis ad Regem Aragonum*, ed. Luis García Ballester and Michael McVaugh (Barcelona: Fundació Noguera-Universitat de Barcelona, 1996), 132 and 590–591.

72 Miquel Forcada, "Astrology in al-Andalus during the 11th and 12th Centuries: Between Religion and Philosophy," in *From Masha'allah to Kepler: The Theory and Practice of Astrology in the Middle Ages and the Renaissance*, ed. Charles Burnett and Dorian Gieseler Greenbaum (Ceredigion: Sophia Centre Press, 2015), 149–176.

73 Luis García Ballester and María de la Concepción Vázquez de Benito, "Los médicos judíos castellanos del siglo XIV y el galenismo árabe: *El Kitāb al-ṭibb al-qašṭalī al-malūkī* (*Libro de medicina castellana regia*) (c. 1312)," *Asclepio* 42 (1990): 119–147. The book has been edited and translated into Spanish by María de la Concepción Vázquez de Benito, "Medicina castellana regia," in *Ciencias de la Naturaleza en al-Andalus. Textos y estudios VI*, ed. Camilo Álvarez de Morales (Granada: Consejo Superior de Investigaciones Científicas, 2001), 11–91 (Spanish Trans.), and "El *Kitāb al-ṭibb al-qašṭalī al-malūkī* (*Libro de medicina castellana regia*) (c. 1312)," in *Ciencias de la Naturaleza en al-Andalus. Textos y estudios VII*, ed. Expiración García Sánchez and Camilo Álvarez de Morales (Granada: Consejo Superior de Investigaciones Científicas, 2004), 11–107 (ed.).

74 *Kitāb al-ṭibb al-qašṭalī*, 71–73 (ed.)/54–55 (Spanish trans.); cf. Samsó, *Las ciencias*, 559–560.

that Galen cannot be wrong, the only conclusion is that the average size of human body has diminished since Antiquity. In support of this claim, he provides supposed evidence that the bones of the men of ancient times are larger than those of his contemporaries. For this idea to make sense, the author questions the invariability of the celestial world and divides it into three sections, although he only mentions two. These are "the nearest world"—the sublunary world—and the "intermediate world" or "world of the orbs"—the celestial world with the exception of the orb that would correspond to the prime mover, the world that the author could have named "the superior world."[75] The main argument is religious in character: philosophers err about the immutability of the universe because they hold that the world is eternal, whereas the Jews, and any other men who profess a religion as well, know that the world was created by God and may therefore change according to his will.[76] The only argument given to justify the mutability of the "intermediate world" is the variation of the solar year since Hipparchus, and the fact that it is at times longer or shorter. The author goes on to criticise Ibn al-Zarqālluh and others for failing to provide a model with a physical entity that explains this variation, possibly because he wants to divert his readers' attention away from the weakness of the solar year argument. Next, the argument of the physical inconsistency is expanded to include a general criticism of the astronomers and their models which encompasses the motion of the fixed stars, the obliquity of the ecliptic and the models for the Moon, Venus, and Mercury, although the author does not claim that the variability found in the underlying phenomena of these models is proof of the variability of the celestial world. However, he takes it for granted that what the astronomers describe is highly inaccurate.

Conclusions

The Toledan astronomers of the eleventh century, led by Ṣāʿid al-Andalusī, made great efforts to improve, refine, or correct the astronomical lore that had come down to them. Ibn al-Zarqālluh was the most prominent member of this group. They were good scientists whose work largely changed the course of Ptolemaic astronomy in al-Andalus. Later scholars who were not astronomers received not only their works but also an idealised image of their achievements.

75 Samsó, *Las ciencias*, 559.

76 Interestingly enough, the author mentions here (though only in passing) *Tahāfut al-falāsifa* (the incoherence of philosophers), the most famous refutation of philosophy written by the Muslim theologian al-Ghazālī (cf. *Kitāb al-ṭibb al-qašṭalī*, ed., 71:14).

The fact that the Toledan astronomers performed their own observations and made complex instruments seems to have contributed to their reputations. For the Ptolemaic astronomers, the contributions of the Toledan circle became a reference as important as the works of the authorities of the past. Even the Aristotelian philosophers—whose aim was to reform Ptolemaism, and who could therefore have merely ignored them—took them into consideration, albeit into critical consideration. For example, Ibn Bājja, who seems to have wanted to undermine their reputation as astronomers, draws attention to their inconsistencies with regard to Ptolemy's paradigm and to the Aristotelian scientific method. For his part, Ibn Rushd adopted a more balanced stance and made the findings of the Toledan circle an object of debate, possibly a collective reflection inspired by Ibn Ṭufayl. However, as a dogmatic Aristotelian Ibn Rushd held that the heavens had to be as they are described in the *Metaphysics*. Paradoxically, the philosophers of the twelfth century made their own interpretation of the *Metaphysics* through the solutions provided by Ibn al-Zarqālluh and others to the problem of precession. It seems that most scholars in twelfth-century Almohad al-Andalus took a realistic stance against the instrumentalism of the Toledan astronomy, which is reflected in the responses of a physician of note and of another physician of a later period. For them, the different descriptions that the astronomers give of one and the same phenomenon—and, as a result, their inability to provide a consistent explanation of reality—is proof that physicians' conclusions based on medical experience are not invalidated by the findings of astronomy. A second implicit conclusion is that medical astrology should not be taken as dogma.

Acknowledgements

The research for this paper has been funded by the Spanish Ministry of Economy, project "La evolución de la ciencia en la sociedad de al-Andalus desde la Alta Edad Media al pre-Renacimiento y su repercusión en las culturas europeas y árabes (siglos X–XV)," ref. FFI2011-30092-C02-01. I wish to thank Julio Samsó and the anonymous referees for their helpful remarks, and Josep Puig Montada for providing me with a copy of Oxford, Bodleian Library, MS. Pococke 206.

CHAPTER 8

Astrolabes on Parchment: The Astrolabes Depicted in Alfonso X's *Libro Del Saber De Astrología* and Their Relationship to Contemporary Instruments

Laura Fernández Fernández
Department of Medieval Art History, Complutense University of Madrid, Ciudad Universitaria, 28040 Madrid, Spain
lfernand@ucm.es

Abstract

This article analyses the relationship between text and image in one of the most important scientific manuscripts commissioned by Alfonso X the Learned, *El Libro del saber de astrologia* (Ms. 156 BH Complutense University, Madrid). The main topic deals with the depiction of astrolabes in the four treatises devoted to this instrument in the codex, and the connexion established between the real astrolabes, the depicted ones, and the written sources used by the *Alfonsine* team.

Keywords

Alfonso X – Libro del Saber de Astrologia – medieval astrolabes – scientific illuminated manuscripts

* I am deeply grateful to Charles Burnett and Josefina Rodríguez for inviting me to participate in the conference *Astrolabes in Medieval Cultures*, as well as for their kind help during this article's revision. I also would like to thank Antonio Momplet and the anonymous reader of this paper for their suggestions and refinements. Finally, I would like to gratefully acknowledge the permission of the Biblioteca Histórica, Universidad Complutense de Madrid, especially its director, Dr. Marta Torres, to reproduce the images of MS 156, BH.

 | DOI:10.1163/9789004387867_011

Introduction

The *Libro del saber de astrologia* (*Book of the Knowledge of Astrology*, hereafter LSA) is one of the scientific books commissioned by King Alfonso X, known as *el sabio*, "the Learned" (1221–1284). The book is a key piece of evidence for the monarch's political program and for the circulation of scientific knowledge during the Middle Ages, especially in the Iberian Peninsula.[1]

The contents of the manuscript, now at the Complutense Library in Madrid, were arranged in sixteen sections, each divided into several chapters of various lengths according to the complexity of their subject matter.[2] The codex was compiled with the aim of gathering all existing knowledge related to the celestial bodies, and thus the essential information required for the study and practice of astrology.[3]

Each treatise is introduced by a prologue giving detailed information about the writing process and the production of the book: the different phases of the

1 For outlines of this matter see Juan Vernet, *Lo que Europa debe al Islam de España* (Barcelona: Acantilado, 2006; 1977); and Julio Samsó, *Las ciencias de los antiguos en Al-Andalus* (Almería: Fundación Ibn Tufayl de Estudios Árabes, 2011; 1992); Josep Casulleras and Julio Samsó (eds.), *From Baghdad to Barcelona: Studies in the Islamic Exact Sciences in Honour of Prof. Juan Vernet*, 2 vols. (Barcelona: Instituto Millás Vallicrosa, 1996); and Emilia Calvo, Mercè Comes, Roser Puig, and Mònica Rius (eds.), *A Shared Legacy. Islamic Science East and West* (*Homage to Professor J.M. Millàs Vallicrosa*) (Barcelona: Universitat de Barcelona, 2008).

2 Madrid, Biblioteca Histórica, Universidad Complutense de Madrid, MS 156. For a detailed study of MS 156, see Anthony Cárdenas, *A Study and Edition of the Royal Scriptorium Manuscript of El Libro del Saber de Astrologia by Alfonso X el Sabio* (unpubl. Ph.D. diss.: University of Wisconsin, Madison, WI, 1974), and "Hacia una edición crítica del Libro del Saber de Astrología de Alfonso X: estudio codicológico actual de la obra regia (mutilaciones, fechas y motivos)" in *Homenaje a Pedro Sainz Rodriguez* (Madrid: Fundación Universitaria Española, 1986), 2:111–20.; Laura Fernández, *Arte y Ciencia en el scriptorium de Alfonso X el Sabio* (Sevilla-Puerto de Santa María: Universidad de Sevilla-Cátedra Alfonso X, 2013), 213–279. The manuscript can be consulted in its entirety through the World Digital Library web portal: https://www.wdl.org/en/item/15401/ (accessed 23 June 2018).

3 For a study on the content of the different treatises included in the *Libro del saber de astrologia*, see Manuel Rico y Sinobas, *Libros del saber de astronomia, del rey D. Alfonso X de Castilla*, I–V (Madrid: Tipografía de don Eusebio Aguado, Impresor de Cámara de S.M y de su Real Casa, 1863–1867); Cárdenas, *A Study and Edition*; Mercè Comes and Julio Samsó (coord.), *Ochava espera y Astrofísica. Textos y estudios sobre las fuentes árabes de la astronomía de Alfonso X* (Barcelona: AECI, 1990); and *De Astronomia Alphonsi Regis: actas del Simposio sobre Astronomía Alfonsí celebrado en Berkeley (agosto 1985), y otros trabajos sobre el mismo tema* (Barcelona: Universitat de Barcelona, Instituto Millás Vallicrosa de Historia de la Ciencia Árabe, 1987); Fernández, *Arte y Ciencia*, 213–279.

work, the aims and purposes set out in the planning of the royal project, and the challenges that emerged during its development, such as the need to add to and reorganize the content, and to incorporate visual aids in order to make the subject matter clearly understood. These prologues provide us with names of scholars who participated in the production of the texts and connect the monarch with the creative process, underscoring his role as the promoter of the project and occasionally suggesting that he participated in it directly.[4] The prologues also reveal that the royal codex was the result of a lengthy project that was initiated in the mid-1250s, when the king commissioned the translation of a few Islamic books on astronomical subjects, and finished in 1278 in the city of Burgos after a long period of translations, revisions, and finalization of these materials.

Among the sixteen treatises are four texts on the construction and use of the astrolabe, each concerning a specific sort of instrument. All these treatises adapted Islamic sources, and in those cases in which previous written materials did not exist, new texts were expressly commissioned by the king.[5] They all

4 The role of the king as "editor" or "corrector" of his texts, and his deliberate involvement in the definition of the Castilian language, has been one of the most frequently mentioned and debated subjects in the field of Alfonsine studies. The question is directly related to the prologue of the *Libro de las estrellas fixas*, which relates, according to Rico y Sinobas's edition, how the king "tollo las razones que eran sobejanas et dobladas, et que non eran en castellano drecho" ("removed the phrases that were superfluous and duplicated, and that were not in correct Castilian"). On this interesting issue, see Antonio García Solalinde, "Intervención de Alfonso X en la redacción de sus obras," *Revista de Filología Española* 2 (1915): 283–288; Rafael Cano Aguilar, "Castellano drecho?," *Verba* XII (1985): 287–306; Anthony Cárdenas, "Alfonso X nunca escribió *castellano drecho*," in *Actas del X Congreso de la Asociación Internacional de Hispanistas*, ed. Antonio Vilanova (Barcelona: PPU, 1992), 1:151–159; Fernández, *Arte y Ciencia*, 36–38.

5 On medieval Hispanic astrolabes in this period, see Ramón Martí and Mercè Viladrich, "En torno a los tratados hispánicos sobre construcción de astrolabios hasta el siglo XIII," in *Textos y Estudios sobre Astronomía Española en el siglo XIII* (Barcelona: CSIC, 1981), 79–99; and Ramón Martí and Mercè Viladrich, "En torno a los tratados de uso del astrolabio hasta el siglo XIII en al-Andalus, la Marca Hispánica y Castilla," in *Nuevos Estudios sobre Astronomía Española en el siglo de Alfonso X* (Barcelona: CSIC, 1983), 9–74; Mercè Viladrich, "Astrolabios andalusíes," in *El legado científico andalusí* (Madrid: Ministerio de Cultura, 1992), 53–65; David King, "Andalusi Astronomical Instruments," in *Al Andalus—The Art of Islamic Spain*, ed. Jerrilynn D. Dodds (New York, NY: Metropolitan Museum of Art, 1992), 376–383. For the medieval astrolabes kept in Spain, see Eduardo Saavedra y Moragas, "Astrolabios árabes que se conservan en el Museo Arqueológico Nacional, en la Biblioteca de Palacio y en colecciones particulares," *Museo Español de Antigüedades* 6 (1875): 394–414; Salvador García Franco, *Catálogo crítico de astrolabios existentes en España* (Madrid: Instituto Histórico de la Marina, 1945).

share an identical structure, which is the same as that used in Islamic books on this subject: a first section, frequently illustrated, explaining the parts of the astrolabe and how to construct it, and a second explaining how to use the instrument.

The treatises were richly illuminated, with the images placed prominently in the layout of the manuscript. The attention to detail and accuracy in these images is without doubt one of the most remarkable features of this codex. The images illustrating scientific instruments demonstrate the same educational ambition as is seen in the textual part of the project, and, just as the written text was required to be "bien complido et bien paladino," "well and clearly done," the visual language of the work also presents accurate technical information. Although of unquestionable beauty, the images were designed not only as decorative elements but as indispensable components of the book, functioning as a kind of visual commentary, without which it would have been incomplete.[6] Moreover, we should not forget the images' function as mnemonic elements to facilitate the learning of a subject, one of the most interesting points to consider in the relationship between text and image in scientific books.[7]

---

6 For a study of the role and use of the scientific image in MS 156 see Laura Fernández, "Textual and Pictorial Multilingualism in Alfonso the Learned's *Libro del Saber de Astrologia*: Between Theoretical Necessity and Aesthetic Erudition," in *Multilingual and Multigraphic Documents and Manuscripts from East and West,* ed. Giuseppe Mandalà and Inmaculada Pérez (Piscataway, NJ: Gorgias Press, 2018) 385–406.

7 The use of images as mnemonic devices has been thoroughly studied in connection with the construction of the memory in the ancient world and the Middle Ages. See Mary J. Carruthers, *The Craft of Thought: Meditation, Rhetoric and the Making of Images. 400–1200* (Cambridge: Cambridge University Press, 1998), and Mary J. Carruthers, *The Book of Memory* (Cambridge: Cambridge University Press, 2008). Regarding specifically the history of art, see Michael Camille, "Seeing and Reading: Some Visual Implications of Medieval Literacy and Illiteracy," *Art History* 8 (1985): 26–49; and Michael Camille, *Gothic Art: Glorious Visions* (London: Harry N. Abrams, 1996); and the works of David Freedberg, "Immagini e risposta emotiva: la prospettiva neuroscientifica," in *Prospettiva Zeri,* ed. Anna Ottani Cavina (Turin: Umberto Allemandi & Co, 2009), 85–105; David Freedberg, "Memory in Art: History and the Neuroscience of Response," in *The Memory Process: Neuroscientific and Humanistic Perspectives,* ed. Suzanne Nalbantian, Paul M. Matthews, and James L. McClelland (Cambridge-London: The MIT Press, 2011). For astronomical images, see Moya Carey, "Mapping the Mnemonic: A Late Thirteenth-Century Copy of al-Ṣūfī's 'Book of the Constellations,'" in *Arab Painting: Text and Image in Illustrated Arabic Manuscripts* (Leiden: Brill, 2010), 65–72. Especially stimulating for analyzing the relationship between art and science are the books: Brian S. Baigrie (ed.), *Picturing Knowledge: Historical and Philosophical Problems Concerning the Use of Art in Science* (Toronto, ON: University of Toronto Press, 1996); and Peter L. Galison and Caroline A. Jones (ed.), *Picturing Science, Producing Art* (New York, NY: Routledge, 1998).

However, these images, in common with many other scientific illustrations, have not yet been analysed as sources of information, often being considered superfluous additions or just artistic elements. Their study can offer us new approaches that will complement those already applied in by complementing the current research with new ways of analysis.

The Impact of Islamic Sources in Alfonsine Manuscripts

The use of Islamic scientific manuscripts as sources of information provided the Alfonsine scriptorium with new textual and visual references. It is important to emphasize that in the Islamic scientific sources the images were conceived as an active element in articulating a scientific text, and this feature was immediately inherited by the Alfonsine team.[8]

The question of the direct influence of Islamic manuscripts on the pictorial aspects of the Alfonsine production has been part of the debate on the royal scriptorium ever since Gonzalo Menéndez Pidal established visual connections between a copy of the *Maqāmāt* of al-Ḥarīrī and several scenes of the *Codice Rico* of the *Cantigas de Santa Maria*.[9] Islamic books also influenced other Alfonsine manuscripts, especially those on scientific subjects. In fact, from an iconographic point of view the whole layout of the LSA cannot be understood without bearing in mind the visual impact of Arabic illuminated manuscripts on the Castilian environment. Nevertheless this book surpasses in its richness and complexity any other scientific manuscript from East or West

8 The importance of the image in Islamic scientific texts has been especially underlined by David King, "Illustrations in Islamic Scientific Manuscripts," in *The Book in the Islamic World: The Written Word and Communication in the Middle East,* ed. George N. Atiyeh (Albany, NY: State University of New York and Washington, DC: The Library of Congress, 1994), 149–177; and David A. King, *In Synchrony with the Heavens: Studies in Astronomical Timekeeping and Instrumentation in Medieval Islamic Civilization,* 2 vols. Vol. 1: *The Call of the Muezzin.* Vol. 2: *Instruments of Mass Calculation.* (Leiden: Brill, 2004–2005), volume two. See also Anna Contadini (ed.), *Arab Painting Text and Image in Illustrated Arabic Manuscripts* (Leiden: Brill, 2010).

9 Gonzalo Menéndez Pidal, "Los manuscritos de las Cantigas. Cómo se elaboró la miniatura alfonsí," *Boletin de la Real Academia de la Historia* 150 (1962): 25–51; Francisco Prado-Vilar, *In the Shadow of the Gothic Idol: The Cantigas de Santa María and the Imagery of Love and Conversion* (Ph.D. Diss., Harvard University, Cambridge, MA, 2002); Francisco Prado-Vilar, "The Gothic Anamorphic Gaze: Regarding the Worth of Others," in *Under the Influence: Questioning the Comparative in Medieval Castile,* ed. Cynthia Robinson and Leyla Rouhi (Leiden: Brill, 2005), 67–100.

made at that time. The fact that the codex was a commission of the king distinguishes the book as a luxury product, connecting it to a specific symbolism of power, but at the same time without losing its value as a scientific product.

As I pointed out above, the images of the scientific instruments in the LSA were made with meticulous precision and accuracy. Each sort of astrolabe was illustrated with its particular features, with attention paid both to physical details and to engraved information used for enriching and complementing the knowledge of the instrument. Measurements were carefully made and each part of the astrolabe was drawn in proportion to the rest, so that if someone wanted to build his own instrument, he could do so by simply following the illustrations.

But when we observe these astrolabes depicted on parchment a question immediately arises: Were these astrolabes inspired by real instruments or did they just reproduce the content of the book in a generic way? To investigate this question I would like to analyse these four treatises in order to determine if a direct relationship can be established between those images and actual astrolabes, or if the Alfonsine team used other kinds of visual resources just to create an accurate and proper scientific appearance.

The Treatises on the Astrolabe in the *Libro del saber de astrologia*

The Spherical Astrolabe—El astrolabio redondo

The first astrolabe discussed in our manuscript is the spherical type.[10] According to the prologue to the treatise on the planispheric astrolabe, this was a spherical instrument shaped like the celestial globe that Ptolemy built first, and its initial position in the book could be explained by this connection. The prologue to the treatise on the spherical astrolabe affirms that, in the absence of available books explaining how to build this instrument, the king himself commissioned Rabiçag, one of his Jewish collaborators, to write the book properly and accurately, with the main goal of providing instructions for anyone who wanted to build it.[11] Regarding the second part, on how to use

---

10 Rico y Sinobas, *Libros del saber*, 2:113–222.

11 Rico y Sinobas, *Libros del saber*, 2:113. "Et porque non fallamos libro en que se fable de como se debe fazer de nuevo, por ende nos Don Alfonso el sobre dicho, mandamos a Rabiçag que lo escribiese bien complido et bien paladino, de guisa que lo entendiesen aquellos que ouiesen saber de lo fazer nuevamientre, assicomo lo habemos fecho de los otros libros que hicimos de los otros estrumentos." ("And because we did not find a book that told how it could be made anew, therefore I, the above-mentioned king D. Alfonso,

the instrument, although the text itself gives no information about a possible Islamic source, Mercè Viladrich has demonstrated that it was an adaptation of Ibn al-Samḥ's book on the planispheric astrolabe.[12] It is quite feasible that Rabiçag himself made the adaptation.

The reference to Ptolemy as the inventor of the spherical astrolabe raises another point. Although it is not clear when the instrument was invented, it seems to have been first developed by Islamic scientists.[13] Why then is Ptolemy considered the maker of the spherical astrolabe? The claim can be explained by a confusion of this instrument with the armillary sphere, traditionally attributed to the Alexandrian scientist. In fact, this mistake is present in several Islamic sources, and the fact that the *Libro del saber* reproduces it betrays again its direct dependence on Islamic texts.[14]

The spherical astrolabe was known in al-Andalus at least since the tenth century, although it did not experience great success, probably, as David King affirms, because of the difficulty involved in its construction, and it was apparently not introduced to the Latin West.[15] However, the fact that it was described in the *Libro del saber* implies that the spherical astrolabe was known, at least from a theoretical point of view, in the Castilian kingdom.

Every part of the spherical astrolabe was described and carefully illustrated in the manuscript, from the structure of the wood or brass sphere, to the alidade, to the rete with its star-pointers (Figures 8.1–2).[16] However, the depiction of these elements and the final image of the complete astrolabe presented

---

ordered Rabiçag to do it very well, and quite clearly, so it could be understood by those that needed to know how to make it once again, such as we did in the other books about the other instruments.")

12 Mercè Viladrich, "Una nueva evidencia de materiales árabes en la astronomía alfonsí," in *De Astronomia Alphonsi Regis*, 105–116.

13 Hugo Seemann and Theodor Mittelberger, *Das kugelförmige Astrolab nach den Mitteilungen von Alfonso X. von Kastilien und den vorhandenen arabischen Quellen* (Erlangen: Mencke, 1925); (reprinted in: *Islamic Mathematics and Astronomy*, vol. 88 (Frankfurt, 1998), 359–431).

14 Seemann and Mittelberger, *Das kugelförmige*, 365.

15 King, *In Synchrony*, 2:68.

16 Ms. 156, 1278, BH, UCM, ff. 46v–54v. Thanks to these accurate images, a brass model after the *Libro del saber* has been made by the Istanbul Museum of the History of Science and Technology in Islam, Inventory No. A 1.11. See Fuat Sezgin, *Science and Technology in Islam*, Vol. 2 *Catalogue of the Instrument of the Institute for the History of Arabic and Islamic Sciences* (Frankfurt am Main: Institut für Geschichte der Arabisch-Islamischen Wissenschaften an der Johann Wolfgang Goethe-Universität, 2010), Part III: Astrolabes, 78–172 (129–130), available online at http://www.ibttm.org/ENG/index.html (accessed 23 June 2018).

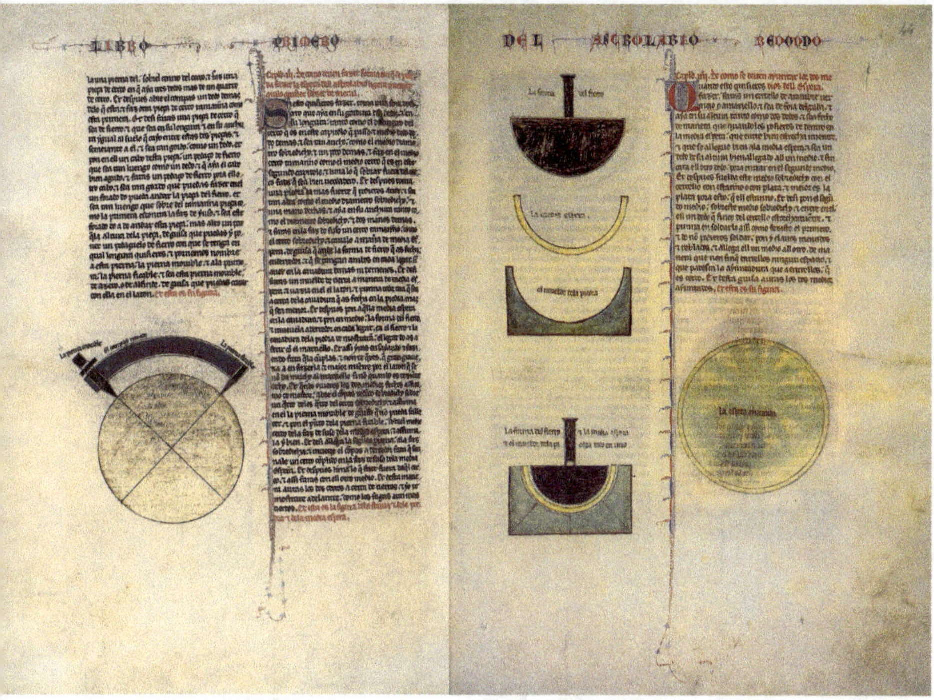

FIGURES 8.1 AND 8.2 *Parts of the spherical astrolabe.* Libro primero del astrolabio redondo, *in the* Libro del saber de astrologia, *MS. 156, BH, UCM, ff. 43v–44r.* © BIBLIOTECA HISTÓRICA MARQUÉS DE VALDECILLA.

pictorial challenges to give the impression of both volume and sphericity. No wonder that at first sight, despite presenting distinctive features, the result appears to be a planispheric rather than a spherical astrolabe (Figure 8.3).

In view of the spherical astrolabe's complexity, it is logical that the illuminator used the image of a planispheric instrument as a starting point. We must remember that this treatise is an adaptation of Ibn al-Samḥ's work on the planispheric astrolabe, so using images of that instrument as a visual reference makes sense. Further, the existence of real spherical astrolabes at the Castilian court seems unlikely, so they would have lacked any other visual model for depicting it.

Analysing these images, we note a new discrepancy. The star-pointers illustrated are a very basic model, without decoration, in the form of an arrowhead. Yet the text mentions a flower shape for the star-pointers, so, although the illustration was conceived as an accurate scientific representation, a disconnection exists between text and image. Additionally, this sort of arrowhead star-pointer

FIGURE 8.3 *Spherical astrolabe.* Libro primero del astrolabio redondo, *in the* Libro del saber de astrologia, *Ms. 156, BH, UCM, f. 55v.*

does not appear in any known Andalusí astrolabe, but it is a motif present in some early oriental astrolabes, such as one kept at the Science Museum in London.[17]

Why did the illuminator use this arrowhead form? Did he perhaps refer to an instrument with this feature as a visual model? Unfortunately, we cannot affirm this. We may plausibly ask if the Alfonsine team used other astrolabe treatises as its main visual source and if the illuminator just reproduced features of early astrolabes from other manuscripts.

The Planispheric Astrolabe—El astrolabio llano

The prologue to the treatise on the planispheric astrolabe provides information about its origin and the significant role of this instrument for knowing the stars.[18] The treatise does not mention any source or any scholar involved in its writing, although the text has been linked to several names.[19]

The images that illustrate the construction of this instrument reproduce every step of its making: how to define the main circles in the rete, place the fixed stars on it, engrave the rete itself, draw the almucantars and azimuth lines, and so on, even how to make the metal pin, until the instrument is perfectly defined (Figures 8.4–6). At the end of this section a portrait of Ptolemy was included, but unfortunately an unscrupulous person cut out this image in the sixteenth century along with other remarkable illustrations.[20]

---

17 Astrolabe, brass, Science Museum, London, 10th century, Syria, INV: 1981–1380, available online at http://www.sciencemuseum.org.uk/ (accessed 23 June 2018). For working on the comparison of astrolabes an online catalogue is very useful: "A Catalogue of Medieval Astronomical Instruments to ca. 1500" developed by David King, available online at http://ciuhct.org/en/wl/medieval-astronomical-instruments, linked to *Epact: Scientific Instruments of Medieval and Renaissance Europe* (Museum of the History of Science, Oxford, British Museum, London, Museo della Scienza, Firenze, Museum Boerhaave, Leiden), available online at http://www.mhs.ox.ac.uk/epact/index.htm (accessed 23 June 2018).

18 Rico y Sinobas, *Libros del saber*, 2:225–92. For a useful and understandable reference on planispheric astrolabes, see Sharon Gibbs and George Saliba, *Planispheric Astrolabes from the National Museum of American History* (Washington: Smithsonian Institution Press, 1984).

19 John D. North, "The Alfonsine Books and Some Astrological Techniques," in *De Astronomia*, 43–50. Julio Samsó, "Alfonso X and Arabic Astronomy," in *De Astronomia*, 23–38 (24). Mercè Viladrich, "Dos capítulos de un tratado de astrolabio en un pergamino del Monasterio de Santa María de Pedralbes," in *Ochava Espera*, 239–246.

20 Fernández, *Arte y Ciencia*, 246.

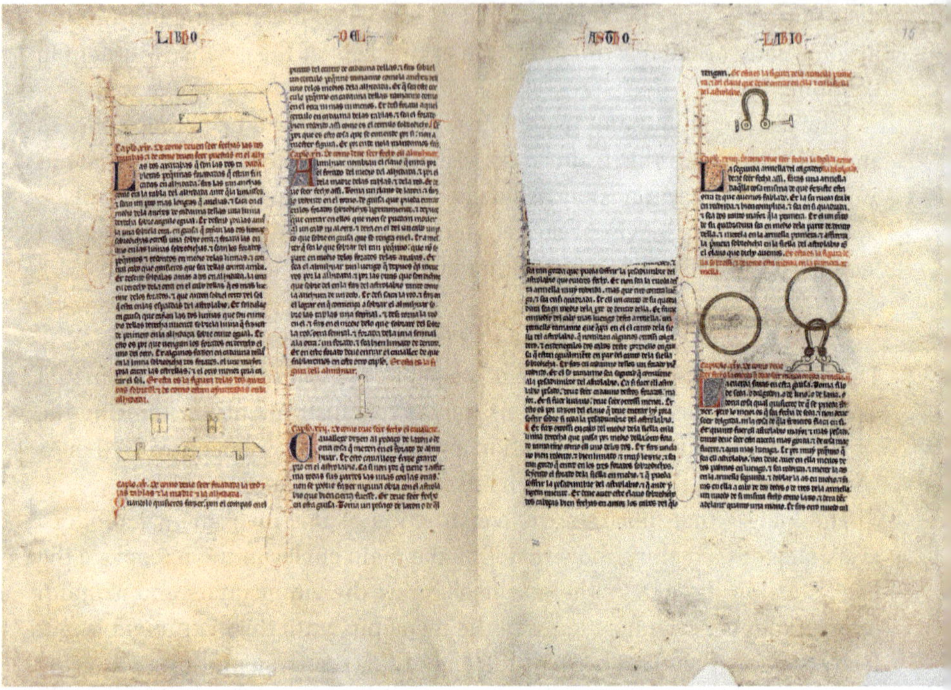

FIGURES 8.4 AND 8.5 *Several parts of the planispheric astrolabe.* Libro del astrolabio llano, *in the* Libro del saber de astrologia, *Ms. 156, BH, UCM, ff. 74v–75r.*
© BIBLIOTECA HISTÓRICA MARQUÉS DE VALDECILLA.

All the common features of Western astrolabes are illustrated: on the back the altitude scales, the zodiac and the calendar with the months, and the shadow square. The latter is depicted only in the bottom-right quadrant (and not in both bottom quadrants). This feature appears in exactly the same position on the astrolabe reproduced in one of the most interesting written sources for documenting the knowledge of this instrument in the Western world, the manuscript Lat. 7412 at the Bibliothèque nationale de France.[21]

21 MS Lat. 7412, BnF, f. 23v, available online at http://gallica.bnf.fr/ark:/12148/btv1b90678942 accessed??. For specific information about this manuscript tradition, see Julio Samsó, "Andalusi Science and the Monastic Scriptoria," in *Islam i Catalunya* (Barcelona: Català de la Mediterrània, 1988), 187–191; Charles Burnett, "King Ptolemy and Alchandreus the Philosopher: The Earliest Texts on the Astrolabe and Arabic Astrology at Fleury, Micy and Chartres," *Annals of Science* 55 (1998): 329–368; Arianna Borelli, *Aspects of the Astrolabe: "Architectonica Ratio" in Tenth- and Eleventh-Century Europe* (Stuttgart: Franz Steiner Verlag, 2008).

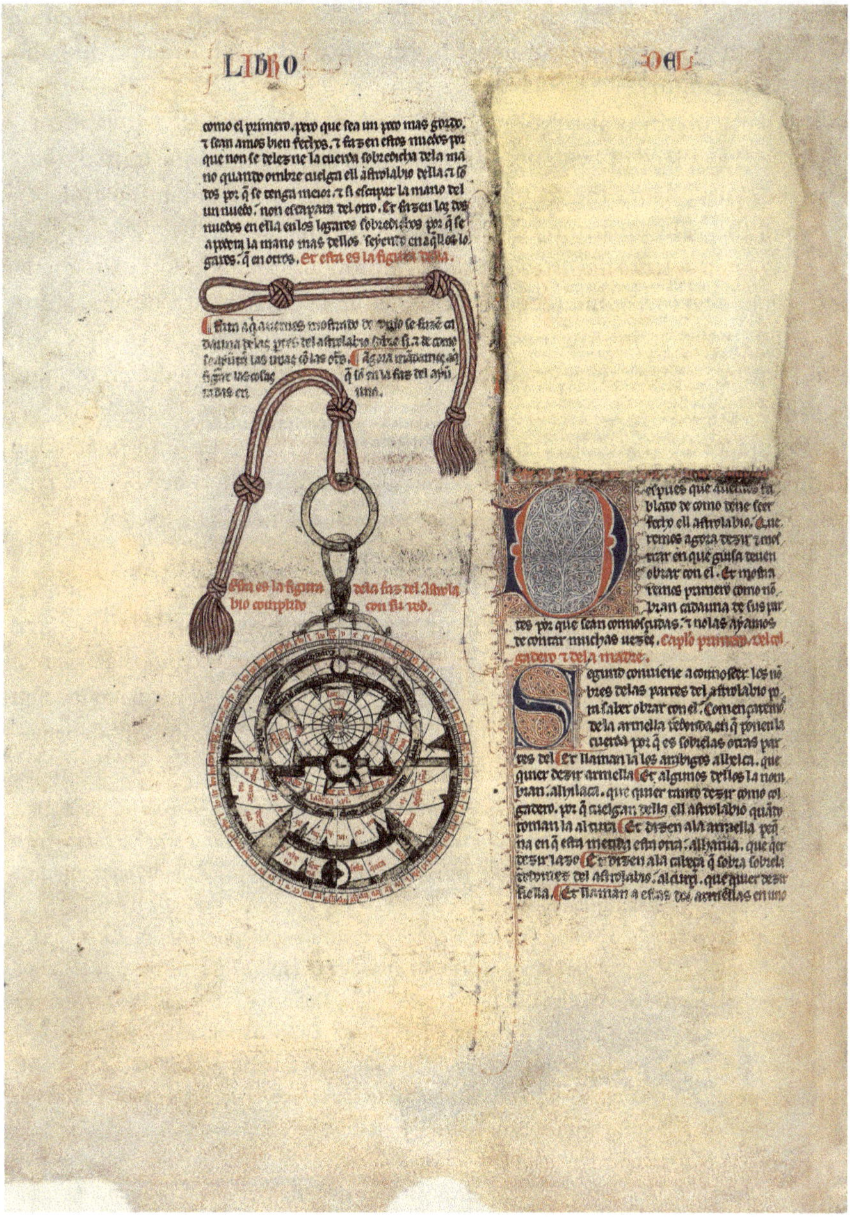

FIGURE 8.6 *Full image of the planispheric astrolabe.* Libro del astrolabio llano, *in the* Libro del saber de astrologia, *Ms. 156,* BH, UCM, *f. 75v.*

According to David King, the image depicts an astrolabe in the tradition of ninth–tenth-century Iraqi astrolabes but clearly made in al-Andalus, probably in the tenth century.[22]

This particularity is also present on the back of the so-called "astrolabe carolingien," a key piece of evidence for how the astrolabe was transmitted to the Christian West,[23] and the feature remained in the early Andalusí astrolabes, as two of the oldest instruments dated in the Peninsula testify.[24] The fact that the image in the *Libro del astrolabio llano* reproduces an early sort of astrolabe once again suggests some reflections. Either an astrolabe with those features was used in the scriptorium as a visual model for this book, or images from Arab written sources were used; or the illuminator may have drawn on both sorts of references.

The images also carefully reproduce how to trace the shape of the plates and all the markings on them. The text insists that the only difference between the plates is the latitude, and specifies one example designed to be used in the city of Toledo, whose latitude is forty degrees (Figure 8.7).

On the front, other particularities can be noted, such as the counterchanged horizontal bar, one of the distinguishing features of Andalusí and Hispanic astrolabes. Azucena Hernández has recently pointed out that the rete design appears to emulate the model of Almohad astrolabes, especially that created by Muḥammad ibn Fattūḥ al-Khamāʾirī, although in a simplified form.[25]

---

22 David King, "Bringing Astronomical Instruments Back to Earth: The Geographical Data on Medieval Astrolabes (to ca. 1100)," in *Between Demonstration and Imagination. Essays in the History of Science and Philosophy Presented to John D. North*, ed. Lodi Nauta and Arjo Vanderjagt (Leiden: Brill, 1999), 3–44.

23 Destombes Astrolabe, Cataluña, tenth century, Institute du Monde Arabe, Paris. Marcel Destombes, "Un astrolabe carolingien et l'origine de nos chiffres arabes," *Archives internationales d'histoire des sciences* XV 58–59 (1962): 3–45. Collected articles in Wesley Stevens, Guy Beaujouan and Anthony Turner, "The Oldest Latin Astrolabe," *Physis* 32 (1995): fasc. 2–3; David A. King and Kurt Maier, "The Medieval Catalan Astrolabe of the Society of Antiquaries, London," in *From Baghdad to Barcelona*, 2:673–678; Paul Kunitzsch, "Traces of a Tenth-Century Spanish-Arabic Astrolabe," *Zeitschrift für Geschichte der arabisch-islamischen Wissenschaften* 12 (1998): 113–120, reprinted in *Stars and Numbers: Astronomy and Mathematics in the Medieval Arab and Western Worlds* (London: Ashgate, 2004).

24 Ibrāhīm ibn Saʿid al-Sahlī, Toledo, 1067, Museo Arqueológico Nacional, Madrid, Inv. 50762 and Ibrāhīm ibn Saʿīd as-Sahlī, Toledo, 1068, Museum of History of Science, Oxford, Inv. 55331.

25 Azucena Hernández, "Astrolabios andalusíes e hispanos: de la precisión a la suntuosidad," *Anales de Historia del Arte*, Actas de las VII Jornadas Complutenses de Arte Medieval *Splendor. Artes suntuarias en la Edad Media Hispánica* (2014): 289–305. I am sincerely

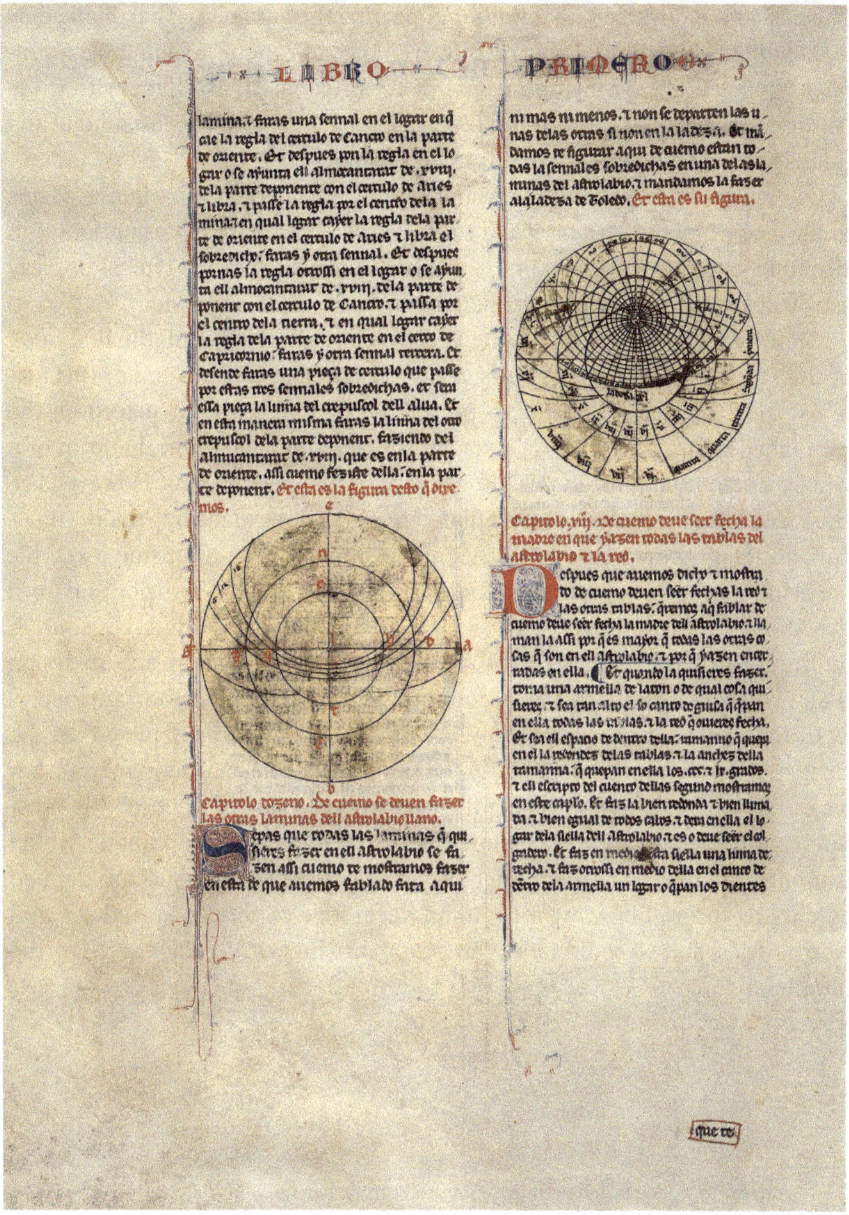

FIGURE 8.7 *Plates for the planispheric astrolabe.* Libro del astrolabio llano, *in the* Libro del saber de astrologia, *Ms. 156, BH, UCM, f. 72v.*
© BIBLIOTECA HISTÓRICA MARQUÉS DE VALDECILLA.

The design includes a circular frame supporting the equinoctial bar, flanked by two elements resembling towers. However, the star-pointers again take a simple arrowhead form, as with the spherical astrolabe. An astrolabe with these features was also depicted in the *Libro del relogio del argento vivo*, one of the treatises on clocks gathered in this manuscript.[26]

Once again the question of visual models arises. The illuminator seems to have combined different visual sources: the common rete design found in Hispanic astrolabes at that time, and the basic star-pointers typical of early astrolabes, both from real instruments and depicted on parchment.

Another example of a planispheric astrolabe was depicted in the first book of the LSA, the *Libro de las figuras de las estrellas fixas*. At the end of this text is a section dedicated to the stars prescribed by Ptolemy to be placed on the astrolabe (Figure 8.8).[27] Those stars are shown on a diagram in the form of a wheel, a common motif in the Alfonsine scientific manuscripts. However, the choice of forty-four stars does not match the selections made by Ptolemy or al-Ṣūfī, although these authors are the main sources for the Alfonsine text. This particularity was interpreted by Mercè Comes as an original contribution from the Alfonsine astronomers, who, as we have just seen, knew different sources for the construction and use of the astrolabe, and were therefore able to design their own star table taking into account their geographical position and needs.

In the centre of this wheel the rete of an astrolabe was depicted, as the treatise asserts, "porque los que lo cataren entiendan en que manera estan las estrellas en el,"[28] emphasizing the pedagogical value of the text. The most interesting feature here is once more the form used for the star pointers, which is totally different from the model discussed above. The pointers are defined by curving forms, following a common decorative pattern in al-Andalus and North Africa, and so could reflect an actual instrument directly known by the illuminator.

---

grateful to Azucena for permitting me to consult her unpublished master's thesis "La difusión del astrolabio andalusí en los reinos cristianos en la España medieval," (Masters thesis, Universidad Complutense, 2014).

26 Mechanism with an astrolabe, *Libro del relogio del argento vivo*, Ms. 156, BH UCM, f. 194v.

27 *Libro de las figuras de las estrellas fixas*, Ms. 156, BH UCM, f. 25v.

28 Rico y Sinobas, *Libros del saber*, 1:142 (for those who used the astrolabe to understand how the stars are distributed on it).

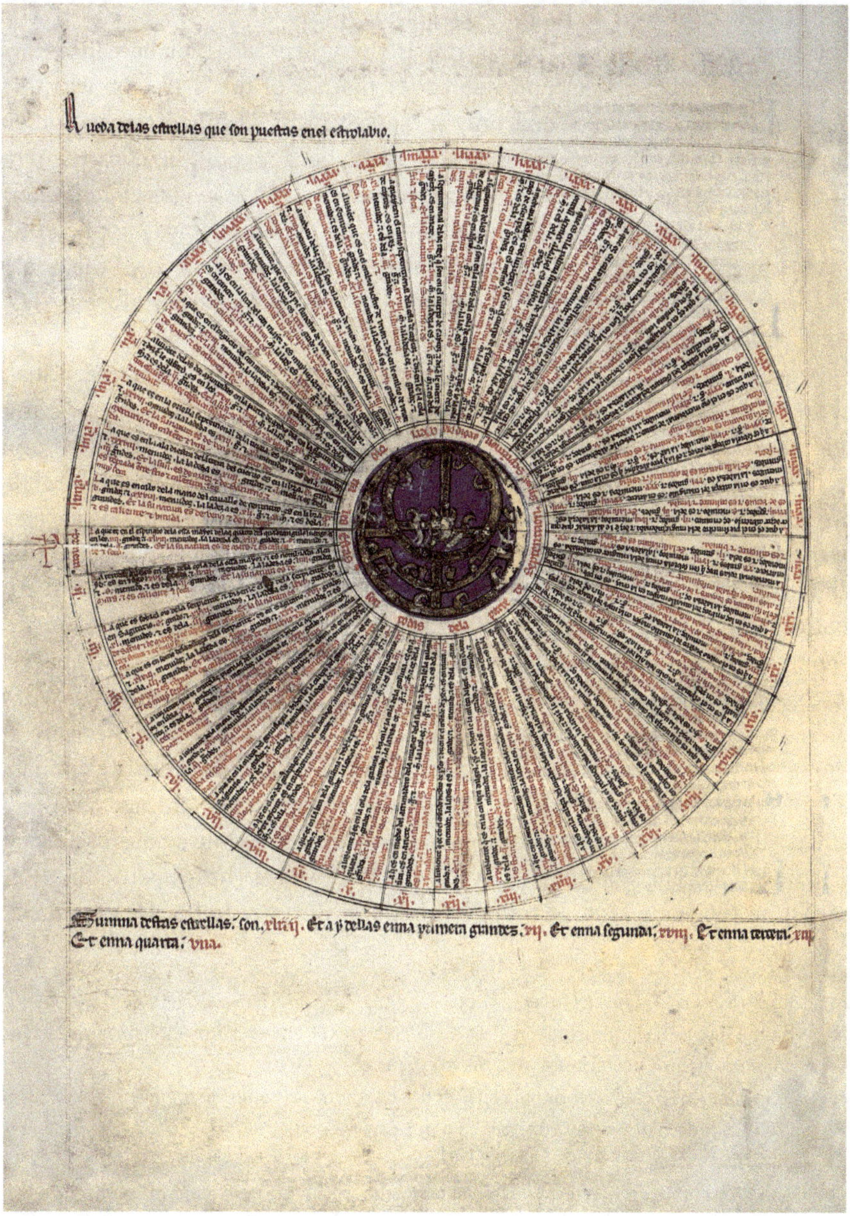

FIGURE 8.8 *Rete of a planispheric astrolabe.* Libro de las figuras de las estrellas fixas, *in the* Libro del saber de astrologia, *Ms. 156, BH, UCM, f. 25v.*

The Universal Astrolabe—La lámina universal

The next astrolabe in our manuscript is the universal astrolabe called the *lámina universal*.[29] The Islamic source for the use of this astrolabe was the book of ʿAlī ibn Khalaf, a text only preserved in its Castilian translation.[30] However, as in the case of the spherical astrolabe, a text on the construction of this instrument did not exist, and the monarch again asked his collaborator Rabiçag to write a new one. As Julio Samsó pointed out, Rabiçag applied the same strategy as he did for the spherical astrolabe treatise: he used the book on the construction of the *azafea*, another universal astrolabe described in this compilation, as the main basis for his new text.[31]

A geometrical figure explaining how to trace the plate opens the visual cycle; three accurate images complete the group, depicting the plate, in front and back, and the rete (Figures 8.9–11).[32] The rete shows a complicated design: the upper half looks like a grid and in the lower half the star-pointers are rendered in a vegetal pattern, more richly decorated than the other images analyzed so far. This striking particularity suggests the possibility that the illuminator was reproducing some real instrument instead of using a pictorial source, but unfortunately no surviving medieval example matches our *lámina universal* illustration, so we lack a real instrument to compare with the images in this manuscript.[33] However, although no medieval *lámina universal* with these features has, to our knowledge, survived, there are late examples with a similar decorative appearance: a precious early seventeenth-century example, the Blagrave-type astrolabe kept at the Adler Planetarium, Chicago, and the universal astrolabe kept at the Museo Galileo (the former Museo di Storia della Scienza) in Florence, both made by Charles Whitwell.[34] These instruments incorporate the same kind of rete used by John Blagrave in his *Mathematical*

---

29 Rico y Sinobas, *Libros del saber*, 3:3–132.

30 Emilia Calvo, "La lámina universal de Alī ibn Jala (s. XI), en la versión alfonsí y su volución en instrumentos posteriores," in *Ochava esfera*, 221–238.

31 Julio Samsó, "La ciencia española en la época de Alfonso el Sabio," in *Alfonso X. Toledo 1984* (Toledo: Museo de Santa Cruz, 1984), 89–101. esp. 99.

32 Ms. 156, BH UCM, ff. 84v and 85v/86v.

33 The only surviving medieval universal astrolabe is kept at the Benaki Museum, Athens, ΓΕ 13178. Images available online at http://www.benaki.gr/ (accessed 23 June 2018). The instrument was made by Ahmad ibn al-Sarraj, in Syria, in 1328–1329, and apparently has no direct relationship to the Toledan universal astrolabe. David A. King, "The Astronomical Instruments of Ibn al-Sarrâj: A Brief Survey," in *Islamic Astronomical Instruments* (London: Variorum, 1987), 9:1–3.

34 Ref. Nº 16, ca. 1620. Roderick and Marjorie Webster, *Historic Scientific Instruments of the Adler Planetarium, Volume I: Western Astrolabes* (Chicago: Adler Planetarium, 1998),

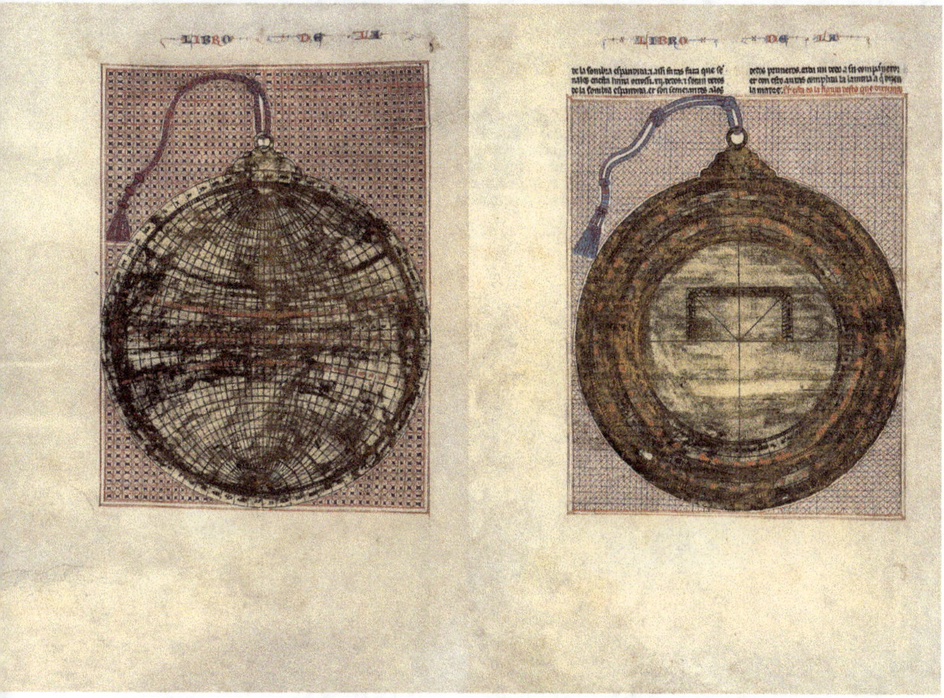

FIGURES 8.9 AND 8.10 *Plate, front and back, of the universal astrolabe.* Libro de la lámina universal, *in the* Libro del saber de astrologia, *Ms. 156, BH, UCM, ff. 84v and 85v.*
© BIBLIOTECA HIS-TÓRICA MARQUÉS DE VALDECILLA.

Jewel, whose woodcuts reproduce an image remarkably close to that in the *Libro del saber*.[35]

According to David King, "[Blagrave's] fine Elizabethan universal astrolabe is a testimony to the unity of astronomical instruments in the Middle

---

85–9 / Inv. 1095, Medici collection, *Museo di Storia della Scienza*, 1595. Images available online at http://catalogue.museogalileo.it/object/Astrolabe_no7.html accessed???.

35 There are several illustrated editions of *Mathematical Jewel*. Especially relevant is the exemplar kept at the Museum of the History of Science, Oxford, with a pen annotation at the bottom: "The Mathematical Jewel of John Blagrave of Reading, Cut by his own hand—1584." Inv. 13599. Image online: http://www.mhs.ox.ac.uk/object/inv/13599 (accessed 23 June 2018). Also very interesting is the copy kept at Cambridge University Library, LE. 28. 5 (London: Walter Venge, 1585): image available in the database "Astronomical Images: Diagrams, Figures, and the Transformation of Astronomy, 1450–1650," University of Cambridge, available online at http://www.astronomicalimages.group.cam.ac.uk/ (accessed 23 June 2018).

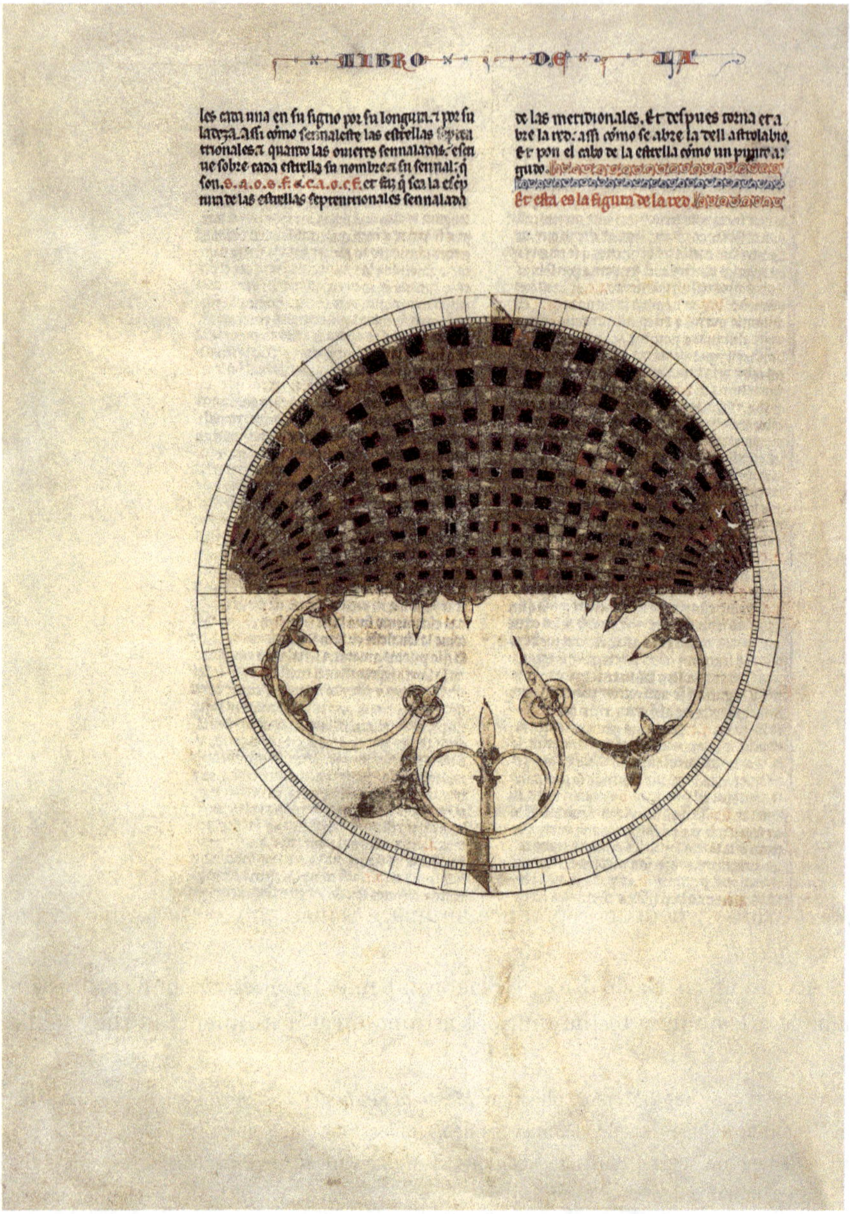

FIGURE 8.11 *Rete of the universal astrolabe.* Libro de la lámina universal, *in the* Libro del saber de astrologia, *Ms. 156, BH, UCM, f. 86v.*

Ages and Renaissance. It shows that the foremost instrument-makers in sixteenth-century England were interested in the same problems as those in eleventh-century Toledo, where the universal astrolabe was invented, and that they favoured the same instrumental solution."[36] Besides the technical issues underlined by King, if we draw our attention to the flourished rete of these English astrolabes a resemblance can be established with the rete depicted in the *Libro del saber*.[37] King himself proposed a connection between the rete of the Adler astrolabe and the image of a rete painted in Biblioteca Nacional MS 3306, a fifteenth-century partial copy of the LSA, f. 174v (Figure 8.12).[38] In fact the rete depicted in this manuscript appears to be a middle step between the design in the original *Libro del saber* and Blagrave's woodcuts.

John Blagrave was probably influenced by a medieval European tradition, but the channel of transmission has yet to be explained. Could Blagrave have known a lost *lámina universal* or a copy of the LSA? The link between the development of the universal astrolabe in modern England and the Alfonsine tradition is difficult to corroborate, but these images may reinforce this connection as a plausible hypothesis.

The Book of the Azafea—Libro de la azafea

The *Libro de la azafea*, another universal instrument, closes the astrolabe section.[39] This treatise could be presented as an example to illustrate the whole elaboration process of the LSA. Its prologue mentions three collaborators who worked at the Alfonsine scientific workshop: first, Fernando of Toledo, a Peninsular Christian who was entrusted with a first translation that the monarch did not find satisfactory; for this reason, according to the text, some years later King Alfonso requested two other members of the workshop to carry out a second translation which proved to be more complete and better executed. One of those scholars was Bernardo, the only contributor of Muslim

36 Webster, *Historic Scientific Instruments*, 88. David A. King, "On the Early History of the Universal Astrolabe in Islamic Astronomy and the Origin of the Term *shakkāzīya* in Medieval Scientific Arabic," *Journal for the History of Arabic Science* 3 (1979): 244–257.

37 Rete, Ms. 156, BH-UCM, f. 86v.

38 This manuscript, formerly L 97, is a remarkable collection of scientific texts that belonged to Juan Fernández de Velasco's library. Among these texts there are several Alfonsine works: the only extant copy of the *Canones de las tablas alfonsies*, and a fifteenth-century copy of the books of the instruments from the *Libro del saber*, including that on the *lámina universal*, Fernández, *Arte y ciencia*, 269–270. The entire manuscript can be consulted at Biblioteca Digital Hispánica, http://bdh-rd.bne.es/viewer.vm?id=0000011059&page=1 (accessed 23 June 2018).

39 Rico y Sinobas, *Libros del saber*, 3:135–237.

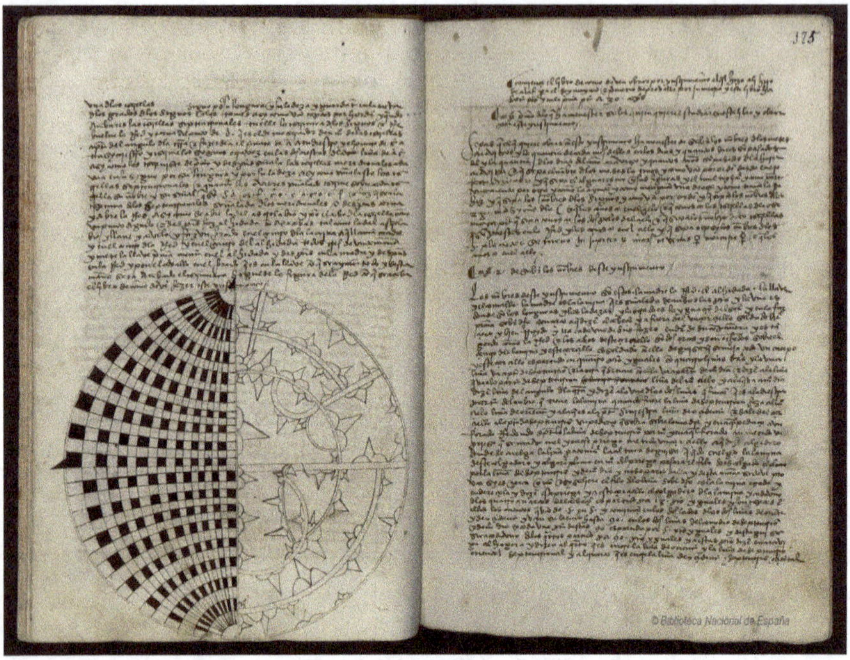

FIGURE 8.12 *Rete of the universal astrolabe.* Libro de la lámina universal, *Ms. 3306*, BNE, *ff. 174v–175r.*

origin documented in the context of the court,[40] and the other was Abraham, a Jewish *alfaquim* of the king.[41] The active participation and cooperation of scholars from different cultural backgrounds is thus perfectly documented.

---

40 In the Italian translation of the *Libro del saber de astrologia*, completed in Seville in 1341 (Ms. 8174, Biblioteca Apostolica Vaticana), Bernardo is described as "Maestro Bernardo Arabico ouero Saracino," a reference that has led to the belief that he was a Morisco (a Muslim converted to Christianity), and therefore the only person of Muslim origin linked to the Alfonsine scriptorium of whom we have documentary information: see Samsó, "Dos colaboradores cientificos musulmanes de Alfonso X."

41 Rico y Sinobas, *Libros del saber*, 3:135. "Este es el libro de la acafeha que es llamada lámina. Dicho auemos fata aqui dell alcora. cuemo es fecha et cuemo deuen obrar por ella. Et otrossi del astrolabio. Cuemo es fecho. et delas huebras que se pueden fazer por el. Mas agora queremos fablar dela açafeha que fizo Azarquiel el sabio astronomo de Toledo á onra del rey Almemun que era entonces sennor dessa cipdat, et nombrola por ende almemonia … Et este libro sobredicho traslado de arabigo en romançe maestre fernando de Toledo por mandado del muy noble Rey don Alfonso, fijo del muy noble Rey don Fernando, et dela Reyna donna Beatriz. et sennor de Castiella. de Toledo. de Leon. de Gallicia. de Seuilla. de Cordoua. de Murçia. de Jahen. et dell Algarue. en el anno quarto

The *azafea* was designed by Azarquiel in Toledo in the middle of the eleventh century. In fact, two versions of this instrument were developed; it was the version called *azafea zarqāliyya* whose description was translated into Old Castilian and that was depicted in the *Libro del saber*.[42] Physically, the *azafea* presents an important difference from the *lámina universal*: the rete has disappeared, and therefore the only images included are of the front and back of the plate, and a rule with the alidade.[43]

On the back of this *azafea*, besides the common scales engraved on the planispheric astrolabe, one may point out, following Roser Puig, three elements: a meridian orthographic projection that occupies three of the four quadrants depicted by parallel lines, a sinus quadrant in the bottom right used as a tool for calculating trigonometric formulae, and a little circle called the "mooncircle." All these features are exactly depicted in our manuscript (Figures 8.13–14).

According to Roser Puig, three Hispanic *azafea zarqāliyya* have been preserved, all of them made by Muḥammad ibn Fattūḥ al-Khamā'irī. Two were

---

que el regno. Et despues mandolo trasladar otra uez en Burgos meior et mas complidamientre á maestre Bernaldo el arauigo. et á don Abrahem su alfaquim. en el XXVI. anno del so regno. que andaua la era del Cesar en mil. CCCXV. Annos." ("This is the book of the *azafea*, which is called plate/lámina. So far we have talked about the alcora, how it is made and how one should use it; and also about the astrolabe, how it is made and about the things one can do with it. But now we want to talk about the *azafea*, made by Azarquiel, the wise astronomer from Toledo, in honor of King Almemun, who was then the sovereign of that city, and therefore named it *almemonía* … And this book was translated from Arabic into Romance by master Fernando de Toledo by order of the very noble king don Alfonso, son of the very noble king don Fernando, and of the queen donna Beatriz, and sovereign of Castile, of Toledo, of Leon, of Galicia, of Seville, of Cordoba, of Murcia, of Jaen, and of the Algarve, in the fourth year that he reigned. And afterwards at Burgos he ordered master Bernaldo the Arab, and Abrahem, his *alfaquim*, to translate it again better and more perfectly, in the 26th year of his reign, that was the year 1315 of Caesar's era.").

42 José María Millás Vallicrosa, *Estudios sobre Azarquiel* (Madrid-Granada: CSIC, 1943–1950); King, "On the Early History of the Universal Astrolabe," 244–257; Roser Puig, "Concerning the Safiha Shakkāziyya," *Zeitschrift für Geschichte der Arabisch-islamischen Wissenschaften* 2 (1985): 123–139; *Los tratados de construcción y uso de la azafea de Azarquiel* (Madrid: Instituto Hispano-Árabe de Cultura, 1987); Julio Samsó, "Sobre el trazado de la azafea y de la lámina universal: intervención de los colaboradores alfonsíes," *Al-Qantara: revista de Estudios Arabes* 8 (1987): 29–43.

43 Front and back of the plate, *LSA*, Ms. 156, BH-UCM, ff. 112v–113r / Rule and alidade, ff. 113v–114r.

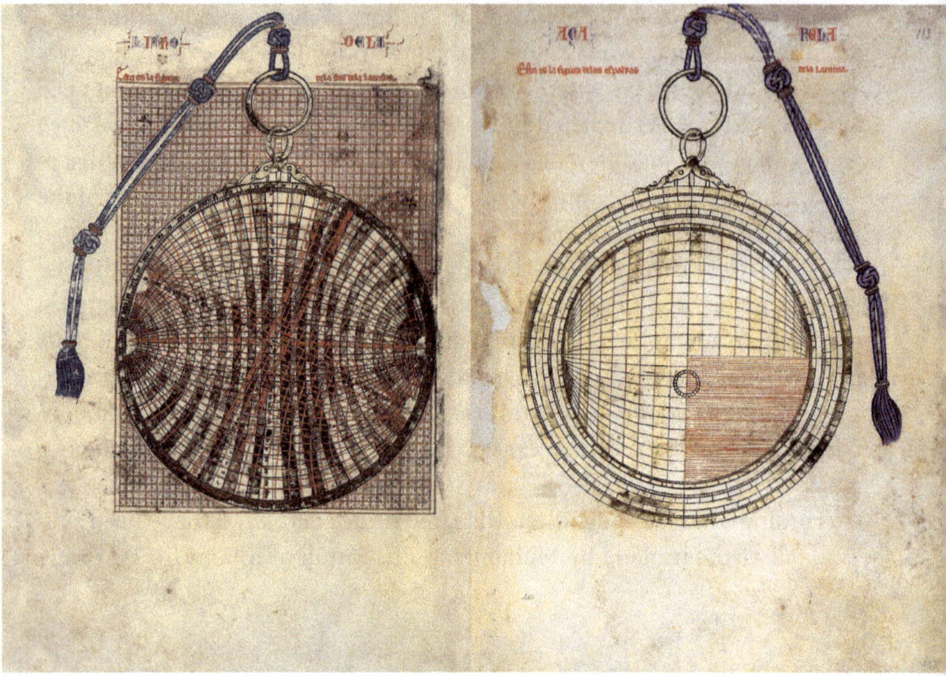

FIGURES 8.13 AND 8.14 *Front and back of the plate of the* azafea. Libro de la azafea, *in the* Libro del saber de astrologia, *Ms. 156,* BH, UCM, *ff. 112v–113r.*
© BIBLIOTECA HISTÓRICA MARQUÉS DE VALDECILLA.

made in 1216 and 1218 in Seville,[44] and one more in Murcia in 1252, thus contemporary to Alfonso X's reign. Following this short list of instruments, Puig suggested that the *azafea* and the *lámina universal* were conceived for theoretical rather than practical use. The instrument made in Murcia in 1252, now at the Real Academia de Ciencias y Artes de Barcelona, was designed using precisely the elements described in the LSA.[45] In this case a close relationship

44 Osservatorio Astronomico di Roma, Referencia n° 694 11/ Almericho da Schio, *Di due astrolabe in caratteri cufici occidental trovati in Valdagno* (F. Ongania, 1880), / Saphée, construite par Mohammed ibn-Fattouh al-Khemairi, à Séville, l'an 615 de l'hégire (1218–1219), Bibliothèque nationale de France, département Cartes et plans, GE A-408, available online at http://gallica.bnf.fr/ark:/12148/btv1b52504946g (accessed 23 June 2018).

45 Muḥammad ibn Fattūḥ al-Khamā'irī, Murcia, 1252, Reial Acadèmia de Ciències i Arts de Barcelona. Jose María Millás Vallicrosa, "Un ejemplar de azafea árabe de Azarquiel," *Al-Andalus* 9 (1944): 118–119; Salvador García Franco, *Catálogo critico*, 311–319; Roser Puig,

can be documented between the real object and the manuscript depiction. We can even argue that the *azafea* now in Barcelona could have been the same instrument known by the Alfonsine scientists.

Conclusion

As we can recognize, the astrolabes depicted in the *Libro del saber de astrología* were designed in close contact with real instruments.[46] However, the use of such other sources as illustrated manuscripts could have provided extra visual information to the Alfonsine astronomers in creating these scientific images. Deeper research into Islamic illuminated manuscripts and their connection with real instruments, with analysis of artistic features, could help us to find new topics of reflection and advance our understanding of the development and spread of medieval science. Both real astrolabes and astrolabes depicted on parchment were used to document the knowledge of astronomy developed in the Castilian kingdom in the thirteenth century, as well as to exhibit the ambition and desires of a monarch who wanted not only to control the world in which he lived but also to understand it.

"L'assafea de la Reial Acadèmia de Ciències i Arts de Barcelona: un exemplar de *zarqalìyya*," *Memories de la Reial Acadèmia de Ciències i Arts* LX-10 (2003): 357–365.

46 It is curious that the astrolabe known as "Astrolabio de Alfonso el Sabio" has apparently nothing to do with the instruments depicted in MS 156, nor even with the figure of the monarch. This astrolabe, kept at the Museo Galileo in Florence, inv. 1113, is a controversial piece. Neither its origin nor its date is clear. It has been linked to Pope Sylvester II, considered a later adaptation of an oriental instrument, and even a nineteenth-century forgery. The reason for relating this astrolabe to Alfonso the Learned is the leather case on which a sixteenth-century hand wrote "Astrolabium arabicum ex Hispania delatum et paratum eo tempore quo aequinoctium vernum haerebat in die 15 martii, id est, anno Christi 1252 quo Alfonsus Rex Hispaniarum restituit motus coelestes." The scholars who have studied the piece insist on dissociating the case from the astrolabe, so that Alfonso X has no connection with the Florentine instrument. In my view, even assuming that the astrolabe has nothing to do with the monarch, the leather case with the inscription has a notable value, and the possibility that the object was altered in order to link it to Alfonso X should be more emphasized. García Franco, *Catálogo de los Astrolabios*, 131–160; Saavedra, "Astrolabios árabes," 396–414; *Alfonso X, Toledo 1984*, 78–79; King, *In Synchrony*, 489–494. Images of the astrolabe and the leather case are available online at http://catalogo.museogalileo.it/oggetto/AstrolabioPiano.html (accessed 23 June 2018).

CHAPTER 9

Fit for a King: Decoding the Great Sloane Astrolabe and Other English Astrolabes with "Quatrefoil" Retes

John Davis
Independent Scholar, Orchard View, Tye Lane, Flowton, Suffolk, IP8 4LD, UK
john.davis51@btopenworld.com

Abstract

The great Sloane astrolabe in the British Museum is the largest and most important of all medieval English instruments and yet its history is completely lost. In this paper, evidence from its various engravings is used to show that it is likely to have been commissioned, ca. 1326, by Richard de Bury as a teaching tool for Prince Edward of Windsor who was soon to become King Edward III. Comparisons are made with two illuminated manuscripts, the "Milemete Treatise" and a copy of the *Secretum Secretorum*, which were also used in the Prince's education. Two other medieval astrolabes, now in Liège and in the London Science Museum, are believed to be closely associated with the Sloane instrument and derived from it.

Keywords

Edward III – Milemete treatise – quatrefoil rete – Richard de Bury – Secretum Secretorum – Sloane astrolabe

Introduction

The Great Sloane astrolabe (Figure 9.1) is one of the treasures of the British Museum.[1] Originally collected by Sir Hans Sloane before 1753, it is a magnificent instrument nearly half a metre in diameter, covered with intricate engraving, complex scales and iconography, and with sculpted zoomorphic features. It is clearly medieval and made for use in England but, as the Museum's website

1 British Museum SLMathInstr 54, "the Sloane astrolabe." Ward 324. Gunther #290. Diameter 462 mm; thickness 12 mm.

 | DOI:10.1163/9789004387867_012

says, it is enigmatic: its date, maker, user(s) and all other details of its history are lost.[2] The size and level of workmanship of the astrolabe indicate that it must

2 The British Museum website (http://www.britishmuseum.org/research/collection_online/collection_object_details.aspx?objectId=54863&partId=1&people=98677&peoA=98677-3-5&sortBy=imageName&page=1) reports that the astrolabe was almost lost completely in its first century at the Museum. The rediscovery of the astrolabe is recorded in the Journals of Sir Frederic Madden (1801–1873), Keeper of Manuscripts at the British Museum 1837–1866, as follows: "Thursday 19th [May 1853] Fine weather. Mr Franks came to my house, and shewed me a very curious Astrolabe of brass, in the most perfect condition, made in England about the year 1330, and lately found in the cellars among refuse, covered with rust!! It formed part

FIGURE 9.1 *Overall views of the back and front of the Sloane astrolabe.* PHOTO COURTESY OF THE TRUSTEES OF THE BRITISH MUSEUM. USED WITH PERMISSION.

have been commissioned by someone with access to considerable funds and at the highest level of society. Its relationship to the small corpus of other medieval

of the Sloane Collection, and has been thrown aside and neglected for just a Century!...." 'Mr Franks' was Augustus Wollaston Franks (q.v.) newly-appointed Assistant in the Department of Antiquities, one of whose early tasks was to identify survivors of the Sloane collection. Madden's journals were bequeathed to the Bodleian Library, Oxford (MS.Eng.hist.c. 140–182). This extract: c. 1866.

English astrolabes is unknown: whilst it clearly has some features in common with a few of these lesser—but still significant—instruments, it is unique in its details. The engraving of the instrument is perhaps the most intricate of all the medieval astrolabes. The features in-between the astronomical scales have sometimes been called "decorative" and are highly reminiscent of illuminated manuscripts. In fact, many, if not all, of these features have been carefully and purposely chosen and are symbols that could have been read by the astrolabe user. This article will propose a reading of the earliest history of this astrolabe and its place in early-fourteenth-century politics, based on a decoding of some of these features and their connections to contemporary manuscripts.

The basic details of the astrolabe were described by Robert Gunther in his monumental catalogue of *Astrolabes of the World* in 1932.[3] A subsequent derivative description is in Frank Ward's *Catalogue of European Scientific Instruments*....[4] Derek de Solla Price's highly critical review of this *Catalogue* described the astrolabe as "probably the most important, interesting and valuable antique instrument in all the world's collections, a jewel in the crown of the British Museum."[5] It was mentioned briefly by David A. King;[6] and it was displayed at the millennium "Time" exhibition at Greenwich.[7] Surprisingly, there have been few scholarly studies of the instrument. Owen Gingerich included it in his study of astrolabes that have retes with zoomorphic star pointers.[8] More recently, Elly Dekker has looked in detail at the actual star names and their positioning to conclude that the astronomical data is closely related to the Kunitzsch Table VI, one of several lists of stars derived ultimately from Ptolemy's catalogue and circulating in manuscript form in the late thirteenth century.[9] Most of the

3 R.T. Gunther, *Astrolabes of the World*, 2 vols., Vol. I: *The Eastern Astrolabes*; Vol. II: *The Western Astrolabes* (London: The Holland Press, first published 1932. Published as a single volume 1976), 462–65.

4 F.A.B. Ward, *A Catalogue of European Scientific Instruments in the Department of Medieval and Later Antiquities of the British Museum* (London: British Museum Publications, 1981).

5 D. de S. Price, "Essay Review: Instruments in the British Museum," *Journal for the History of Astronomy* 13 (1982): 129–32.

6 David A. King, "An Astrolabe from Einbeck Datable ca. 1330," in *Mathematics Celestial and Terrestrial – Festschrift für Menso Folkerts zum 65. Geburtstag*, ed. Joseph W. Dauben et al., *Acta Historica Leopoldina* 54 (Halle: Deutsche Akadamie der Naturforscher Leopodina, 2008), 54–178; reprinted in *Astrolabes from Medieval Europe* (Ashgate: Variorum, 2011).

7 K. Lippincott (ed.), *The Story of Time* (London: Merrell Holberton & NMM, 1999). The Great Sloane Astrolabe appears on p. 105 and is assigned the date "ca. 1295."

8 O. Gingerich, "Zoomorphic Astrolabes: Arabic Star Names Enter Europe" *in The Eye of Heaven: Ptolemy, Copernicus, Kepler* (New York, NY: American Institute of Physics, 1993), 81–101.

9 E. Dekker, "A Close Look at Two Astrolabes and Their Star Tables," in *Sic Itur Ad Astra: Studien zur Geschichte der Mathematik und Naturwissenschaften: Festschrift für den*

manuscripts listing Kunitzsch Table VI state that this list was "verified" in Paris in 1246 with an armillary sphere and one manuscript explicitly gives the observer as John of London: however, the amount of actual observation is unclear.[10]

The above literature gives dates for the astrolabe ranging from "the second half of the thirteenth century" to the 1340s.[11] These "guesstimates" are based (often without giving the reasoning) on a mixture of the star details (including precession), the date of the first point of Aries on its calendar, and the palaeography and general style. A span of nearly a century is perhaps not unreasonable for an undated object but it would be helpful if this could be narrowed.

Proposed Origins of the Sloane Astrolabe

The hypothesis for the astrolabe that the rest of this paper will attempt to justify is that it was commissioned by Richard de Bury somewhere in the period 1325 to 1330. Richard Bury was at that time a tutor and/or mentor to Prince Edward of Windsor, son of King Edward II of England and his Queen, Isabella of France. The prince was in his early teens and soon to be crowned Edward III, first in 1327 as regent with his mother and then, in 1330, to rule in his own right. The astrolabe's purpose was a didactic one, partly to teach Edward the basics of astronomy and astrology but also, possibly more importantly, the essential elements of kingship and the future role that he would be expected to fulfil.

There are two other "quatrefoil" astrolabes that are hypothesized to derive from the Sloane instrument.[12] One, once in the collection of Max Elskamp and now in the Musée de la vie Wallonne, Liège, is very much like a smaller and slightly simplified version of the Sloane instrument.[13] Features of its scales (the stars and the feast days) are extremely close to those on the Sloane

Arabisten Paul Kunitzsch zum 70 Geburtstag ed. M. Folkerts and R. Lorch (Wiesbaden: O. Harrassowitz, 2000), 177–215.

10 P. Kunitzsch, *Typen von Sternverzeichnissen in astronomischen Handschriften des zehnten bis vierzehnten Jahrhunderts* (Wiesbaden: O. Harrassowitz, 1966).

11 A.B. Tonnochy, "Department of British and Medieval Antiquities," *The British Museum Quarterly* 18.1 (1953): 16–17, quoting Gunther's *Astrolabes of the World* (see above, n. 3), 462–65.

12 A third related quatrefoil astrolabe has recently been described in J. Davis, "A Royal English Medieval Astrolabe Made for Use in Northern Italy," *Journal for the History of Astronomy* 48.1 (2017): 3–32.

13 H. Michel, *Les Cadrans Solaires de Max Elskamp* (Liège: Musée de la vie Wallonne, 1966). The astrolabe is illustrated as Plate xxviii. It is described as diameter 146 mm, bronze, ca. 1200—only the first of these "facts" is likely to be correct.

instrument. It will be shown how the Liège astrolabe is likely to have been made soon after the Sloane instrument as a near-copy and as a gift to achieve political favours at the Papal Court (then in Avignon).

The other astrolabe, now in the London Science Museum, has been much altered but, it is speculated, was originally made as a royal gift to John de Stratford, at the time Bishop of Winchester and later Archbishop of Canterbury.[14] Stratford was deeply involved with the political intrigues of Isabella and her lover Roger Mortimer in 1325–7 and then with their overthrow by Edward III in 1330.

Historical Background

There are a great many books describing the histories of the various kings of fourteenth-century England.[15] In addition, the principal characters all have extensive entries in the *Oxford Dictionary of National Biography*. Thus, the intention of this section is only to give the key points in the story as they affect the commissioning of the astrolabes.

The English royal court was in disarray in the middle of the 1320s. Edward II had not lived up to the model of kingship exhibited by his father, Edward I, and had spent his time with inappropriate favourites—first, Piers Gaveston and then the Hugh Despensers, father and son. He had fallen out badly with his barons, lost ground to the Scots and the French and become estranged from his French wife Isabella. She continued to prove useful as an ambassador to France where her brother, King Charles IV, was demanding that Edward pay due homage for his lands in Gascony. Edward, not wanting to leave England, had made his eldest son Prince Edward duke of Aquitaine so that he could go to Paris and perform this duty for him. However, once there the Prince fell under the influence of his mother who refused to allow him to return to England as his father demanded.

In Paris, Isabella met up with Sir Roger Mortimer, who had escaped from the Tower of London where he had been imprisoned for treason. Together, they planned an invasion of England. Support for the invasion, in the form of ships

14 Science Museum, London inv. no. 1880–26. Gunther 293. Tzvi Herz astrolabe.

15 For recent and easily accessible accounts of the relevant histories, see Ian Mortimer, *The Perfect King. The Life of Edward III, Father of the English Nation* (New York, NY: Random House, 2008); and Alison Weir, *Isabella. She-Wolf of France, Queen of England* (London: Vintage, 2012). A more authoritative and academic study is W. Mark Ormrod, *Edward III* (New Haven, CT: Yale University Press, 2011).

and a small army, came from Count William of Hainault as an early dowry for the proposed marriage of Prince Edward to his second daughter Philippa. The invasion occurred in 1326, landing at Orwell in Suffolk, and was remarkably successful as the people and towns of England sided with Isabella and her son. After very little bloodshed, it ended in the capture of Edward II and the hated Despensers. The latter were summarily executed and Edward was "persuaded" to hand the crown to his son, the fourteen-year-old Edward III, in 1327.

The joint reign of Isabella and Edward III started well but rapidly deteriorated as Roger Mortimer, acting in the background, exerted a controlling influence over Isabella. Mortimer showed an increasing self-aggrandisement and appeared to want to rule absolutely. In 1330, Edward III and a small group of friends captured Mortimer in Nottingham Castle: he was swiftly executed and Isabella, for whom Edward retained much affection, was sidelined under nominal house arrest.

It is with this background that the Sloane astrolabe was commissioned, drawing on an existing presence of smaller English astrolabes, mainly with the simpler Y-form of rete.

Richard de Bury

Richard de Bury (his family name was Aungerville, 1287–1345) was born in Suffolk near Bury, famous for its large Benedictine abbey, which housed the relics of St. Edmund, a king of East Anglia killed by the Vikings.[16] (It also had relics of St. Botolph of Iken, another favoured East Anglian saint.) A nominal portrait of Bury is in Thomas Walsingham's *Liber benefactorum of St. Albans Abbey* (Golden Book of St. Albans) but as this was written some thirty-five years after Bury's death it is unlikely to be a convincing likeness.[17] He is also depicted with his head in half-profile on his second seal as Bishop of Durham.[18]

Bury's early education was probably at the local grammar school, in what is now Schoolhall Street, Bury St.Edmunds, just outside the abbey precinct.[19] It is quite likely that he was at least aware of the large library at the Benedictine

16 For a modern biography of Richard de Bury, see the *Oxford Dictionary of National Biography*. For an older but more detailed discussion, see N. Denholm-Young, "Richard de Bury (1287–1345)," *Transactions of the Royal Historical Society* 20 (1937): 135–168.

17 British Library, Cotton MS Nero D VII f. 87r.

18 See http://www.gutenberg.org/files/18938/18938-h/images/05.jpg (accessed 24 March 2018).

19 A.F. Leach, *The Schools of Medieval England* (London: Methuen & Co, 1915), 120–121.

Abbey of St. Edmund at Bury, where there was a collection of over two thousand books.[20] He studied at Oxford (possibly at the Benedictine Durham College, on the site of the present Trinity College) and, although he does not seem to have graduated, he acquired a lifelong interest in books and learning. In 1316 he became a king's clerk where, as chamberlain to Prince Edward's estates as Earl of Chester, he started a long and steady rise in power. The first major step came when he became the mentor/tutor to the young Prince in 1324 at the time when a separate court was established for him at Savoy Palace.

Bury was with the Prince in France in 1325 when, as duke of Aquitane, Edward travelled to Paris to swear allegiance to King Charles IV of France. When Isabella and her lover Sir Roger Mortimer plotted to invade England, Richard was in a position as constable of Bordeaux to raise money illicitly for her cause, at some risk to himself.[21] This allegiance between Bury and Isabella seems to have been a very long-lasting one and important in the raising of Prince Edward. Both were avid collectors of books, in Isabella's case particularly "romances" and stories of King Arthur. As an aside, it is worth noting that she also had some interest in scientific instruments: towards the end of her life, she took delivery of "several quadrants of copper" from a monk of St. Albans.[22]

In his role as tutor/mentor to Prince Edward, Bury made use of the available literature and this included the genre known as the "mirror of princes." One example of this is the *Secretum Secretorum*, widely copied at the period, and which purported to describe the teaching of Alexander the Great by Aristotle. It seems that Bury saw himself in the Aristotle role. Several of the illustrations in the version of the *Secretum Secretorum* presented to Edward on his coronation and now in the British Library feature astrologers and astronomy, such as one showing a king (Alexander) with Aristotle pointing to a diagram of the cosmos.[23] Another, shown in Figure 9.2, has a king surrounded by astrologers

20 E.A. Savage, *Old English Libraries* (London: Methuen & Co., 1911), 63.

21 Weir, *Isabella*, 180.

22 "[to] William Orologer, monk of St Alban's, bringing to the lady queen several quadrants of copper from the like gift by his own hands on the same day: 20s[hillings], 6th January [1358]". The entry was first translated from the original Latin by E.A. Bond, "Notices of the Last Days of Isabella, Queen of Edward II, drawn from an Account of the Expenses of her Household," *Archaeologia*, 35 (1854): 453–69, quoting from the original (and badly fire-damaged) *'Dona'* section of the accounts now British Library, Cotton MS Galba E XIV, f. 15r ll.19–20. The fuller translation here is due to Simon Neal. The quadrants were described as being of "Cup[ro]", but the scribe would not have been a metallurgist so it is more likely that they were of latten (an alloy of copper with additions of zinc, tin and lead). The value of 20 shillings is equivalent to around £500 at today's values.

23 *Secretum Secretorum*, British Library, MS Add 47680, f. 51v.

FIGURE 9.2 *A king (Alexander) with astrologers and a physician. Note the use of an armillary sphere and a quadrant, with the sun and moon in the sky. "Secretum Secretorum," British Library MS Add. 47680 f. 53v.*

and a physician holding instruments including an armillary sphere and a quadrant (held upside-down!). It is easy to visualize an astrolabe held by Bury instead of the armillary sphere.

After Edward II had been deposed, Bury was part of the council of ministers who were supposed to guide the young King Edward III though the council was very much side-lined by Roger Mortimer who heavily influenced all decisions through Isabella. Bury remained very close to Edward and was part of the group that helped him arrange the overthrow of Mortimer in 1330. He was rewarded by being made Bishop of Durham (one of the richest sees in England) in 1333 and ultimately chancellor of the realm in 1334. He remained highly influential for the rest of his life.

Bury met his fellow bibliophile Petrarch whilst on one of his diplomatic missions to Avignon in 1330. Petrarch records that he was "not ignorant of literature and from his youth up curious beyond belief of hidden things."[24] He asked him for information about Thule which Bury promised to answer when he was back at home among his books. Thule (Iceland) is known as the central location of Climate VIII (the Ptolemaic scheme for specifying latitude) and sometimes found as a latitude on astrolabe plates.

Bury assembled a large household of scholars, clerks and chaplains who travelled with him. These came to include a number of scholars from Merton College Oxford including Thomas Bradwardine (who briefly became Archbishop of Canterbury, following Bury's patronage), Walter Burley, and John Maudith. The household is described as "representing the single most notable circle or sequence of scholars under the patronage of one person in fourteenth-century England."[25] It seems likely that the subject of astrolabes would feature in any conversations with these "Oxford Calculators."

Bury let it be known that no present was more appealing to him than one of books and in his lifetime he accumulated a very large library. In his will, he bequeathed most of these to the Benedictine Durham College in Oxford, though due to his debts the "five cartloads" of books had to be sold off by his executors instead. One source of books was from Richard of Wallingford, Abbot of St. Alban's Abbey, builder of the famed astronomical clock and probably the best astronomer of the century, who made a gift of four books and sold Bury a further thirty-two from the Abbey, to the outrage of the monks. The two knew each other from being members of Edward's parliaments, though they may have met much earlier in Paris.[26]

Towards the end of his life, in 1344, Bury wrote a book—the *Philobiblon*—on the importance of book collecting and which is considered to be one of the earliest books to discuss librarianship.[27] It has been described as an "enthusiastic rant about the virtues of books" in which Bury declares: "In books I find the dead as if they were alive; in books I foresee things to come; in books warlike affairs are set forth; from books come forth the laws of peace."[28]

24 N. Denholm-Young, (see above, n. 16), particularly p. 148; see also the Oxford Dictionary of National Biography entry for Richard Bury, written by W. J. Courtenay.

25 *Oxford Dictionary of National Biography*: entry for Richard de Bury.

26 J. North, *God's Clockmaker: Richard of Wallingford and the Invention of Time* (London: Hambledon & London, 2005), 110–11.

27 R. Bury, *Philobiblon*. A modern transcription/translation is *The Love of Books: The Philobiblon of Richard De Bury*, trans. E.C. Thomas (Oxford: Basil Blackwell for The Shakespeare Head Press, 1960), available online at http://www.philobiblon.com/philobiblon.shtml.

28 Mortimer, *Perfect King*, (see above, n. 15), 36.

Bury's interest in astronomy—as well as many other topics—is made explicit in chapter seven of the *Philobiblon* (translated from the Latin):[29]

> There [in books] is most excellent in doctrine, so far as relates to this passing sublunary world; there Ptolemy measures epicycles and eccentric apogees and the nodes of the planets by figures and numbers.

And in chapter fifteen:

> But if we please to visit the heavenly inhabitants, Taurus, Caucasus, and Olympus are at hand, from which we pass beyond the realms of Juno and mark out the territories of the seven planets by lines and circles. And finally we traverse the loftiest firmament of all, adorned with signs, degrees, and figures in the utmost variety. There we inspect the Antarctic pole, which eye hath not seen, nor ear heard; we admire the luminous Milky Way and the Zodiac, marvellously and delightfully pictured with celestial animals.

In chapter ten he discusses the need to understand languages other than Latin:

> One thing, however, we conclude from the premises, that the ignorance of the Greek tongue is now a great hindrance to the study of the Latin writers, since without it the doctrines of the ancient authors, whether Christian or Gentile, cannot be understood. And we must come to a like judgment as to Arabic in numerous astronomical treatises.

From this, it is clear that Bury was familiar with the existence of Arabic astronomy and probably had these works, including details of astrolabe construction and use, in his library. We do not know how many of these astronomical books he had actually read and understood but clearly he was aware of what was contained in his library. Together with the presence of Oxford astronomers including Thomas Bradwardine and John Maudith in his close circle, it certainly suggests that he would have at least been aware of the importance of the astrolabe and probably knew the basics of operating one.

29 Thomas, *The Love of Books* (see above, n. 27).

Features of the Sloane Astrolabe

The Rete

The most striking feature of the rete of the Sloane astrolabe is the set of quatrefoils (and trefoils and semi-quatrefoils) supporting the star pointers. These design elements were already quite common on both Islamic and Latin astrolabes.[30] There is a small group of English medieval astrolabe retes that feature quatrefoils, but the layout of the Sloane is unique in having a row of three large quatrefoils inside the ecliptic circle. These form the most important single clue as to the meaning of the whole instrument. Its large size allows room for much fine detail in the features that the maker has exploited by including many small items, rather than simply scaling up those found on smaller astrolabes.

Medieval symbolism was often a multi-layered story with parallel meanings being read into each element. Here, the usual religious icon of the quatrefoil representing a Christian cross (or the four evangelists) is supplemented by a secular meaning as a four-leaved clover, for centuries a sign of good luck where, traditionally, the lobes stand for faith, hope, love, and luck. The medieval name for this plant was *corona regia* or the king's crown. It is this symbolism that, it is proposed, was behind their choice for the astrolabe. Herbariums of the period say that the symbol is used because "the yellowe flowers doe crown the top of the stalkes as with a chaplet of gold."[31] Support for this interpretation is given in the detailed shape of the Sloane astrolabe quatrefoils where the inside tips of the lobes are intricately shaped to give them a close interpretation of the plant (Figure 9.3). This very fine workmanship is not generally present on other, smaller instruments with quatrefoils.

A contemporary and highly suggestive use of the cloverleaf as an analogy for the king's crown is shown in an illustration in the *Milemete Treatise*, a spectacular and sumptuous illuminated manuscript developed by the King's Clerk Walter de Milemete and presented to Edward III on his coronation in 1327.[32]

30 For astrolabes with quatrefoil/trefoils, see, for example, David A. King, *The Ciphers of the Monks: A Forgotten Number-Notation of the Middle Ages* (Stuttgart: F. Steiner, 2001). Appendix J (The quatrefoil on medieval astrolabe retes), p. 380. Also, David A. King, "In Synchrony with the Heavens: Studies in Astronomical Timekeeping and Instrumentation," in *Medieval Islamic Civilization*, 2 vols. Vol. 1: *The Call of the Muezzin*; Vol. 2: *Instruments of Mass Calculation* (Leiden: Brill, 2004–5), 2:XVII.

31 Parkinson, *Theatre of Botany* (1643), 720.

32 The "Milemete Treatise" is entitled *De Nobilitatibus, sapientiis et prudentiis regum*, Christ Church, Oxford MS 92. Available online at the Christ Church library website https://digital.bodleian.ox.ac.uk/inquire/p/575d3fbd-21d9-4521-a938-301756bf88d7. Originally conceived for Edward II, it was reworked in 1326/7 by Walter de Milemetre as

FIGURE 9.3 *Close-up of part of a Sloane quatrefoil.*
THIS AND ALL SUBSEQUENT PHOTOS OF THE SLOANE ASTROLABE ARE BY THE AUTHOR, WITH THE PERMISSION OF THE TRUSTEES OF THE BRITISH MUSEUM.

In the introductory section (added after the main treatise had been largely completed) is a full page miniature of Edward III and his mother Isabella on thrones being simultaneously handed green—the colour is significant—quatrefoils by an angel of god. The meaning is clearly that Edward and Isabella were divinely destined to rule together (Figure 9.4).[33]

There exists a similar depiction of another enthroned young and clean-shaven king with his mother (Louis IX of France and Blanche of Castile, ca. 1230)[34] as well as a later image of King Henry VI of England being handed

---

a coronation gift for Edward III and as a partner to an equally magnificently illuminated copy of the pseudo-Aristotelian *Secretum Secretorum* (Brit Lib MS Additional 47680). For a discussion of these manuscripts see Libby Karlinger Escobedo, *The Milemete Treatise and Companion Secretum Secretorum: Iconography, Audience, and Patronage in Fourteenth-Century England* (New York, NY: Edwin Mellen Press, 2011); and Michael Michael, "The Iconography of Kingship in the Walter of Milemete Treatise," *Journal of the Warburg and Courtauld Institutes* 57 (1994): 35–47, which states: "These two manuscripts that Walter de Milemete prepared for Edward III in 1326/7 fit within the medieval tradition of the *speculum principis*, or mirror of princes. Such texts were intended to educate rulers in statecraft, ethics, war and courtly behaviour."

33 *Milemete Treatise*, (see above, n. 32), f. 4v.

34 *Bible Moralisée*, Pierpont Morgan Library, MS M.240, f. 8r.

FIGURE 9.4
Edward & Isabella receiving corona regia from an angel. Milemete Treatise, Christ Church, MS *92, f. 4v.*

his crown (rather than symbolic quatrefoils) by angels that support the interpretation of the *Milemete* image.[35]

A second instance of the *corona regia* is to be found in the half-page miniature on folio 18v of the *Milemete Treatise* (Figure 9.5). Here, the King (Edward III) is being crowned by Christ while three angels hand him a cross, a shield and a cloverleaf. The illumination is part of the heading of chapter five, "On the nobility of the king and the prosperity of his government; and on the causes of his success."

The next question is why there is a set of *three* quatrefoils and this can be answered by referring to a powerful and long-lasting piece of political prophecy current at the time. This is generally known as *The Six Kings who will follow King John* and describes the characters and deeds of the kings Henry III, Edward I, Edward II, Edward III, and two as-yet unborn kings (Richard II and

35 British Library, Hargrave MS 274 f. 204v, historiated initial "A," c. 1488.

FIGURE 9.5 *King Edward being presented with a cross, a shield and a corona regis (cloverleaf) by angels. Milemete Treatise, Christ Church, MS 92, f. 18v.*

Henry IV).[36] It is also sometimes known as *The Last Kings of the English.*[37] The prophecy claims its origins going back to King Arthur and Merlin, as recounted in Geoffrey of Monmouth's *Historia Regum Britanniae*, and was widely believed and studied at the time, including by the educated elite. It is thought initially to have been written in Anglo-Norman prose soon after Edward of Windsor's birth in 1312 and circulated verbally but many slightly later written versions exist in English couplets and in Latin: it also appears in the *Brut*.[38] Part of this "prophecy" was written with hindsight and hence was rather accurate but the future parts became increasingly vague. Each king was represented by an animal: Henry III as the Lamb of Winchester, Edward I as the Dragon, Edward II as the despised Goat of Carnarvon, Edward III as the boar, to be followed by two lesser kings, the Ass with Leaden Feet and the Accursed Mole. Edward III, the "boar who will come out of Windsor" (his birthplace), was prophesised to

36 For the Six Kings Prophecy see, for example, Lesley Ann Coote, *Prophecy and Public Affairs in Later Medieval England* (Woodbridge: Boydell & Brewer, 2000). See chapter 3, "Expectation and disappointment 1307–1340," 83–120. Page 107 refers to the Three Crowns at Cologne and of England/Scotland/France. Or see Roy Martin Haines, *King Edward II: His Life, His Reign, and Its Aftermath, 1284–1330* (Montreal, QC: McGill-Queens Press, 2003), especially 26, "Prophecy in retrospect."

37 W. Mark Ormond, *Edward III*, (see above, n. 15). Chapter 4.

38 T. M. Smallwood, "The Prophecy of the Six Kings," *Speculum* 60.3 (1985): 571–592.

be the *King who will wear three crowns*, regaining the lands won by his grandfather, the fearsome Edward I, but lost by his ineffectual father. Many great things were predicted for him. The choice of the boar has significance because the "Boar out of Cornwall" was King Arthur, whom Edward was to spend his life trying to emulate, culminating in 1348 with the foundation of the Order of the Garter as his equivalent to the Round Table.

There are at least two possible roles that can be identified with the "king with three crowns." One is with the Holy Roman Emperor who traditionally wore crowns of iron, silver, and gold. His seat in Cologne, which held the shrine of the Magi (the Three Kings of the Nativity), was visited in 1338 by Edward, who made a very large offering there.[39] Much later, he made a solemn oath that he would not be buried amongst the Kings in Cologne as the prophecy had predicted and as he had at one time seriously contemplated, clearly he had taken the prophecy to heart.[40]

The other set of "three crowns" that the king might wear were those of England, France and Scotland. Edward I (reg. 1272–1307) had fought for the Scottish crown for many years but Edward II (reg. 1307–1327) had shamefully lost control at the battle of Bannockburn in 1314. Edward III (reg. 1327–1377) was the duke of Aquitaine and had a claim to the crown of France through his mother Isabella. It was this claim that was one of the underlying causes of the Hundred Years War. Later in his reign, after the triumphant battle of Poitiers (1356), Edward held both King John II of France and David II of Scotland captive in London so, in some sense, the prophecy could have been said to have come true.

The three large quatrefoils are supported from the ecliptic ring by two smaller trefoils. These were usually taken in the Gothic period to represent the Trinity (Father, Son and Holy Spirit) and are another sign that the three crowns were to be given to the king by divine right.[41]

There is a second, quite different, botanical link to the quatrefoil shape in the form of *Paris quadrifolia* (herb paris), which the herbalist John Gerarde described thus:

> Herbe Paris riseth up with one small tender stalke two hands high; at the very top come forth foure leaves directly set one against another in the manner of a Burgundian Crosse or True-love knot: for which cause among the Antients it has been called Herbe True-love.[42]

---

39 Mortimer, *Perfect King*, (see above, n. 15), 21.

40 Ormrod, *Edward III*, (see above, n. 15), 467.

41 F. Rest, *Our Christian Symbols* (Cleveland, OH: Pilgrim Press, 1954).

42 John Gerarde, *Grete Herbal of Generall Historie of Plants* (1597), p. 328. See also the Middle English Herbal at Oxford, Bodleian, MS Ashmole 1443.

Thus in the medieval mind the leaves of the herb paris can stand for both the cross of Christ and the true-love knot. Chaucer used "trewe-love" as an abbreviation for "true-love knot" in the last stanza of his *Court of Love*: some of these knots had four loops giving rise to the association.[43]

A popular Middle English rhyme alludes to the custom of placing the spring-blooming herb paris in hats, hoods, or purses as a charm for luck in love, while stressing how quickly such love is lost.[44] The quatrain was set down in the fourteenth-century *Fasciculus Morum*, a handbook for Franciscan preachers.[45]

Again, we can look to the *Milemete Treatise* for an example of the quatrefoil pattern used to indicate true love. The introductory section has a major double-folio illumination of the "Storming of the Castle of Love" (Figure 9.6).[46] This is a common medieval allusion to the art of courtly love and chivalric behaviour often found on contemporary ivory caskets and mirror backs as well as in other manuscripts such as the Luttrell and Peterborough psalters.[47] It is clear that this symbol would appeal to a love-sick Edward, separated from betrothed Philippa. A further interpretation of the *Three Crowns* is that it is the heraldic banner of St. Edmund, of whom there is more to report below.[48]

In addition to the set of three large quatrefoils, the rete also has a set of four semi-quatrefoils at the ends of two diametrical bars. These could be regarded as the representation of God controlling the cosmos, depicted in many manuscripts as him holding an orb in his hand. The lobes of these semi-quatrefoils have a different geometry to those of the large quatrefoils: the latter are pointed and look leaf-like whereas the former are circular arcs. These two forms of quatrefoil are nicely illustrated in the same figure in a marginal illustration of the *Milemete Treatise* as seen in Figure 9.7.

The other major feature of the Sloane rete is that many (thirty-four out of forty) of the star pointers have a zoomorphic form. Again, this is quite common,

---

43 H.A. Kelly, *Chaucer and the Cult of St. Valentine* (Leiden: Brill, 1986), 75.

44 See http://d.lib.rochester.edu/teams/text/fein-moral-love-songs-and-laments-four-leaves-of-the-truelove-introduction; or http://blog.metmuseum.org/cloistersgardens/2010/06/23/herb-paris/ (Last accessed 23 June 2018).

45 See http://www.psupress.org/books/titles/0-271-00642-0.html (Last accessed 23 June 2018).

46 *Milemete Treatise*, ff. 3v–4r. Here, a group of young knights on horseback is attacking a castle defended by girls with longbows and crossbows firing true-love flowers instead of arrows.

47 Escobedo, *The Milemete Treatise* (see above, n. 32), 152.

48 Azure, three imperial crowns or, two and one. In some later depictions, the crowns each contain two arrows (Edmund's attribute), and occasionally a wolf's head (for the myth of the finding of his decapitated head) is also present.

FIGURE 9.6 *"Storming of the Castle of Love." Milemete Treatise, Christ Church* MS *92, ff. 3v–4r.*

FIGURE 9.7 *Nested quatrefoils with both rounded and pointed lobes. Milemete Treatise, Christ Church* MS *92, f. 6r.*

FIGURE 9.8 *Montage of selected zoomorphic star pointers on the rete of the Sloane astrolabe. Note the variety of animal forms.*

even usual, on English medieval astrolabes but here the detailing is at the highest level. Often, the pointers use animals that form a rebus or *aide-memoire* for the star name, e.g., a dog's head for Alhabor or Sirius (the Dog Star) and a bird for Corvus. Gingerich points out that whereas Gunther describes the sickle-shaped pointers as issuing from the mouths of dogs, at least twelve of these animals have wings and might best be described as dragons.[49] As the close-ups in Figure 9.8 show, the animals are actually quite varied and form a complete menagerie. It could be suggested that at least some of them represent the animals in the *Six Kings* prophesy. Exotic animals of all sorts appear in the illuminations of the *Milemete Treatise* and real animals were kept in menageries such as that in the Tower of London which, in 1289, housed lions, leopards, and a bear. Edward III himself moved his menagerie around the country with him.[50] Other hybrid animals were frequently shown in the bestiaries of the time.[51]

A detailed feature of the rete that is easily overlooked is a small pair of human faces at either end of the equatorial arc (Figure 9.9). The one on the left

49 Gingerich, "Zoomorphic Astrolabes," n. 8, p. 86.

50 Ormrod, *Edward III*, (see above, n. 15), 103.

51 Escobedo, *The Milemete Treatise* (see above, n. 32), p. 91.

FIGURE 9.9 *Close-up of the two faces on either end of the equatorial arc on the rete of the Sloane astrolabe.*

seems to be of a young girl and hence it is suggested it is actually a depiction of Philippa of Hainault, betrothed to Prince Edward. The one on the right is more masculine and could be Edward himself. Of course, the images are far too crude to be certain of this identification but further support for the suggestion is provided below and, if it is believed, then it could establish a *terminus post quem* of 1325 for the astrolabe.

The Plates

The Sloane astrolabe has a set of three double-sided plates with stereographic projections labelled for:

(i) 53 GRAD^{9} and 54 GRAD^{9} [52]
(ii) 51 GRAD^{9} and 48 GRAD^{9} 30 m^{a}
(iii) 55 GRAD^{9} and 52 GRAD^{9} (marked for LVNDONIARVM)

(Here, the ^{9} superscript shows the common medieval abbreviation for the -us word-ending. m^{a} is the abbreviation for "minuta.")

In addition, the womb (inside of the mater) has a further projection for 42 GRAD, marked for ROMA. With the exception of the plate for 48° 30′, which is clearly intended for Paris, the latitudes of the plates are for England and cover virtually the whole country. The possibility that they were intended for locations in mainland northern Europe can be discounted. There is only a relatively small amount of

52 Gunther, "Astrolabes of the World," p. 464.

lettering on the plates but it appears that the one for London has received the most attention (for example, it includes a set of lines for the "Houses of Heaven" not on the other plates) and was thus probably the one in main use. In general, there is relatively little that can be deduced about the origins of the astrolabe from the plates.

The choice of 42° (Rome) for the womb is rather unexpected as Avignon had become the location for the Papal court at this time. Neither the almucantars nor the unequal-hour lines are numbered, as they are on the plates. On the basis of the engraving of the "G" of Grad, though, it is possible that this was added by a different hand and hence it may originally have been blank.

The Back of the Sloane Astrolabe

The back of the astrolabe, like most medieval English instruments, is mainly devoted to various calendrical scales. What sets the Sloane astrolabe apart from its contemporaries is the high level of additional detail that fills all of the spare space around the numerical scales. This detail, previously dismissed as decoration, includes numerous groteseques, dragons, and floral elements, and is extremely reminiscent of the style seen in illuminated manuscripts of the time, including the *Milemetre Treatise* and the British Library copy of the *Secretum Secretorum*.

Amongst the memorable images on the back of the astrolabe are two pairs of anthropomorphic birds facing each other, one pair near the throne and the other near the bottom of the instrument (Figure 9.10). Similar pairs of creatures (Figure 9.11) can be found in the British Library *Secretum Secretorum* amongst the over sixty hybrid creatures that it contains.[53] For the astrolabe creatures, it is suggested that the birds are intended to represent falcons and the faces—a male and a female in each pair—are real individuals. The presence of portrait busts, possibly representing Edward III himself, have been noted in the *Milemete Treatise* and the British Library *Secretum Secretorum*.[54]

The royal sport of falconry was an enduring passion of Edward III and of Isabella.[55] He was depicted in an introductory illumination of the *Milemete Treatise* with a falcon on his arm and he was similarly depicted in a book that Philippa gave him as a wedding present.[56] Later in life, he maintained a staff of up to twenty

53 Escobedo, *The Milemete Treatise*, 75.

54 Escobedo, *The Milemete Treatise*, 191 references *Milemete* f. 13v and *Secretum Secretorum* 14r.

55 Ormrod, *Edward III*, 103.

56 *Milemete Treatise*, f. 1r shows Prince Edward sitting on a bed with a falcon on his arm.

FIGURE 9.10 *Photographs of the two pairs of anthropomorphic falcons on the back of the Sloane astrolabe.*

FIGURE 9.11 *Anthropomorphic beasts from (top) the Milemete Treatise, Christ Church, MS 92, f. 2r; and (bottom) Secretum Secretorum, British Library, MS Add 47680 f. 9v.*

COURTESY OF THE TRUSTEES OF THE BRITISH LIBRARY.

falconers and there are many other accounts of him rewarding his favourites and of taking falcons with him on trips abroad.[57] Thus the appearance of falcons on the astrolabe would be very attractive and personal to the young prince.

I suggest that the faces of the upper pair of falcons are meant to represent Prince Edward himself with, facing him, his betrothed Philippa of Hainault, aged about fourteen and with her hair in a very fashionable "crispinette," thought to have been introduced to England from Paris by Isabella.[58] There are few portraits of Philippa as a young woman but one can be found of her at her coronation (Figure 9.12). The female face in the other pair of falcons is clearly an older married woman, as shown by the wearing of a "torque" with a veil under the chin.[59] This, then, is likely to be an image of Queen Isabella, Edward's mother, herself. Identification of the matching clean-shaven male, wearing a "hood and liripipe" is more difficult. The first thought is that it might be Isabella's husband, King Edward II but since the pair were estranged and the King was soon to be deposed (or had already been deposed) this is unlikely. Also, King Edward was known to wear a long beard that he did not shave off until he was captured by Isabella's and Mortimer's followers.[60] It does not seem conceivable that the Queen's lover, Roger Mortimer, would be shown on an astrolabe for her son and it would have been highly presumptuous for Richard Bury to have had his own image shown facing a member of royalty. This leads to the conclusion that perhaps it is a second representation of the Prince, facing his mother and co-ruler.

The *Six Kings* was not by any means the only political prophecy in circulation at the beginning of the fourteenth century. Another was the *Prophecy of the Bull*, a long prose poem later attributed to John of Bridlington that also foresaw an heroic future for a new king, mirroring the deeds of earlier generations.[61] Like others of the genre, this also told the story by means of the actions of animals representing the key players and in this case Edward III was represented by a bull, a common choice for him (as well as the boar and the leopard/lion) at the beginning of his reign. The prophecy states (in Latin): "The

57 Ormrod, *Edward III*, 103.

58 See http://www.kats-hats.co.uk/crispinette.shtml (Last accessed 23 June 2018).

59 See http://www.kats-hats.co.uk/torque.shtml (Last accessed 23 June 2018). See also Weir, *Isabella*, Plate 1, for a picture of Isabella with similar headgear.

60 Mortimer, *Perfect King*. The forcible shaving of Edward II's beard in a muddy puddle features in Christopher Marlowe's play of the same name.

61 Caroline Shenton, "Edward III and the Symbol of the Leopard" in *Heraldry, Pageantry and Social Display in Medieval England*, ed. P. R. Coss and M. Hugh Keen (Woodbridge: Boydell, 2002), 69.

FIGURE 9.12 *Illuminated miniature of the coronation of Queen Philippa, from Jean Froissart, Chronicles, BnF MS Française 2863.*
COURTESY OF THE BIBLIOTHÈQUE NATIONALE DE FRANCE.

Bull will be strong, fearing not at all the sadness of death; sober and chaste, upright without any crime, fortunate. Not slow to do good, bold as the leopard."[62] Two faces of bulls can be found on the back of the Sloane astrolabe, on the north-south line, below the shadow square and above and below the lower pair of falcons (Figure 9.13). They look out of the astrolabe and must have reminded the prince what characteristics were expected of him.

Elsewhere on the astrolabe (Figure 9.13), in the corners of the shadow square, are overhead images of what could well be heraldic lions, known as

62 Quoted in Shenton, "Edward III," 122.

FIGURE 9.13 *Close-up of the bulls and leopards(?) on the back of the Sloane astrolabe.*

leopards. The "three [heraldic] lions of England" were already well established at this time but Edward III also took the leopard as his personal symbol, particularly early in his reign.[63] He used the name "Sir Lionel" in jousts and named his third son Lionel (1338–1368). When he produced the first English coinage in gold for many years in 1344, one of the coins was the now-famous and extremely rare "double leopard."[64] On the obverse, the coin shows King Edward in the form of Solomon with leopards on either side and on the reverse it has a frame of a large quatrefoil (or *corona regis*) around a cross with crowned quatrefoils and with four leopards outside it. Edward also had real leopards in his menagerie: in 1333 a payment to a man described as "keeper of the king's leopards" is recorded and later, in 1365, the king's son, the Black Prince, sent him a gift of

63 The coin, once exhibited at the Fitzwilliam Museum, Cambridge, can now be seen at http://british-coin-price-guide.homelinux.com/England-Coinage-1272-1399/england-king-edward-iii-rare-gold-double-florin-coinage-1344-1351-british-coin-price-guide-page15.html (accessed 26 September 2018).

64 See http://www.fitzmuseum.cam.ac.uk/dept/coins/exhibitions/CoinOfTheMoment/doubleleopard/ (accessed 23 March 2015).

a lion and a leopard to replenish his menagerie.[65] Richard Bury was also fond of the lion/leopard symbol: the English three lions shield appears twice on his aforementioned seal as Bishop of Durham. If the identification of the creatures on the astrolabe is correct, it sheds new light on the origins of Edward's lifelong affection for leopards.

There are two major images (Figure 9.14) featuring dragons (strictly, wyverns, as they have wings and only one pair of legs) on the back of the astrolabe, positioned either side of the shadow square. That on the left is of a large dragon being attacked and bitten by two much smaller creatures. This is likely to represent Prince Edward's grandfather, the fearsome Edward I (depicted as the dragon in the *Six Kings* prophecy) in the battles he fought throughout his reign with the Welsh and the Scots. Edward III was to spend much of his life reclaiming the inheritance of his grandfather, especially against the Scots, and in 1346 he used a dragon standard at the battle of Crécy as a sign that the battle would be without quarter (i.e., to the death), the equivalent of the revered French *Oriflamme* standard, standing for a "*guerre mortelle*."[66]

The dragons on the right of the shadow square form a different picture: a pair of near-mirror-image creatures are bound together by two belts around their torsos. I suggest that they stand for the combined kingdom of England and Wales with the historic red dragon of the Celts being joined to the white dragon of the Anglo-Saxons.[67] This raises a question of whether the engraving of the astrolabe was coloured in any way. Certainly, there is no trace of it today but it does remain a possibility. Once again, Prince Edward was being encouraged to emulate the feats of his grandfather.

Feast Days

A feature of the calendar scales on many medieval English astrolabes is that they include a set of feast days, taken from the Christian calendar and usually accompanied by their corresponding Dominical Letters. This feature is not normally found on astrolabes from other cultures, locations or periods. The number of feasts included varies with the size of the astrolabe and hence the space available. On a very small astrolabe it might be as few as twenty-four

65 Shenton, "Edward III," 77.

66 Ormrod, *Edward III*, 281.

67 In Geoffrey of Monmouth, *Historia Regum Britanniae* (c. 1130), Merlin prophesises long struggles between the two dragons. Edward I incorporated Wales into England in the Statute of Rhuddlan in 1284 and in 1301 made his son (later Edward II) the Prince of Wales.

FIGURE 9.14 *Close-up of the dragons located on either side of the shadow square on the back of the Sloane astrolabe.*

(two per month);[68] at the other extreme, the Science Museum astrolabe has fifty-five: the Sloane and the Liège instruments each have (the same) forty-eight. A popular collection of descriptions of the saints' lives by Jacobus de Voragine, writing sometime before 1267, lists 182 saints.[69] Clearly, not all of these saints were celebrated throughout Christendom and no astrolabe, or written calendar, had space for them all. Even quite comprehensive *Kalendaria*, such

68 J. Davis and M. Lowne, "An Early English Astrolabe at Gonville & Caius College, Cambridge, and Walter of Elveden's Kalendarium," *Journal for the History of Astronomy* 46.3 (2015): 257–290.

69 Jacobus de Voragine, *Legenda Aurea* or *Golden Legend*. Voragine (1231–1298) was Archbishop of Genoa, beatified 1896. His *Legenda Aurea* exists in about 900 manuscripts and achieved a dominance in Western hagiographical literature. Between 1470 and 1530 it was the most-printed book in Europe.

as those of Walter de Elveden and, later, John Somer and Nicholas of Lynn, omit some of the names.[70]

The choice of which feast days are included on an individual astrolabe provides an extremely important clue to its origins, one that has been largely overlooked until now, even though catalogue entries for astrolabes usually list them. An exception is Eagleton's study of several "Chaucerian" astrolabes but her limited section on the calendars focuses on trying to find whether the sets of feast days on three astrolabes can be associated with particular manuscript *kalendaria* of the period, with a negative conclusion.[71]

An early comparative listing of the saints on just three astrolabes was published in 1888 by Chancellor Ferguson.[72] This was increased to seven in 2012 by Falk and this has been extended to ten for the present study, as shown in Table 9.1.[73] Many of the choices are of the major feasts (for example, the Nativity, the apostles, All Saints, etc.) and appear on all or nearly all of the astrolabes inspected. These do not provide any significant insights. But when a rather minor saint appears, it is clearly of interest to wonder why they have been selected as they are very likely to have been specially chosen by the commissioner (or possibly the maker) of the instrument. It might be that it is a local saint, or one with a particular localised following or cult, or simply that the feast day is a date of particular, personal, importance—perhaps a birth date or anniversary, an important battle, or just a convenient day of the month to provide an evenly-spaced calendar.

Table 9.1 can be used for several purposes. Firstly, it shows that there are a number of saints who are particularly English (for example, ss. Dunstan, Cuthbert, Augustine of Canterbury, Thomas Becket) thus confirming that these are English astrolabes, or at least ones made for use in England. Secondly, the table shows the remarkably high level of correlation between the selected three astrolabes (the Sloane, the Liège, and the Science Museum instruments) when compared to the "other" seven instruments that also fall into the "medieval English" category and have feast days on their calendars. In contrast, the totals at the bottom show that there is a significant number (over twenty)

70 For Nicolas of Lynn, see Sigmund Eisner (ed.), *The Kalendarium of Nicholas of Lynn* (Atlanta, GA: University of Georgia Press, 1980). For John Somer see Linne R. Mooney (ed.), *The Kalendarium of John Somer* (Atlanta, GA: University of Georgia Press, 1998).

71 Catherine Eagleton, "'Chaucer's Own Astrolabe': Text, Image and Object," *Studies in History and Philosophy of Science* 38 (2007): 303–326.

72 Chancellor Ferguson, 'On an Astrolabe Planisphere of English make', *Archaeologia*, 52, 75–84, 1888.

73 Seb Falk, *Astrolabes in Context. A Reappraisal of Medieval Astronomical Instruments* (MPhil. Diss., Cambridge University, Cambridge, 2012). The total number of astrolabes with saints' days calendars has now reached 14—see Postscript.

TABLE 9.1 *The Feast Days on the Sloane Group and other Medieval English Astrolabes*

Date	Feast	Sloane	Liège	SciMus	Other[a]
January 1	Circumcisio domini	X	X	X	5
6	Epiphania domini	X	X	X	7
13	Hilarius	X	X	X	1
22	Vincentius of Saragossa	X	X	X	2
25	Paulus, conversion	X	X	X	6
February 2	Maria, purification	X	X	X	6
14	Valentinus	X	X	X	1
22	Petrus in cathedra in Antiochia	X	X	X	7
24	Matthias	X	X[b]	X	5
March 7	Perpetua			X	0
12	Gregorius	X	X	X	6
21	Benedictus	X	X	X	2
25	Maria, annunciation	X	X	X	7
April 4	Ambrosius	X	X	X	6
14	Tiburtius et Valerianus	X	X	X	1
23	Georgius			X	4
25	Marcus	X	X	X	7
May 1	Philippus et Jacobus	X	X	X	3
3	Inventio sancta crucis	X	X	X	5
6	Johannis ante portam latinam	X	X	X	3
19	Dunstanus	X	X	X	4
26	Augustinus, Anglorum ap	X	X	X	5
June 5	Bonefacius			X	1
11	Barnabus	X	X	X	7
17	Botulfus			X	2
24	Johannes Bapt., nativitas	X	X	X	6
29	Petrus et Paulus	X	X	X	5
July 4	Martinus, ordination et translatio	X	X		1
7	Translatio S. Thome			X	5
15	Swithinus			X[c]	1
20	Margareta	X	X	X	4
25	Jacobus	X	X	X	5
August 1	Petrus ad vincula	X	X		3
10	Laurentius	X	X	X	7
15	Maria, assumption	X	X	X	5
24	Bartholemew	X	X	X	6

Date	Feast	Sloane	Liège	SciMus	Other[a]
29	Johannes Bapt., decollation aut natalis			X	3
September 1	Egidius (St Giles)	X	X	X	0
8	Maria, nativitas	X	X	X	7
14	Exaltatio Sancte Crucis	X	X	X	5
21	Mattheus	X	X	X	6
29	Michael, archangel	X	X	X	5
October 1	Remigius, translation	X	X		0
9	Dionysius (St Denys)	X	X	X	6
18	Lucas	X	X	X	7
19	Frideswida			X	0
28	Simon et Jude	X	X[d]	X	7
November 1	Omnium Sanctorum	X	X	X	5
6	Leonardi confessoris et abbis			X	1
11	Martinus	X	X	X	6
20	Edmundus, rex	X	X	X	0
25	Katerina	X	X	X	4
30	Andreas	X	X		3
December 6	Nicholas	X	X	X	5
13	Lucia	X	X	X	2
21	Thomas	X	X	X	6
25	Nativitas domini	X	X	X	6
27	Johannes, Apost.			X	0
29	Thom (Becket)			X	0
TOTAL		48	48	55	76

a "Other": how many other English medieval astrolabes (out of seven) does this feast also occur on. The other astrolabes studied are: (i) Oxford MHS, 47869 (the Painswick astrolabe), (ii) Adler Planetarium, Chicago, M-26, (iii) Brit. Mus. 1914, 0219.1, (iv) Gonville & Caius College 'B', (v) Brit. Mus. 1909, 0617.1 (dated 1326), (vi) Innsbruck #2579, (vii) Whipple Museum 1264. This table is a modified version of one originally compiled by Seb Falk, *Astrolabes in Context*.

b Mistakenly shown on Jan 25.

c Mistakenly shown on 1 July.

d Mistakenly shown on 24 October.

of feast days chosen by the makers of these other seven instruments that do not appear on the selected trio. It is worth highlighting that the other seven astrolabes include instruments with both quatrefoil and Y-form retes and some may be as late as the early fifteenth century.

The details of the lives of the saints in the following paragraphs come from several standard sources.[74]

Of particular interest are the feast days that appear on the Sloane group but not on the others, or perhaps only on one later (and derivative?) instrument. A key example is St. Edmund whose feast day is on 20 November. Edmund was a king of East Anglia in the late ninth century. He had been killed by arrows in a battle with invading Vikings: his usual attribute was an arrow. His relics were kept at the large Benedictine monastery at Bury (now called Bury St. Edmunds) that became very rich as a result. Not only was this the hometown of Richard de Bury but it was also the first overnight stopping place for the invasion force of Prince Edward and Queen Isabella when they landed on the Orwell in 1326. Later, as king, Edward III would prove a frequent visitor and supporter of St. Edmund, the first patron saint of England before St. George took over the role. Throughout his life, Edward regularly patronised the Anglo-Saxon royal saints, as did Isabella: one of his first independent progresses, in September 1328, included Bury St. Edmunds and several other East Anglian religious sites. Later, Edward had pennants and flags made for his army with St. Edmund's heraldic badge of *three crowns*.[75] The crowns in this instance are said to represent the three principal elements of his saintly character: King, Virgin, and Martyr.[76] An example of the arms, produced in the St. Edmundsbury scriptorium c. 1465, is shown in Figure 9.15.

St. Edmund had been the major patron saint of England since the Conquest (amongst several others, including St. Cuthbert) and although St. George was added to the list by Edward I, it was not until the middle of the reign of Edward III that St. George became the main English patron saint, at least amongst the core of knights surrounding the king. But St. George's feast day of 23 April does not occur on the Sloane astrolabe's calendar (nor on the Liège one), which suggests that it was engraved no later than the first three decades

74 The standard reference is A.B. Butler, *Lives of the Saints* (Dublin, 1884), available online at http://www.sacred-texts.com/chr/lots/. Other sources are http://www.catholic.org/saints/ (last accessed 23 June 2018) or Wikipedia entries for individual saints.

75 Ormrod, *Edward III*, 101 & 158.

76 George Henry Preble, *Origin and History of the American Flag and of the Naval and Yacht-Club Signals, Seals and Arms, and Principal National Songs of the United States, with a Chronicle of the Symbols, Standards, Banners, and Flags of Ancient and Modern Nations*, 2 vols. (Philadelphia, PA: Nicholas L. Brown, 1917), 1:123.

FIGURE 9.15
An historiated letter I with the arms of St. Edmundsbury Abbey, in a set of poems by John Lydgate produced there c. 1465. British Library, Harley MS 2255, f. 1r.
COURTESY OF THE TRUSTEES OF THE BRITISH LIBRARY.

of the thirteenth century, before Edward III began to adopt St. George to represent him, preferring a triumphant knight to a king who had been killed by invaders.[77] Nevertheless, Edmund remained a favorite with King Edward who gave this name to his fifth son Edmund of Langley (1341–1402), later the first Earl of Cambridge—the nearest sizeable town to Bury.

St. Edmund also remained a favourite of the people and, even at the end of the century, of the later Plantagenets. In the famous Wilton Diptych, Edward III's grandson Richard II is depicted with three patrons, one of whom, representing St. Edmund (with his arrow attribute) is said to resemble Richard's father, Edward the Black Prince.[78] St. Edmund's cloak is painted a vivid blue with gold crowns, reflecting his arms shown in Figure 9.15. Still later, in 1434 Henry VI stayed for many weeks at St. Edmundsbury Abbey where the Abbot presented him with a magnificent illuminated copy of John Lydgate's "Life of St. Edmund," which includes another version of the Abbey's "three crowns" arms.[79]

---

77 Ormond, *Edward III*, 101, 158.

78 Margaret Galway, "The Wilton Diptych: A Postscript," *The Royal Archaeological Institute of Great Britain and Ireland* 107 (1950): 9–14.

79 For a history of St. Edmundsbury Abbey, including Henry VI's visit, see William Page (ed.), *A History of the County of Suffolk*, 2 vols. (Folkstone: Institute of Historical Research by Dawson, 1975), 2:56–272. The presentation copy of John Lydgate's Metrical lives of Saints Edmund and Fremund is now British Library, MS Harley 2278; the three crowns are illustrated on f. 3v.

St. Valentine's feast, celebrated on 14 February, is today well known for being the day for romantic love. His feast appears on the three "Sloane group" astrolabes and on just one other (thought to be rather later, and with a Y-form rete).[80] He was a widely celebrated third-century Roman martyr and had an obscure history but his association with the tradition of *courtly love* is generally believed to date only from the time of Chaucer through his *Parlement of Foules* in 1382.[81] Chaucer, though, often got his ideas from earlier writers so there is a reasonable chance that the connection between St. Valentine and courtly love could well have been established at the beginning of the century. Certainly, St. Valentine is listed on 14 February in Walter Elveden's *Kalendarium* of 1327, even though the feast is not on the calendar of the astrolabe associated with him.[82] If it was already associated with courtly love, it would certainly have appealed to Prince Edward in 1326, looking forward to his marriage with Philippa, to whom he showed genuine affection throughout their long marriage.

The rather obscure SS. Tiburtius and Valerianus (14 April) appear on the Sloane, Liège, and Science Museum astrolabes and only one other instrument (the same one that also has St. Edmund). This fact on its own serves to strengthen the connection between the trio of quatrefoil astrolabes though it is initially a puzzle to know why they were selected. Valerian was a Roman senator in the third century, the brother of Tiburtius and married to the more famous martyr St. Cecilia (who does not appear on any of the English astrolabes). Together with the Roman soldier Maximus, who was charged with beheading them but converted instead, they were martyred. However, a commentary regarding them by the nineteenth-century writer St. John Bosco reads:

> If you keep with good companions, you can be assured that you will one day rejoice with the blessed in heaven, whereas if you keep with those who are bad, you will become bad yourself, and you will be in danger of losing your soul.[83]

---

80 British Museum, inv. no. 1914, 0219.15; Gunther 298; Ward 327 (c. 1450?).

81 J.E. Jost, "Chaucer's Parlement of Foules as a Valentine Fable: the Subversive Poetics of Feminine Desire," *Parenthesis* 1 (1999): 53–82, available online at http://www.yorku.ca/inpar/inpar001_jost.pdf (accessed 23 June 2018).

82 Elveden's *Kalendarium* can be found, for example, in Corpus Christi College, Cambridge, MS 37. The association with the Gonville and Caius astrolabe is explored by Davis and Lowne, n. 68.

83 Quoted in A.J.M. Mousolfe and J.K. Mousolfe, *Saint Companions for Each Day, St. Pauls BYB* (1986), 135.

For Richard Bury, who had seen Edward II suffering the bad advice of the favourites Piers Gaveston and the Despensers, the need to inculcate in his son and successor the need for good advice and strong comradeship would be a key message for him: the joint martyrdom of the brothers would have been an ideal vehicle to use, adding a religious dimension to the military one provided by the Round Table knights. Edward III learned this lesson well and came to regard loyalty and comradeship as the highest personal attribute of his own "band of brothers," which later bore fruit at Crécy and elsewhere and led to the foundation of the Order of the Garter in 1348.

Another saint on the three "Sloane group" astrolabes, this time exclusively, is St. Egideus (St. Giles, on 1 September). Falk has suggested that this choice was made simply to identify the first day of the month and this is certainly possible.[84] St. Giles was a French Benedictine, much celebrated by all Benedictine monasteries, of which Richard Bury seems to have had a particular liking—Durham College Oxford was located on St. Giles' Street. (St. Benedict's own feast on 21 March is also on the Sloane trio and just two other instruments.) As Bishop of Durham, Bury built a new gatehouse for the Kepier Hospital that was dedicated to St. Giles: the Hospital is currently in ruins but it the gatehouse still exists and retains Bury's arms (now illegible) carved above the window.[85] In the medieval period, the feast day was also celebrated by a large and important St. Giles' Feast in Winchester at which many traders from London and internationally would attend.[86] In Norfolk, the dedication to St. Giles on churches ranks as the twentieth most popular so its inclusion on the astrolabes is certainly not an obvious choice.[87]

St. Hilarius (St. Hilary, on 13 January) is another feast shown on the Sloane trio (and again, on the British Museum instrument).[88] A fourth-century "Doctor of the Church" from Poitiers, he seems an unusual choice but in this instance it seems to have been made on practical calendrical grounds. Hilary Term is the second of the academic year at Oxford (and Dublin) University and it is also the second of the legal terms of the English judicial courts. Richard

84 Falk, *Astrolabes in Context.*

85 W. Page (Ed.), *A History of the County of Durham* Vol 3: *Parishes: St. Giles*. Victoria County History (1928) 182–190.

86 See http://www.british-history.ac.uk/report.aspx?compid=42048.

87 Rev. C.L.S. (Charles Lawrence Scruton) Linnell, *Norfolk Church Dedications*, Issue 21 (York: Borthwick Publications, 1962). Note that this source is not completely reliable for medieval church dedications as there are many differences to modern ones (see, for example, N. Orme, *English Church Dedications* (Exeter, 1996), especially p. 57). But the basic point remains true.

88 See BM astrolabe, inv. no. 1914,0219.15 (see n. 80).

Bury had continuing connections at the University and the legal terms would have been part of everyday life for the royal court too.

St. Remigius is an obscure saint who occurs only on the Sloane and Liège astrolabes—they show the date of the translation of his relics on 1 October rather than his modern feast day of 13 January. A bishop of Reims (438–533), St. Remigius looks a strange choice for an English astrolabe but the reason could well be a misidentification. A later Benedictine monk, Remigius de Fécamp (d. 1092) came to England with the invasion of William I and became Bishop of Lincoln, responsible for building the cathedral on the current site. In the late twelfth and thirteenth centuries, there were attempts to have him canonized (Gerald of Wales wrote a *Vita* in 1196–99) that were ultimately unsuccessful although Matthew Paris's *Chronica Vita* for 1254–1259 does describe him as a saint.[89] (The canonization of Robert Grosseteste, another Bishop of Lincoln, was also unsuccessful though that of Bishop Hugh did succeed and he appears on one slightly later astrolabe calendar.[90]) Lincoln Cathedral was of particular importance to Edward I and his descendants as it was the site of the northernmost Eleanor Cross and it remained important to the household of John of Gaunt. There are relatively few English churches dedicated to St. Remigius, with just four in the county of Norfolk (the 26th most frequent dedication there).[91] One of these is at Hethersett in Norfolk and reputedly was begun in 1320 by Remigius de Hethersete. It has as a patron Gonville and Caius College Cambridge, founded by Norfolk churchmen in the early fourteenth century and with another astrolabe connection.[92]

The main feast day for St. Martinus (St. Martin of Tours) is on 11 November and this occurs on almost all medieval English astrolabe calendars of this type—he was a favourite martyr. But the feast for his *ordination et translatio* is on a separate date (4 July) and this occurs only on the Sloane and Liège astrolabes—clearly he was of particular interest to the commissioner of these instruments. The cult of St. Martin was particularly strong in East Anglia, as was noted in an article on the Gonville & Caius astrolabe, which is believed to date from the same decade (i.e., the 1320s).[93]

89 Diana Webb, *Pilgrimage in Medieval England* (London, Hambledon & London; 2000), 69. See also Rachael Koopermans, *Wonderful to Relate: Miracle stories and miracle collecting in high medieval England* (Philadephia, PA: University of Pennsylvania Press; 2011), 201; and Corpus Vitreum Medii Aevi, medieval stained glass in Great Britain—St. Remigius Church, Dunston, Norfolk, available online at http://www.cvma.ac.uk/publications/digital/norfolk/sites/dunston/history.html (accessed 26 February 2017).

90 Oxford Museum of the History of Science, inv. no. 47869, "the Painswick astrolabe."

91 Linnell, Norfolk Church Dedications.

92 Davis and Lowne, "An Early English Astrolabe."

93 Davis and Lowne, "An Early English Astrolabe."

The Liège Astrolabe

The Liège astrolabe, shown in Fig. 9.16, is, at 145 mm diameter, much smaller than the Sloane and its rete only has a single quatrefoil inside the ecliptic ring. There are, though, several strong links between the two instruments. As well as an exact correspondence in the list of saints' days noted above (with two dates seemingly mis-transcribed), the selection of the stars on the rete and their spelling is also very similar, as noted by Gingerich: all 34 of the Liège's stars also appear on the Sloane, with all but three having the same spellings.[94] There are a number of stylistic similarities as well, including the riveted construction with a narrow limb and the small throne.

The Liège astrolabe was published by Henri Michel in 1947 where it was described as '*le plus ancien astrolabe gothique esistant*' and given the ludicrously early date of 1190 on the spurious grounds that the calendar did not include St Thomas Becket.[95] Michel opined that it could be the work of a Jew who left Spain to take refuge in Guienne and made for an English lord, description repeated twenty years later though this is now known to be totally incorrect.[96] He did, though, rather perceptively note the similarity with the London Science Museum astrolabe (described below) though the link to the Sloane astrolabe escaped him.

The rete of the Liège astrolabe features many zoomorphic star pointers, most in the form of birds' heads with their beaks showing the star position. They are not well-modelled, in contrast to those on the Sloane, and not up to the general standard of craftsmanship of the instrument as a whole. The reason becomes clear when the pointers are examined more closely, as in Fig. 9.17. The shapes have been merely roughed-out with chisels: the markings for the final finished shape to which they should have been filed are still visible in places. The instrument has then been seriously over-polished, rounding-over the edges of the pointers and giving them their current rather amorphous appearance. Thus it seems that the astrolabe maker was not given sufficient time to finish his work properly: the delivery of the instrument on time was more important than refining the finish. Why?

---

94 Gingerich, n. 8. Also, Lowne and Davis, n. 68.

95 H. Michel: *Traitè de l'Astroblabe*, (Paris, Gauthier-Villars; 1947. Reprinted Librairie Alain Brieux, Paris, 1976. Translated, edited and revised James E. Morrison 2007) pp. 145–6 & 169 (Plate III). In an Erratum to the 1976 edition, Michel amends this to "*un des plus ancien astrolabe gothique existant*" and but revises the date only to c. 1200.

96 H. Michel: *Les Cadrans Solaires de Max Elskamp*, Musée de la vie Wallonne, (1966). Plate 28.

The instrument has four double-sided plates for latitudes of 39/40; 42/45; 47/49 and 52/54 degrees: no place names are given. This is a rather widely spaced set and covers from Athens to the north of England, including most of Europe. Latitudes of places were not, in general, accurately known in the fourteenth century so it is not certain for what locations the astrolabe has been designed. As a typical example, a list of latitudes written in the early fifteenth century by William of Worchester gives alternative latitudes, sometimes much more than a degree apart, of many cities.[97] For example, the plate for 45° could eas-

97 William of Worchester writing *c.* 1438, (Bodleian MS Laud Misc. 674, ff. 73r–74r, quoted in John D. North: *Horoscopes and History*, (London: The Warburg Institute, 1986), pp. 186–195.

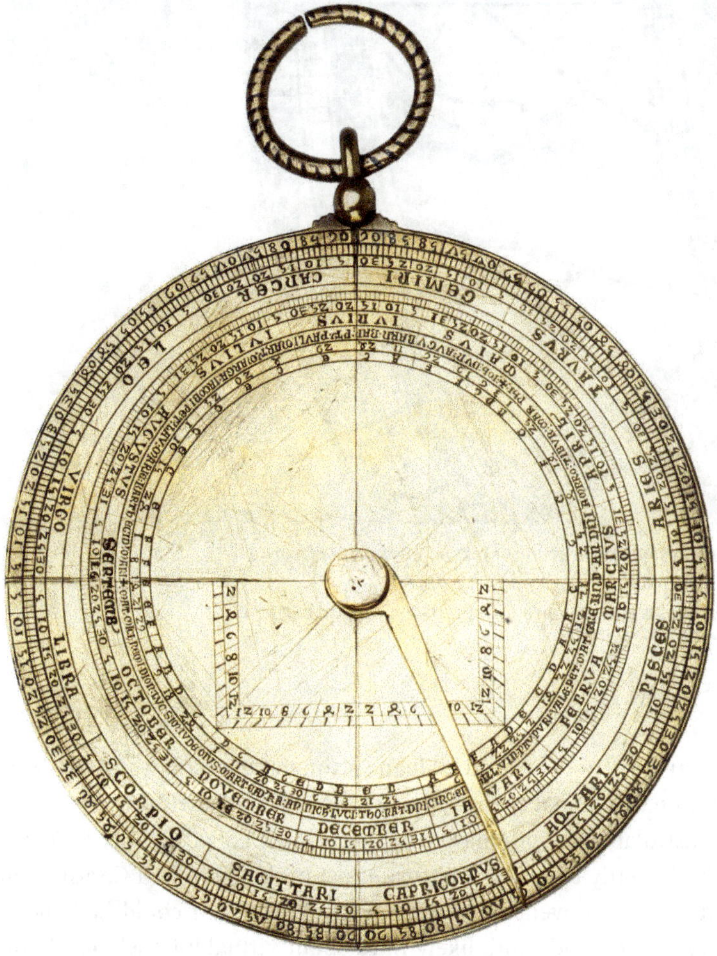

FIGURE 9.16 *Photographs of the front and back of the Liège astrolabe, inv. no. 400.*
COURTESY OF THE MUSÉE DE LA VIE WALLONNE, LIÈGE.

ily operate for locations such as Avignon or Milan, both cities with which the English Royal Court had strong connections. Avignon at this period (1309–77) was the seat of the Pope and Milan's ruling dynasty of the Visconti family had links which would eventually see Edward III's third surviving son Lionel of Antwerp (Duke of Clarence) marry into in 1367.[98] The closeness of the features

98 Ormrod, Edward III, n. 15, p. 446.

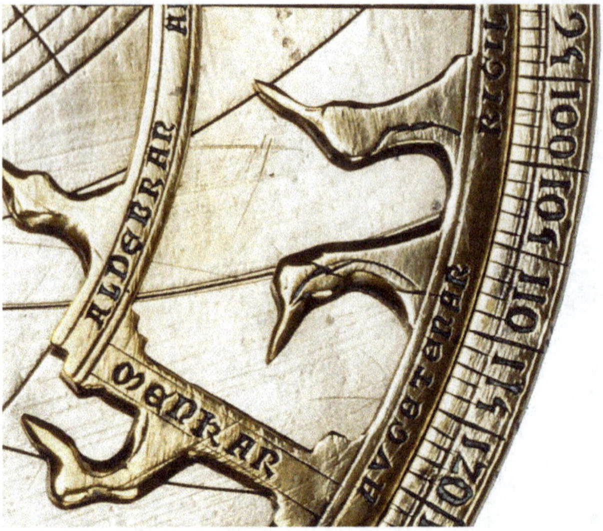

FIGURE 9.17 *Close-up photograph of star pointers on the Liège astrolabe.*
COURTESY OF THE MUSÉE DE LA VIE WALLONNE, LIÈGE.

of the Liège astrolabe to those of the Sloane strongly suggest it was made very close to it in time which makes the Milan link look less likely.[99]

The Liège astrolabe was purchased by the Belgian collector Max Elskamp (1862–1931) in the early twentieth century, most probably from a Continental dealer, but its earlier provenance is unknown. Although it could have been recently imported it could more likely have spent virtually its whole life on the Continent. One scenario that can be proposed is that it was made for an English churchman who was seconded to the Papal Court for a significant period. Alternatively, it could have been made as a present (bribe!) for an Englishman to give to a member of the Papal Court for any one of a number of favours, from appointment to an important papal post (such as an English bishopric) to an audience with the Pope.

If we speculate that the Liège astrolabe was based on the Sloane instrument and also commissioned by Richard Bury as a means of furthering a royal cause with the papacy, the question is what that cause might have been. The first occasion which can be suggested is when a papal dispensation was required for

99 Another of the 'quatrefoil' astrolabes, currently in Milan, is more likely to be associated to the Visconti family and will be discussed in a forthcoming paper.

Edward to marry Philippa of Hainault on account of their being second cousins. This occurred in 1327 when a deputation was sent to Avignon to present the case. At first, Pope John XXII refused but he relented after Edward sent a personal letter in August.[100] The date does seem a little early, not leaving much time after the making of the Sloane instrument with its images of Philippa.

A more likely suggestion is in the autumn of 1329 when Edward was preparing to overthrow the control of the self-aggrandised Roger Mortimer, who had developed into an overbearing tyrant. To obtain the Pope's support, Richard Bury drafted a secret letter from Edward to the Pope which bore the code words of "*Pater sancta*" in Edward's own hand to show that it, and any future correspondence, was truly from him.[101] The plan was hatched by a very small group around Edward which included Sir William Montague who took the letter to Pope John XXII in Avignon. What better proof that the letter came from Edward than a customised version of his own astrolabe as a present to the Pope? John XXII is not known particularly for an interest in astronomy but he has been described as "one of the most liberal patrons of education and of science in history" particularly in the medical field: medicinal astrology was one of the main uses of astrolabes at this time.[102]

The Science Museum Astrolabe

The Science Museum astrolabe (Figure 9.18) is best-known for an owner's name, Tzvi Herz, in Hebrew characters ("Ẓvi Herṣ"), which is engraved on both sides of the throne, and the fact that the scales have some Hebrew inscriptions added to the Latin ones. However, these features are not original and the underlying instrument matches the Sloane and Liège astrolabes in terms of the stars, the feast days and several of the other key features. It currently has a single double-sided plate engraved and labelled for the latitudes of 49° and 52° but the style of the numerals, as well as its metallurgy,[103] indicate that although the plate is one of the original set the numerals were added in the mid

100 Weir: *Isabella*, n. 15, pp. 265–6 & p. 274.

101 Weir, *Isabella*, n. 15, p. 327.

102 J.J. Walsh, *The Popes and Science: the history of the papal relations to science during the middle ages and down to our own time*, (New York, Fordham University Press; 1915), especially pp. 130–140.

103 The metallurgy of the various components have been measured by the author using XRF as part of a project on the Metallurgy of Medieval Scientific Instruments (MoMSI), which will be reported elsewhere.

sixteenth century. The womb in the mater is quite deep and could originally have housed at least four or five plates: today, the extra depth is occupied by four rectangular blocks of a modern composition brass soldered to the inside of the mater as spacers. The original plates, now lost, could easily have covered the whole of England at intervals of 1° in latitude, like the Sloane instrument.

The Oxford Museum of the History of Science has in its collections a rete—the so-called Baillie's rete—which is so close in layout, style, and star names to the Science Museum one that it is certainly related to it and may well have been produced in the same workshop.[104] As the rest of the astrolabe is lost, it

104 Oxford MHS inv. 45133; Gunther #294 (471 and Fig. 185).

FIGURE 9.18 *Front and back of the London Science Museum "Tzvi Herz" astrolabe.* PHOTOGRAPHS BY THE AUTHOR, WITH THANKS TO THE SCIENCE MUSEUM, SOUTH KENSINGTON.

is not possible to speculate on its possible ownership but the possibility that a small group of astrolabes was made for Edward III's closest and most trusted supporters must be considered.

The Science Museum rete, together with Baillie's one and like that on the Liège instrument, features a single quatrefoil inside the ecliptic ring and a set of four half-quatrefoils at the ends of the diametrical bars. Only some of the star pointers are zoomorphic, with others being finely modelled floral designs. Many of the star names have been re-engraved: only those in capitals are original. They are a subset of those used on the Sloane and

Liège instruments but with one extra star, CORVUS, which has a bird as a pointer.

One feature of Baillie's and the Science Museum retes that is not found on the Sloane or Liège instruments is that the star pointers for Delfin, Aquila, and Mucida Equi spring from a single base that has the appearance of the face of a bull (though Gingerich calls it a "satyr(?)").[105] It is difficult not to see a link here to *The Prophecy of the Bull* mentioned earlier and thus that it is a representation of Edward III. The star pointers give the impression of a forked beard, a feature of Edward's adult appearance though admittedly his tomb effigy shows two rather than three points.

The calendar on the Science Museum astrolabe includes fifty-five feast days, the most of any of the ten English medieval astrolabes studied (see Table 9.1). As well as having most of the feasts of the Sloane and Liège instruments (missing are St. Martin, St. Petrus ad vincula, St. Remigius, and St. Andreas), it has ten extra, and unusual, ones that they do not have. The first of these is the unique St. Frideswide (19 October). She is the patron of both the city and university of Oxford, which suggests that the owner was a very keen Oxford man, almost certainly educated there.

The next unusual saint is St. Swithun on 15 July though the Science Museum calendar has him on 1 July, surely a transcription error by the engraver (St. Swithun has the correct date on the only other instrument on which he occurs). St. Swithun's relics are at Winchester Cathedral where he is naturally the patron saint. Also present is the apostle St. John (27 December) who is surprisingly missing from all the other English astrolabes, perhaps because of the surfeit of feast days in late December. This does not seem to offer a strong clue to the owner.

Two days later (29 December), the feast for the martyrdom of Thomas Becket is also uniquely shown on the Science Museum calendar. The link here is clearly to Canterbury Cathedral where he was murdered and his relics were kept. As an important English saint, the feast for the translation of his relics on 7 July (which occurred in 1220 with much ceremony) appears on around half of the astrolabes with saints' days calendars but not on the other Sloane group instruments. Perhaps a reason for this, as well as the pressure of space, is that it was not a good idea to remind a king about a murder carried out on behalf of one of his predecessors! (Becket (c.1118–70) was initially Henry II's Chancellor but was killed by four of his knights in Canterbury Cathedral.) More likely is that the list of saints was originally compiled by Richard de Bury who saw Becket's spiritual home of Canterbury as a rival to his beloved Bury St. Edmunds. (There was considerable confusion in the fourteenth century as

105 Gingerich, "Zoomorphic Astrolabes," 86.

to the correct day to celebrate St. Thomas Becket and several contemporary calendars show at least one of the two possible days erased.[106])

On 7 March there is another unique and rare saint, St. Perpetua, usually celebrated with another female saint, St. Felicite. It is difficult to find a connection here, unless it is that the date is also the date of death of Thomas Aquinas.

St. Boniface (7 June) is a further "extra" saint on the Science Museum calendar. Born Winfrid (or Wynfrith) in Wessex, he was a Benedictine who spent most of his working life in Germany where he became the Archbishop of Mainz. It is difficult to find a plausible connection to the astrolabe, other than the co-incidence (surely?) that Edward III's fifth son, Edmund of Langley, was born on this day in 1341.

One person who matches the interests of this set of saints, as well as being a close confidant of Edward III, is John de Stratford (ca. 1275–1348).[107] He had studied at Oxford and had been a dean at Canterbury Cathedral before being made Bishop of Winchester in 1323, rather against Edward II's wishes as he had fallen foul of the king's favourites, the Despensers. He had taken up Isabella's cause in Paris in 1325/1326 and, after the capture of Edward II following her invasion in 1327, he had been a leading member of the group that arranged for the king to hand the crown to his son. Stratford was part of the council of ministers advising Edward III during his co-reign with Isabella but then fell out with her and transferred his allegiance to Edward. Immediately after Roger Mortimer had been arrested, Bishop Stratford was elected as Chancellor, a post that he was to hold three times before becoming Archbishop of Canterbury in 1333. Stratford briefly fell out with Edward in a financial crisis in 1340/41 but he was later reconciled.[108]

Edward had much to be thankful to Stratford for in achieving the smooth transfer of power to him and in his political support in England, Avignon, and Paris. Stratford was an independent and forthright thinker so, if the Science Museum astrolabe was a gift made to him, the changes to the set of feast days from the original set on the Sloane instrument would be well within his character and at his behest.[109]

---

106 Examples where the dates have been altered are the 'Queen Mary Psalter' (Brit. Lib. Royal MS Harley 2332) and the Old Sarum Processional, Salisbury Cathedral Library MS 148. Additionally, the *Kalendaria* of Walter Elveden, Nicholas of Lynn and John Somer all show alterations or erasures to their 7 July and/or 29 December entries on at least some of their copies.

107 See Stratford's entry in the Oxford Dictionary of National Biography for details of his life.

108 Weir, *Isabella*, 114.

109 See the ODNB or Weir, Isabella, for an assessment of his character.

If the hypothesis that the astrolabe belonged to Stratford is correct, it begs the questions of when it was made and whether it was a gift from the royal court or whether Stratford himself had it made to show his continued allegiance to Edward. It is the only one of the trio of instruments to include the feast of St. George on its calendar and that might suggest a date a few years into Edward's reign, when he was beginning to supplant St. Edmund.

Final Thoughts

It is accepted that the association of the Sloane, Liège, and Science Museum astrolabes with a short period of time early in the reign of Edward III that has been proposed here is based almost exclusively on circumstantial evidence. There are, however, numerous separate threads that all lead in the same direction and have a remarkable fit with the known historical facts of the period. The sheer size and quality of the Sloane astrolabe indicate strongly that it was designed to express kingly magnificence, and to stress to a young prince that he was expected to take on the role of leading his country to the head of Europe. Its unique mirroring of the appearance of illuminated manuscripts that are known to have been used in the education of the prince is a point that is difficult to explain by other means. The threads include the three large quatrefoils that are now interpreted as standing for the king's crowns, handed down by God; the human faces on both the rete and the back of the instrument; the animals that match those used to describe King Edward III in contemporary political prophesies, and the carefully selected set of saints whose feast days are given on the calendar and who fit with the known interests of Richard Bury—some of which were followed by Edward later in his reign.

The earliest dated English astrolabe known is another instrument in the British Museum.[110] It is of the "Chaucerian" design with a Y-form rete and is dated 1326. The small astrolabe at Gonville & Caius College, Cambridge, has also been given a likely date very close to this.[111] Since the Sloane astrolabe's most probable date is the same, it seems highly likely that there was some connection between the instruments, though which influenced which has not been established. The Sloane's date is believed to lie in the period when Richard Bury was acting as tutor to Prince Edward of Windsor, running from his betrothal to Philippa of Hainault in 1325 to his crowning in 1327 or, possibly, his overthrow of Isabella and Mortimer in 1330.

110 British Museum 1909,0617.1. Ward 325, Gunther #291, dated 1326.

111 Davis and Lowne, "An Early English Astrolabe."

It is clear that there is a number of questions that have not yet been tackled. The first two are who actually made the instruments, and where? There are some early indications, such as their palaeography and the metallurgy, that suggest that they may have been made in Europe—Paris, perhaps, or in the Low Countries—but this requires much further research. An English workshop, perhaps influenced by skilled immigrant artisans from Europe, is not ruled out. It is a reflection on the large gaps in our knowledge that we have little idea where they could have been made.

A proper mathematical analysis of the geometries of the astrolabes has still to be performed. The accuracy of the stereographic projections and the calendars needs to be established and may lead to interesting conclusions about the way in which the engraving was performed and the underlying datasets that were employed.

Other English "quatrefoil" astrolabes that are not quite such a good fit to the Sloane trio also require further study as these may well have been influenced by them even if they are not direct, contemporary, copies. And the connection between these instruments and the "Chaucerian" designs, which may be both earlier and later than the Sloane group, still needs to be established.

Acknowledgements

I am grateful to the Trustees of the British Museum, and particularly to the curator Oliver Cooke, for access to the Sloane astrolabe. Rebecca Storr at the London Science Museum kindly arranged access to the Tzvi Herz instrument. Marie-Claude Thurion and Anne Stiernet at the Musée de la vie Wallonne kindly provided photographs of the Liège astrolabe. Sebastian Falk graciously allowed use of the calendrical information in Table 9.1 from his unpublished MSc thesis. I am also grateful to Silke Ackermann, Elly Dekker, Michael Lowne and Harriet James for their help.

Postscript

Since the text for the original *Medieval Encounters* article was written in 2014, another 'Sloane group' quatrefoil astrolabe has been identified in an American museum and its details published.[112] In addition, a further four quatrefoil

112 J. Davis: 'A Royal English Medieval Astrolabe Made for Use in Northern Italy', *J. Hist. Astron.*, 48(1), 3–32 (Feb 2017). DOI: 10.1177/0021828616681214.

astrolabes incorporating calendars with English saints have been identified in various Continental museums bringing the total to ten: these are currently being studied further. Two of these astrolabes have the same set of saints' days as the Sloane and Liège calendars but in others some different saints are included, unique to individual instruments and thus providing clues to their possible original owners. Also, an evolution in the design of the rete and the stars can be identified leading to the belief that a chronology of the instruments, all from the first half of the fourteenth century, may be identifiable. Details of the mechanical construction lead to the view that all the members of this Sloane group of astrolabes were made in a single English workshop which was operating over several decades. These other members of the group are:

1) Adler Planetarium, Chicago, 'Mensing-26' IIL #200 and #295
2) Kugel Collection, Paris, IIL #546
3) Belgian private collector, IIL #4518 (with a Chaucerian-style mater)
4) Istanbul, IIL #4508
5) Pinacoteca Ambrosiana, Milan inv. 1006 (very close to the Sloane's rete design but with three trefoils instead of three quatrefoils inside the ecliptic)

Further possible members are also being investigated, including an instrument sold by Sotherby's in 1986, possibly a later copy or an unfinished example.

As well as these quatrefoil instruments, 'Chaucerian' astrolabes with Y-shaped strapwork on the rete also often have calendars with feast days—at least six examples have been recognised. They can be shown mainly to be slightly later (second half of the fourteenth century or early fifteenth) though they continue in the same English tradition and at least some of them were probably made in the same workshop which produced the Sloane Group quatrefoil astrolabes. The connection between the 'quatrefoil' and 'Chaucerian' astrolabes has been made much clearer by the existence (again in Continental Europe) of two 'hybrid' astrolabes with saints' days calendars which share features of the two design groups.

CHAPTER 10

European Astrolabes to ca. 1500: An Ordered List

David A. King
Frankfurt
d.a.king@orange.fr

Abstract

Research on medieval European astrolabes has hitherto been somewhat haphazard. Most pieces are unsigned and undated, many difficult to assign to a specific region. Some early ones cannot be understood without reference to the Islamic tradition from which they derive. What are perhaps the most important pieces from a historical point of view—the earliest-known astrolabe, from 10th-century Catalonia, and the astrolabe made by the leading astronomer of 15th-century Europe, Regiomontanus,—were declared fakes or suspicious before they could be studied seriously. A detailed study of groups of related instruments, for example, those with Hebrew inscriptions, is a most welcome contribution. A survey of the clearly-identifiable astrolabes made in medieval England (or France or Italy) has never been undertaken; maybe this list might encourage somebody willing to learn the language of instruments to undertake such a task.

Keywords

astrolabes – medieval – Europe – Destombes – Chaucer – Fusoris – Regiomontanus – museums

Introduction

This list is arranged chronologically by provenance. It dates from the early 1990s and has barely been modified since then. It was generated as the table of contents of the European astrolabe component of a catalogue of medieval instruments (to ca. 1500) in the making, which for various reasons was not completed. The German Research Foundation (DFG) generously funded the basic research for several years. This list corresponds to what was to be Section 6 of the catalogue, and references to individual instruments should

 | DOI:10.1163/9789004387867_013

be in the form "6.m.n", where "m.n" is the number assigned in the list. (See my *In Synchrony with the Heavens*, Leiden, 2005, XVIII, for a list of Islamic instruments from Sections 1–5.) Subsequent sections dealing with instruments other than astrolabes to ca. 1500 have been suppressed here. I present this list in the hope that it may be useful to future researchers. It was already published in my *Astrolabes from Medieval Europe* (Ashgate: Variorum, 2011), XII.

Note

The symbol # precedes a unique number assigned to each instrument, a tradition started by Derek de Solla Price in his *Astrolabe Checklist* first circulated in 1973.

Abbreviations

Antwerp NSM, National Maritime Museum (Het Steen); Barcelona OF, formerly preserved in the Observatorio Fabra; Barcelona MM, Museu Marítim; Bernkastel-Kues NKS, Nikolaus-Kusanus-Stift; Boston, Mass., MFA, Museum of Fine Arts; Brescia MEC, Museo dell'Età cristiana, now Museo di Santa Giulia; Brugine CISST, Centro Internazionale di Storia dello Spazio; Cambridge CC, Caius College; Cambridge WMHS, Whipple Museum of the History of Science; Chicago, Ill. AP, Adler Planetarium & Astronomy Museum; Cracow CZ, Czartoryski Museum; Cracow JM, Jagiellonian Museum; Cologne SM, Stadtsmuseum; Darmstadt HLM, Hessisches Landesmuseum; Edinburgh RSM, Royal Scottish Museums; Florence MNB, Museo Nazionale Bargello; Florence MSS, Museo di Storia della Scienza, now Museo Galileo; Frankfurt am Main HM, Historisches Museum; Ghent BM, Bijloke Museum; Greenwich NMM, National Maritime Museum; Kassel HLM, Hessisches Landesmuseum; Leiden MB, Museum Boerhaave; Istanbul BTTM, Bilim ve Teknoloji ve Teknik Müzesi (Museum of Science, Technology and Technics); Liège MVW, Musée de la vie wallonne; London BM, The British Museum; London NG, Nur Gallery, collection of Dr David Khalili; London RGS, Royal Geographical Society; London SA, Society of Antiquaries; London SM, Science Museum; London VA, Victoria and Albert Museum; Luneburg MFL, Museum Fürstenturm Lüneburg; Milan MCAA, Museo Civico d'Arti Applicati; Milan PA, Pinacoteca Ambrosiana; Milan TT, former collection of Dr. Tullio Tomba; Moulins (Allier) MD, former Musée départementale, now Musée Anne de Beaujeu; Munich BNM, Bayerisches Nationalmuseum; Munich DM, Deutsches Museum; New York N.Y. CU, Columbia University, Special Collections; Nuremberg GNM, Germanisches Nationalmuseum; Oldenzaal (NL) HM, Historisch Museum "Het Palthe-Huis"; Oxford MC, Merton College; Oxford MHS, Museum of the History of Science;

Oxford OC, Oriel College; Paris BNF, Bibliothèque nationale de France; Paris CNAM, Centre national des arts et métiers; Paris IMA, Institut du monde arabe; Pavia MSU, Museo per la Storia dell'Università di Pavia; PC, private collection; PLU, present location unknown; Prague NTM, National Technical Museum; Rockford, Ill. TM, formerly Time Museum, now dispersed; Rome OA, Osservatorio astronomico, now Museo Copernicano; Salem, Mass., PM, Peabody Museum; Salzburg CA, Museum Carolino-Augusteum; Salzburg PS, Museum des Stiftes St. Peters; Schweinfurt SM, Stadtsmuseum; Stockholm NM, Nordiska Museum; Turin MPMT, Museo Palazzo Madama de Torino; Strasbourg MH, Musée historique; Utrecht UM, University Museum; Washington, D.C., NMAH, National Museum of American History; Wroclau OBS, Observatory, now in Cracow JM.

1 Miscellaneous Early European Astrolabes (I)

1 The earliest European astrolabe, from tenth-century Catalonia (#3042—Paris IMA)
2 An astrolabe for the seven climates (#161—London BM)
3 An Italian astrolabe (#3037—PLU < Milan TT)
4 An unsigned, undated (?) astrolabe (#4504—PLU)
5 A Spanish (?) astrolabe (for Saragossa and other latitudes, with later additions for Paris) (#191—Oxford MHS)
6 An English or French astrolabe for latitudes between 24° and 60°, including for [Paris] and [London] (#300—Oxford MHS)
7 A French astrolabe (#428—Greenwich NMM)
8 An Italian (?) astrolabe with a zoomorphic rete (#2041—Oxford MHS)
9 An Italian (?) astrolabe with plates for the climates (#166—Oxford MHS)
10 An Italian (?) astrolabe for the climates (#167—London BM)
11 A French astrolabe for the upper climates (#420—Greenwich NMM)
12 An English astrolabe missing its rete (#303—Oxford MHS < MC)
13 A mater and plates owned by Nikolaus of Cusa (#589—Bernkastel-Kues NKS)
14 An astrolabe of uncertain provenance for latitudes between 16° and 52° (#558—Nuremberg GNM)
15 An astrolabe dated 1374 (#3114—PLU)
16 A medieval Italian astrolabe with a rete design incorporating a frame shaped like a bishop's mitre (#4556—PLU < Christie's 1994)

2 Miscellaneous Early European Astrolabes (II)

1 A Catalan astrolabe with a V-shaped frame on the rete (#416—Greenwich NMM)

2 An Italian astrolabe with a distinctive rectangular frame on the rete (#493—Florence MSS)

3 A related astrolabe, signed "KP" and made in Urbino (?) in 1462 (#4506—Moulins MD—stolen)

4 An Italian (?) astrolabe (#441—London SM)

5 An astrolabe with an equatorial ring on the rete (#567—Cracow JM < Wroclau OBS)

6 An astrolabe with an equatorial ring on the rete, dated 1375 and signed by Petrus Raimondus (#3053—Boston MFA)

7 A composite astrolabe with Hebrew inscriptions on the back of the mater (#621—Munich DM)

8 An astrolabe with a latitude grid around the ecliptic on the rete and zodiacal figures on the back (#213—PLU)

9 A fifteenth-century astrolabe from the Low Countries with a rete in an *art nouveau* style (#536—Leiden MB)

10 An unsigned, undated astrolabe with an unusual rete (#3096—Salem PM)

11 An Italian astrolabe with degenerate quatrefoil markings and a *shakkâziyya* grid (#2044—Oxford MHS)

12 An Italian astrolabic plate for the second climate and a simplified rete combining north and south projections (#169—Oxford MHS)

13 *deest*

14 A late Italian astrolabe copied from an earlier type (#4507—Florence MNB)

15–16 Two related astrolabes, one Italian made to look like a Dutch one, and a Dutch copy

15 A composite Italian astrolabe with a dubious signature by Henricus de Hollandia (#548—Nuremberg GNM)

16 An unsigned 15th-century (?) Dutch astrolabe in imitation of the style of the Nuremberg astrolabe "signed" by Henricus de Hollandia (#566—Prague NTM)

17 Two "lost" astrolabes (#625 and #626—PLU, formerly Salzburg PS)

3 Astrolabes with Quatrefoil Decoration (I)

1 An astrolabe with a rectangular frame inside the ecliptic on the rete and plates for latitudes down to that of [Palestine] (#162—London SA)

2 A French astrolabe with gear mechanism (#198—London SM)
3 An English astrolabe dated 1342 and signed by Blakeney (#292—London BM)
4 The "Great Sloane astrolabe" (#290—London BM)
5 An Italian astrolabe with zoomorphic representations on the throne and back (#547—Nuremberg GNM)
6 An Italian astrolabe (#4515—PLU < Milan TT)
7 An Italian astrolabe dedicated to (Alfonsus ?) Bonhominus (#4516—Milan MCAA)
8 An Italian astrolabe (#3203—Turin MPMT)
9 An Italian astrolabe (#2043—Oxford MHS)
10 A German astrolabe owned by Ludolf de Scicte of Einbeck *ca.* 1330 (#2027—Cracow JM)
11 An unsigned rete (#294—Oxford MHS)
12 An English astrolabe (#200, #295—Chicago AP)
13 An Anglo-French (?) astrolabe with an owner's mark in Hebrew script (#293—London SM)
14 An Italian (?) astrolabe (#2062—Pavia MSU)

4 Astrolabes with Quatrefoil and Trefoil Decoration (II)

1 A French (?) astrolabe (#4751—Rome OA—stolen)
2 An astrolabic plate for [Paris] (#190—London VA)
3 An astrolabe with a replacement quatrefoil rete (#627—Salzburg CA)
4 An astrolabe with a "broken" half quatrefoil (#4752—Cambridge WMHS)
5 A German astrolabe with a half quatrefoil on the solstitial bar of the rete (#461—Ghent BM)
6 A fourteenth-century Picard astrolabe with a half-quatrefoil on the solstitial bar, featuring monastic cipher numerals, dedicated by the Benedictine monk Berselius to Amerotius, his teacher of Greek, at Louvain in 1522 (#202—PC < London Christie's 1991)
7 A French or English astrolabe owned by John of London (?) (#3058—Greenwich NMM)
8 A North Italian astrolabic plate with a single trefoil on the rete (#4753—Milan MCAA)
9 A French or English astrolabe with half-quatrefoils on the supports of the equatorial bar, supposedly dated 1501 (#4517—PLU)
10 A French (?) astrolabe (#2042—Oxford MHS)
11 An English astrolabe (#4518—PC)
12 An English astrolabe (#457—Liège MVW)

13 A French (?) astrolabe (#546—PC)
14 An English astrolabe (#4508—Istanbul BTTM)
15 A "cosmopolitan" Italian astrolabe with a replacement plate from a Byzantine astrolabe and additional markings in Hebrew (#4509—PC)
16 An Italian (?) astrolabe (#4754—Milan PA)
17 "Sir John Findlay's astrolabe" (#238—PLU)
18 A debased trefoil Gothic astrolabe (#573—PLU formerly Needham Collection)

5 Astrolabes with Inscriptions in Judaeo-Arabic or Hebrew

1 An astrolabe with a rectangular frame inside the ecliptic on the rete, dating from ca. 1300 and with inscriptions in Judaeo-Arabic, including a corrupt poem on the nature of an astrolabe (#3915—London NG)
2 An astrolabe with Hebrew inscriptions from the mid-fourteenth century (#158—London BM)
3 A fifteenth-century astrolabe with Hebrew inscriptions (#3906—PC)
3 A fifteenth-century astrolabe with Hebrew inscriptions on the back (#621—Munich DM)
4 A sixteenth-century astrolabe with quatrefoil markings on the rete and Hebrew inscriptions (#159—Chicago AP)
5 Miscellaneous bits and pieces with Hebrew inscriptions and instruments made by Jewish craftsmen:
(a) A rete with messy Hebrew inscriptions (#2572—Washington NMAH)
(b) A Maghribi astrolabe with Arabic inscriptions made by a Jewish craftsman (#1122—PC)
(c) The back of a fourteenth-century "cosmopolitan" astrolabe (#4520—PC)
(d) An owner's mark in Hebrew script (#293)

6 English Astrolabes with a Y-shaped Frame on the Rete in the Tradition of Geoffrey Chaucer

1 An astrolabe for Oxford or London (#301—Cambridge CC)
2 An Anglo-French astrolabe for Montpellier, Oxford and elsewhere, dated 1326 (#291—London BM)

3 "The Painswick astrolabe" for localities between Toledo and [Thule] (#299—Oxford MHS)
4 An astrolabic plate for Oxford fitted with an equatorium (#297—Oxford MC)
5 An astrolabic plate for Oxford with a universal horary quadrant (#296—Oxford OC)
6 An astrolabe for localities from Dover to Berwick (#298—London BM)
7 "Parnel's astrolabe" (# 304—Washington NMAH)
8 An English astrolabe (#4755—Oxford MHS)
9 An astrolabe with plates for Paris, Cologne and London (#4521—Milan TT)
10 An English zoomorphic astrolabe (#2006—Washington NMAH)

7 Early French and Italian Astrolabes with an Upper Frame on the Rete

1 An early Italian astrolabe with a bowl-shaped frame (#4756—Milan PA)
2 A Franco-Italian astrolabe for the climates (#4512—PLU < Rockford TM)
3 A Franco-Italian astrolabe (#168—Oxford MHS)
4 A Franco-Italian (??) astrolabe similar to the previous one (#410—Oxford MHS)
5 A Franco-Italian astrolabe (#4522—PLU)
6 An astrolabe similar to the previous one (#609—Luneburg MFL)
7 An astrolabe with an unfinished rete (#4757—Rome OA—stolen?)
8 An Italian astrolabe dated 1420 (#4523—PC)
9 A French astrolabe for Nevers (#4524—Frankfurt HM)
10 An Italian astrolabe with an upper equatorial frame (#186—Washington NMAH)
11 A German astrolabic plate with a tulip-shaped frame, dated 1468 and serving a latitude in N. Italy, and said to have been owned by Regiomontanus (#550—Nuremberg GNM)
12 An unsigned Italian astrolabe / astrolabic plate with a thistle-shaped frame on the rete, dated 1500 (#4758—Milan PA)
13 An Italian astrological plate for the latitude of [Palestine] (#170—Oxford MHS)
14 A late Italian astrolabe with a bowl-shaped upper frame on the rete (#576—Cambridge WMHS)
15 A late Italian astrolabe with an upper frame on the rete (#247—London SM)

16 An Italian astrolabe with two short upper bars (#479—Florence MSS)
17 An Italian astrolabe from ca. 1500 (#174—London RGS)
18 An astrolabe of uncertain provenance (#431—Greenwich NMM)
19 An astrolabe from the Low Countries or from Germany influenced by the "Fusoris" tradition but with a prominent bowl-shaped frame on the rete (#4525—Munich DM)
20 An anomalous late-fifteenth-century German astrolabe exhibiting French influence (#601—Cologne SM)

8 Fifteenth-Century French Astrolabes in the Tradition of Jean Fusoris

1 The "Thornoe" astrolabe (#337—Greenwich NMM)
2 An astrolabe related to the Greenwich Thornoe one (#476—Barcelona OF)
3 A French (?)/ Anglo-French (?) astrolabe (#163—Oxford MHS)
4 A "Fusoris-type" astrolabe without an upper frame on the rete (#598—Kassel HLM)
5 A simplified "Fusoris-type" astrolabe (#4526—PC)
6–20 "Fusoris-type" astrolabes
6 A "Fusoris-type" astrolabe (#196—Oxford MHS)
7 A "Fusoris-type" astrolabe (#194—Oxford MHS)
8 A "Fusoris-type" astrolabe, maybe slightly later than the rest (#192—Oxford MHS)
9 The "Fusoris-type" astrolabe of Robert Gordon of Straloch (#3082—Edinburgh RSM)
10 A "Fusoris-type" astrolabe (#3080—Paris CNAM)
11 A "Fusoris-type" astrolabe with rete and plates reworked by an Egyptian or a Syrian in the nineteenth (?) century (#460—Antwerp NMM)
12 "Sir John Findlay's "Fusoris-type" astrolabe" (#193—Chicago AP)
13 A "Fusoris-type" astrolabe (#199—Chicago AP)
14 A "Fusoris-type" astrolabe (#195—London BM)
15 A "Fusoris-type" astrolabe (#568—Cracow JM < Wroclaw OBS)
16 A "Fusoris-type" astrolabe (#4527—Stockholm NM)
17 A simplified "Fusoris-type" astrolabe (#193B, #3070—PLU ???)
18 A "Fusoris-type" astrolabe (?) (#3083—PLU)
19 A "Fusoris-type" astrolabe (?) (#3084—Barcelona MM)
20 An astrolabe in the "Fusoris" tradition (#518—Paris BNF)
21 Two imitation "Fusoris" astrolabes

(a) Undated (#407—Cambridge WMHS)
(b) An early copy of a "Fusoris-type" astrolabe dated 1447 or 1487 (#623, #3208—Munich BNM)

22 An astrolabe in the "Fusoris" tradition signed by Hooft (#197—Cambridge CC)

23 An astrolabe in the "Fusoris" tradition (#4528—Brescia MEC)

24 An unsigned imitation "Fusoris-type" astrolabe (#3081—PLU)

25 Another imitation "Fusoris-type" astrolabe (#4529—Brugine CISST)

26 Three sixteenth-century Dutch astrolabes, imitations of "Fusoris-type" astrolabes, signed by Johannes Bos
(a) Undated (#539—Utrecht UM)
(b) Dated 1591 (#430—Greenwich NMM)
(c) Undated (#534—Leiden MB)

9 Fifteenth-century German Astrolabes in the Tradition of Regiomontanus and Hans Dorn

1–5: Some mid-fifteenth-century German astrolabes with similar ("Regiomontanus-type") retes

1 Two astrolabes by the same maker (?) with a lunar volvelle on the back
(a) Dated 1457 (#549—Nuremberg GNM)
(b) Undated (#452—PC)

2–3: Two astrolabes with an *organum Ptolemaei* on the back

2 An unsigned astrolabe with later additions on the blank side of the plate of the organum (#256—PC)

3 An astrolabe signed by Ioannes (Regiomontanus) and dedicated to Cardinal Bessarion in 1462 (#640—PC, England < London Christie's 1989)

4 Three unsigned astrolabes with similar basic features and inscriptions in Gothic script
(a) Dated 1466 (#592—Darmstadt HLM)
(b) Dated 1481 (#530—Strasbourg MH)
(c) Undated (#620—Munich DM)

5 Two astrolabes with unusual plates and three different scripts
a) An astrolabe with plates for tropical latitudes (#516—Paris CNAM)
b) An astrolabe with universal plates (#545—New York CU)

6–8: Some related astrolabes, some signed by Johannes/Hans Bamman of Schmalkalden

6 An astrolabe with a "Regiomontanus-type" rete dated 1482 (#4501—Cracow CZ)
7 Some simplified astrolabes
  (a) Dated 1484 (#633—Schweinfurt SM)
  (b) Dated 1487 (#604—PLU < lost from Schloß Kreuzenstein)
  (c) A mater dated 1490 (#248—Oxford MHS)
  (d) Dated 1491 (#249—Munich BNM)
  (e) A mater dated 1491 (#540—Oldenzaal HM)
8 An astrolabe with an unusual rete signed "hb" and dated 1493 (#4502—PLU < Milan TT)

9–13: Some instruments signed by Hans Dorn and some others unsigned but probably also due to him

9 An *organum Ptolemaei* on the back of an astrolabic plate attached to a celestial sphere signed by Hans Dorn in 1480 (#4510—Cracow JM)

10–12: Three astrolabes representing the generation after Regiomontanus, with a single half quatrefoil and a rectangular frame developed from the "Regiomontanus-type" rete

10 An astrolabe owned by (?) Ioannes Broscius (#643 or #644—Cracow JM)
11 An unsigned astrolabe dated 1483 and fitted with an *organum Ptolemaei* and an eclipse calculator (#492—Florence MSS)
12 An unsigned astrolabe owned by Martin Bylica (#642—Cracow JL)
13 A compendium with an astrological plate signed by Hans Dorn in 1491 (#250—London BM)
14 An unsigned, undated astrological plate and an equatorium (#4511—Milan PA)
15 An astrolabe on paper and vellum by Hanns Herghamer dated 1492 (#403—Oxford MHS)

CHAPTER 11

Too Many Arabic Treatises on the Operation of the Astrolabe in the Medieval Islamic World: Athīr Al-Dīn Al-Abharī's *Treatise on Knowing the Astrolabe* and His Editorial Method

Taro Mimura
Graduate School of Integrated Arts and Sciences, Hiroshima University, 1-7-1 Kagamiyama, Higashi-Hiroshima 739-8521, Japan
taromimu@hiroshima-u.ac.jp

Abstract

In the medieval Islamic world, many scholars engaged in astronomy composed books on astrolabe, especially treatises on how to operate it (called "Treatises on the Operation of the Astrolabe") such as *Book of the Astrolabe* (*Kitāb al-Asṭurlāb*) by Kūshyār ibn Labbān (fl. second half of the tenth century CE); however, most of them had similar contents. One might ask why these scholars sought to write their own treatises of this kind, even though they differed little from existing works on the subject? To answer this question, I compare Kūshyār's *Book of the Astrolabe* (*Kitāb al-Asṭurlāb*) and Athīr al-Dīn al-Abharī's (d. 1262 or 1265) *Treatise on Knowing the Astrolabe* (*Risāla fī maʿrifat al-Asṭurlāb*), which was written under the strong influence of Kūshyār's *Book*: a comparison between them reveals that they presupposed the use of their own astrolabes. This analysis shows that many Arabic treatises on the operation of the astrolabe had their novelty at least guaranteed by the uniqueness of a specific astrolabe presupposed by each author.

Keywords

Athīr al-Dīn al-Abharī – Book of the Astrolabe – Kūshyār ibn Labbān – Treatise on Knowing the Astrolabe

 | DOI:10.1163/9789004387867_014

Introduction

The astrolabe was one of the most popular astronomical instruments in the medieval Islamic world, and many scholars who were engaged in astronomy composed books on it. In these works, they described not only how to make and improve it, but also explained the geometrical principles necessary for its construction. In particular, treatises on how to operate it (normally called *Risālāt fī al-ʿamal bi-'l-asṭurlāb*, "Treatises on the Operation of the Astrolabe") were written by quite a few scholars, including Muḥammad ibn Mūsā al-Khwārizmī (ca. 780–ca. 850), ʿAlī ibn ʿĪsā al-Asṭurlābī (fl. 832), Abū al-Ḥusayn ʿAbd al-Raḥmān ibn ʿUmar al-Ṣūfī (903–986), Ibn al-Ṣaffār al-Andalusī (d. 1035), Abū al-Rayḥān Muḥammad ibn Aḥmad al-Bīrūnī (973–ca. 1050), Ibn al-Samḥ al-Gharnāṭī (979–1035), and Abū al-Ṣalt al-Dānī al-Andalusī (ca. 1068–1134).[1] Through the work of these and other writers, the "Treatise on the Operation of the Astrolabe" became one of the popular genres of astronomical works in Arabic. To show a typical table of contents of this genre, I will give an overview of Kūshyār ibn Labbān's (fl. second half of the tenth century CE) *Book of the Astrolabe* (*Kitāb al-Asṭurlāb*).[2]

Kiyā Abū al-Ḥasan Kūshyār ibn Labbān Bāshahrī al-Jīlī, an eminent scholar in the Iranian region, wrote several works on mathematics (e.g., *Principles of Hindu Reckoning*), astronomy (e.g., *Zīj al-Jāmiʿ*), and astrology (e.g., *Introduction*

---

1 On the astrolabe in the medieval Islamic world see the comprehensive studies by David A. King, *In Synchrony with the Heavens: Studies in Astronomical Timekeeping and Instrumentation in Medieval Islamic Civilization*, 2 vols., Vol. 1: *The Call of the Muezzin*; Vol. 2: *Instruments of Mass Calculation* (Leiden: Brill, 2004–2005).

2 On Arabic manuscripts of this work see Appendix A of this paper, where I also discuss the title of this work, the textual relationship between these manuscripts, and how to establish the text by using them. In this paper, I establish the text by using the four oldest manuscripts from Family α: MSS G, A, P1, and S. When some manuscripts have a different reading, I record it at the bottom of the text with the sigla defined in Appendix A, i.e.:

:	Separates variant and manuscript sigla
+	Added in
–	Missing from
( )	My comments
أ	MS A
ب	MS P1
جـ	MS G
ص	MS S
شا	مشطوب (crossed out in)
فا	فوق السطر في (above the line in)
ها	هامش (margin)

to Astrology), each of which was popular in the medieval Islamic world as a textbook.[3] One of his works, the *Book of the Astrolabe* was also widespread, so that today Arabic manuscripts of it are extant all over the world.

At the beginning of the *Book of the Astrolabe* (*Kitāb al-Asṭurlāb*, henceforth "K-*Book*"), Kūshyār explains the importance of the instrument for astronomical research, and then gives the details of its parts and plates. This introduction is followed by four sections:[4]

- Section one: on indispensable matters among the subjects (in twenty-four chapters)
- Section two: on the remaining subjects rarely needed (in twelve chapters)
- Section three: on examining the parts, circles, and lines and [on] knowing the rectification of their defects (in twelve chapters)
- Section four: on constructing the astrolabe (in twelve chapters).

الفصل الأوّل في مهمّات من الأبواب لا يستغنى عنها
الفصل الثاني في بقية أبواب يحتاج إليها في الندرة
الفصل الثالث في امتحان الآلات والدوائر والخطوط[1] ومعرفة صحّتها من سقمها[2]
الفصل الرابع في صنعة الأسطرلاب[3].

[1] الآلات والدوائر والخطوط] الات ودواير وخطوطَ : ب [2] من سقمها] وَسقمها : ب [3] الأسطرلاب] + بطريق سهل : أ = + بطريق وجيز : ب .

Among them, section one is formed by the following twenty-four chapters:[5]

- Chapter one: on knowing the altitude of the Sun and stars and what is connected to that [topic]
- Chapter two: on knowing the ascendant from the altitude of the Sun and stars

3 Mohammad Bagheri, "Kūshyār Ibn Labbān," in *The Biographical Encyclopedia of Astronomers*, ed. Thomas Hockey et al. (New York, NY: Springer, 2007), 560–561, gives a useful overview of his scientific biography.

4 The section titles are taken from MS A: folio 10a, lines 11–15; MS P1: folio 4b, lines 5–13; MS G: folio 363b, lines 3–7; MS S: folio 39a, lines 5–10.

5 The chapter headings are taken from MS A: folio 10a, line 17–folio 11a, line 4; MS P1: folio 4b, line 15–folio 5b, line 6; MS G: folio 363b, lines 10–25; MS S: folio 39a, line 11–folio 39b, line 13.

- Chapter three: on knowing the adjustment of the Sun's degree[6] when it does not fit the division of the astrolabe[7]
- Chapter four: on knowing the adjustment of the almucantars' degrees when they do not fit the division of the astrolabe
- Chapter five: on knowing the adjustment of the ascendant's degree when it does not fit the division of the astrolabe
- Chapter six: on knowing the rotation of the celestial sphere from the rising of the Sun and a star or from their setting until the moment of the observation and [on] the equal hours that have passed of a day from their rising or setting
- Chapter seven: on knowing the seasonal hours that have passed of a day or a night until the moment of the observation and [on] knowing the degrees of these hours
- Chapter eight: on knowing the existence of fractions of seasonal hours when the hours are neither exact nor complete
- Chapter nine: on knowing the cardines and the rest of the places[8] of the celestial sphere
- Chapter ten: on knowing the Sun's declination
- Chapter eleven: on knowing the distance of a star from the equator
- Chapter twelve: on knowing the degree that culminates with the planet and the [degree] rising and setting with it
- Chapter thirteen: on knowing the terrestrial latitude of a region and the altitude of midday[9]
- Chapter fourteen: on knowing the day-arc of the Sun and a star as well as their night-arc[10] and what is connected to that [topic]
- Chapter fifteen: on knowing the equation of the daytime of the Sun and stars
- Chapter sixteen: on knowing the rising-times of the zodiacal signs in the terrestrial equator and in the region assumed on the plate

---

6 The celestial longitude of the Sun at the moment of the observation.

7 The calibration of the astrolabe's plates.

8 The twelve astrological places on the ecliptic; on the concept "place" in the astrological context, see Charles Burnett, Keiji Yamamoto, and Michio Yano eds., *Al-Qabīṣī (Alcabitius): the Introduction to Astrology* (London: The Warburg Institute, 2004), 47, footnote 31.

9 The Sun's altitude at midday.

10 The day-arc or the night-arc is the amount of the rotation of a celestial object during a day or a night.

- Chapter seventeen: on knowing the transfer of the rising-times' degrees into equal degrees[11]
- Chapter eighteen: on knowing the transfer of years of the world and nativities
- Chapter nineteen: on knowing the rising of dawn and the setting of dusk
- Chapter twenty: on knowing the meridian line
- Chapter twenty-one: on knowing the shadow [length] from the altitude
- Chapter twenty-two: on knowing the beginning of *al-aṣr*[12] and its end
- Chapter twenty-three: on knowing the altitude of an object whose stone's place[13] can be reached
- Chapter twenty-four: on knowing the altitude of an object whose stone's place cannot be reached.

ا في معرفة ارتفاع الشمس والكواكب وما يليق بذلك

ب في معرفة الطالع من ارتفاع الشمس وارتفاع[1] الكواكب[2]

جـ في معرفة تعديل جزء الشمس إذا لم يكن موافقاً لقسمة الأسطرلاب[3]

د في معرفة تعديل أجزاء المقنطرات إذا لم تكن موافقة[4] لقسمة الأسطرلاب[5]

ه في معرفة تعديل[6] جزء الطالع إذا لم يكن موافقاً لقسمة الأسطرلاب[7]

و في معرفة الدائر من الفلك لطلوع الشمس والكوكب[8] أو لمغيبهما إلى وقت القياس والساعات المستوية الماضية من طلوعهما أو غروبهما

ز في معرفة الساعات الزمانية الماضية من النهار أو الليل إلى وقت القياس ومعرفة أجزاء هذه الساعات

ح في معرفة وجود[9] كسور الساعات الزمانية إذا لم تكن الساعات صحيحة تامّة

ط في معرفة الأوتاد وسائر بيوت[10] الفلك

ي في معرفة ميل الشمس

يا في معرفة بعد الكوكب عن[11] معدّل النهار

يب في معرفة الدرجة التي تتوسّط السماء مع الكوكب[12] والتي تطلع وتغيب معه

يج في معرفة عرض البلد وارتفاع نصف النهار

يد في معرفة قوس نهار الشمس والكوكب[13] وقوس ليلهما وما يليق بذلك

يه في معرفة تعديل نهار الشمس والكواكب[14]

11 Degrees with respect to the ecliptic.

12 Afternoon prayer.

13 The landing place of a plumb line of the object.

يو في معرفة مطالع البروج بخطّ[15] الاستواء وبالبلد المفروض في الصفيحة
يز في معرفة تحويل درج المطالع إلى درج السواء
يح في معرفة تحويل سني العالم والمواليد[16]
يط في معرفة طلوع الفجر ومغيب الشفق
ك في معرفة خطّ نصف[17] النهار
كا في معرفة الظلّ من الارتفاع
كب في معرفة أوّل[18] العصر وآخره
كج في معرفة ارتفاع شخص يمكن الوصول إلى مسقط حجره
كد في معرفة ارتفاع لا يمكن الوصول إلى مسقط حجره .

[1]وارتفاع] او ارتفاع : ب [2] الكواكب] أ ، ب = الكوكب : جـ ، ص [3] الأسطرلاب] الاصطرلاب : ص [4] تكن موافقة] أ ، ص = تكن موافقاً : ب = يكن موافقا : جـ [5] الأسطرلاب] ب ، جـ = الاصطرلاب : أ ، ص [6] تعديل] –أ [7] الأسطرلاب] أ ، جـ = الاصطرلاب : ب ، ص[8] والكوكب] جـ ، ص = او الكواكب : أ = والكواكب : ب [9] وجود] –ف [10] الأوتاد وسائر بيوت] اوتاد بيوت : ب [11] عن] من : أ [12] الكوكب] جـ ، ص = الكواكب : أ ، ب [13] والكوكب] جـ = والكواكب : أ ، ب = او الكوكب : ص [14] والكواكب] + الثابته : جـ = او الكوكب : ص [15] بخطّ] أ ، جـ = لخط : ب ، ص [16] العالم والمواليد] المواليد والعَالم : أ [17] نصف] أ ، ب ، جـ ، فاص = –ص [18] أوّل] + وقت : جـ ، ص .

The structure of K-*Book* suggests that Kūshyār aims to write a comprehensive work on the astrolabe: a detailed explanation of its parts and plates (introduction), basic and non-basic operations (sections one and two), how to maintain it (section three), and how to draw lines on the plates (section four). However, the majority of the treatises on the operation of the astrolabe cover only an explanation of its parts and plates and its essential operations, just like the contents found in the introduction and section one of K-*Book*. As a result, every treatise has similar contents. Thus, one might ask why these scholars sought to write their own treatises of this kind, even though their content differed little from that of existing works on the subject.

In this paper, I will focus on Athīr al-Dīn al-Mufaḍḍal ibn ʿUmar ibn al-Mufaḍḍal al-Samarqandī al-Abharī's (d. 1262 or 1265) *Treatise on Knowing the Astrolabe* (*Risāla fī maʿrifat al-Asṭurlāb*), a hitherto unstudied work, to answer the above question. As will be elucidated, this treatise is written under the strong influence of K-*Book*. Thus, by comparing the two works, I will illustrate how Abharī composes his work by using K-*Book*, and will show his purpose of writing it, which will lead to clarifying the *raison d'être* of many treatises on the operation of the astrolabe in the medieval Islamic world.

Athīr al-Dīn al-Mufaḍḍal ibn ʿUmar ibn al-Mufaḍḍal al-Samarqandī al-Abharī

Abharī[14] was especially known as an author of philosophical works such as *Īsāghūjī* (an introduction to Aristotelian logic)[15] and *Hidāyat al-ḥikma* (a summary of Aristotelian philosophy).[16] His philosophical books became popular, and many scholars wrote commentaries on them. He was also known as a famous teacher on philosophy, having many disciples, including Shams al-Dīn Abū al-ʿAbbās Aḥmad Ibn Muḥammad Ibn Khallikān (1211–1282), Najm al-Din ʿAlī Dabīrān-i al-Kātibī al-Qazwīnī (d. 1277) and Shams al-Dīn Muḥammad ibn Ashraf al-Ḥusaynī al-Samarqandī (d. 1302).

The scholarship of Abharī was not limited to philosophy; for it extended also to mathematical sciences.[17] In fact, he wrote quite a few works on mathematics and astronomy, such as *Treatise on a Compass for Conic Sections, Recension of Euclid's* Elements, and *Utmost Attainment in Comprehending the Orbs.* One of his astronomical works is *Treatise on Knowing the Astrolabe.*[18]

---

14 On his biographical information see Heidrun Eichner, "Al-Abharī, Athīr al-Dīn," in *Encyclopaedia of Islam, THREE*, ed. Kate Fleet, Gudrun Krämer, Denis Matringe, John Nawas, and Everett Rowson. Last accessed 21 March 2017; Dag Nicolaus Hasse, "Mosul and Frederick II Hohenstaufen: Notes on Athiraddin al-Abhari and Siragaddin al-Urmawi," in *Occident et Proche-Orient: Contacts scientifiques au temps des Croisades. Actes du colloque de Louvain-la-Neuve, 24 et 25 mars 1997*, ed. I. Draelants, A. Tihon, and B. van den Abeele (Turnhout: Brepols, 2000), 145–163, explores his philosophical connection to Europe.

15 On this work, see Hans Daiber, *Bibliography of Islamic Philosophy*, 2 vols. (Leiden: Brill, 1999), vol. 1, nos. 108, 109, 954, 1783, 2810, 5711, 5994, 5997, and 6000.

16 On this work, see Daiber, *Bibliography of Islamic Philosophy*, vol. 1, nos. 107, 3553, and 6148.

17 On his astronomical achievement see Hüseyin Sarıoğlu, "Abharī," in *Biographical Encyclopedia of Astronomers*, 7–8. A list of his scientific works is in Boris Rosenfeld and Ekmeleddin Ihsanoğlu, *Mathematicians, Astronomers & Other Scholars of Islamic Civilisation and their Works (7th–19th c.)* (Istanbul: Research Centre for Islamic History, Art and Culture (IRCICA), 2003), 209–210.

18 On Arabic manuscripts of this work, see Appendix B of this paper, where I also discuss its title and my editorial method of establishing the text. In this article, I establish the text by using MS F. When I emend a reading, I note the original at the bottom of the text, where I

The Treatise on Knowing the Astrolabe

The *Treatise on Knowing the Astrolabe* (henceforth "A-*Treatise*") consists of the following fourteen chapters:[19]

- Chapter one: on the definition of technical terms
- Chapter two: on taking the altitude of the Sun and stars
- Chapter three: on knowing the ascendant from the altitude of the Sun or stars
- Chapter four: on knowing the rotation of the celestial sphere from the rising of the Sun and its setting and [on] knowing the hours that have passed of a night and a day
- Chapter five: on knowing the cardines and the rest of the places
- Chapter six: on knowing a day-arc and a night-arc and their hours
- Chapter seven: on knowing the excess of a night and a day
- Chapter eight: on knowing the rising-times of degrees of the zodiacal signs and the transfer of the rising-times into equal degrees
- Chapter nine: on knowing the transfer of years of the world and nativities
- Chapter ten: on knowing the Sun's declination approximately
- Chapter eleven: on knowing the terrestrial latitude of a region
- Chapter twelve: on knowing the beginning of *al-aṣr* and its end
- Chapter thirteen: on knowing the height of an object whose base can be reached
- Chapter fourteen: on knowing the height of an object whose base cannot be reached.

الباب الأوّل في تعريف الاصطلاحات
الباب الثاني في أخذ الارتفاع للشمس والكواكب
الباب الثالث في معرفة الطالع من ارتفاع الشمس أو الكواكب
الباب الرابع في معرفة الدائر من الفلك لطلوع الشمس وغروبها ومعرفة الساعات الماضية من الليل والنهار
الباب الخامس في معرفة الأوتاد وسائر البيوت
الباب السادس في معرفة قوس النهاروالليل وساعاتهما

---

denote MS F as "ف" with the same notation defined in Appendix A; in the text, words in < > are additions made by me to restore the text.

19 MS F has no table of contents. The listed headings are collected from each chapter.

الباب السابع في معرفة الفضل من الليل والنهار
الباب الثامن في معرفة مطالع أجزاء البروج وتحويل المطالع[1] إلى درج السواء
الباب التاسع في معرفة تحويل سني العالم والمواليد
الباب العاشر في معرفة ميل الشمس بالتقريب
الباب الحادي عشر في معرفة عرض البلد
البال الثاني عشر في معرفة أوّل العصر وآخره
الباب الثالث عشر في معرفة <ارتفاع شخص> يمكن الوصول إلى أصله
الباب الرابع عشر في معرفة ارتفاع شخص لا يمكن الوصول إلى أصله .

[1] المطالع] الطّالع : ف .

Chapter one is an introduction to the names of parts and plates of the astrolabe with brief explanations of what they are. In the rest of the chapters, the author explains how to operate the instrument for solving specific problems.

The headings of chapters two to fourteen impress on us that they are very similar to those of the chapters in K-*Book*, section one, in terms of their order and wording. By comparing the texts of A-*Treatise* and K-*Book*, we realize that the core of A-*Treatise*, chapters two to fourteen is a copy from K-*Book*, section one, although he never mentions the source. To clarify Abharī's dependence on K-*Book*, I will present the whole of chapter two of A-*Treatise* with its source, namely the entire text of K-*Book*, section one, chapter one:

A-*Treatise*, chapter two[a]	**K-*Book*, section one, chapter one**[b]
[1] The astrolabe hangs to the right side, and the astrolabe's back is made facing the Sun, and you turn the alidade on the altitude degrees until the Sun's ray enters the upper hole of the alidade's edge and departs from the lower hole. So the altitude degrees between the alidade's edge and the east point are the Sun's altitude; what is between the meridian line and the alidade's edge is the complement of the altitude.	[1] You hang the astrolabe to your right side, with the mater's back facing the Sun, and you turn the alidade on the altitude degrees until you place its two sighting vanes facing the Sun so that the Sun's ray enters the upper hole and passes through the lower hole. So the altitude degrees between the alidade's head and the east-west line[c] are the Sun's altitude at that moment.

(cont.)

A-*Treatise*, chapter two[a]	K-*Book*, section one, chapter one[b]
[2] If the day is cloud (sic!), and the Sun's body is apparent, while its rays are hidden, we hang the astrolabe facing one of the eyes, and turn the alidade until the Sun's body is seen by this eye through a hole [of the sighting vanes] of the alidade. So [the place] where the alidade's head has fallen is the altitude. Likewise, the altitude of a star at night is taken. It is better to put a tube between the two holes [of the sighting vanes] of the alidade in order to prevent the visual rays from spreading.	[2] If a day is cloud (sic!), and the Sun's body is apparent, while its rays are hidden, we hang the astrolabe facing one of the eyes, and turn the alidade until we see by this eye the Sun's body through the two holes [of the sighting vanes] of the alidade. So [the place] where the alidade's head has fallen is the altitude. Likewise, the altitude of a star at night is equally taken.
[3] If we wonder whether the altitude is eastern or western, we take the altitude and leave for a little while, and then take the altitude again. If the second altitude increases from the first altitude, it [the first altitude] is eastern; if it decreases, it is western. This question only occurs when the Sun or stars are in the vicinity of the midday.	[3] If we wonder whether the altitude is eastern or western, we pause for a little while after taking the altitude, and then we take the altitude again. If the second altitude increases, the first altitude is eastern; if the second altitude decreases, the first altitude is western. This question does not occur except when the Sun or a star is in the vicinity of the midday.
[4] If we want the utmost altitude of the Sun or a star, we take its altitude, in the condition that it is in the vicinity of the midday, moment after moment. So when we find it beginning to decrease, the altitude that is before it is the utmost altitude.	[4] If we want the utmost altitude, we observe the Sun or a star until, when it is in the vicinity of the midday, that is, it is approaching it, we take its altitude moment after moment. So when we see it beginning to decrease, the altitude adjacent to it beforehand is the utmost altitude.

A-*Treatise*, chapter two[a]	K-*Book*, section one, chapter one[b]
[1] يعلّق الأصطرلاب باليمين ويجعل ظهر الأصطرلاب إلى الشمس ، وتدير العضادة على أجزاء الارتفاع حتّى يدخل شعاع الشمس في الثقبة العليا من طرف العضادة ويخرج من الثقبة السفلى ، فما بين طرف العضادة ونقطة المشرق من أجزاء الارتفاع هو ارتفاع الشمس . وما بين خطّ نصف النهار وطرف العضادة تمام الارتفاع .	[1] تعلّق[1] الأسطرلاب[2] بيمينك[3] وظهر[4] الأمّ إلى الشمس ، وتدير[5] العضادة على أجزاء الارتفاع حتّى تقيم[6] دفّتيه[7] بإزاء الشمس فيدخل الشعاع في الثقبة[8] العليا وينفذ من الثقبة[9] السفلى ، فيكون ما بين رأس العضادة وبين خطّ المشرق والمغرب من أجزاء الارتفاع هو ارتفاع الشمس في ذلك الوقت .
[2] وإن كان اليوم غيم (!) وجرم الشمس كان ظاهراً وشعاعها[1] خفياً ، علّقنا الأصطرلاب بإزاء إحدى العينين ، وأدرنا العضادة حتّى يرى بتلك العين جرم الشمس من ثقبة العضادة ، فحيث وقع رأس العضادة فهو الارتفاع . وهكذا يؤخذ[2] ارتفاع الكوكب . والأولى أن يركّب فيما بين ثقبتي العضادة أنبوبة ليحفظ شعاع البصر عن الانتشار.	[2] فإن[10] كان يوم غيم (!) وجرم الشمس ظاهر وشعاعها خفي ، علّقنا الأسطرلاب بإزاء إحدى العينين ، وأدرنا العضادة حتّى نرى بتلك العين جرم الشمس من[11] ثقبتي العضادة[12] ، فحيث[13] وقع رأس العضادة فهو الارتفاع . وهكذا[14] سواء[15] يؤخذ ارتفاع الكوكب بالليل .
[3] وإن شككنا أنّ الارتفاع شرقي أوغربي[3] ، نأخذ[4] الارتفاع ويخلّى زماناً يسيراً ثمّ نأخذ[5] الارتفاع ثانياً . فإن زاد الارتفاع الثاني من الارتفاع الأوّل شرقي ، وإن نقص فغربي . وإنّما يقع هذا الشكّ إذا كانت الشمس أو الكواكب قريباً من نصف النهار .	[3] فإن[16] شككنا في أنّ الارتفاع شرقي[17] هو أم[18] غربي ، توقّفنا بعد أخذ الارتفاع قليلاً ثمّ أخذنا الارتفاع ثانياً . فإن زاد الارتفاع الثاني فالارتفاع الأوّل شرقي ، وإن نقص الارتفاع الثاني فالارتفاع الأوّل[19] غربي[20] . ولا يقع هذا الشكّ إلّا إذا كانت[21] الشمس أو الكوكب[22] بالقرب من نصف النهار .

(cont.)

A-*Treatise*, chapter two[a]	K-*Book*, section one, chapter one[b]
[4] وإذا أردنا غاية الارتفاع للشمس أو للكوكب ، أخذنا ارتفاعه حال كونه قريباً من نصف النهار لحظة بعد لحظة . فإذا وجدناه بحيث يأخذ من النقصان ، الارتفاع[6] الذي قبله هو غاية الارتفاع .	[4] فإن[23] أردنا غاية الارتفاع ، رصدنا الشمس أو الكوكب حتّى إذا كان بالقرب من نصف النهار وهو سائر[24] إليه ، أخذنا[25] ارتفاعه لحظة بعد لحظة . فإذا رأيناه[26] قد أخذ في النقصان ، كان الارتفاع الذي يليه من قبل هو غاية الارتفاع .
[1] وشعاعها] وشعاعُها : ف [2] يؤخذ] يأخذ : ف [3] أوغربي] وغربيّ : ف [4] نأخذ] يأخذ : ف [5] نأخذ] يأخذ : ف [6] الارتفاع] والارتفاع : ف .	[1] تعلّق] ب = نعَلِق : أ = نعلق : جـ ، ص [2] الأسطرلاب] ب ، جـ = الاصطرلاب : أ ، ص [3] بيمينك] أ ، ب = يميننا : جـ = يمىىا : ص [4] وظهر] ظهر : جـ [5] وتدير] أ ، ب = وندىر : جـ ، ص [6] تقيم] ب = ىقيم : أ ، جـ ، ص [7] دفّتيه] دفىها : أ [8] في الثقبة] بالىقىه : جـ [9] الثقبة] ىقبه : جـ [10] فإن] فصل فان : جـ [11] من] عن : ص [12] العضادة] أ ، ب = الدفىين حميعا : جـ = الدفىين («جمىعا» زائد فوق السطر) : ص [13] فحيث] + الدفتىن جمىعا: شاص [14] وهكذا] جـ = هكذا : أ = وها كدي : ب = وهكدى : ص [15] سواء] سوي : جـ [16] فإن] فصل فان : جـ [17] شرقي] اشرقى : أ («الـ«ا»» مشطوب) [18] هو أم] او : أ [19] فالارتفاع الأوّل] أ ، جـ = فالاول : ب [20] وإن نقص الارتفاع الثاني فالارتفاع الأوّل غربي] –ص • غربي] + وَان ىقص الارتفاع الثانى فالاول غربى : أ [21] كانت] أ ، ب = كان : جـ ، ص [22] الكوكب] الكواكب : ب [23] فإن] فصل فان : جـ [24]

(cont.)

A-*Treatise*, chapter two[a]	K-*Book*, section one, chapter one[b]
	سائر] أ ، ص = صار : ب ، جـ 25 أخذنا] واخذنا : ب 26 رأيناه] راينَا : ب .

a MS F: folio 22a, line 14–folio 23a, line 2.

b MS A: folio 101a, line 5–folio 101b, line 3; MS P1: folio 5b, line 6–folio 6a, line 6; MS G: folio 364a, lines 2–13; MS S: folio 39b, line 15–folio 40a, line 10.

c One of the two central lines on the plate drawn from east to west; on this technical term see K-*Book*, introduction edited in Appendix C of this paper.

This example illustrates that chapter two is a copy of K-*Book*, section one, chapter one with some modification. Especially, the first part of [2] ("If a day is cloud" vs. "If the day is cloud") shows Abharī's strong dependence on Kūshyār's text. First, this sentence written by Kūshyār must be "*in kāna yawm ghayman*," not "*in kāna yawm ghaym*," where he fails to make *ghaym* accusative owing to *kāna*. And the expression of *ghaym* is odd in this context because this term means "cloud"; he should use *ghāʾim*, which denotes "cloudy." But Abharī copies this awkward sentence with this grammatical mistake.[20]

On the other hand, Abharī sometimes modifies the source text. In the last sentence of [3] ("This question does not occur except when the Sun or a star is…"), he changes its structure from "…does not occur…except when…(*wa-lā yaqaʿu…illā idhā…*)" to "…only occurs when…(*innamā yaqaʿu…idhā…*)." Moreover, he puts additional information in the last parts of [1] and [2].

In each of chapters two to fourteen, Abharī utilizes a chapter (or two chapters) of K-*Book*, section one. The following is the list of K-*Book*, section one's chapters from which Abharī copies the text:

- A-*Treatise*, chapter two: K-*Book*, section one, chapter one;
- A-*Treatise*, chapter three: K-*Book*, section one, chapter two;
- A-*Treatise*, chapter four: K-*Book*, section one, chapters six and seven;
- A-*Treatise*, chapter five: K-*Book*, section one, chapter nine;
- A-*Treatise*, chapter six: K-*Book*, section one, chapter fourteen;
- A-*Treatise*, chapter seven: K-*Book*, section one, chapter fifteen;

20 This point is owed to the anonymous referee of this article.

- A-*Treatise*, chapter eight: K-*Book*, section one, chapters sixteen and seventeen;
- A-*Treatise*, chapter nine: K-*Book*, section one, chapter eighteen;
- A-*Treatise*, chapter ten: K-*Book*, section one, chapter ten;
- A-*Treatise*, chapter eleven: K-*Book*, section one, chapter thirteen;
- A-*Treatise*, chapter twelve: K-*Book*, section one, chapter twenty-two;
- A-*Treatise*, chapter thirteen: K-*Book*, section one, chapter twenthy-three;
- A-*Treatise*, chapter fourteen: K-*Book*, section one, chapter twenty-four.

The above list shows Abharī's skipping K-*Book*, section one, chapters three to five, eight, eleven, twelve, and nineteen to twenty-one. It is remarkable that while Kūshyār gathers basic operations in section one entitled "on indispensable matters," Abharī further chooses fundamental ones from them.

According to Abharī's aim of selecting essential operations, the omission of chapter twenty (on how to draw the meridian line) is reasonable, because Kūshyār does not use the astrolabe in it. Among the rest of the deleted chapters, we find that Kūshyār explains similar problems in chapters two to five and eight: on operating for the adjustment of degrees (e.g., the Sun's degree) not fitting the calibration of the astrolabic plate. His removal of them suggests that he thinks that the adjustment is not essential. And chapters eleven and twelve can be classified in one category: on operating concerning the fixed stars; this category might contain chapter nineteen, where Kūshyār determines the rising of dawn and the setting of dusk by using the pointer of a fixed star. In their omission, we detect his hesitation in keeping subjects on the fixed stars as fundamental, although he mentions two operations using a fixed star's pointer in chapters three and four.

In some chapters of A-*Treatise*, Abharī uses a part of the source chapter(s), while skipping some topics not relevant (or fundamental) to A-*Treatise*. For example, he composes chapter four from K-*Book*, section one, chapters six and seven, and omits the last two topics of chapter seven: how to determine the degrees of the equal hours from the altitude of a celestial object and how to determine the seasonal hours by using special curves engraved on a type of alidade; in chapter eight, he utilizes K-*Book*, section one, chapters sixteen and seventeen, and he skips the case of the equatorial region; in chapter nine, he copies the most part of K-*Book*, section one, chapter eighteen, but he deletes the last topic: on how to determine the rotation of the celestial sphere from the Sun's rising until the transfer of years; in chapter ten, he skips Kūshyār's explanation about the fact that the astrolabe only gives the Sun's declination approximately; in chapter eleven, he copies the first part of K-*Book*, section one, chapter thirteen, but skips the second part, where Kūshyār presents how to know the terrestrial latitude from the altitude of a fixed star; and in chapters thirteen and fourteen, he uses K-*Book*, section one, chapters twenty-three and twenty-four, respectively, but in each chapter he omits Kūshyār's explanation of his operation by using a diagram.

As shown above, Abharī made his own "Treatises on the Operation of the Astrolabe" by selecting topics from K-*Book*, section one. Moreover, he sometimes inserts extra information not found in Kūshyār's text. A-*Treatise*, chapter twelve contains an interesting example of his addition and heavy editorial efforts.

Of the two chapters on shadows in K-*Book*, section one, namely chapters twenty-one and twenty-two, Abharī skips twenty-one (on knowing the shadow length from the altitude) in his treatise, but he uses twenty-two (on knowing the beginning of the Aṣr and its end) in his chapter twelve. Since he omits the explanation of how to obtain the shadow length from the Sun's altitude by using the astrolabe given in K-*Book*, section one, twenty-one, he expands a part of Kūshyār's text of twenty-two. The following is the whole of A-*Treatise*, chapter twelve as well as K-*Book*, section one, twenty-two:

A-*Treatise*, chapter twelve[a]	K-*Book*, section one, chapter twenty-two[b]
You put the alidade on the [Sun's] altitude at midday. So the shadow degrees on which the other edge of the alidade has fallen is the digits of the shadow [length] corresponding to the [Sun's] altitude at midday.	You know the digits of the shadow corresponding to the [Sun's] altitude at midday.
You add twelve digits to them, and put an edge of the alidade on these degrees of the shadow. So the altitude degrees on which the other edge of the alidade has fallen is the [Sun's] altitude at the beginning of the time *Aṣr*.	You add twelve digits to them, and take its altitude. So what has resulted is the [Sun's] altitude at the beginning of the *Aṣr*.
You add twenty-four digits to the shadow [length] of the [Sun's] altitude at midday, and put the alidade on that altitude. So the altitude degrees on which the other edge of the alidade has fallen is the degrees of the [Sun's] altitude at the end of the time.	You add twenty-four digits to the shadow [length] of the [Sun's] altitude at midday, and take its altitude. So what has resulted is the [Sun's] altitude at the end of the Aṣr.

(cont.)

A-*Treatise*, chapter twelve[a]	K-*Book*, section one, chapter twenty-two[b]
	If the shadow [length] is [given] in feet, seven and fourteen are [added] instead of twelve and twenty-four, respectively. Some Islamic people make the beginning of the time *Aṣr* at the end of its time we have described, and the end of its time before the Sun's setting. This is a topic based on the *Sharīʿa* and the variety of their doctrines.
تضع[1] العضادة على ارتفاع نصف النهار ، فما وقع عليه الرأس الآخر للعضادة[2] من أجزاء الظلّ فهو أصابع الظلّ لارتفاع[3] نصف النهار.	تعرف[1] أصابع ظلّ الارتفاع[2] لنصف[3] النهار.
وتزيد عليه[4] اثني عشر إصبعاً ، وتضع[5] رأس العضادة على تلك الأجزاء من الظلّ ، فما وقع عليه الرأس الآخر للعضادة من أجزاء الارتفاع فما كان فهو ارتفاع أوّل وقت العصر.	وتزيد[4] عليه اثني عشر إصبعاً ، وتأخذ[5] ارتفاعه ثانياً[6] ، فما كان فهو ارتفاع أوّل العصر.
وتزيد[6] على ظلّ ارتفاع نصف النهار أربعة[7] وعشرين إصبعاً ، وتضع[8] العضادة على ذلك الارتفاع فما وقع عليه الرأس الآخر من العضادة من أجزاء الارتفاع فهو أجزاء ارتفاع آخر الوقت.	وتزيد[7] على ظلّ ارتفاع نصف النهار كد[8] إصبعاً ، وتأخذ[9] ارتفاعه ، فما كان فهو ارتفاع آخر العصر.

(cont.)

A-*Treatise*, chapter twelve[a]	K-*Book*, section one, chapter twenty-two[b]
	فإن كان الظلّ أقداماً ، كانت السبعة والأربعة عشر بدلاً من الاثني عشر[10] والأربعة العشرين[11] . ومن أهل الإسلام[12] من يجعل أوّل وقت العصر آخر وقته الذي ذكرناه[13] ، وآخر وقته قبل[14] مغيب الشمس . وهذا[15] مبني على الشريعة واختلاف مذاهبهم[16] فيه[17] .
1 تضع] يضع : ف 2 للعضادة] العضادة : ف 3 لارتفاع] ارتفاع : ف 4 وتزيد عليه] ويزيد على : ف 5 وتضع] ويضع : ف 6 وتزيد] ويزيد : ف 7 أربعة] اربعًا : ف 8 وتضع] ويضع : ف .	1 تعرف] أ = نعرف : ب ، جـ = نعرف : ص 2 الارتفاع] أ ، ب = ارتفاع : جـ ، ص 3 لنصف] ب ، ص = نصف : أ ، جـ 4 وتزيد] وبزيد : أ ، ص = ونزيد : ب ، جـ 5 وتأخذ] وناخذ : أ ، ب ، جـ = وباخذ : ص 6 ثانياً] أ ، ب = –جـ ، –ص 7 وتزيد] ونزيد : أ ، ب ، جـ = وبرىد : ص 8 كد] أ ، ب = اربعه وعشرن : جـ = اربعه وعسرون : ص 9 وتأخذ] وناخذ : أ ، ب ، جـ = وباحد : ص 10 الاثني عشر] سب : أ 11 والأربعة العشرين] و كد: أ 12 أهل الإسلام] أ ، ب = الفقها : جـ 13 ذكرناه] أ ، ب = ذكرنا : جـ 14 قبل] أ ، ب =قبيل : جـ 15 وهذا] + شي : جـ 16 مذاهبهم] أ ، ب = مذاهب العلما : جـ 17 ومن أهل . . . مذاهبهم فيه] –ص .

a MS F: folio 25b, line 12–folio 26a, line 5.

b MS A: folio 108b, line 14–folio 109a, line 3; MS P1: folio 14b, lines 5–11; MS G: folio 369b, lines 3–8; MS S: folio 46a, lines 16–20.

Here he tries to keep the consistency of Kūshyār's text in his treatise by adding information on how to obtain the shadow length from the altitude degrees and *vice versa*; however, he skips the last part of twenty-two, probably because of its being not relevant to his work.

Another example is found in A-*Treatise*, chapter five, in the beginning of which we also detect Abharī's addition to the text of K-*Book*, section one, chapter nine, as follows:

A-*Treatise*, chapter five[a]	**K-*Book*, section one, chapter nine**[b]
You put the ascendant degree on the eastern horizon. So among the degrees of the zodiacal signs, that which has fallen on the line of the mid-heaven, i.e., the upper half of the meridian line, is the beginning of the tenth place.	When the ascendant degree is put on the eastern horizon, the degree falling on the line of the mid-heaven is the beginning of the tenth place.
That which has fallen on the western horizon is the beginning of the seventh [place].	The degree falling on the western horizon[c] is the beginning of the seventh place.
That which has fallen on the line of the cardine of the earth, i.e., the lower half of the meridian line, is the beginning of the fourth place.	The degree falling on the line of the cardine of the earth is the beginning of the fourth place.
These are the four cardines.	These four places are the cardines.
تضع[1] جزء الطالع على أفق المشرق ، فما وقع من أجزاء البروج على خطّ وسط السماء وهو النصف[2] الأعلى من خطّ نصف النهار فهو أوّل البيت[3] العاشر .	إذا وقع جزء الطالع على أفق المشرق ، كان الجزء الذي يقع على[1] خطّ وسط السماء هو أوّل البيت العاشر[2] .
وما وقع[4] على أفق المغرب أوّل السابع .	والجزء الذي يقع على أفق المغرب[3] هو أوّل[4] البيت[5] السابع[6] .
وما وقع على خطّ وتد الأرض وهو النصف الأسفل من خطّ نصف النهار وهو أوّل البيت[5] الرابع .	والجزء الذي يقع على خطّ وتد الأرض[7] هو[8] أوّل[9] البيت الرابع .

(cont.)

A-*Treatise*, chapter five[a]	K-*Book*, section one, chapter nine[b]
فهذه هي الأوتاد الأربعة	فهذه البيوت الأربعة هي الأوتاد .
1 تضع] يضع : ف 2 وهو النصف] والنصف : ف 3 البيت] بيت : ف 4 وقع] قربه : ف 5 البيت] بيت : ف .	1 يقع على] نقطع : جـ 2 هو أوّل البيت العاشر] أ ، ب = جزو وسط السما : جـ = وهو البت العاشر : ص 3 المغرب] + جزو المغرب : جـ ، ص 4 هو أوّل] أ ، ب = وهو : جـ 5 البيت] أ ، جـ = -ب 6 والجزء الذي . . . البيت السابع] -ص 7 الأرض] + جزو وتد الارض : جـ ، ص 8 هو] أ ، ب = وهو : جـ ، ص 9 أوّل] أ ، ب = -جـ ، -ص .

a MS F: folio 23b, line 13–folio 24a, line 4.

b MS A: folio 104b, line 17–folio 105a, line 4; MS P1: folio 9b, line 14–folio 10a, line 2; MS G: folio 366a, lines 16–19; MS S: folio 43a, lines 2–4.

c MSS G and S add the following phrase: “i.e., the degree of the west.”

In this part, he inserts descriptions of the two lines (the line of the mid-heaven and the line of the cardine of the earth) on the astrolabic plate, while Kūshyār uses these technical terms without explanation. That is because Kūshyār has already explained them when describing two central lines of the plate in the introduction, as follows:[21]

والخطّان المتقاطعان في الصفائح : أمّا الذي يخرج من عند العلاقة فيقسم الصفيحة بنصفين فهو خطّ نصف النهار ، ويسمّى النصف الذي يلي العلاقة منه خطّ وسط السماء ، والنصف الآخر خطّ وتد الأرض . وأمّا الذي يقطع هذا الخطّ بنصفين على زوايا قائمة عند مركز الصفيحة فهو خطّ المشرق والمغرب .

(The two lines cutting each other on the plates: As for that which is extended from the cord and halves the plate, it is the meridian line; the half which is next to the cord is called the line of the midheaven, and the other half is [called] the

21 The whole Arabic text of the latter half of K-*Book*'s introduction, where Kūshyār defines astrolabic parts, is edited in Appendix C of this paper with an English translation.

line of the cardine of the earth. As for that which halves this line perpendicularly at the center of the plate, it is the east-west line).

On the other hand, Abharī explains these two lines in chapter one, as follows:[22]

خطّ نصف النهار هو الخطّ الذي يأخذ من الجانب الكرسي إلى الجهة المقابلة .
خطّ الاستواء هو خطّ مقاطع لخطّ نصف النهار على زوايا قائمة .

(The meridian line is the line taken from the side of the throne to the other side. The line of the equator is the line cutting the meridian line perpendicularly.)

This quotation shows that in chapter one, he does not introduce the line of the mid-heaven and the line of the cardine of the earth, so he needs to define them when mentioning them for the first time. Moreover, a comparison between the two texts on the two central lines makes clear that unlike chapters two to fourteen, he writes their explanations without using Kūshyār's text; in fact, his terminology "the line of the equatorial region" is different from "the east-west line" found in K-*Book*. Thus, we must ask: Why does he compose a new explanation of astrolabic parts? To elucidate his aim, I will next compare the introduction of astrolabic parts by Abharī and that by Kūshyār.

A striking difference between them is found in their texts on the horizon circle and the almucantars, as follows:

From A-*Treatise*, chapter one	From K-*Book*, introduction
The horizon circle is the circle passing through the two intersection points between the equator and the line of the equatorial region. The almucantars are the circles above the earth.	The almucantar circles are the circles drawn on the plates around point C; some of them are complete circles, whereas others are incomplete; between them, the numbers of their degrees are written according to the astrolabe's calibration: if it is a one-sixth [astrolabe], every six degrees are [written]; if it is a one-third, every three degrees are [written]; if it is a half, every two degrees are [written]; and if it is a complete [astrolabe], every degree is [written].

22 A-*Treatise*, chapter one is edited in Appendix D of this paper with an English translation.

From A-*Treatise*, chapter one	From K-*Book*, introduction
	The first of these almucantar circles is the eastern and western horizon,[a] and "east" and "west" are written on it. The almucantars adjacent to the eastern horizon towards the meridian line are the eastern almucantars, and the [almucantars] adjacent to the western horizon towards the meridian line are the western almucantars. What is between the center of the plate and the first almucantar on the line of the cardine of the earth is the terrestrial latitude of the region [supposed in] the plate.
دائرة الأفق الدائرة التي هي بنقطتي التقاطع بين معدّل النهار وخطّ الاستواء . المقنطرات هي الدوائر التي فوق الأرض .	ودوائر المقنطرات هي الدوائر المخطوطة في الصفائح محيطة بنقطة ص ، منها دوائر تامّة ومنها غير تامّة ، وفيما بينهما مكتوب عدد أجزائها بحسب قسمة الأسطرلاب : إن كان سدساً فستّة ستّة ، وإن كان ثلثاً فثلاثة ثلاثة ، وإن كان كان نصفاً فجزآن جزآن ، وإن كان تامًّا فجزء جزء . وأوّل دوائر هذه المقنطرات أفق المشرق والمغرب مكتوب عليه المشرق والمغرب . والمقنطرات التي تلي الأفق الشرقي إلى خطّ نصف النهار هي المقنطرات الشرقية ، والتي تلي الأفق الغربي إلى خطّ نصف النهار هي المقنطرات الغربية . وما بين مركز الصفيحة وبين أوّل المقنطرات على خطّ وتد الأرض هو عرض بلدان الصفيحة .

a MSS G and S add: "and the directions of east and west."

When describing the almucantars, Abharī only indicates their place, whereas Kūshyār gives more detailed information about them and mentions several types of plates with different calibrations. In this part, Abharī's text is shorter than Kūshyār's just like in A-*Treatise* chapters two to fourteen, but Abharī writes it in his own words. What is remarkable in his wording is that we can detect the necessity for the readers to refer to an astrolabe. Here he only denotes the position of the almucantars with respect to the earth, i.e., the horizontal circle; if the readers do not have an astrolabe, his definition would not make any sense to them.

As for the text by Kūshyār, we must note that he sets the almucantars around point C. In this context, point C means the point of ninety degrees[23] given on the zenith, but this point is not always marked by C in every astrolabe. That is because the plate of each astrolabe is designed individually, so only some astrolabes have a plate marked with C. Kūshyār's usage of point C impresses on us that, while he refers to several types of astrolabe, he supposes in K-*Book* a plate having point C. Moreover, in his description of the horizon circle, he mentions the East and West marks; these are also not always engraved on every plate, so his reference to them shows that the plate he is thinking of has particular marks for East and West. These three marks confirm that he composes the introduction to the astrolabic parts with respect to a specific astrolabe at his disposal.

On the other hand, we find that "the column (*'amūd*) of Aries and Libra," one of Abharī's parts, is not mentioned in K-*Book*. Abharī explains it, as follows:

والعمود الذي يقطع العنكبوت على نقطتي الحمل والميزان يسمّى عمود الحمل والميزان .

(The column cutting the rete at the two points Aries and Libra is called the supporter of Aries and Libra.)

This description indicates that it is a part of the rete between the vernal equinox point and the autumnal equinox point; however, this part is not always found in every rete, since there exist various types of rete according to their designs. Thus, Abharī's inclusion of it confirms that he supposes a specific rete that has it, while Kūshyār's rete probably does not contain it.

The contents of A-*Treatise*, chapter one, and K-*Book*, introduction, show that Abharī and Kūshyār have their specific astrolabes in mind when introducing astrolabic parts: while Kūshyār's astrolabe has the marks of C, east, and west, without a supporter of the rete, Abharī's astrolabe does not have these

23 That is because C (*Ṣād*) is 90 in abjad numerals.

three marks, but has the supporter. That their supposed astrolabes are different can be detected in both works after the introductory part. For example, K-*Book*, chapter two, gives an example of how to obtain the Sun's altitude with the astrolabe, as follows:[24]

(For example: on a plate for the latitude thirty-six degrees, we suppose the [Sun's] altitude at that time to be twenty-four degrees east, and suppose the Sun's degree to be Taurus twelve degrees.[25] Then we put Taurus twelve degrees on the twenty-four degrees almucantar,[26] so we find Gemini at eighteen degrees on the eastern horizon. We call it the ascendant degree.

If this altitude is, for example, the altitude of Aquila, we put its pointer on the twenty-four degrees almucantar. So we find Capricorn at twenty-two degrees on the eastern horizon. We call this the ascendant degree).

مثال ذلك في صفيحة عرض لو نفرض[1] ارتفاع الوقت كد[2] درجة شرقية ونفرض[3] جزء الشمس في يب من الثور[4] ، فوضعنا الدرجة يب[5] من الثور على مقنطرة كد[6] ، فوجدنا الدرجة يح[7] من الجوزاء واقعة على أفق[8] المشرق فقلنا هو الجزء الطالع .

فإن كان هذا الارتفاع هو ارتفاع[9] النسر الطائر مثلاً ، وضعنا شظيته على مقنطرة كد[10] فنجد الدرجة كب[11] من الجدي بالتقريب واقعة على أفق المشرق فقلنا هو الجزء الطالع .

[1] نفرض] ب = بفرض : أ ، جـ ، ص [2] كد] أ = اربعًا عشرن : ب = اربعه وعشرن : جـ ، ص [3] ونفرض] ب ، جـ = وبفرض : أ = بفرض : شاص [4] ونفرض جزء الشمس في يب من الثور] أ ، ب («من الثور» زائد في الهامش) = ونفرض جزو الشمس فى اثنى عشر درجه من الثور : جـ = نفرض السمس فى اىى عشر : ص = + ليافقا («لبوافقان» في نسخة أ) اقسام الاسطرلاب («الاصطرلاب» في نسخة ص لكنّه مشطوب و«نطاق البروح» زائد في الهامش مع رمز «ص») الى ان سن العمل فما لا يوافقها : جـ ، ص [5] فوضعنا الدرجة يب] هاب = فرضنَا الدَّرجه الىانيه عشر : أ = –ب = فوصعنا الدرجه الىانيه عشر : جـ ، ص [6] كد] أ ، ب = اربعه وعشرن : جـ = على اربعه وعسرن : ص = + فى الناحيه الشرقه : أ ، ص [7] يح] ب = الىامنة عشر : أ ، جـ

24 MS A: folio 101b, line 8–14; MS P1: folio 6a, line 11–folio 6b, line 2; MS G: folio 364a, lines 19–26; MS S: folio 40a, line 15–folio 40b, line 4.

25 MSS G and S add: "in order that both fit the division of the astrolabe until we will explain how to operate with that which does not fit them."

26 MSS G and S add the following phrase: "on the eastern side."

، ص 8 أفق] الافق : جـ 9 ارتفاع] الاربماع : جـ 10 كد] أ ، ب = اربعه وعشرن :
جـ ، ص 11 كب] ب = النانية والعشرن : أ ، جـ ، ص .

Here he uses a plate designed for the latitude of thirty-six degrees, which corresponds to the place of his activity, Rayy. Thus, in this example he clearly refers to his specific astrolabe. Moreover, the setting out of the example suggests that the almucantars of his astrolabe are calibrated at six degrees, which means that he has a one-sixth astrolabe.[27]

As already shown, in A-*Treatise*, chapter two, Abharī uses K-*Book*, section one, chapter one; however, he deletes this example. It seems odd because his treatise is aimed to present fundamental operations of the astrolabe, so a specific example like this will help the readers to understand the process of operation. Thus, his omission indicates that this example is not applicable to his astrolabe designed for a latitude other than thirty-six degrees.

The comparison of their contents illustrates that both of them presuppose the use of their own astrolabes throughout the works. Suggested by the table of contents of K-*Book*, however, Kūshyār tries to compose a treatise that applies to a wider context than his specific astrolabe. In fact, he mentions several features found in other astrolabes: in the introduction, he mentions various types of astrolabe such as the one-sixth astrolabe; in section one, chapter seven, he describes a type of alidade engraved with curves denoting seasonal hours, although they are not included in the introduction; in section one, twenty-one, he introduces the two types of shadow unit, digits and feet, and he explains how one knows which unit is used in his astrolabe. These examples confirm his tendency of composing a comprehensive work, which led his work to be one of the most popular treatises on the operation of the astrolabe in the medieval Islamic world.

Unlike Kūshyār, Abharī never mentions another type of astrolabe; when copying K-*Book*, section one, chapter seven, he omits the topic of the special curves on a type of alidade. As for the two units of shadow, he skips the whole of K-*Book*, section one, twenty-one, and when using the shadow length for determining the *Aṣr*, he supposes digits as the unit without giving any note on the other unit feet. His selection of topics from K-*Book* impresses on us that he limits the explanation to the astrolabe at his disposal.

The above analysis shows the reason why Abharī rewrites an explanation of astrolabic parts: since he presupposes an astrolabe different from Kūshyār's, he

27 In fact, MSS G and S add the explanation that Kūshyār chooses twelve degrees for adjusting the example to his astrolabe; see footnote 25 of this paper.

cannot use Kūshyār's text, and writes the introduction afresh. As regards the operations, most of them are relevant to any astrolabe, so he duplicates a part of Kūshyār's text with some editorial changes for adjusting the extracted words to his own astrolabe. His editorial work illustrates that every single astrolabe has a distinct design that leads a scholar write a new handbook. Although the major part of A-*Treatise* is a copy of K-*Book*, it is original because Abharī writes it for his own astrolabe.

In the medieval Islamic world, scholars proposed many innovations on the astrolabe, and instrument makers made a number of astrolabes with new layouts. As a result, each astrolabe had distinctive features, so theoretically every astrolabe needed its own handbook. Indeed, many Islamic scholars wrote treatises on the operation of the astrolabe, and these treatises would obviously have been useless without reference to an astrolabe. As the case of Abharī's treatise reveals, many Arabic treatises had their novelty at least guaranteed by the uniqueness of the specific astrolabe that each author had at his disposal.

Appendix A: Arabic Manuscripts of Kūshyār's Book of the Astrolabe

The Arabic manuscripts of K-*Book* at my disposal are the following thirteen manuscripts (in the order of the dates of their copying; the first letter is the siglum of each manuscript):

G: Istanbul, Yeni Cami MS 784, folios 362b–376b (copied around the sixth century AH)[28]
The copy was written in *naskh* without vocalization. As Krause notes, the last three folios have damage in the lower part. The copyist did not enter any correction. The chapter headings, tables, and diagrams are in rubric.

A: Istanbul, Ayasofya MS 2671, folios 80b–103a (copied in 621 AH = 1224/5 CE)[29]
The copy was written in *naskh* with occasional vocalization. The copyist entered a few corrections in the margin. The chapter headings are in rubric.

28 Based on Krause's estimation; see Max Krause, "Stambuler Hanschriften islamischer Mathematiker," *Quellen und Studien zur Geschichte der Mathematik, Astronomie und Physik*: Abteilung B: Studien 3 (1936): 437–531 (472).

29 See Krause, "Stambuler Hanschriften," 472.

P1: Paris, Bibliothèque nationale de France, MS *arabe 2487, folios 2a–31b* (*copied on 24th Shawwāl 679* AH *= 23 February 1281* CE)[30]
The copy was completed on 24 Shawwāl 679 AH, as the copyist noted in the front page (folio 2a) and the colophon (folio 30b):

Front Page:

كتاب
الاصطرلاب للشيخ الامام العالم ابو الحسن كوشيار
ابن لبان باشهري الجيلي رحمه الله عليه يحتوي
على معاني واعمال وتدقيق حساب وذلك في نوبه
العبد الراحمى رحمه ربه وغفرانه
وذلك بحكم النسخ في منتصف شوال
من سنه تسعه وسبعين وسمايه عربيه

(Book of the Astrolabe by the master, leader, and scholar Abū al-Ḥasan Kūshyār ibn Labbān Bāshahrī al-Jīlī, may God have mercy upon him, comprising the concepts, the operations, and the exact calculation. That is the responsibility of the merciful servant, may God have mercy upon him and forgive him.
That [copy] is by virtue of the copying in the middle of Shawwāl of the Arabic year 679).

Colophon:

ووافق الفراغ من نسّخه يوم الاحد الرابع والعشرين من شهر شوال
من سنه تسع وسبعين وستمايه

(The completion by the person who copied it happened on Sunday 24 Shawwāl of the year 679).

It was written in *naskh* with occasional vocalization. The copyist entered one correction in the margin of fol. 6a.

S: Istanbul, Ayasofya MS *2672, folios 37b–54b* (*copied in 684* AH*=1285/6* CE)[31]
The copy was written in *naskh* without vocalization. The copyist collated it with the exemplar as well as with another copy; he entered many corrections in the margin (frequently with the mark "correct (صح)" or "another copy (خ)").

30 De Slane, *Catalogue des manuscrits arabes* (Paris: Bibliothèque Nationale, 1883–1895), 441 erroneously records the copying date as 699 AH.

31 See Krause, "Stambuler Hanschriften," 472.

P2: Paris, Bibliothèque nationale de France, MS *arabe 5972, folios 2b–36b* (*copied in 1641* AG*=1329* CE)[32]
The copy was written in *naskh* without vocalization (except a few *tanwīns*). The copyist collated it with the exemplar as well as with another copy; he entered many corrections in the margin (frequently with the mark "correct (صح)" or "another copy (خ)"); moreover, he added a long extract from another copy on folios 30b–31b.

C: Dublin, Chester Beatty Library MS *Ar. 5254, folios 163a–186b* (*copied around the tenth century* CE)[33]
The copy was written in *naskh* with occasional vocalization. The copyist collated it with the exemplar as well as with another copy; he entered many corrections in the margin (sometimes with the mark "correct (صح)" or "another copy (خ)").

Te: Tehran, Tehran University MS *2092, folios 3b–25b* (*copied in 1057* AH *= 1647/8* CE)[34]
The copy was written in *naskh* without vocalization. The copyist entered a few corrections in the margin (sometimes with the mark "correct (ص)").

Se: Istanbul, Selim Aga MS *730, folios 23b–42a* (*copied in 1107* AH *= 1695/6* CE)[35]
The copy was written in *naskh* without vocalization. The copyist did not copy section four. He entered a few corrections in the margin (sometimes with the mark "correct (ص)"). The names of astrolabic parts in the introduction are in rubric; the chapter numbers are in rubric.

Ma: Qom, Mashhad Astan Quds MS *5529, pp. 7–21* (*copied around the eleventh century* AH)[36]
The copy was written in *naskh* without vocalization. The copyist copied the text in the framed space as well as in the margin. He did not enter any corrections.

---

32 See E. Blochet, *Catalogue des manuscrits arabes des nouvelles acquisitions (1884–1924)* (Paris: Bibliotheque Nationale, 1925), 152.

33 Based on Arberry's estimation; see Arthur J. Arberry, *The Chester Beatty Library: A Handlist of the Arabic Manuscripts* (Dublin: Hodges, Figgis & Co., 1964), 7:81. Note that Fuat Sezgin, *Geschichte des arabischen Schrifttums* 9 vols. (Leiden: Brill, 1978), 6:248 records the date 891 AH without any reference.

34 See Mustafa Derayati, *Fehrestwara-ye dastneweshtha-ye Iran*, 12 vols. (Tehran: Library, Museum and Document Center of Islamic: 1389 AH solar-year), 1:796.

35 See Krause, "Stambuler Hanschriften," 472.

36 See Derayati, *Fehrestwara-ye*, 1:796.

M1: Tehran, Malik MS *2242, folios 26b–44b (copied around the eleventh century* AH*)*[37]
The copy was written in *naskh* without vocalization. The copyist entered several corrections in the margin (sometimes with the mark "correct (صح)").

Y: Princeton, Princeton University Library MS *(Yehuda collection) 1168, folios 69b–91b (copied around the twelfth century* AH*)*[38]
The copy was written in *naskh* without vocalization. The copyist entered many corrections in the margin.

M2: Tehran, Malik MS *3455, pp. 320–378 (no date)*
The copy was written in *naskh* without vocalization. The copyist entered some corrections in the margin.

Mu: Mumbay, Mulla Firuz MS *86, folios 2b–33b (no date)*[39]
The copy was written in *naskh* without vocalization. The copyist entered quite a few corrections in the margin (sometimes with the mark "correct (صح)").

From catalogue entries,[40] the following Arabic manuscripts are also known as witnesses of this book, although I could not examine them:

(1) Kazimiya, the copy registered in Mahfuz's catalogue no. 176 (copied in 868 AH)[41]
(2) Tehran, Tehran University MS 8412/1 (copied in the tenth century AH)
(3) Tehran, a manuscript in Nasiri Private Library (copied in 1057 AH)[42]
(4) Qom, Mashhad Astan Quds MS Hayat 520/2 (copied around the eleventh century AH)
(5) Cairo, Dār al-Kutub MS Fādil mīqāt 158m (copied around the eleventh century AH)

---

37 See Derayati, *Fehrestwara-ye*, 1:796.

38 Based on Mach's estimation; see Rudolph Mach, *Catalogue of Arabic Manuscripts (Yehuda Section) in the Garrett Collection in the Princeton University Library* (Princeton, NJ: Princeton University Press, 1977), 423.

39 Edward Rehatsek, *Catalogue Raisonné of the Arabic, Hindustani, Persian and Turkish Manuscripts in the Mulla Firuz Library* (Bombay: Managing committee of the Mulla Firuz library, 1873) 43–44 records that this copy has no date; however, Sezgin, *Geschichte*, 6:249 notes the date sixth century AH.

40 See Sezgin, *Geschichte*, 6:248–249; Rosenfeld and Ihsanoğlu, *Mathematicians*, 118; Derayati, *Fehrestwara-ye*, 1:796.

41 See Sezgin, *Geschichte*, 6:249.

42 See Sezgin, *Geschichte*, 6:248.

(6) Qom, Mashhad Shaykh Alī Haydar MS 77/32 (copied in 1243 AH)
(7) London, British Library MS ADD 7490/11 (copied around the seventeenth century CE)
(8) London, British Library MS DEL AR 1665/b (copied around the seventeenth century CE)
(9) Princeton, Princeton University Library MS (Yehuda collection) 4382 (copied in 1925)[43]
(10) Cairo, Dār al-Kutub MS miqat 895/2 (no date)
(11) Cairo, Azhar MS falak 50 (no date)
(12) Tehran, Majlis shurā MS 3281 (no date)

Since the beginning and ending of the text do not include its title, this work has not a definitive title. As far as I know, none of the manuscripts gives the title except MS P1, where the copyist adds the above-mentioned front page containing the title "Book of the Astrolabe." Thus, I use it as the title.

As for the textual relationship between these manuscripts, there are two families according to the contents of section one, chapters fifteen to seventeen:

Family α: MSS G, A, P1, S, C, Se, M2, and Mu;
Family β: MSS P2, Te, Ma, M1, and Y.

I have already published the whole Arabic text of this work by using MSS P1, P2, C and Y.[44] In this edition, I presented the texts of both families; however, the fact that Family α contains the oldest four manuscripts confirms that Family α represents the early stage of the transmission of K-*Book*, whereas manuscripts in Family β are affected by later modification. Hence, in this paper I establish the text by using the four oldest manuscripts from Family α: MSS G, A, P1, and S.[45] When some manuscripts have a different reading, I record it at the bottom of the text. Throughout this paper, I use the following notation for denoting variant readings (also given in note 4):

[Separates reading in the text from the variant
: Separates variant and manuscript sigla
+ Added in

---

43 This is a modern copy of Dār al-Kutub MS Fāḍil mīqāt 158; see Mach, *Catalogue*, 423.

44 The Arabic text edited by me with a Japanese translation is included in Mohammad Bagheri ed., *Risālah-yi Usṭurlāb-I Kūshyār Gīlānī* (Tehran: Miras-e Maktoob, 2014).

45 I am planning to edit this work based on these four manuscripts with an English translation.

–	Missing from
( )	My comments
أ	MS A
ب	MS P1
جـ	MS G
ص	MS S
شا	مشطوب (crossed out in)
فا	فوق السطر في (above the line in)
ها	هامش (margin)

The variant readings show the proximity of MS A to MS P1 against MSS G and S. Here we must remark that section four, chapters ten to eleven are missing in MSS G, S, C, and M2; this confirms the closeness of MS G and MS S. Since the tables of contents of MSS G and M2 after the introduction indicate section four "in ten chapters,"[46] while those of MSS S and C mention section four "in twelve chapters,"[47] the original of section four obviously had twelve chapters just like in MSS A, P1, and so on. Therefore, I will give priority to the readings from the group of MSS A and P1.

Appendix B: Arabic Manuscripts of Abharī's Treatise on Knowing the Astrolabe

The following two Arabic manuscripts of this treatise are known to me:

F: Paris, Bibliothèque nationale MS arabe 2544, folios 21b–26b (copied in 1009 AH = 1600/1 CE).
Ca: Istanbul, Carullah MS 1468, folio 1b (no date).

The codex BnF MS 2544 is a *majmū'* (collection) of 15 treatises on astronomical instruments. Item 5 is a copy of Abharī's *Treatise*.[48] The beginning of the copy (folio 21b) gives the title "Treatise on Knowing the Astrolabe":

Beginning:

46 MS G: folio 363b, line 7; MS M2: p. 335, line 5.
47 MS S: folio 39a, line 10; MS C: folio 164b, line 20.
48 The detailed information about this codex, see De Slane, *Catalogue*, 455–6.

بسم الله الرحمن الرحيم
هذه رسالة فى معرفة الاصطرلاب لمولانا الامام اثير الدين الابهرى
نوّر الله مضجعه وهو اربعة عشر باباً

(In the name of God the Merciful and Compassionate.
This is a treatise on knowing the astrolabe by our master and leader Athīr al-Dīn al-Abharī, God illuminate him by resting him, which is in fourteen chapters.)

It was completed in 1009 AH in Larende, namely Karaman in Turkey, as recorded in the colophon (folio 26b).

Colophon:

الرساله بعون الله الملك
الوهاب فى سنه سبع والف
من الهجرة النبوية
فى لارنده
المحروسه

(The treatise is by the help of God the King and Donor in the Hijra and Prophetic year 1009 in Larende, the protected city.)

It was written in *naskh* with occasional vocalization. The copyist did not enter any corrections. Another hand entered one correction in the margin of folio 25b.

Ca: Istanbul, Carullah MS *1468, folio 1b* (*no date*)[49]
As Krause notes, this copy has only a part of chapter one. The rest of this codex (folios 2a–16b) is an extraction from anonymous *zīj*s copied by other two hands.

Beginning:

بسم الله الرحمن الرحيم
هذه رساله في معرفه الاصطرلاب لمولانا الامام اير الدين نوّر الله مضجعه وهو اربعة عشر بابا

(In the name of God the Merciful and Compassionate.
This is a treatise on knowing the astrolabe by our master and leader Athīr al-Dīn al-Abharī, God illuminate him by resting him, which is in fourteen chapters.)

49 Krause, "*Stambuler Hanschriften*," 493 comments that folio 1b is by a modern hand.

As one can see from the beginning, the copy is very similar to MS F. By comparing MSS F and Ca, I cannot find any significant variant between them.

I also found a copy of a recession of this work, described below.

Mi: Tehran, Millī MS arabī 1171, pp. 273–282 (copied around the tenth century AH)[50]
The major part of this codex (pp. 1–272) is a copy of Abharī's *Utmost Attainment in Comprehending the Orbs*, his largest astronomical work.[51] The last part of it (pp. 273–282) is a treatise on the operation of the astrolabe. A comparison between this copy and MS F reveals that this is a recession of the text transmitted in MS F. Whereas MS F has fourteen chapters, a redactor divides some chapters, so that this version has sixteen chapters.

From catalogue entries,[52] there exist at least three manuscripts of this recension, although I have had no opportunity of examining them:

(1) Cairo, Dār al-Kutub MS Taymur riyada 165/3
(2) Qom, Mashhad Astan Quds MS Hayat 700/2 (copied around the eleventh century AH)
(3) Qom, Mar'ashī MS 4060/1 (no date)

In this article, I establish the text by using MS F, the only manuscript containing the entire number of chapters. When I emend a reading, I note the original at the bottom of the text, where I denote MS F as "ف" with the same notation defined in Appendix one of this paper; in the text, words in < > are additions made by me to restore the text.

Appendix C: Kūshyār's introduction to Astrolabic Parts: Text and Translation

As already described, Kūshyār's introduction consists of two parts: on the purpose of using the astrolabe, and on astrolabic parts and plates. To compare

50 Based on the estimation by Derayati, *Fehrestwara-ye*, 7:749.

51 On the importance of this work in the history of Islamic astronomy, I gave a paper "*Summary of the Almagest*: Another Astronomical Tradition in the Days of Naṣīr Dīn Ṭūsī and its Significance in the History of Solving Problems Concerning the Diurnal Motions," SAW Seminar "Exploring mathematical practices around an astronomical question one: Mathematical approaches to the Diurnal motions of the Sun, the Stars, and the Moon," on 21 February, 2014 in Paris.

52 See Rosenfeld and Ihsanoğlu, *Mathematicians*, 209; Derayati, *Fehrestwara-ye*, 1:797.

Kūshyār's explanation of the parts and plates with A-*Treatise*, chapter one, I will present the text of the second part of K-*Book*'s introduction[53] with an English translation.[54]

فالأمّ[1] هي الصفيحة الظاهرة التي فيها العروة وحلقة العلاقة يعلّق[2] بها الأسطرلاب[3] . وعلى ظهرها خطّان يتقاطعان عند مركز الصفيحة[4] على زوايا قائمة ، أحدهما يخرج من عند العلاقة فيقطع الصفيحة بنصفين يسمّى[5] خطّ العلاقة ، والآخر يسمّى خطّ المشرق والمغرب .

والحجرة[6] هي الحلقة المركّبة على باطن هذه الصفيحة مقسومة ثلاث مائة وستّين جزءاً[7] ، وابتداء قسمتها من عند خطّ العلاقة[8] . وسمّيت حجرة لأنّها تتضمّن الصفائح كلّها والعنكبوت .

والعضادة[9] هي التي على ظهر الأمّ . ولها رأسان محدّدان ، وعليها دفّتان مركّبتان[10] في كلّ واحدة[11] منهما ثقبة صغيرة[12] .

وأجزاء الارتفاع[13] هي الأجزاء التسعون التي عن إحدى جنبتي العلاقة . وربّما كان الربعان[14] اللّذان[15] عن جنبتيها[16] كلّ واحد منهما مقسوماً[17] بتسعين جزءاً[18] .

وأجزاء الظلّ[19] هي الأجزاء التي في الربع المقابل لأجزاء الارتفاع على ظهر الأمّ .

والخطّان المتقاطعان في الصفائح[20] : أمّا الذي يخرج من عند العلاقة فيقسم[21] الصفيحة بنصفين فهو خطّ نصف النهار ، ويسمّى النصف الذي يلي العلاقة منه خطّ وسط السماء ، والنصف الآخر خطّ وتد[22] الأرض . وأمّا الذي يقطع هذا الخطّ بنصفين[23] على زوايا قائمة عند مركز الصفيحة فهو خطّ المشرق والمغرب .

والدوائر الثلاث[24] في الصفائح[25] : أصغرها مدار رأس السرطان ، وأوسطها مدار رأس الحمل والميزان[26] ، وأعظمها مدار رأس الجدي .

ودوائر المقنطرات[27] هي الدوائر المخطوطة في الصفائح محيطة بنقطة[28] ص ، منها[29] دوائر تامّة ومنها غير تامّة ، وفيما بينهما[30] مكتوب عدد أجزائها بحسب قسمة الأسطرلاب[31] : إن كان سدساً[32] فستّة ستّة ، وإن كان ثلثاً[33] فثلاثة ثلاثة ، وإن كان كان نصفاً[34] فجزآن جزآن ، وإن كان تامًّا فجزء جزء . وأوّل دوائر هذه المقنطرات أفق المشرق والمغرب[35] مكتوب عليه[36] المشرق والمغرب . والمقنطرات[37] التي تلي الأفق الشرقي إلى خطّ نصف النهار هي المقنطرات الشرقية ، والتي تلي الأفق الغربي إلى خطّ

53 MS A: folio 99a, line 7–folio 100a, line 10; MS P1: folio 3a, line 10–folio 4b, line 4; MS G: folio 362b, line 22–folio 363b, line 2; MS S: folio 38a, line 5–folio 39a, line 4.

54 The whole text of the introduction can be found in Bagheri ed., *Risālah-yi*, 115–117.

نصف النهار هي المقنطرات[38] الغربية . وما بين مركز الصفيحة وبين أوّل المقنطرات على خطّ[39] وتد[40] الأرض هو عرض بلدان[41] الصفيحة .

وخطوط الساعات الزمانية[42] هي الخطوط[43] التي عن جنبتي وتد الأرض إحدى نهايتيها[44] مدار رأس السرطان والأخرى[45] مدار رأس الجدي ، وفيها تقويس قليل . وخطّ وتد الأرض مأخوذ معها .

والعنكبوت[46] هو[47] الصفيحة العليا[48] المخرّقة[49] ، وفيها شظايا محدّدة .

ودائرة نطاق البروج[50] هي الدائرة التامّة في العنكبوت عليها[51] أسماء البروج وأجزاؤها وأقسامها بحسب صنعة الأسطرلاب[52] ستّة ستّة أو ثلاثة ثلاثة أو جزآن جزآن أو جزء جزء كما[53] قلنا[54] في المقنطرات .

والشظايا المحدّدة[55] في العنكبوت[56] هي الكواكب[57] الثابتة ، وعلى كلّ واحد منها اسم الكوكب[58] .

ومري الأجزاء[59] هو[60] الزيادة التي في رأس الجدي تعدّ[61] به أجزاء المجرة[62] .

وما سوى ذلك[63] كالقطب والفلس والفرس والعروة وحلقة[64] العلاقة والشطبة النابتة[65] عن وجه العنكبوت لإدارة العنكبوت[66] ، فكلّها ظاهر الفعل وسبب الاحتياج إليها بيّن لا يحتاج إلى تفسيرها[67] .

[1] فالأمّ] الام فالام : جـ [2] يعلّق] أ = تعَلق : ب = يعلق : جـ = يعلف : ص [3] الأسطرلاب] ب ، جـ = الاصطرلاب : أ ، ص [4] الصفيحة] –ب [5] يسمّى] تسمى احَدَنحما : ب [6] والمجرة] المُجره والمجره : جـ [7] وستّين جزءاً] وستون قسمًا : جـ . جزءاً] + وفيها العروه وحلقه العلاقه ايضا : جـ ، ص [8] العلاقة] + الى الحهه اليمنى : هاص (مع رمز «ص») [9] والعضادة] العضاده والعضاده : جـ [10] مركّبتان] مركان : جـ [11] واحدة] أ ، ب = واحد : جـ ، ص [12] صغيرة] + وتسمى الهدفان : هاص (مع رمز «ص») [13] وأجزاء الارتفاع] اجزا الارتفاع واجزا الارىفاع : جـ [14] الربعان] المربعان : جـ [15] اللّذان] ص ، جـ = الّذان : أ [16] العلاقة ، وربّما كان الربعان اللّذان عن جنبتيها] –ب [17] مقسوماً] مقسوم : ب [18] جزءاً] قسما : ب [19] وأجزاء الظلّ] اجزا الظل واحزا الظل : جـ [20] والخطّان المتقاطعان في الصفائح] الخطان المتقاطعان في الصفايح والخطان المىقاطعان ى الصفايح : جـ [21] فيقسم] فقسم : جـ [22] وتد] («نصف» مشطوب) وتد : جـ [23] بنصفين] أ ، ب = –جـ ، –ص [24] الثلاث] الثلثه : أ [25] والدوائر الثلاث في الصفائح] الدواير الىلث في الصفايح والدواير اللث ى الصفايح : جـ [26] والميزان] –ص [27] ودوائر المقنطرات] دواير المقىطرات ودواير المصطرات :

جـ 28 بنقطة] ب ، جـ = («ننصفن» مشطوب) سقطه : أ = حول ىمطه : ص 29
منها] فمنها : ب 30 بينهما] أ ، ب = ىىها : جـ ، ص 31 الأسطرلاب] الاصطرلاب :
ص 32 سدساً] سُدسيّا : أ 33 ثلثاً] ثلثا : أ 34 نصفاً] نصفيّا : أ 35 والمغرب] + وجهه
المسرق والمغرب : جـ ، ص 36 عليه] علها : ص 37 والمقنطرات] فالمقنطرات : ب 38
الشرقية ، والتي تلي الأفق الغربي إلى خطّ نصف النهار هي المقنطرات] –أ 39 خطّ]
+ نصف النهار : هاص (مع رمز «ص») 40 وتد] («و» زائد تحت السطر في نسخة
ص) 41 بلدان] أ ، ب = بلد : جـ ، ص 42 وخطوط الساعات الزمانية] خطوط
الساعات زمانيه وخطوط الساعات الزمانيه : جـ 43 الخطوط] –جـ 44 إحدى نهايتيها]
أ = احدي نهايته : ب = من : جـ = ىىدى عند : ص 45 والأخرى] ب = الاخر :
أ = وىنتهى عند : جـ ، ص 46 والعنكبوت] العنكبوت : جـ 47 هو] هي : ب 48 العليا
] أ ، ب = الاعلى : جـ ، ص 49 المخرّقة] + وهى الشكه : هاص (مع رمز «ح») 50
ودائرة نطاق البروج] دايره نطاق البروج ودايره نطاق البُروج : جـ 51 عليها] وعليها :
جـ 52 الأسطرلاب] ب ، جـ = الاصطرلاب : أ ، ص 53 كما] وكلّما : أ 54 قلنا] أ ،
ب ، جـ ، هاص = –ص 55 المحدّدة] المحدُودَة : أ 56 والشظايا المحدّدة في العنكبوت]
الشطايا المحدده والشطايا المحدده فى العنكبوت : جـ . في العنكبوت] –أ 57 الكواكب
] أ ، ب = للكواكب : جـ ، ص 58 الكوكب] كوكب : جـ 59 ومري الأجزاء]
مري الاجزا ومرى الاحرا : جـ 60 هو] ص = هى : أ ، ب ، جـ 61 تعدّ] يُعدّ : جـ 62
المجرة] + عند الحاجه الىها : ص 63 ذلك] + من هذه الالات : جـ = + من الالاب
: ص 64 وحلقة] والحلَقه : أ 65 النابتة] الناينة : أ = النايته : ب = النابته : جـ = النايته
: ص 66 لإدارة العنكبوت] لادارته : هاص = –ص 67 تفسيرها] أ ، ص = التفسير
: ب = تفسير : جـ .

(The mater is the outside plate having the handle and the ring for the cord from which the astrolabe hangs. On the back, there are two lines cutting each other at the center of the plate perpendicularly: one of them is drawn from the place of the cord and halves the plate, which is called the cord line; and the other is called the east-west line.

The rim is the ring set on the inside of this plate, being divided into 360 degrees.[55] The beginning of its division is on the cord line. It is called the rim because it embraces all the plates as well as the rete.

55 MSS G and S add: "It has also the handle and the ring for the cord."

The alidade is that which is on the back of the mater. It has two pointed edges. There are two sighting vanes erecting on it, each of which has a small hole.

The altitude degrees are the ninety degrees on one side of the cord; sometimes there are two quadrants on both sides of the cord, each of which is divided into ninety degrees.

The shadow degrees are the degrees in the quadrant opposite to the altitude degrees on the back of the mater.

The two lines cutting each other on the plates: As for that which is extended from the cord and halves the plate, it is the meridian line; the half [of the line] next to the cord is called the line of the midheaven, and the other half is [called] the line of the cardine of the earth. As for that which halves this line perpendicularly at the center of the plate, it is the east-west line.

The three circles on the plates: the smallest is the circuit of the head of Cancer; the middle is the circuit of the heads of Aries and Libra; and the largest is the circuit of Capricorn.

The almucantar circles are the circles drawn on the plates around point C; some of them are complete circles, whereas others are incomplete; between them, the numbers of their degrees are written according to the astrolabe's calibration: if it is a one-sixth [astrolabe], every six degrees are [written]; if it is a one-third, every three degrees are [written]; if it is a half, every two degrees are [written]; and if it is a complete [astrolabe], every degree is [written].

The first of these almucantar circles is the eastern and western horizon,[56] and "East" and "West" are written on it. The almucantars adjacent to the eastern horizon towards the meridian line are the eastern almucantars, and the [almucantars] adjacent to the western horizon towards the meridian line are the western almucantars. What is between the center of the plate and the first almucantar on the line of the cardine of the earth is the terrestrial latitude of the region [supposed in] the plate.

The lines of seasonal hours are the lines on both sides of the cardine of the earth, [each of which has] one edge [on] the circuit of the head of Cancer and the other [on] the circuit of the head of Capricorn. They are slightly bent. The line of the cardine of the earth is taken with them.[57]

The rete is the upper hollowed plate, in which there are sharpened pointers.

The circle of the zone of the zodiacal signs is the complete circle on the rete. On it are the names of the zodiacal signs and their degrees and divisions

56 MSS G and S add: "and the directions of east and west."

57 The line of the cardine of the earth is taken as one of the lines of seasonal hours.

according to the construction of the astrolabe: per six degrees or three degrees or two degrees or one degree, as we mentioned in [the section on] the almucantars.

The sharpened pointers in the rete are the fixed stars. On each of them is the name of a star.

The indicator of the degrees is the index on the head of Capricorn, by which the degrees of the rim are counted.

As for other [parts] such as the pin, the small ring, the horse, the handle, the ring for the cord, the small branch protruding from the rete's front[58] for rotating the rete, the function of all of them is obvious and the reason why they are needed is clear, not requiring explanation.)

Appendix D: Abharī's Treatise on Knowing the Astrolabe, Chapter One: Text and Translation

The following is the text of A-*Treatise*, chapter one,[59] with an English translation.

أمّ الصفائح هي الصفيحة التي يحيط بها حلقة مقسومة بثلاث مائة وستّين جزءاً تسمّى[1] هذه الحلقة بالمجرة .

والزيادة التي فيها من الجانب الأعلى تسمّى[2] الكرسي .

والتي يتعلّق بها الأصطرلاب من الحلقة يقال لها العروة .

خطّ نصف النهار هو[3] الخطّ الذي يأخذ من الجانب الكرسي إلى الجهة المقابلة .

خطّ الاستواء هو خطّ مقاطع لخطّ نصف النهار على زوايا قائمة .

والعنكبوت هي الصفيحة المخروقة .

منطقة البروج الدائرة المقسومة بالبروج على العنكبوت .

المري هو الزيادة التي في رأس الجدي .

مدار <رأس> الجدي هو[4] الدائرة التي يدور[5] عليها رأس الجدي وهي أعظم المدارات في الصفيحة في الأصطرلاب الشمالي .

مدار رأس السرطان هو[6] الدائرة التي يدور عليها رأس السرطان .

58 A small handle on the rete.

59 MS F: folio 21b, line 4–folio 22a, line 12.

دائرة معدّل النهار هي الدائرة التي يدور عليها رأس الحمل والميزان .
دائرة الأفق الدائرة التي هي بنقطتي التقاطع بين معدّل النهار وخطّ الاستواء .
المقنطرات هي الدوائر التي فوق الأرض .
والعمود الذي يقطع العنكبوت على نقطتي الحمل والميزان يسمّى عمود الحمل والميزان .
خطوط الساعات هي الخطوط التي تقع[7] بين معدّل النهار ومدار المنقلبين تحت الأفق .
والذي في مركز الصفائح من العنكبوت يقال له القطب .
والذي فيه القطب يسمّى الفلس .
العضادة مسطرة من جوهر الأصطرلاب على ظهره عليها هدفتان فيها ثقبتان .
دائرة الارتفاع هي الدائرة المخطوطة على ظهر الأصطرلاب محيطة به .
أجزاء الارتفاع على ظهر الأصطرلاب في محيط دائرة الارتفاع .

[1] تسمّى] يسمّى : ف [2] تسمّى] يُسمّى : ف [3] هو] وهو : ف [4] هو] هى : ف [5] يدور] تدُور : ف [6] هو] هى : ف [7] تقع] يقع : ف .

(The mater of the plates is the plate surrounded by a ring divided into 360 degrees; this ring is called the rim.

The addition to it in the upper side is called the throne.

The part of the ring from which the astrolabe hangs is called the handle.

The meridian line is the line taken from the side of the throne to the other side.

The line of the equator is the line cutting the meridian line perpendicularly.

The rete is the incised plate.

The belt of the zodiacal signs is the circle on the rete divided into the zodiacal signs.

The indicator is added to the head of Capricorn.

The circuit of the head of Capricorn is the circle on which the head of Capricorn rotates, being the largest circuit in the northern astrolabe's plate.

The circuit of the head of Cancer is the circle on which the head of Cancer rotates.

The equator circle is the circle on which the heads of Aries and Libra rotate.

The horizon circle is the circle passing through the two intersection points between the equator [circle] and the line of the equator.

The almucantars are the circles above the earth.

The column cutting the rete at the two points Aries and Libra is called the column of Aries and Libra.

The lines of hours are the lines between the equator and the circuits of the tropics below the horizon.

The part of the rete in the center of the plates is called the pin.[60]

That in which there is the pin is called the small ring.[61]

The alidade is the ruler belonging to the astrolabe's body on the back; on it are two sighting vanes having a hole.

The altitude circle is the circle drawn on the astrolabe's back, which embraces it [i.e., the back].

The altitude degrees on the astrolabe's back are in the circumference of the altitude circle.)

---

60 Note that the pin is a separate component of the astrolabe, rather than a part of the rete.

61 The small ring is the one lying between the horse and the rete, preventing the latter from being damaged.

CHAPTER 12

Changing the Angle of Vision: Astrolabe Dials on Astronomical Clocks

Günther Oestmann
Institut für Philosophie, Literatur-, Wissenschafts- und Technikgeschichte, Technische Universität Berlin, Straße des 17. Juni 135, 10623 Berlin, Germany
oestmann@nord-com.net

Abstract

From the middle of the fourteenth century until the Early Modern period, several monumental astronomical clocks were erected in Europe, and on many of them astrolabe dials were placed. On a group of earlier clocks, "southern astrolabes" (i.e. with stereographic projection from the North Pole) were employed, whereas later examples show a "northern astrolabe" (i.e., a stereographic projection from the South Pole), which is commonly used on portable astrolabes. The material and textual evidence as well as reasons for this change shall be examined. Moreover, the question of transmission of special variants of stereographic projection from East to West will be discussed.

Keywords

Astronomical clocks – astrolabes – astrolabe dials – stereographic projection

Introduction

The Earth's axis points towards the Pole Star, and the stars circulate once every twenty-four hours in an anticlockwise direction (Figure 12.1). According to the pattern of the "celestial dial," this should logically be the case on clock dials also. But the hands of clocks and watches move "backwards," because the sense of rotation has been adopted from the astrolabe, which preceded the mechanical clock as a device for time measuring.[1]

1 Henri Michel, "Pourquoi les horloges tournent-elles à l'envers?" *Journal suisse d'horlogérie* 71 (1946): 127–134; Siegfried Wetzel, "Astrolabium, Uhr und Uhrzeiger-Sinn," *Chronométrophilia* 53 (2002): 45–53.

 | DOI:10.1163/9789004387867_015

FIGURE 12.1 *Sense of rotation of the celestial sphere for a terrestrial viewer (Henri Michel, "Pourquoi les horloges tournent-elles à l'envers?," 127).*

Rotating star maps show the celestial sphere from the interior according to our natural impression (Figure 12.2). But in case of the astrolabe the location of the viewer is outside the cosmos, and he looks down from the northern celestial pole to the plane of the celestial equator (Figure 12.3): in his line of vision the celestial sphere rotates clockwise, and the positions of stars appear mirror-inverted (according to ancient practice celestial globes display this "outside perspective" also, i.e., figurative representations of the constellations are depicted in dorsal view).[2]

In portable astrolabes the tropic of Capricorn usually coincides with the outer circumference of the tympanum, and the centre of projection of the celestial sphere is located in the southern celestial pole (also denominated as "northern astrolabe"). However, the North Pole can also constitute the centre of projection, so that the tropic of Cancer is on the outside ("southern astrolabe") (Figure 12.4). If the astrolabist's traditional view from outside of the celestial sphere is abandoned in favour of a terrestrial angle of vision, the direction of rotation remains clockwise. The "southern astrolabe" (northern projection) with its convex horizon line can be found on astronomical dials of early monumental clocks erected until the mid-fifteenth century.

2 Elly Dekker, *Illustrating the Phaenomena: Celestial Cartography in Antiquity and the Middle Ages* (Oxford: Oxford University Press, 2013), 34–35.

FIGURE 12.2 *Rotating star-map of the nineteenth century* (*Modern reproduction,* © ASTROMEDIA VERLAG, NEUSTADT IN HOLSTEIN).

The clock in the cathedral of Lund (Figure 12.5) is one of those early examples, allegedly dating to 1380 (although there is no archival confirmation of this).[3] Then follow Doberan (1390) (Figure 12.6), Stralsund (1394), Berne, Lübeck[4]

3 Theodor Wåhlin, *Horologium mirabile Lundense: Det Astronomiska Uret i Lunds Domkyrka* (Lund: C. W. K. Gleerup, 1923), 178–179.

4 Gustav Schaumann, Friedrich Bruns and Fritz Hirsch, *Die Marienkirche* (= *Die Bau- und Kunstdenkmäler der Freien und Hansestadt Lübeck, II.2*) (Lübeck: Nöhring, 1906), pp. 248–56.

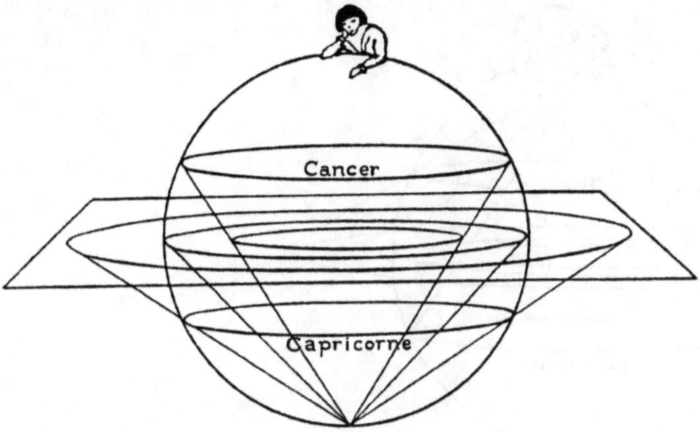

FIGURE 12.3 *Stereographic projection from the south pole–"Northern Astrolabe" (Henri Michel, Traité de l'astrolabe [Paris: Librairie Alain Brieux, 1976], 17).*

(both 1405), Chartres (1407),[5] Prague (1410),[6] Wismar (ca. 1421),[7] and Bourges (1423–1424) (Figure 12.7).[8] The astronomical clock of Villingen (1401, destroyed in 1714) very likely had an astrolabe dial with northern projection, too. This can be concluded from a certificate crediting the clockmaker Claus Gutsch with the construction of an astronomical dial which showed the course and seasonal height of the Sun:

> We, the Sheriff, Mayor and Council confirm publicly, that master Claus Gutsch from Rottweil [...] has made a calendar disk and a sphere [an astrolabe] for our Cathedral in Villingen [...], on which the Sun rises and

5 Adolphe Lecocq, "Notice historique et archéologique sur les horloges de l'église Notre-Dame de Chartres," *Mémoires de la société archéologique d'Eure et Loir* 4 (1867): 328–31.

6 Zdeněk Horský and Emanuel Procházka, "Pražký orloj," *Acta historiae rerum naturalium nec non technicarum* 9 (1964): 83–146.

7 Manfred Schukowski, "Die astronomische Uhr in Wismars Marienkirche," *Wismarer Beiträge: Schriftenreihe des Archivs der Hansestadt Wismar, H.* 13 (1998): 21–33.

8 Emmanuel Poulle, *Un constructeur d'instruments astronomiques au XVe siècle: Jean Fusoris (= Bibliothèque de l'École des Hautes Études, IVe section, Fasc. 308)* (Paris: Librairie Honoré Champion, 1963), 27–40.

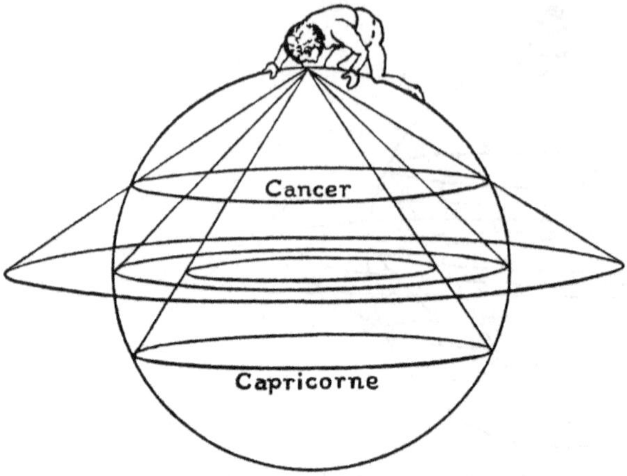

FIGURE 12.4 *Stereographic projection from the north pole—"Southern Astrolabe" (Michel, Traité de l'astrolabe, 16).*

FIGURE 12.5 *Astrolabe dial of the astronomical clock in Lund Cathedral in its condition before 1916 (Wåhlin, Horologium mirabile Lundense, 45).*

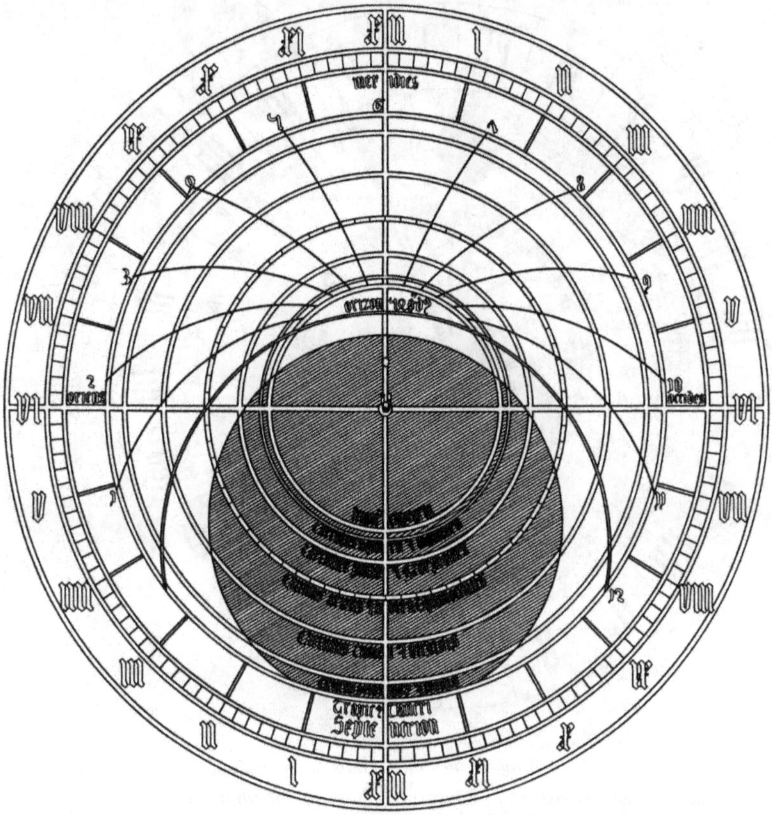

FIGURE 12.6 *Tympanum of the astrolabe dial of Doberan, 1390. Drawing by Hans Wåhlin, 1923; (Wåhlin, Horologium mirabile Lundense, 80).*

sets and moves high and low according to the time of the year, and the Moon rises and sets and rises and sets and is waxing and waning [...].[9]

9 "Wir der Schulthaiss, der Burgermaister vnd der rautt gemainlich der Statt ze Vilingen kunden meniglichem mit disem brieff vnd vergehen offenlich, daz der erber man maister Claus Gutsch von Rotwil, zöger dis briefs vns in vnserm münster ze Vilingen ain Sper geordnot vnd gemachot hat, die also wiset vnd betutet, alz hie nauch an disem brieff geschriben stat: Dez ersten ain rade, daz des jars ainost vmb gat, vnd ist daz kalendarium daran vnd die zwölf manot gemalot mit den geberden, alz denn ze mal gewonhait ist ze tund; vnd sicht man die lengi zwuschent wihennechten vnd vasnaht allu jar, vnd sitzt ain mennly dauor, daz zaiget vff alle tag durch das jar, vnd vindet man alle virtage in dem jar. Ob dem rad ist die sper geordnot, daran ist bezaichent luft, wasser vnd ertrich vnd Sunne vnd man vnd die zwölf zaichen, vnd gat die sunne vff vnd ze gnaden vnd hohet vnd nidret nauch dem, alz es in dem iare ist; der man gat ouch vff vnd ze gnaden vnd hohet vnd nidert in ie dem schin alz vil, alz

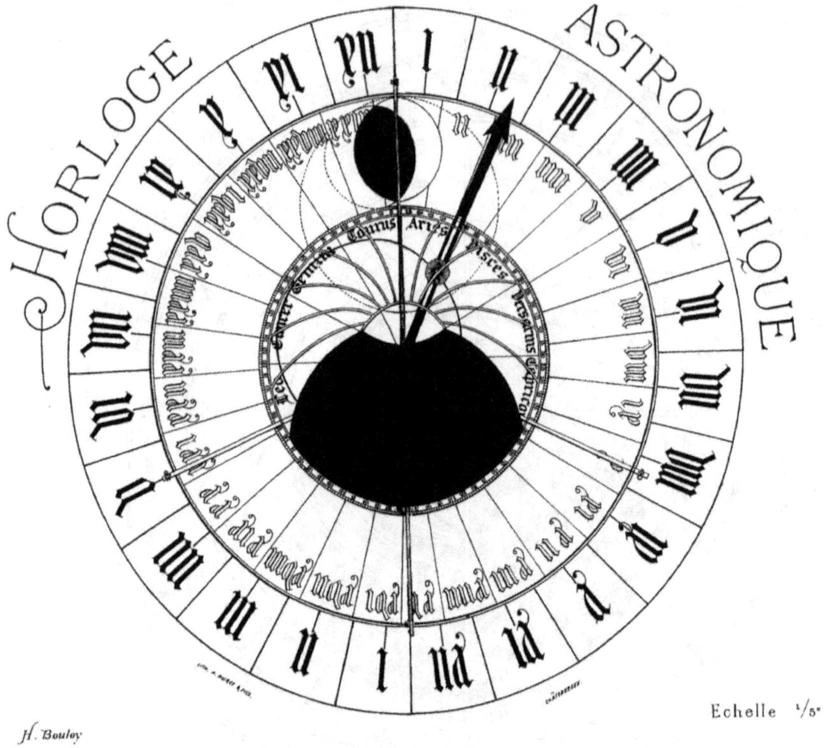

FIGURE 12.7 *Astrolabe dial of Bourges, 1423. (Jules-Louis Barreau, Description de la cathédrale, des vitraux de Bourges et des autres monuments de la ville [Châteauroux: A. Majesté, 1885], n.p.).*

Shortly before (in 1398) the magistrates of Rottweil had sent a request to the councillors of Strasbourg to show the calendar disk of the astronomical clock

---

die sunne ains jars tut, vnd lait zu vnd nimpt ab an dem schin vnd wirt nuwe in der stund, alz man liset, daz er an dem himel nuwe sig; vnd stand beide, Sunne vnd man, taeglichen in irem staffel vnd zaichen, alz man liset, daz si an dem himel standin; vnd sicht man in der sunnen alle stunden dez tags vnd der naht, vnd sint darob die hailgen drig kung vnd drig kneht, die gand vmb zu ainer tur vss vnd zu der andren tur wider in; wisent vnser frowen vnd kerent sich vmb gen ir vnd nigent ir, vnd kert sich daz kindli vmb gen inen, vnd schwept ain engel ob inen mit ainem sternen, vnd stand zwein engel vff dem gehus vnd blasent, so die kung wellent gan; vnd ist ain positiff daran, daz singt, die wil si gant. Dis ordnung vnd dis werk hat der obgenant maister Claus mit holtzwerk wol geziert. Vnd dis alles zu ainem waren vnd stätem vrkund so haben wir vnser Statt Insigel offenlich gehenkt an disen brieff, der ze

in Strasbourg Cathedral to Gutsch.[10] Hence it can be assumed that the first clock of Strasbourg Cathedral (which had been erected in 1352–1354) served as a model for the clock in Villingen.

The astronomical clock in Frankfurt Cathedral, completed in 1372–1373, was largely rebuilt in 1384 by Johann Schnitzler from Haguenau in Alsatia. It is likely that he also adopted the first Strasbourg clock as a model.[11] On an engraving of 1612 showing the interior of the cathedral the astronomical clock is visible in the background, but only a tripartite structure (calendar plate, astronomical dial, and automata) can be discerned from the depiction (Figure 12.8). The clock was demolished in 1855, and no other pictorial records or descriptions are extant.

But there are at least two exceptions. The astronomical clock constructed by Richard Wallingford (1292–1336) in the monastery of St. Albans had an astrolabe with southern projection. It can be concluded that Wallingford followed the ancient tradition of the anaphoric clock[12] which also had southern projections (this is proved by archaeological findings).[13] The "Astrarium" of Giovanni

---

Vilingen geben ist an dem nechsten fritag nauch dem maye tag, do man zalt nauch Cristy geburt vierzehenhundert Jare vnd darnauch in dem ersten Jare" (Villingen-Schwenningen, Municipal Archive, MS GG 4a).

10 "Vnser williger dienst stevor. Besundern lieben fründ, als ir in uwerm münster ze Straszburg hand daz jar rade ist vns geseit, daz daz ze mäl ain hübsch gut werke sig, vnd daz wir ouch ze mäl gern hettent in vnser Statt Rotwil. Vnd vf daz so habent wir zu ûch geschikt Clausen Gutschen, zöger des briefs, dasselb werck ze beschowen vnd darinne rat ze suchen, das vns daz ouch gemachet werde. Darvmb uwer guten fruntschaft bitten wir mit ernst vnd flissentlich, daz ir schaffen wellent gegen dem Custos oder schafner oder gen wenne dz notdurftig sige, daz man den egenennten Clausens Gutschen daz egenant uwer werk aygenlich beschowen lasse vnd daz Im darinne beraten vnd geholffen werde, vnd daz Im weg darzu geben werdint daz er vns dasselbe werk ouch vollbringen muge. Daz wellent wir in allen sachen allezijt willetlich vmb uch gedienen. Geben am mitwochen noch sant Geonen [?] tag Anno dni. M.CCC.LXXXX.VIII. Von vns dem Rat ze Rotwil" (Archives de la Ville et Communauté urbaine de Strasbourg, Sér. II, 46, Nr. 32).

11 Ernst v. Bassermann-Jordan, "Die einstige Monumentaluhr im Dome zu Frankfurt am Main," *Die Uhrmacherwoche: Handelszeitung für den Uhren- und Bijouteriewaren-Markt* 36 (1929): 441; Igor A. Jenzen, "Die Monumentaluhr im Frankfurter Dom," in: *Uhrzeiten: Die Geschichte der Uhr und ihres Gebrauches*, ed. Igor A. Jenzen (Marburg: Jonas, 1989), 40–41.

12 John David North, "Monasticism and the First Mechanical Clocks," in *The Study of Time II: Proceedings of the Second Conference of the International Society for the Study of Time*, ed. Julius Thomas Fraser and Nathaniel Lawrence (Berlin: Springer, 1982), 388; John David North, *Richard of Wallingford* (Oxford: Clarendon Press, 1976), 1:486–489; 2:342.

13 On anaphoric clocks of Late Antiquity see L. Maxe-Werly, "Note sur les objets antiques découvertes à Gondrecourt (Meuse) et à Grand (Vosges)," *Mémoires de la Société Nationale*

FIGURE 12.8 *The astronomical clock in the northern transept of Frankfurt Cathedral. Detail of an engraving in the "Crowning Diary" of Emperor Matthias, 1612 (Carl Wolff, Der Kaiserdom in Frankfurt am Main: Eine baugeschichtliche Darstellung, [Frankfurt a. M.: Jügel, 1892], 22).*

de'Dondi (1318–1389) built in 1348–1364 also had an astrolabe dial with southern projection commonly used on portable astrolabes[14] (Figure 12.9).

The southern projection has been applied on astrolabe dials of monumental clocks from the early sixteenth century onwards: Winterthur (1529),[15] Münster (1540–1542),[16] Saint-Omer (1555–1558) (Figure 12.10), Ulm (1581), Olomouc (1573–1575),[17] Strasbourg II (1571–1574) (Figure 12.11), Augsburg (1594),[18] and Lyon (1598).[19]

The change from northern to southern projection was recognized first by Theodor Wåhlin (1864–1948), the architect of the Cathedral of Lund,[20] and in 1967, when the Czech science historian Zdeněk Horský (1929–1988) presented a typology of astronomical dials.[21] This change was the result of the growing

---

des Antiquaires de France 48 (1887): 170–178; Otto Benndorf, Ernst Weiss, and Albert Rehm, "Zur Salzburger Bronzescheibe mit Sternbildern," *Jahreshefte des Österreichischen Archäologischen Instituts in Wien* 6 (1903): 32–49; Marcel Nordon, "Sur un objet en bronze découvert à Grand en Lorraine en 1886," *Bulletin d'ANCAHA* 57 (1990): 27–42; Anthony John Turner, "The Anaphoric Clock in the Light of Recent Research," in *Sic itur ad astra: Studien zur Geschichte der Mathematik und Naturwissenschaften. Festschrift für den Arabisten Paul Kunitzsch zum 70. Geburtstag*, ed. Menso Folkerts and Richard Lorch (Wiesbaden: Otto Harrassowitz, 2000), 536–547.

14 *Giovanni Dondi dall'Orologio: Tractatus Astrarii, Biblioteca Capitolare di Padova, Cod. D. 39*, ed. Antonio Barzon, Enrico Morpurgo et al. (Vatican City: Biblioteca Apostolica Vaticana, 1960), fols. 7^{v}, 10^{v} of the facsimile.

15 Adolf Schenk, *Die Uhrmacherfamilie Liechti von Winterthur und ihre Werke* (Winterthur: Schönenberger, 1970), 26–29.

16 See the postscript to this essay.

17 Günther Oestmann, "Die astronomische Uhr am Rathaus zu Olmütz–Olomouc," in *Mittelalterliche astronomische Großuhren: Internationales Symposium in Rostock 25. bis 28. Oktober 2012*, ed. Manfred Schukowski, Uta Jahnke and Wolfgang Fehlberg (Leipzig: Akademische Verlagsanstalt, 2014), 125–154.

18 Ernst v. Bassermann-Jordan, "Die verlorene Monumentaluhr im Dome zu Augsburg," *Die Uhrmacherkunst* 48 (1929): 181.

19 François Branciard and Charles Morat, *L'horloge astronomique de la Cathédrale Saint-Jean de Lyon* (Saint-Étienne: Dumas, 1993).

20 Theodor Wåhlin, "Astrolabe Clocks and Some Thoughts Regarding the Age and Development of the Astrolabe," in Robert Theodore Gunther, *The Astrolabes of the World: Based upon the Series of Instruments in the Lewis Evans Collection in the Old Ashmolean Museum at Oxford, with Notes on Astrolabes in the Collections of the British Museum, Science Museum, Sir J. Findlay, Mr. S. V. Hoffman, the Mensing Collection, and in other Private Collections* (London: The Holland Press, 1976), 540–541.

21 Zdeněk Horský, "Astronomy and the Art of Clockmaking in the Fourteenth, Fifteenth and Sixteenth Centuries," *Vistas in Astronomy* 9 (1967): 25–34.

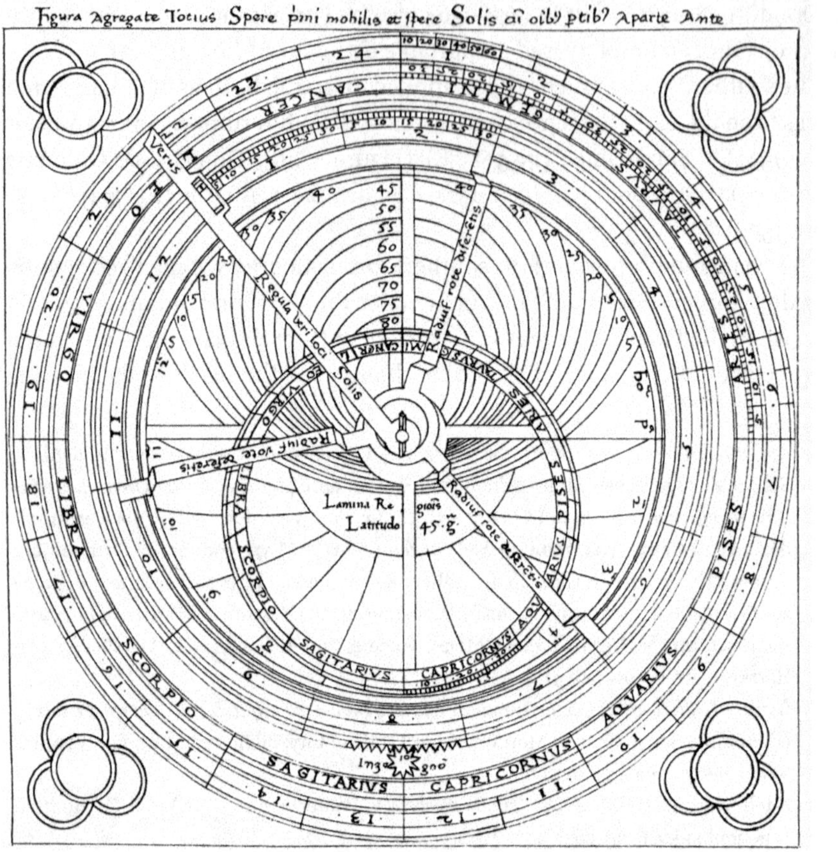

FIGURE 12.9 *Astrolabe dial of Giovanni de'Dondi's "Astrarium," 1348–1364* (*Granville Hugh Baillie, H. Alan Lloyd and F. A. B. Ward, The Planetarium of Giovanni de Dondi* [*London: The Antiquarian Horological Society, 1974*], *60*).

proliferation of portable astrolabes, in which the southern projection was always used. There is only one notable exception: a portable European astrolabe with northern projection, which originated in the thirteenth-century France. This instrument has a rete equipped with a gear train for representing the movements of sun and moon, but the engravings on the tympanum and on the back have been left unfinished. In all probability this instrument was derived from the dial of an astronomical clock.[22]

22 London Science Museum inv. no. 1880–32; Sharon Gibbs, Janice A. Henderson and Derek de Solla Price, *A Computerized Checklist of Astrolabes* (New Haven, CT: Department of History and Science and Medicine, Yale University, 1973), # 198; Gunther, *The Astrolabes of*

FIGURE 12.10 *The astronomical clock of Saint-Omer, 1555–1558 (historical postcard in possession of the author).*

To sum up, three variants of projection of the celestial sphere are to be distinguished:

---

the World, 347; John David North, "Opus quarundam rotarum mirabilium," *Physis: Rivista Internazionale di Storia della Scienza* 8 (1966): 369–371; Johann Hügin, "Astrolabium mit Räderwerk aus dem 13. Jahrhundert," *NUZ Neue Uhrmacherzeitung: Schmuck und Uhren* 10 (1978): 6–10.

FIGURE 12.11 *Astrolabe dial of the astronomical clock of Strasbourg Cathedral, 1571–1574.* © MUSÉE DES ARTS DÉCORATIFS DE STRASBOURG; PHOTO MUSÉES DE STRASBOURG.

- Type I has the centre of projection and position of the observer in the South Pole. The appearance of the sky is terrestrial (inside the celestial sphere), the outer limit of the projection is the tropic of Capricorn, the sense of rotation counterclockwise. It represents the "normal" view of the observer on the earth and is used for rotating star maps.
- Type II has the centre of projection and position of the observer in the North Pole. The appearance of the sky is also terrestrial (inside the celestial sphere), the outer limit of the projection is the tropic of Cancer, and the sense of rotation clockwise. This variant is found on astrolabe dials of early monumental clocks.

- Type III has the centre of projection in the South Pole, but the observer is in the North Pole. The appearance of the sky is mirror-inverted (external to the celestial sphere), the outer limit of the projection is the tropic of Capricorn, and the sense of rotation also clockwise. Contrary to type II, star positions between the tropic of Capricorn and the North Pole can be shown. It is encountered in the vast majority of portable astrolabes and late monumental astronomical clocks (i.e. from the early sixteenth century onwards).

In all three variants there is no difference in the accuracy and the ease of reading, but one may wonder why clockmakers deviated from the common astrolabe when constructing astrolabe dials and employed the unusual projection from the northern celestial pole? Reproducing the Sun's motion in the zodiac was of foremost importance when designing an astronomical dial, and this projection enabled the display of the annual course of the Sun in a most natural way, because here the daily arc above the convex shaped horizon line is longer in summer than in winter and thus corresponds to real conditions. If the projection center is located in the South Pole, the Sun runs through a short daily arc above the concave shaped horizon line in summer, and the winter arc is very long. But notwithstanding this advantage the older representation was discarded in the sixteenth century with the proliferation of the common planispheric astrolabe.

Is the northern projection a specific European development coinciding with the construction of the first mechanical clocks in the thirteenth to fourteenth century? It is well known that knowledge about stereographic projection and astrolabe construction mainly spread from Muslim Spain to Europe, and so the northern projection may be a transfer from Arabic-Islamic culture also.[23]

Al-Farghānī (ca. 820–after 861), al-Sijzī (second half of the tenth century) and al-Bīrūnī (973–1048) worked on new projection types and forms of the astrolabe. The construction of the southern astrolabe was described by al-Farghānī, but the projection probably already existed before his time.[24] The mathematician and astronomer al-Sijzī dealt exhaustively with the theory

---

23 According to Wåhlin, "Astrolabe Clocks," 542 and 547, the southern projection is older, and he postulated that the change to northern projection commonly used on portable astrolabes might be due to Hipparchus, but this is a very speculative assumption and rather unlikely.

24 Muḥammad ibn Kathīr al-Farghānī, *On the Astrolabe*, ed. and tr. Richard Lorch (Wiesbaden: Steiner, 2005), 341–361, 434–438; Josef Frank, "Zur Geschichte des Astrolabs," *Sitzungsberichte der physikalisch-medizinischen Sozietät Erlangen* 50–51 (1918–1919): 280.

of the astrolabe[25] and composed a treatise on mixed forms of the northern and southern astrolabe with a wide variety of shapes of the rete. The text however, is not extant.[26] Numerous constructions of "mixed astrolabes" were adopted by the polymath al-Bīrūnī,[27] who wrote a "Comprehensive Treatment of the Possible Methods of the Production of the Astrolabe,"[28] where he treated the "northern" and "southern" astrolabe[29] as well as many hybrid forms.[30] He also mentioned these in his "Instructions to the Elements of the Art of Astrology" composed in 1029 CE/AH 420.[31]

It may be assumed that the construction of those variants was only regarded as an intellectual challenge and exercise of mathematical theory. But it was not merely a "coquetterie géometrique," as Henri Michel once thought.[32] Though the utility of such instruments was limited and they were very difficult to handle compared to the ordinary astrolabe, we have occasional examples

---

25 Fuat Sezgin, *Geschichte des arabischen Schrifttums*, 17 vols. (Leiden: Brill (vols. 1–9); Frankfurt am Main: Institut für Geschichte der Arabisch-Islamischen Wissenschaften an der Johann Wolfgang Goethe-Universität (vols. 10–17), 1967–2000), 5:329–334 and 6:224–226; Max Krause, "Stambuler Handschriften islamischer Mathematiker," *Quellen und Studien zur Geschichte der Mathematik, Astronomie und Physik*, Abt. B, vol. III (1936): 437–532, at 468–469.

26 *Kitāb fī Qawānīn mizājāt al-asṭurlāb al-shimālī maʿ al-janūbī*, as quoted by al-Bīrūnī (Sezgin, *Geschichte des arabischen Schrifttums*, 6:226).

27 Sezgin, *Geschichte*, 261–276.

28 *Istīʿāb al-wujūh al-mumkina fī ṣanʿat al-asṭurlāb*; for a German translation of the introduction see Eilhard Wiedemann, "Einleitungen zu arabischen astronomischen Werken," *Das Weltall* 20 (1919): 24–6; also Heinrich Suter and Eilhard Wiedemann, "Über al Bīrūnī und seine Schriften," *Sitzungsberichte der phys.-med. Sozietät in Erlangen* 52–53 (1920–1921): 83.

29 *Maqāla fī tashīl al-tasṭīḥ al-asṭurlābi wa-l-ʿamal bi-murrakabātihī min al-shimālī wa-l-janūbī* (On an easily comprehensible account of the projection of the astrolabe and the use of the compound astrolabe made of the northern and southern astrolabe); Sezgin, *Geschichte des arabischen Schrifttums*, 6:269–270.

30 Eilhard Wiedemann, "Beiträge zur Geschichte der Naturwissenschaften, LXI: Allgemeine Betrachtungen von al Bīrūnī in einem Werk über die Astrolabien," *Sitzungsberichte der phys.-med. Sozietät in Erlangen* 52–53 (1920–21): 97–115; Suter and Wiedemann, "Über al Bīrūnī," 74; François Charette, *Mathematical Instrumentation in Fourteenth-Century Egypt and Syria: The Illustrated Treatise of Najm al-Dīn al-Miṣrī* (Leiden: Brill, 2003), 66–83. According to a quote of the teacher and patron of al-Bīrūnī, Abū Naṣr Manṣūr b. ʿAlī b. ʿIraq, the astronomer Ḥabash (living in Baghdad c. 900) had written a guide for the construction of the northern and southern astrolabe (*Ṣanʿat al-asṭurlāb al-shimālī wa-l-janūbī*; Sezgin, *Geschichte des arabischen Schrifttums*, 6:175).

31 *The Book of* Instruction *in the Elements of the Art of Astrology*, ed. and tr. Robert Ramsay Wright (London: Luzac & Co., 1934), 198.

32 Michel, *Traité de l'Astrolabe*, 69.

of their construction.[33] But the manuscripts on astrolabe construction by al-Farghānī, al-Bīrūnī, and al-Sijzī did not reach Europe, as far as is known.[34]

However, there is evidence for the transmission of special astrolabe types, namely the "southern" astrolabe. The Jewish scholar Abraham ben Meir ibn Ezra (1089–1092/1164–1167) lived in Spain until 1140, and on his subsequent restless *Wanderleben* he travelled to Italy, France, and England. He wrote three treatises on the astrolabe in Hebrew (1146 and 1148) and one in Latin (1154). In his Hebrew treatises Ibn Ezra is much concerned with the handling of a "southern" astrolabe in a northern latitude[35] and describes the different layout of tympana of "northern" and "southern" astrolabes.[36]

A Latin treatise on the astrolabe (*De forma astrolabii*), extant in two copies of the mid-twelft century (London, British Library: MS Cotton Vesp. A II 13, fols 37v–40v and MS Arundel 377, fols 63r–68v) has been attributed to Ibn Ezra by José M. Millás Vallicrosa (1897–1970).[37] This was at first questioned,[38] but recent research tends to confirm the authorship of Ibn Ezra.[39] Contrary to Millás Vallicrosa (who assumed London as place of composition), now Rouen,

33 Charette (n. 30), 73–74; Alain Brieux, "Notes sur un astrolabe nord-sud," *XIVth International Congress of the History of Science, Tokyo and Kyoto, Japan, 19–27 August, 1974*, 4 vols. (Tokyo: Science Council of Japan, 1974), 3:282–283; Anthony John Turner, *Astrolabes, Astrolabe-Related Instruments* (The Time Museum, Catalogue of Collection, Vol. I, Part 1) (Rockford: The Time Museum, 1985), 74–83.

34 Dominique J. Boilot, "L'Oeuvre d'al-Beruni: Essai bibliographique," *Mélanges de l'Institut Dominicain d'Études orientales du Caire* 2 (1955): 191–200; Francis J. Carmody, *Arabic Astronomical and Astrological Sciences in Latin Translation: A Critical Bibliography* (Berkeley–Los Angeles, CA: University of California Press, 1956), 113–116.

35 Josefina Rodríguez Arribas, "Medieval Jews and Medieval Astrolabes: Where, Why, How, and What For?" in *Time, Astronomy, and Calendars in the Jewish Tradition*, ed. Sacha Stern and Charles Burnett (Leiden: Brill, 2014), 226–227.

36 Rodríguez Arribas, "Medieval Jews", 232.

37 José M. Millás Vallicrosa, "Un nuevo tratado de astrolabio, de R. Abraham Ibn ᶜEzra," *Al-Andalus: Revista de las escuelas de estudios árabes de Madrid y Granada* 5 (1940): 1–29.

38 Raphael Levy, "The Authorship of a Latin Treatise on the Astrolabe," *Speculum* 17 (1942): 566–569.

39 Shlomo Sela, *Abraham Ibn Ezra and the Rise of Medieval Hebrew Science* (Leiden: Brill, 2003), 31–36; Renate Smithuis: "Science in Normandy and England under the Angevins: The Creation of Abraham Ibn Ezra's Latin Works on Astronomy and Astrology," in *Hebrew to Latin, Latin to Hebrew: The Mirroring of two Cultures in the Age of Humanism, Colloquium held at the Warburg Institute, London, October 18–19, 2004*, ed. Giulio Busi (Berlin–Torino: Nino Aragno Editore, 2006), 25, 31, 38–39.

Verona, or Mantova are considered.[40] The text follows Ibn Ezra's Hebrew treatises on the astrolabe very closely,[41] and in it "right" and "left" astrolabes are mentioned: the largest circle of the latter, which is more frequent, is the tropic of Capricorn, the smallest circle the tropic of Cancer and *vice versa.* In both variants the celestial equator is in the middle, i.e., between the tropics.[42]

This source confirms that knowledge of variations of the stereographic projection and special astrolabe constructions must have reached Europe from Islamic areas (probably Spain) in the twelfth century, thus providing essential knowledge for the construction of astrolabe clock dials.

Acknowledgement

This article is a revised and enlarged form of a paper given at the XXXIIIth Scientific Instrument Symposium of the Scientific Instrument Commission, Tartu, 2014.

Postscript

The astronomical clock of Münster Cathedral represents an exceptional case. It was constructed c. 1396–1400 by Fridericus, monk of the Cistercian monastery Hude near Bremen[43] and damaged during the sectarian government

40 Sela, *Abraham Ibn Ezra*, 36; Shlomo Sela and Gad Freudenthal: "Abraham Ibn Ezra's Scholarly Writings: A Chronological Listing," *Aleph: Historical Studies in Science & Judaism* 6 (2006): 40–42.

41 Sela, *Abraham Ibn Ezra*, 32; Rodríguez Arribas, "Medieval Jews," 249–250.

42 "Genera astrolabiorum duo sunt: dextrum et sinistrum. Numerus sinistrorum maior quam dextrorum est. Dupliciter ea discernimus. Primo quod in sinistris stelle fixe septentrionales intra zodiacum, australes extra figurantur. In dextris e contrario. Secundo uero quod in sinistris trium rotarum maxima est Capricorni, minima Cancri, in dextris e contrario. Equinoctialis in utriusque media scripta est": Millás Vallicrosa, "Un nuevo tratado," 9.

43 "Temporibus suis Fredricus monachus cisterciensis de Huda complevit horalogium in ecclesia monasteriensis In qua quidem horalogio habentur hore equales et inequales ascendens signum motus medii omnium planetarum regimine eorundem in horis et kalendarium cum festis mobilibus"; Chronicle of the Cistercian abbey Marienfeld; Staatsarchiv Münster: Msc. VII nr. 1340, fol. 17—see also Das Chronicon Campi s. Mariae in der ältesten Gestalt (1185–1422), ed. Friedrich Zurbonsen (= Münsterische Beiträge zur Geschichtsforschung, 5) (Paderborn: Schöningh, 1884), 59. The date 1408 given later

of Anabaptists with its iconoclasm in 1534/35.[44] In 1540 the Mathematician and printer Dietrich Tzwyvel, Joannes Aquensis (Johannes from Aachen), Minorite friar and minister of Münster Cathedral, and the blacksmith Nikolaus Windemaker joined for a refurbishment of the clock. The paintings were carried out by Ludger tom Ring the Elder and (probably) his son Hermann. In its present state the astrolabe dial has a southern projection connected with a rete constructed from the interior of the celestial sphere, and therefore its sense of rotation is counterclockwise. The tympanum of the astrolabe dial showing a map of the world (viewed from the interior of the globe) was painted in 1662.[45] A dendrochronological analysis showed that the trees used for the clock plate with its astrolabe dial were cut down between 1519 and 1542, and x-ray examinations covering large areas conducted before commencing an extensive restoration in 2017/18 revealed, that in 1540/42 originally a northern projection with lines for the temporal hours named literally counterclockwise had been applied.[46] The dial with its convex horizon line looked similar to the dial in Doberan (fig. 6) and was most likely a copy of the original painting of the late 14th century. Perhaps the desire for a proper orientation of the astronomical clock within the church building gave rise to provide for a counterclockwise rotational sense.[47]

(„Horologium opus istud artificiosissimum à monacho quodam Cœnobii Hudensis ordinis Cisterciensis Comitatus Delmenhorstensis Anno 1408 perfectum, malleis securibusque pulsant, singulari vero odio et furore Baptisterium, reliquiarumque repositoria comminuunt"; Hermann Kock, Series episcoporum monasteriensium, eorundemque vitae ac gesta in ecclesia 3 [Münster: Koerdinck, 1802], 30) is not confirmed by any original source, however.

44 On the sources see Max Geisberg, Die Stadt Münster (= Die Bau- und Kunstdenkmäler von Westfalen 41, part 5) (Münster: Aschendorff, 1937), 121.

45 Theodor Wieschebrink, *Die astronomische Uhr im Dom zu Münster* (Münster: Aschendorff, 19832), 36.

46 The investigation was made by Hans Portsteffen (Cologne Institute of Conservation Sciences) and restorer Marita Schlüter (Everswinkel).

47 For the dial of the astronomical clock of Berne see Markus Marti, *Das Astrolabium am Zeitglockenturm in Bern: Die Neuentdeckung eines Kulturguts von europäischer Bedeutung* (Berne: Stämpfli, 2018).

CHAPTER 13

Astrolabes for the King: The Astrolabe of Petrus Raimundi of Barcelona

Azucena Hernández
Department of Medieval Art History, Complutense University of Madrid, Ciudad Universitaria, 28040 Madrid, Spain
azucena.hernandezperez@gmail.com

Abstract

The astrolabe of Petrus Raimundi, made in Barcelona in 1375, occupies a significant position in the set of medieval Spanish astrolabes with Latin inscriptions, as it is the only one signed and dated that has survived to the present day. A full description and study of the astrolabe is presented in the context of the support given to the manufacturing of scientific instruments by King Peter IV of Aragon. Although the astronomical and time reckoning features of the astrolabe are fully detailed, special attention is given to its artistic and decorative features. The relationships between Petrus Raimundi's astrolabe and those manufactured in al-Andalus, the region under Islamic rule within the Iberian Peninsula in the Middle Ages, are highlighted, as well as the links with astrolabe production in other European Christian kingdoms. The role played by astrolabes in medicine is considered and first steps are taken towards discovering the identity of Petrus Raimundi.

Keywords

Dated astrolabes – King Peter IV of Aragon – Latin astrolabes – Petrus Raimundi

Introduction

The interest that political and religious powers took in astrolabes in the Middle Ages, and the aristocratic patronage of those who manufactured these fascinating artistic and scientific objects, explain the good condition of conservation and the significant number of them that have survived to the present day, despite their small size and the ease of reusing brass, the alloy of which they

 | DOI:10.1163/9789004387867_016

are made. The use of the astrolabe by astronomers cannot alone account for its proliferation in the Middle Ages; we must also consider the political, social, and technological aspects of the culture that produced such instruments.[1]

The astrolabe of Petrus Raimundi is a witness to the society of the second half of the fourteenth century in the kingdom of Aragon[2] under King Peter IV *el Ceremonioso* who was a supporter of science, mainly astronomy, astrology, medicine, and, above all, the manufacturing of scientific instruments.

The Astrolabe of Petrus Raimundi of Barcelona

The planispheric astrolabe of Petrus Raimundi presented here (Figure 13.1) occupies an important position among the medieval Spanish astrolabes with Latin inscriptions, as it is the only one signed and dated that has survived to the present day, as far as we are aware.[3] It bears an inscription engraved on the edge of the rim that gives the name of the maker or owner, and the year and the city of its manufacture. The inscription will be carefully analyzed below, but it is relevant to highlight now that Petrus Raimundi identifies himself as a "member of the House of the King of Aragon." The term "member of the House of the King" appears in the documents to identify individuals who lived where the king lived and therefore moved from one city to another with the king, made for him what he ordered and were paid from the royal court budget.[4]

---

1 A complete study of the fourty-nine extant astrolabes made in the Iberian Peninsula during the Middle Ages, both in Al-Andalus and the Hispanic Christian Kingdoms including the related political, social and technological aspects in Azucena Hernández, *Astrolabios en al-Andalus y los reinos medievales hispanos* (Madrid: La Ergástula, 2018) and *Catálogo razonado de los astrolabios de la España medieval* (Madrid: La Ergástula, 2018).

2 The Kingdom of Aragon under Peter IV included the present regions of Aragon, Catalonia, and Valencia in the Iberian Peninsula, plus the Balearic Islands and the southeast of France up to Montpellier.

3 The astrolabe is held in the Museum of Fine Arts in Boston (accession number 88654). It is identified as #3053 in the International Instrument Checklist Number established by David A. King (this list is reproduced in this special issue—see section two, #6, in King's article—and is also publicly available online at http://www.davidaking.org/instrument-catalogue TOC.htm). This astrolabe #3053 is not included in Derek J. Price's checklist, Derek J. Price "An International Checklist of Astrolabes," *Archives Internationales d'Histoire des Sciences* 34 (1955): 243–263, 363–380.

4 For example, The Valencian silversmith Guillem de Binaya is identified as "member of the House of the King" in two documents signed by King Peter IV in 1345 and 1346 for the payment for a silver astrolabe: "… un strelabi de argent que en Guillem de Bisaya, de casa sua, motla en Perpenya en lo present mes de octubre a ops del dit senyor rey …" ("… a silver astrolabe

FIGURE 13.1 *Astrolabe of Petrus Raimundi, front and back.*
PHOTOS BY THE AUTHOR WITH THE COURTESY OF THE MUSEUM OF FINE ARTS, BOSTON.

Various hypotheses emerge from this fact: either Petrus Raimundi was one of the astrolabe makers protected and paid by King Peter IV or he was a scholar (e.g., astronomer, astrologer, physician) who ordered the astrolabe to be made by a third person, for his own use or to be given to the king as a present. All these hypotheses will be taken into consideration within this paper, paying attention to the role that astronomy and astrolabes played in medieval medicine. The astrolabe was already mentioned by David A. King in 1995 and in 2004, including a brief description of the instrument, as one of the five medieval astrolabes made in the Kingdom of Aragon.[5]

cast by Guillem of Binaya, a member of his house. He cast the astrolabe in Perpignan in the present month of October at the request of the king ..."); "... fas vos saber que en G de Binaya, argenter de Valencia de casa del senyor rey, li es degut per [...] de i estalabre d'argent del senyor rey qui pesa v unzes quart ..." ("... we have a debt with Guillem of Binaya, silversmith in Valencia, member of the house of the king. The debt is due to [...] and one silver astrolabe for the king whose weight was five and a quarter onzas ..."). For more details see Antoni Rubio i Lluch, Documents per l'Historia de la Cultura Catalana Mig-eval, 2 vols. (Barcelona: Institut d'Estudis Catalans, 1908–1921), 2:79–80.

5 David A. King, "Astrolabis de la Catalunya medieval," in *La Ciència en la Historia dels Països Catalans*, ed. J. Vernet and Ramon Parés, 3 vols. (Valencia: Institut d'Estudis Catalans,

The astrolabe is made of gilt brass, has a diameter of 109 mm, a weight of 420 grams, and serves ten latitudes (from 0° to 66°N) in five plates engraved on both faces.[6] Its back bears the usual zodiac and calendrical scales and an additional scale with the twenty-eight houses of the moon for astrological purposes.[7] It has a decorative throne with a symmetrical pair of foliated volutes. Its rete (Figure 13.2) bears a complete equatorial ring that crosses the ecliptic, a rare feature in medieval astrolabes.

The Mater

The womb of the mater itself is not engraved. Its limb carries a scale graduated from 0° to 360° labelled for each 10° interval in Gothic numerals and subdivided for each 5° and then each 1°. The rim is damaged, and the parts of the scale from 356° to 360° and from 0° to 5° are missing. The digit 2 in the number 20 of the scale is engraved as a mirror image of a regular 2, or rather, as a mirror image of letter Z. This inversion occurs only once on the rim but more times on the back of the astrolabe. The digits 4, 5, and 7 are displayed in their standard medieval forms.[8]

2004–2009), 1:198–199; David A. King, "The Earliest Known European Astrolabe in the Light of Other Early Astrolabes," *Physis: Rivista internazionale di storia della scienza* 32 (1995): 359–404 at 380, 396; David A. King and Kurt Maier, "The Medieval Catalan Astrolabe of the Society of Antiquaries of London," in *From Baghdad to Barcelona. Essays on the History of the Islamic Exact Sciences in Honour of Prof. Juan Vernet*, ed. Josep Casulleras and Julio Samsó, 2 vols. (Barcelona: Universidad de Barcelona "Instituto Millás Vallicrosa," 1996), 2:673–718 at 697.

6 Full details of the diameter, thickness and weight of each piece of the astrolabe in Hernández, *Catálogo razonado de los astrolabios*, 362–369.

7 David A. King, "Astronomical Instrumentation in the Medieval Islamic World," in David A. King, *In Synchrony with the Heavens: Studies in Astronomical Timekeeping and Instrumentation in Medieval Islamic Civilization*, 2 vols. Vol. 1: *The Call of the Muezzin*. Vol. 2: *Instruments of Mass Calculation* (Leiden: Brill, 2004–2005), 2:52. King considers that the first mention of the procedure for the construction of solar and calendrical scales on the back of an astrolabe is found in the Treatise of the Astrolabe by the Andalusī astronomer Abū al-Ṣalt (ca. 1000); Julio Samsó, *Las Ciencias de los Antiguos en al-Andalus*, 2nd ed. (Almería: Fundación Ibn Tufayl de Estudios Árabes, 2011), 33. The presence of the solar and zodiac calendar scales on the back of an astrolabe establishes a bi-univocal correspondence between each date on the Julian calendar and the actual solar longitude.

8 Charles Burnett, "Indian Numerals in the Mediterranean Basin in the Twelfth Century, with Special Reference to the 'Eastern Forms,'" in *From China to Paris: 2000 Years Transmission of Mathematical Ideas*, ed. Y. Dold-Samplonius et al. (Stuttgart: Steiner, 2002), 237–288 at 241

FIGURE 13.2 *Rete of the astrolabe of Petrus Raimundi.*
PHOTO BY THE AUTHOR WITH THE COURTESY OF THE MUSEUM OF FINE ARTS, BOSTON.

The edge of the rim carries an elegant Latin inscription in Gothic script that includes the name and affiliation of the astrolabe maker or patron, place and date of manufacture, and the latitude and longitude of Barcelona. The inscription reads (Figure 13.3):

> HOC ASTRALABIUM BARCHINONE: ANNO XPI 1375 PER PETRUM RAIMŨDI: DE DOMO REGIS ARAGONUM: LATI BARCHINONE: 41: LONGI 39:

---

and 265; Richard Lemay, "The Hispanic Origin of our Present Numeral Forms," *Viator* 8 (1977): 435–459.

FIGURE 13.3 *Details of the inscription engraved in the edge of the rim of the astrolabe with the name of the author, date, place, latitude and longitude.*
PHOTOS BY THE AUTHOR WITH THE COURTESY OF THE MUSEUM OF FINE ARTS, BOSTON.

This may be translated as:

> This Astrolabe <made> in Barcelona, in the year of Christ 1375, by Petrus Raimundi of the House of the King of Aragon: at the Latitude of Barcelona 41, Longitude 39.

The city, Barcelona, and the year, 1375, ascribe this astrolabe to the Crown of Aragon under King Peter IV who ruled Aragon, Valencia, Catalonia, and Mallorca from 1336 until his death in 1387.[9] With regards to the geographic coordinates indicated in the inscription, the latitude 41° is the value assigned to Barcelona in the Middle Ages, whilst the longitude 39° requires clarification.[10] The present value for the longitude of Barcelona in relation to the Greenwich meridian is 2°12′E but Greenwich was not established worldwide as the prime meridian until 1844. Various places, such as Alexandria, Arin, Mecca, Paris, Cadiz, the Canary Islands, Naples or the Azores, were used as a zero point to measure longitude in the Middle Ages. The value 39° for the longitude of Barcelona probably refers to the water meridian, a prime meridian identified in *Andalusī* historical sources. Such a point corresponds to a position in the Atlantic Ocean, 17°30′W of the most westerly point in the Canary Islands.[11]

---

9 The designation of Peter IV as King of Mallorca took place several years before, in 1344.

10 King, *Astrolabis de la Catalunya medieval*, 198. King reads "Longitude 28" in the inscription rather than "Longitude 39" as engraved, probably due to the poor quality of the photographs he worked with. Consequently, the explanation given to justify the value 28° for the longitude of Barcelona is no longer applicable.

11 Julio Samsó, *Las Ciencias de los Antiguos en al-Andalus*, 90, 469–470. The position of the "water meridian" was set at 17°30′ west of the Canary Islands by Azarquiel in his Toledan Tables (ca. 1080 AD). For more details about this issue, see Mercé Comes, "The 'Meridian of the

Considering that the island Hierro is the westernmost island of the Canary archipelago and that the modern longitude of Roque del Guincho, the westernmost cape within the island Hierro, is 18°10′W, we get a longitude of 35°40′W for the *Andalusī* water meridian. Barcelona would have had a longitude related to that prime meridian of 37°52′E. Although the number in the inscription of the astrolabe is 39°, it differs by little more than 1° from the calculated value, an acceptable error. This astrolabe is thus a testimony to the use of the *Andalusī* water meridian to establish geographical longitudes in the Iberian Peninsula during the fourteenth century.

The Back

The back bears four altitude scales divided from 0° to 90° for every 5° and subdivided for each degree. Each quadrant is numbered from the east and west points to the south and north points. The two scales in the upper quadrants are labelled for each 5° and the digit 2 in numbers 20 and 25 is engraved as a mirror image of a letter Z as on the mater. The two scales in the lower quadrants are labelled for each 10° and the mirror image of digit 2 appears also in the number 20.

The usual zodiac and calendrical scales, non-concentric with each other, are the next ones to occur. The zodiac scale is divided for each 5° and subdivided for each degree and labelled for each 10°. The names of the zodiac signs are engraved as follows:

> ARIES TAURUS GEMINI CANCER LEO UIRGO LIBRA
> SCORPI9 SAGITARI9 CAṖCORN9 AQUARI9 PISCES

The engraver has replaced the ending in "-us" of four of the signs by a character that looks like the cipher "9" which is quite common within the repertoire

Water' in the Tables of Geographical Coordinates of al-Andalus and North Africa," *Journal for the History of Arabic Science* 10 (1994): 41–49; Mercè Comes "Tables of Geographical Coordinates and the Size of the Mediterranean According to the Astronomers of al-Andalus," in *Al-Andalus. The Scientific Legacy*, ed. Rodrigo Gutiérrez (Granada: Proyecto Sur de Ediciones, 1995), 22–36; E.S and M.H. Kennedy, *Geographical Coordinates of Localities from Islamic Sources* (Frankfurt am Main: Institut für Geschichte der Arabisch-Islamischen Wissenschaften and der J.W. Goethe-Universität, 1987), 67. Kennedy includes the value of the longitude for Barcelona related to other meridians but not to the water meridian.

of medieval abbreviations. The little dot over the "P" in Capricornus indicates also an abbreviation, namely "RI."[12]

The calendrical scale is divided for each five days and subdivided for each day and then labelled for each ten days (10–20–31 for the months with 31 days and 10–20–28 for February). The names of the months are engraved as follows:

> IANUARIUS FEBRUARI9 MARCIUS APRILIS MAYYUS IUNIUS IULIUS AUGUSTUS SEPTEMBER OCTOBER NOUEMBER DECEMBER

In these month names, the replacement of the ending in "-us" by a "9" has only been made in February. The name MAYYUS is an orthographical variant of the Latin name MAIUS.[13] The vernal equinox is set at 12 March.[14]

An additional scale has been included following the calendar. It carries the twenty-eight mansions of the moon labelled with the numbers one to twenty-eight. Since the lunar mansions are used more for astrological judgements, one might consider that the astrolabe's first owner practiced astrology. The first mansion corresponds to the twenty-fifth degree of Aries.

A double horary quadrant has been located in the upper central space, whilst the lower part features a double shadow square with scales divided into twelve digits and labelled for each three.

---

12 David A. King, "Rewriting History through Instruments: The Secrets of a Medieval Astrolabe from Picardy," in *Making Instruments Count: Essays on Historical Scientific Instruments Presented to Gerard L'Estrange Turner*, ed. R.G.W. Anderson, J.A. Bennett, and W.F. Ryan (Farnham: Ashgate Variorum, 1993), 59–60. During the Middle Ages the names of the signs were written in various ways, including numerous abbreviations. The abbreviation of Capricornus with a dot over the "p" is attested on various European medieval astrolabes such as the French astrolabes #202, #337 and the Dutch astrolabe #566, among others. For the abbreviation of "-us" see Adriano Capelli, *Dizionario di Abbreviature Latine ed Italiani* (Milano: Ulrico Hoepli, 1912), 25.

13 David A. King, "The Astrolabe Depicted in the Intarsia of the studiolo of Archduke Federico in Urbino," in *Astrolabes from Medieval Europe* (Farnham: Ashgate Variorum, 2011), X:21. The variant MAYUS for May is found on the calendrical scale featured on the back of astrolabe #493 preserved in the Museo di Storia della Scienza, Florence.

14 King, *Astrolabis de la Catalunya medieval*, 199. The vernal equinox in two astrolabes whose date and place of manufacture is attributed by King to the fourteenth century and the Kingdom of Aragon is set at 12 March (astrolabe #162 preserved at the Society of Antiquaries in London) and 11,5 March (astrolabe #416 preserved at the National Maritime Museum, Greenwich).

The Rete

The rete is unusual in design. Its distinctive feature is a complete equatorial ring that crosses the circle of the ecliptic at the two equinoxes, Aries and Libra (see Figure 13.2). The outer and equatorial rings of the astrolabe's rete both carry scales graduated from 0° to 360° labelled for each ten-degree interval, in Gothic numerals, and subdivided for each two degrees. The outer ring is damaged and the part of the scale from 356º to 360º is missing. The zero of the scale coincides with the zero degree of Capricorn in the outer ring and with the zero degree of Aries in the equatorial ring.[15] The star-pointers are flame-shaped, the equinoctial bar has a single counter-change outside the ecliptic, and the Gothic inscriptions are all carefully executed in bold script.

Each zodiac sign of the ecliptic is divided into three unlabelled ten-degree intervals and subdivided for each two degrees. The names of the zodiac signs are engraved as follows:

> ARIES , TAURUS , GEMIN , CANCE , LEO , UIRGO , LIB RA , SCORPI9 , SAGITARI9 , CAPRICORN9 , AQARI9 , PIS CES.

The engraver has also replaced the ending in "-us" of four signs by a "9". The reduced space available in the scale of the ecliptic results in the shortening of the names Gemini and Cancer to GEMIN and CANCE. The names LIBRA and PISCES are split in the middle with a space between the two parts to allow room for the joining of the ecliptic and the equatorial rings. The sign Aquarius has been engraved "Aqari9" which is one of the various medieval abbreviations for that name.[16]

The rete features twenty-two star pointers, with the star names inscribed in Gothic script. Table 13.1 identifies each of the stars.[17]

---

15 King, *Astrolabis de la Catalunya medieval*, 199. The scale on the outer ring measures RA in the way defined in Antiquity whilst the scale on the equatorial ring measures it in the present standard way.

16 King, *Rewriting History through Instruments*, 59. The abbreviation of Aquarius as "Aqari" or "Acari" is attested in various European medieval astrolabes such as the Spanish astrolabes #416, #1148 and #3622 and the French astrolabe #163. "Acarius" is the term used by Ramon Llull in his Treatise on Astronomy (1297) when he introduces the Zodiac signs: see Jordi Gayà, with Lola Badia, "Ramon Llull. Tractat d'Astronomia (segons el Ms. Add. 16.434 del British Museum)," in *Textos y estudios sobre Astronomía española en el siglo XIII*, ed. Juan Vernet (Barcelona: Universidad Autónoma de Barcelona, 1981), 205–323 at 222.

17 King, *Astrolabis de la Catalunya medieval*, 199. King provides the names of twelve of the twenty-two stars without identification; Burkhard Stautz, *Untersuchungen von*

TABLE 13.1 *Identification of the Stars in the Rete of the Petrus Raimundi Astrolabe*

No.	Star-name in the rete	Current star-name	Arab star-name	Identification
1	DENEBCAYTO	Diphda	*dhanab al-qayṭūs*	β Cet
2	PANTACAYTAC	Baten Kaitos	*baṭn al-qayṭūs*	ζ Cet
3	ALGETENAR	Achernar	*ākhir al-nahr*	α Eri
4	ALDEBARAN	Aldebaran	*al-dabarān*	α Tau
5	RIGIL	Rigel	*al-rijl*	β Ori
6	ALHABOR	Sirius	*al-ʿabūr*	α CMa
7	ALGOME	Procyon	*al-ghumayṣā*	α CMi
8	ALFARAR	Alphard	*al-fard*	α Hya
9	COR LEO	Regulus	*qalb al-asad*	α Leo
10	CORUUS	Gienah Corvi	*janāḥ al-ghurāb al-yaman*	γ Crv
11	ALGORAB	Algorab	*al-ghurāb*	δ Crv
12	AZIMECH	Spica	*al-simāk* (*al- aʿzal*)	α Vir
13	ALRAMECH	Arcturus	*al-rāmiḥ*	α Boo
14	BMENAX	Denebola	*dhanab al-asad*	β Leo
15	URSA	Dubhe	*ẓahr al-dubb al-akbar*	α UMa
16	ALAY	Alkaid	*al-qā'id banat al-naʿsh*	η UMa
17	ALPHAC	Alphecca	*nayyir al-fakka*	α CrB
18	UEGA	Vega	*al-wāqiʿ*	α Lyr
19	ATAYR	Altair	*al-ṭā'ir*	α Aql
20	DELFIN	Deneb Dulfin	*dhanab al-dulfīn*	ε Del
21	MIRAD	Mirach	*al-mi'zar*	β And
22	YOUD	Capella	*al-ʿayyūq*	α Aur

There are many construction marks engraved on the reverse of the rete, namely all the scale lines and some dots made to fix the position of the lines in the three rings (Figure 13.4).

mathematisch-astronomischen Darstellungen auf mittelalterlichen Astrolabien islamischer und europäischer Herkunft (Bassum: GNT, 1997), 81–110 and 238–273; Paul Kunitzsch, *Arabische Sternnamen in Europa* (Wiesbaden: Otto Harrassowitz, 1959), 65–224.

FIGURE 13.4 *Reverse of the rete of astrolabe of Petrus Raimundi with workshop construction marks.*
PHOTO BY THE AUTHOR WITH THE COURTESY OF THE MUSEUM OF FINE ARTS, BOSTON.

The Plates

The astrolabe has five plates engraved on both sides. They all bear altitude circles for every five degrees, azimuth circles for each ten degrees, and the lines of the twelve unequal hours. Neither the circles nor the lines are labelled.

The inscriptions on the plates read as shown in Table 13.2. The fifth plate features the stereographic projection for latitude 0°, the equator, on the side where the inscription EQUINOCCIUM stands, which is quite unusual (Figure 13.5). The other side serves latitude 66°N which corresponds to the limit of the inhabited world in the medieval sources. The last two plates were not meant for observation, but only for teaching.

The rest of the latitudes served by the astrolabe are quite common in the Spanish medieval astrolabes. The word "Valecia" appears rudely scratched below the well-engraved label "LA 39" and would have been added by a later user of the astrolabe (very likely living in Valencia) in order to select the plate for his city more easily (Figure 13.6). A scratch with the word "Paris" appears below the label "LA 48" in the corresponding plate.

TABLE 13.2 *The Inscriptions on the Plates*

Plate No.	Inscription
1a	LA 30
1b	LA 36
2a	LA 39 Valecia
2b	LA 40
3a	LA 41
3b	LA 42
4a	LA 45
4b	LA 48 Paris
5a	EQUINOC CIUM
5b	LA 66

FIGURE 13.5 *Detail of the plate for latitude 0° with the engraved word EQUINOC CIUM.* PHOTO BY THE AUTHOR WITH THE COURTESY OF THE MUSEUM OF FINE ARTS, BOSTON.

The possible cities intended to be served by the four plates are given in Table 13.3.[18]

18 David A. King, World Maps for Finding the Direction and Distance to Mecca. Innovation and Tradition in Islamic Science (Leiden: Brill, 1999), 27. It is worth noting that latitudes

FIGURE 13.6 *Detail of the word VALECIA scratched in the plate for latitude 39 of the astrolabe of Petrus Raimundi.*
PHOTO BY THE AUTHOR WITH THE COURTESY OF THE MUSEUM OF FINE ARTS, BOSTON.

TABLE 13.3 *Possible Cities Served by the Astrolabe*

Plate No.	City	Plate No.	City
1a	30° = Cairo	3a	41° = Barcelona or Zaragoza
1b	36° = Algeciras	3b	42° = Girona or Perpignan
2a	39° = Valencia	4a	45° = Toulouse
2b	40° = Toledo	4b	48° = Paris

The Remaining Elements of the Astrolabe

The alidade is counter-changed; it has no scale and each of the pinnulae has one open hole. The front ruler is identical in shape to half of the alidade.

The suspensory system is original and encompasses a remarkable throne fixed onto the rim at four points of contact. It presents a pair of symmetrical

30°, 36°, 41°, 45°, and 48° correspond to the central bands of five of the seven geographical climates of Antiquity and they were commonly included on medieval astrolabes.

foliated volutes whose shape is analyzed later in this paper. The handle is trefoil-shaped and fitted with a circular ring. The pin seems to be original whilst the wedge is a modern addition.

Brass Analysis Report

A full set of tests done on each piece of the astrolabe using the X-ray fluorescence (XRF) method confirms that all its elements are made of medieval brass.[19] The mater, the rete, and all the plates are gilt but it has been possible to analyze the brass under the gilt. The alloy for the mater, the rim, the rete, the throne, and two of the plates (those for latitudes 30°/36° and 41°/42°) is very similar, with around 83% copper, 15% zinc and less than 1% tin, lead, and silver. The other three plates (those for latitudes 39°/40°, 45°/48° and 66°/equinox) have a slightly higher percentage of zinc, with around 80% copper, 19% zinc, and less than 1% tin, lead, and silver. The additional pieces (alidade, ruler, pin, horse, handle, and ring) have widely different compositions with percentages of zinc from 0.24% to 21.2%.[20]

Decorative Dimension of the Astrolabe of Petrus Raimundi

Medieval astrolabes concentrate their ornamental elements in the rete and the throne. In Islamic al-Andalus and in the Christian kingdoms of the Iberian Peninsula, the decorations are moderate and reveal a preference for austerity. As already noted, the rete of the astrolabe of Petrus Raimundi is distinctive in design due to the presence of the complete equatorial ring which somehow determines the shape and location of the decorative elements, namely a full circle in the vertical axis and two semicircles in the horizontal axis. They are all placed inside the upper half of the equatorial ring.

The presence of circular frames as ornamental elements in the rete is quite common in the astrolabes from al-Andalus, which were the source of

---

19 The XRF analysis of each part of the astrolabe was made on 31 October 2013 by Michele Derrick of the Scientific Research Laboratory of the Museum of Fine Arts, Boston, MA, USA. My gratitude to Abigail Hykin and Rebecca Tilles for making the analysis report available to me.

20 Full details of the results of XRF analysis in Azucena Hernández, *Astrolabios en la España Medieval: de al-Andalus a los reinos cristianos*, (Madrid: Universidad Complutense de Madrid, 2017), 603–604.

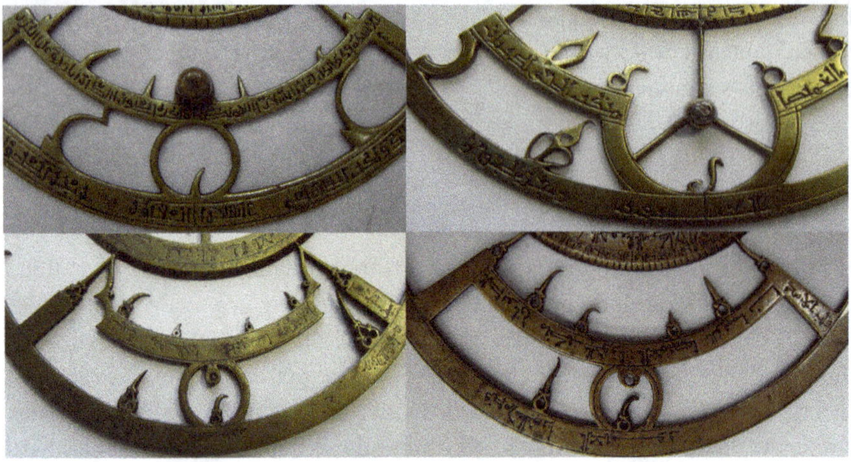

FIGURE 13.7 *Detail of decorative circles in Andalusí astrolabes #116 (top left), #2527 (top right), #129 (bottom left) and #136 (bottom-right).*
PHOTOS BY THE AUTHOR WITH THE COURTESY OF THE STAATSBIBLIOTHEK BERLIN AND THE MUSEUM OF THE HISTORY OF SCIENCE, OXFORD.

inspiration for the medieval astrolabes with Latin, vernacular, and/or Hebrew scripts manufactured in the Christian kingdoms of the Iberian Peninsula. As examples of such use of the circular frame it is worth mentioning astrolabes #116 (Berlin Staatsbibliotek), #2527 (Museum of the History of Science, Oxford), #136 (Spanish Royal Academy of History), and eight of the fourteen *Andalusī* Almohad astrolabes that Muḥammad ibn Fattūḥ al-Khamāʾirī made in Seville between 609AH (1212–1213 CE) and 634AH (1236–1237 CE), all of which feature a circle between the celestial equator frame below the ecliptic and the outer rim of the rete (Figure 13.7).

The interest in circles as decorative elements moved from al-Andalus to the Christian kingdoms of the Iberian Peninsula and the small circle below the celestial equator frame is present in astrolabes #3915, #4560, and #416, all unsigned but considered by David King to be Hispanic work of the fourteenth or fifteenth century.[21] Particularly relevant with regard to the use of circular

21 For astrolabe #3915, see Mohamed Abuzayed, David A. King and Petra Schmidl, "From a Heavenly Arabic Poem to an Enigmatic Judaeo-Arabic Astrolabe," *Suhayl* 10 (2011): 85–142 at 86 and 115. For astrolabe #4560, see David A. King, "An Astrolabe from Medieval Spain with Inscriptions in Hebrew, Arabic and Latin," in *In Synchrony with the Heavens*, ch. 15, at 2:845, and Sophie Makariou, "Ficha de astrolabio," in *Los mundos del Islam en la colección del Museo Aga Khan. Catálogo de exposición* (Barcelona: Fundación la Caixa, 2009), 92. For astrolabe #416 see King and Maier, *The Medieval Catalan Astrolabe*, 697.

FIGURE 13.8 *Retes of the astrolabe of Petrus Raimundi (left, photo by the author with the courtesy of the Museum of Fine Arts, Boston) and of Mamluk astrolabe #4164, Cairo Museum of Islamic Art nr. 15368 (right, photo courtesy of Alain Brieux, Paris).*

frames is astrolabe #416 (Greenwich Maritime Museum), which features two semicircles, one opening downwards and the other upwards, in the equinoctial bar between the ecliptic and the outer rim. This feature, together with the flame-shaped star pointers of this astrolabe, identifies it as the closest one, within medieval Spanish production, to the Petrus Raimundi astrolabe from the decorative point of view.

With regards to the two semicircles in Petrus Raimundi's rete, they both open downwards and are laid in the two segments of the equinoctial bar inside both the equatorial and the ecliptic rings. Among those studied by previous scholars, the only astrolabe that offers a similar feature is #4164 (Cairo Museum of Islamic Art, nr. 15368), an unsigned and undated Mamluk astrolabe examined by David King in 2005[22] (Figure 13.8). Whether or not the similarities between the Mamluk astrolabe and the one from Barcelona are a result of direct relationships between their respective manufacturers is under study. In any case, the intermediary role played by the Nasrid kingdom of Granada should not be underestimated.[23]

22 David A. King, "Some Astronomical Instruments form Medieval Syria," in *In Synchrony with the Heavens*, ch. 14b, at 2:705–709.

23 For relationships between astronomers from the Mamluk sultanate and those from the al-Andalus Nasrid kingdom, see Emilia Calvo, "Ibn Bāso's Astrolabe in the Maghrib and East," in *From Baghdad to Barcelona*, 2:755–767 at 763.

All of the star pointers in the rete of Petrus Raimundi's astrolabe are flame-shaped. Flame-shaped star pointers do not come from the *Andalusī* astrolabe tradition but became popular in Christian Europe in the fourteenth and fifteenth centuries.[24]

Our astrolabe has a throne with a remarkable design of curved leaves resembling volutes symmetrically placed and falling onto the rim with an elegant fishtail-shaped end. These ends seem to carefully hold and frame the inscription engraved on the edge of the rim. The front and back of the throne are identical in shape but the front bears more engraved lines. The design may be related to the *Andalusī* Aḥmad ibn Ḥusayn ibn Bāṣo astrolabe thrones[25] but with an additional point of sophistication, evolving from Ibn Bāṣo's geometrical patterns to a more naturalistic style of leaves and fishtail shapes (Figure 13.9). The circulation within the Crown of Aragon of the only treatise on the planispheric astrolabe written in vernacular Castilian, the *Libro del Astrolabio Llano* (one of the sixteen parts of the *Libro del Saber de Astrología* by King Alfonso X of Castile), explains the visual relationship between the throne depicted in the manuscript and the one in the Barcelona astrolabe, probably via Ibn Bāṣo's

FIGURE 13.9 *Throne of the astrolabe of Petrus Raimundi compared with the one of the andalusí astrolabe #132 of Aḥmad ibn Ḥusayn ibn Bāṣo.*
PHOTOS BY THE AUTHOR WITH THE COURTESY OF THE MUSEUM OF FINE ARTS, BOSTON, AND THE SPANISH ROYAL ACADEMY OF HISTORY.

24 The well-known French astrolabe maker Jean Fusoris (ca. 1365–1436) used flame-shaped star pointers in the retes of his astrolabes, and other unsigned and undated pieces also include these elegant star pointers.

25 In particular astrolabe #132 (Spanish Royal Academy of History) and #144 (Museum of Islamic Art, Doha, Qatar).

thrones.[26] The Petrus Raimundi throne design may have influenced future astrolabes such as the so-called "Urbino astrolabe" probably made in 1462.[27]

Having described and analyzed the astronomical and ornamental dimensions of the Petrus Raimundi astrolabe, it is time to approach the context that helps to understand why this object was made.

Historical and Social Context of this Astrolabe: King Peter IV of Aragon and the Astrolabe Workshops Patronized by Him

The development of science was significant in the Kingdom of Aragon during the fourteenth century. A whole dynasty of kings supported scientific research, focusing mostly on its practical dimension: the construction of scientific instruments. Among these, James II (1291–1327), Alphonse IV (1327–1336), Peter IV *el Ceremonioso* (1336–1387), John I (1387–1396), and Martin I (1396–1410) were the greatest promoters of instrument-making.[28] Additionally, they recorded most of the activities in their kingdom that had any economic impact, including the construction of astrolabes and other astronomical instruments. These documents, which have survived in large numbers, are an excellent source for the social and political context in which an astrolabe such as that of Petrus Raimundi might have been constructed.

The written sources identify three main centres of production of astrolabes in the second half of the fourteenth century in the Kingdom of Aragon: Mallorca, Barcelona, and Perpignan, with Mallorca being the most important. The throne supported the workshops and showered the manufacturers, both Jewish and Christian, with honours. Despite the good number of documented astrolabe makers and goldsmiths, the name of Petrus Raimundi has not yet been found.

---

26 King, *Astrolabis de la Catalunya medieval*, 198. King notes the relationship between the throne of Petrus Raimundi's astrolabe and that of the astrolabe depicted in the Libro del Astrolabio Llano, part of Alfonso X of Castile's *Libro del Saber de Astrología*, and extends this relationship to other elements in the rete.

27 For details of this astrolabe #4506 which was stolen in 1977 and whose current location remains unknown, see King, *The Astrolabe Depicted in the Intarsia*, 22–26.

28 Miguel Batllori Munné, "La cultura catalano-Aragonesa durant la dinastía de Barcelona 1162–1410: Corrents actuals de la investigación," in *VII Congreso de Historia de la Corona de Aragon*, ed. J. Zurita, 3 vols. (Barcelona: Universidad de Barcelona, 1962), 1:329–407.

Who was Petrus Raimundi?

As said before, research has not yet yielded a clear result on the identity of Petrus Raimundi, the person whose name is engraved on the edge of the rim of the astrolabe.[29] Nevertheless, a person with the same name, *Petrus Raimundi, magister in artibus et bacallarius in medicina* has been found in the archives as the author of an interesting manuscript of 1338, *Tractatus primus Rosae Philosophiae* (The First Treatise of the Rose of Philosophy'), which has survived in a copy made in 1490 (Figure 13.10).[30] The manuscript is kept in the Biblioteca Colombina y Capitular in Seville.

The *Tractatus* is a study of the relationship between the science of the stars and medicine, and its author, who declares himself to be a physician, demonstrates throughout the book knowledge of both astronomy (planets, moon, stars, comets, etc.) and medicine (the human body, diseases, etc.), two disciplines very closely linked in the Middle Ages.[31] The book was dedicated

29 I want to thank Dr Antonio Momplet for drawing my attention to Pere Ramon and for all his support.

30 Josep Perarnau i Espelt, "Nous autors i textos catalans antics: Pere de Puigdorfila, Fogatges, Guillem Aldomar, Pere Ramon," *Arxiu de textos catalans* 17 (1998): 540–569. Incipit: "Tractatus primus Rosae Philosophiae, factus et compilatus per magistrum Petrum Raymundi, magistrum in artibus et bacallarium in medicina, quem compilavit in villa Trempi." Explicit: "Compositus et compilatus fuit liber iste qui Rosa Philosophiae apellatur, a magistro Petro Raymundi, magistro in artibus et bacallario in medicina in loco de Trempo, anno Domini 1338, 13 die mensis januarii. Et presens scriptura fuit scripta per manum mei Jacobi Herla, studentis Balagarii et finita 14 die septembris anno a Nativitate Domini 1490." See Carmen Álvarez Márquez, "El itinerario de adquisiciones de libros de mano de Hernando Colón," *Historia, Instituciones, Documentos* 30 (2003): 55102 at 57, 99. Tractatus primus Rosae Philosophiae was one of the 15 344 registered books in the private library of Ferdinand Columbus (1488–1539), the second son of Christopher Columbus (1451–1506).

31 I present some quotations from the manuscript Seville, Biblioteca Capitular y Colombina, Ms 5-2-23 as examples: "Hic tractatur de quadam costellatione, que fuit non est diu, scilicet anno Domini M° CCC° XXX'V', scilicet de coniunctione Martis cum Saturno…." (f. 12r); "Hic ostenditur quot modi sunt nutrientium corpus nostrum. Hic ostenditur quem aspectum habet venenum ad corpus humanum, et quot sunt genera venenorum." (f. 18v); "Hic ostenditur quod a quatuor mundi partibus est concitatio ventorum et motus, ratione planetarum et signorum. Et primo, de planetis, post de signis …" (f. 32v);. "Capitulum octavum, in quo tractatur de stella comata, ubi ostenditur que est materia stelle comate, et que causa efficiens, et modus generationis ipsius, et quot sunt stele comate, et locus magis conveniens sue generationi, et regio, quia septentrionalis. Et an atestetur supra mortem principum, et magis principum quam aliorum. Et de significatione ipsius. Ultimo, de scintillatione stellarum, quare ipse scintillant" (f. 41r). All these references show that this is not a technical astronomical or astrological text; it is rather a cosmological or philosophical

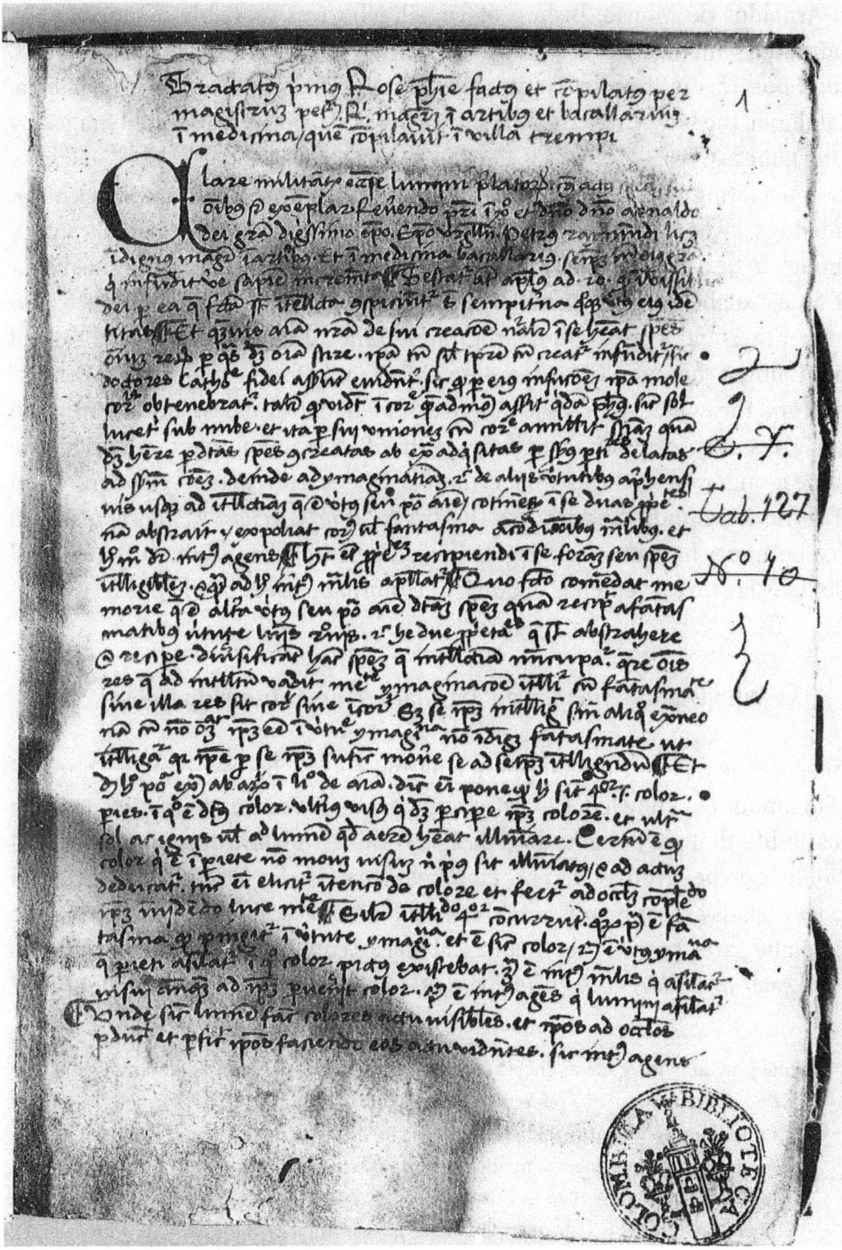

FIGURE 13.10 *First folio of Tractatus primus Rosae Philosophiae by Petrus Raimundi.* PHOTO COURTESY OF THE BIBLIOTECA CAPITULAR AND COLOMBINA OF SEVILLE, BCC MS 5-2-23.

to Arnaldus de Manso, Bishop of Urgell, who was probably a supporter of Raimundi's medical career and particularly of the production of this book. The book was written in Tremp, a small village eighty-five km from Lleida, Catalonia, the city where the first university of Aragon was founded in 1300 by King James II.[32]

Considering that Peter IV ruled his kingdom from 1336 to 1387, the book, dated 1338, was written during his reign. It is very unlikely that a man mature enough to be a physician and to write this book could have also been the maker of an astrolabe dated 1375, but it is possible that the two were relatives, perhaps father and son or grandfather and grandson. The Spanish tradition of maintaining the same Christian name in a family from generation to generation, and the evidence that the practice of medicine also passed from father to son, gives us some grounds to propose that Petrus Raimundi, the man whose name is engraved on the astrolabe, may have been a physician, son or grandson of Petrus Raimundi, the author of the *Tractatus primus Rosae Philosophiae* and that both may have got their degree in medicine from the Studium Generale of Lleida.[33] Further research is ongoing to confirm or reject this hypothesis.

Who Qualifies to Sign an Astrolabe? The Authorship Debate

The documents confirm that physicians and doctors had astrolabes. They had a fair amount of knowledge of astronomy, geometry, and mathematics, and the possibility that they themselves designed the astrolabes they commissioned should not be dismissed. The craftsmanship of the instruments required specific skills and only brass-smiths would have been able to manufacture the astrolabe parts, but the first theoretical phases could have been undertaken by a *magister in artibus et medicina,* the title held by Petrus Raimundi. It should

one providing the general encyclopedic knowledge needed for medieval medicine. The writer does not make any reference to astrolabes.

32 The Studium Generale of Lleida was founded in Lleida (Lerida) by King Jaime II of Aragon (1291–1327) and the studies of medicine took place there during the reign of King Peter IV. The two villages mentioned in this manuscript, Tremp, where Petrus Raimundus wrote his book, and Balaguer ("studentis Balagarii"), the birthplace of the student who made the copy, are located in the province of Lleida.

33 Astronomy, one of the seven liberal arts and medicine were taught in the Studium Generale of Lleida. For details see Jon Arrizabalaga, "Les Universitats," in *La Ciència en la Historia dels Països Catalans* 1:373–374; also Elías Serra Ráfols, *Una Universidad medieval. El Estudio General de Lérida* (Madrid: Librería General Victoriano Suárez, 1931), 42–44 and J.J. Busqueta (ed.), *Llibre de les Constitucions i Estatuts de l'Estudi General de Lleida* (Principio del formulario; Lleida: Universitat de Lleida, 2000).

be noted that an identical title was held by the French astrolabe maker, Jean Fusoris (1365–1436).[34]

The process of making an astrolabe involves at least three sets of skills:

- Skills for the stereographical projection of different circles, spheres, and points and for the graduation of the different scales that the astrolabe displays.
- Metalworking skills to cast or hammer the brass into circular sheet-metal plates and to make the rete and throne with their ornamental elements.
- Engraving skills to draw the lines into the brass pieces, keeping their correct position to the maximum extent in order to ensure the accuracy of the instrument, as well as writing the names and numbers properly.

Could a single person have enjoyed all these skills together in the Middle Ages? Perhaps, but probably not, for the skills are too diverse and would have been acquired at dissimilar institutions, each with specific degrees regulated by the appropriate guilds.

A fourth personality should be added to the already identified trio and this is the patron: the king, a courtier, a prelate, or a scholar who actually ordered the manufacture of the astrolabe and paid for it. The question then arises: whose name among the four people possibly involved in the astrolabe production should sign it, if any?[35] This question has already been asked by many art historians when studying the few medieval artworks that happen to have been

34 King John I of Aragon received in 1396 an astrolabe made by Jean Fusoris and the document which records the event refers to him as "dilecto nostro Johanni Fusorii in artibus et medicina magistro, Rubio i Lluch," *Documents* 1:84.

35 Most of the historiography of astrolabes links each astrolabe to a single person, the astrolabe maker, if the astrolabe is signed, or an anonymous person if it is unsigned. A single signature does not mean a single hand was involved in the whole process from the theoretical conception of the object to the final adjustments of the metal pieces, and this principle applies to any kind of manufactured object, whether simple or complex, whether scientific, artistic, or both. The division of labor in the manufacturing of astrolabes is supported by scholars (i.e., Robert B. Gordon, "Sixteenth-Century Metalworking Technology used in the Manufacture of two German Astrolabes," *Annals of Science* 44 (1987): 71–84 and B.D. Newbury, M.R. Notis, B. Stephenson, G.S. Cargill III and G.B. Stephenson, "The Astrolabe Craftsmen of Lahore and Early Brass Metallurgy," *Annals of Science* 63 (2006): 201–213). On the subject of medieval astrolabes manufacturing process with references to medieval documents see Azucena Hernández "Del concepto al dibujo y del trazo al latón. Manufactura de un astrolabio en un taller medieval," in *Afilando el pincel, dibujando la voz. Prácticas pictóricas góticas*, M. Miquel, O. Pérez and P. Martínez (eds.) (Madrid: Editorial Complutense) 199–220.

signed. A good example of how complex the establishment of authorship can be is the "Bobrinski Bucket" (Hermitage Museum, St Petersburg), a magnificent cast brass cauldron inlaid with silver and copper signed by four people: the patron who ordered it, the metalworker who chased it, the artist (artisan) who made the inlaid decoration and finally the person who received it as a present.[36] Which name should have been kept if not enough space for all the names was available? Who plays the most relevant role in the process?

The usual anonymity of art pieces applies also to the astrolabes and this is why this astrolabe signed by Petrus Raimundi is unique among those in Latin script from medieval Spain. The efforts to identify Raimundi and his role in the manufacture are ongoing. The presence of an astrological feature such as the scale with the twenty-eight mansions of the moon in a preferential position on the back of the astrolabe, suggests that its first owner was particularly interested in astrological predictions. All research directions remain active, but at the moment we are focusing our attention on the community of physicians in Aragon around 1375 rather than the metalworking guilds. Following this approach will, one hopes, yield a clearer answer.

Acknowledgements

This paper is dedicated to David A. King. My gratitude to Dr. Charles Burnett and Dr. Josefina Rodríguez Arribas for their invitation to submit this paper. The study of this astrolabe would have been impossible without the help of the Museum of Fine Arts of Boston, in particular of Abigail Hykin (curator), Michele Derrick (associate research scientist), and Marietta Cambareri (curator). This study is part of the Spanish R&D Project HAR2013–45578-R: "Al-Andalus, los reinos hispanos y Egipto: arte, poder y conocimiento en el Mediterráneo medieval. Las redes de intercambio y su impacto en la cultura visual," led by Susana Calvo and Juan Carlos Ruiz Souza. My thanks to them for their support and also to Gerry Lawrence, Herbert González, David Fisac and A.D. Morrison-Low for their help.

36 The inscription on its rim may be translated: "Ordered by ʿAbd-al-Raḥmān ibn ʿAbdallāh al-Rashīdī. Chased by Moḥammad ibn ʿAbd-al-Wāḥid. Decorated by Ḥājib Masʿūd ibn Aḥmad, the decorator [al-naqqāsh] in Herat for its owner, the exalted Ḵhwāja Rukn-al-Dīn, pride of the merchants, the most trustworthy of the Muslims, grace of the pilgrims and the two sanctuaries [the holy places Mecca and Medina], Rashīd-al-Dīn ibn ʿAzīzī ibn Abū-al-Ḥusayn al- Zanjānī, may his glory last." The handle provides the date "in the month of Muḥarran in year 559 AH." equivalent to December 1163 CE.

CHAPTER 14

A New Approach to the Star Data of Early Planispheric Astrolabes

Giorgio Strano
Museo Galileo, Piazza dei Giudici 1, 50122 Florence, Italy
g.strano@museogalileo.it

Abstract

During the twentieth century, scholars attempted to date early astronomical instruments by the scientific information discernable in their design. In particular, they devised "methods" to date planispheric astrolabes by the position of the vernal equinox on the calendar scale or by the precessional effect over the star coordinates. Such methods usually overlooked the many problems behind the instruments' design and construction and the vicissitudes in a particular object's history. Criticisms of the naiveté of the scientific approach did not persuade scholars with an astronomical background. They remained convinced of the supreme explanatory power that the present knowledge has for deciphering the past. The criticisms, instead, dissuaded scholars with a historical background from paying attention to the scientific data. They felt comfortable in analysing scientific relics on the basis of other types of evidence, such as textual sources, iconography, and style. The star coordinates of early planispheric astrolabes, as witnessed in their star pointers, can neither be regarded nor treated as the data of present science. Even if they do not describe the sky at the very moment of the making of the instruments, they may, nevertheless, be considered as non-textual sources that contain precious information. The star coordinates on the astrolabe retes may in fact connect groups of instruments with early star tables.

Keywords

Catalan astrolabe – Geoffrey Chaucer – Data interpretation – English medieval astrolabes – Pseudo-Messahalla star table

 | DOI:10.1163/9789004387867_017

Introduction

Planispheric astrolabes have fascinated both historians and scientists for centuries. The two scholarly disciplines, however, seldom interact fruitfully. Rather, they evolve specific methods for studying astronomical instruments. When the two scholarly groups meet, each party prefers its own "historical" or "scientific" methods of investigation and tries to convince the other party that these are the most reliable. In order to provide assistance, other people are occasionally involved in the studies, including art historians, philologists, materials experts, and 3D modellers.

In this article, I will focus on two traditional "scientific" methods for handling astrolabes' data. Notwithstanding their relevant drawbacks, they tend to reappear from time to time in scholarly studies. I will successively outline an intermediate approach to the astrolabes' star data. The new approach was tentatively developed during the comparative study of the Tomba-Koelliker astrolabe (Figure 14.1),[1] and five other early English "Chaucer-type" astrolabes.[2] The new approach starts from the gathering of star data from the instrument's pointers, as an astronomer would do, but uses such data to establish possible relationships with other available sources, as a historian or a philologist would do.[3] The preliminary results are encouraging enough to test the approach on

1 Tomba-Koelliker astrolabe, Museo Galileo: Institute and Museum of the History of Science, Florence, inv. no. 3931 (King no. 4521). For King's numbers see: David A. King, "An Ordered List of European Astrolabes to ca. 1500," in King, *Astrolabes from Medieval Europe* (Farnham: Ashgate, 2011), XII/1–11 (XII/7). See also: David King, *A Catalogue of Medieval Astronomical Instruments: Provisional Table of Contents* (Version May 2002), available online at http://www.davidaking.org/instrument-catalogue-TOC.htm (accessed 13 April 2015).

2 "Chaucer" astrolabe, dated 1326, British Museum, London, inv. no. MLA 1909, 6–17.1 (Gunther no. 291, Ward no. 325, EPACT no. 21); Painswick astrolabe, Museum of the History of Science, Oxford, inv. no. 47869 (Gunther no. 299, EPACT no. 67); Astrolabe, British Museum, London, inv. no. MLA 1914,0219.1 (Gunther no. 298, Ward no. 327, EPACT no. 91); Parnel astrolabe, Smithsonian National Museum of American History, Washington DC (Gunther no. 304); Gunther astrolabe, Museum of the History of Science, Oxford, inv. no. 49359 (King no. 4755, EPACT no. 40). For Gunther, Ward and EPACT numbers see respectively: Robert W.T. Gunther, *The Astrolabes of the World*, 2 vols. (Oxford: University Press, 1932), 2: 465–483; Francis A.B. Ward, *A Catalogue of European Scientific Instruments in the Department of Medieval and Later Antiquities of the British Museum* (London: British Museum, 1981), 112–113; EPACT, available online at http://www.mhs.ox.ac.uk/epact/catalogue.php (accessed 13 April 2015).

3 See Jim Bennett and Giorgio Strano, "The So-Called 'Chaucer Astrolabe' from the Koelliker Collection, Milan," *Nuncius: Journal of the Material and Visual History of Science* 29.1 (2014): 179–229 (205–220).

FIGURE 14.1 *The Tomba-Koelliker astrolabe, front (Museo Galileo: Institute and Museum of the History of Science, Florence, inv. no. 3931).*

other instruments as, for example, the Catalan astrolabe with a Chaucer-type rete, preserved at the National Maritime Museum of Greenwich.[4]

Dating Early Astrolabes by Astronomical Methods

Since the last century, professional and amateur scientists use the habit of dating planispheric astrolabes astronomically. Among the "scientific" methods to

4 Catalan astrolabe, National Maritime Museum (Caird Collection), Greenwich, London, inv. no. AST0552 (Gibbs no. 416). For Gibbs's numbers see: Sharon L. Gibbs, Janice A. Henderson, Derek John de Solla Price, *A Computerized Checklist of Astrolabes* (no locality: no printer indication, 1973), 16.

accomplish the purpose, two are largely preferred. They both rely upon the implicit assumption that astronomers in the past acted in the same way that they do today and that they always used reliable, freshly gathered data. By following this line of reasoning, the astronomical information represented on an instrument would approximate the real sky at the moment of the instrument's creation. Any discrepancy between the sky and the instrument would be due to random errors in observations and the manufacturing process.

The "First Point of Aries" Method

The "inventor" of the most common method of dating early astrolabes is unknown. As the mathematics involved is elementary, the definition of the operational procedure would have been within the ability of anyone aware of what prompted the Gregorian reformation of the calendar.

The purpose of the reformers was to fix the exact date of the vernal equinox on 21 March and keep it from drifting over the course of centuries. However, it was not clear how to do this. The work continued over a period of more than a century and involved first-class mathematicians such as Johannes Regiomontanus, Nicolaus Copernicus, and Egnazio Danti. The most difficult part of the work was the new definition of the solar year cycle of 365 and 366 days.

In the old Julian calendar, based on a simple four-year cycle, the average length of the year (365.2500 days) was too long in comparison to the tropical year, based on two consecutive transits of the Sun at the vernal equinox (365.2422 days, present figure). The reformers outlined three rules to determine if a year would be bissextile or not. The years whose figure is divisible by four are usually bissextile, as in the Julian calendar. If the figure is also divisible by 100, the year is, however, not bissextile. But if the figure is also divisible by 400, the year returns to being bissextile. The resulting average year of 365.2424 days represented an excellent approximation to the tropical year. The difference (ca. seventeen seconds) will require a further correction only after another 4500 years.

Until the Gregorian reform, the excessive length of the average Julian year had an important consequence: the date of the vernal equinox progressively shifted backwards. As a result, by the end of the sixteenth century, the equinox fell on 10 March. The calendar reformers, therefore, had to reposition the equinox to the correct date. In 1582, Pope Gregory XIII decreed, with the papal bull *Inter gravissimas*, the suppression of ten days. In Catholic countries, 4 October was followed by 15 October. Other countries adopted the new calendar several years or even a few centuries later. As an important corollary, the equinox day of the new calendar did not start at noon, as in the astronomical praxis, but at midnight. This fact explains the half-day difference between the Gregorian and the Julian count of the days, which is still used by astronomers in long-term calculations.

The so-called "First Point of Aries method" relies upon the average difference of 0.0075 days between the Julian and the Gregorian average years. Before the reform, the vernal equinox shifted back at the rate of one day every 133 years (or 128 years by referring to the present figure of the tropical year). As the day and time of the vernal equinox can be found by examining the calendar scale on the back of the astrolabe, the displacement of the first point of Aries represents a time indicator. The difference from the ten days corrected by the calendar reform, multiplied by the shift rate, would indicate the so-called "age of the instrument."

In 1923, this technique was favoured by Robert Gunther: "The epoch of construction [of an astrolabe] is accurately indicated by the day of the month of the vernal equinox, when the Sun enters the first point of Aries."[5] The opinion of such an authoritative scholar, who in 1932 completed the first extensive work on planispheric astrolabes, was not immune from criticism. In 1947, Henry Michel re-examined the method and pointed out two flaws in this way of dating an astrolabe by the position of the first point of Aries. According to Michel: (1) the hour system of early astrolabes is the astronomical one (time is measured in equinoctial hours from noon); (2) the bissextile cycle produces a mean oscillation of the equinox's date of ±12 hours.[6] Because of such drawbacks, Michel disagreed with the outcomes based on the method. Not only was Gunther's opinion "completely wrong,"[7] and "the date of the vernal point does not allow us to determine the fabrication year,"[8] but the application of the method to check a dated astrolabe demonstrated that "also in this case, for which we have at our disposal a happy combination of information, the date of the vernal point does not give us a modicum of precision on the fabrication year."[9] Further criticism was offered in 1956 by Emmanuel Poulle.[10]

These warnings went either unheeded or misunderstood. On the one hand, in 1973, the method was applied to the undated astrolabes included in the *Computerized Checklist of Astrolabes* by Derek de Solla Price and Sharon

5 Robert W.T. Gunther, *Early Science in Oxford*, 14 vols. (Oxford: Clarendon Press, 1923–1945); facsimile reprint (London: Dawsons of Pall Mall, 1968–1969), 2:187.

6 Henry Michel, *Traité de l'Astrolabe* (Paris: Gauthier-Villars, 1947); anastatic reprint (Paris: Librairie Alain Brieux, 1976), 135–141.

7 Henry Michel, *Traité*, 137.

8 Henry Michel, *Traité*, 139.

9 Henry Michel, *Traité*, 141.

10 Emmanuel Poulle, "Peut-on dater les astrolabes médiévaux?," *Revue d'Histoire des Sciences et de Leurs Applications* 9 (1956): 301–322 (304–309).

Gibbs,[11] and from time to time it has reappeared in studies by scholars with a scientific background.[12] On the other hand, Michel's *Traité* has been interpreted as a defence, instead of a criticism of the first point of Aries method.[13]

Therefore, it is worth noting again the complete unreliability of this technique. This unreliability should be apparent to scientists much more than to historians. In fact, if a scientist takes the trouble to examine the method carefully, he or she will realize that there are a number of factors that make it problematic to locate the first point of Aries within the calendar scale:

- Nobody knows if an early astrolabe was made for a bissextile or another year of the Julian cycle. Moreover, was such a distinction important to an astrolabe maker? This uncertainty leads to an *a priori* error of about ±12 hours;
- There is no guarantee that the time system adopted on the back of the early astrolabes is the astronomical one. Depending on the maker, the customer, local tradition, and other factors, an astrolabe might use the Babylonian (count of hours from sunrise), Italian (count of hours from sunset), civil (count of hours from midnight), or an other system. This uncertainty produces an *a priori* error of about ±12 hours;
- Instruments are not exempt from errors. In particular, the position of the vernal equinox is determined by the smallest division of the calendar scale. For a scale divided by one day, the *a priori* error amounts, at best, to half a division, that is ±12 hours;
- Nobody has yet noticed the random error of determining the exact instant of the equinox. The error estimated by Claudius Ptolemy in the *Almagest* on the ability to distinguish 1/10° on a vertical graduated ring remained the same up to the age of Regiomontanus. Ptolemy's optimistic estimate was ±6 hours.[14]

Even discounting other uncertainties—as, for example, the fact that the maker might have relied upon old astronomical tables or instruments—the result of the dating process is very vague. The total error on the position of the vernal

11 Gerald L'Estrange Turner, "A Critique of the Use of the First Point of Aries in Dating Astrolabes," in *Sic Itur ad Astra: Studien zur Geschichte der Mathematik und Naturwissenschaften: Festschrift für den Arabisten Paul Kunitzsch zum 70. Geburtstag*, ed. Menso Folkerts and Richard Lorch (Wiesbaden: Otto Harrassowitz, 2000), 548–554 (548).

12 Guglielmo Righini, "Il grande astrolabio del Museo di Storia della Scienza di Firenze," *Annali dell'Istituto e Museo di Storia della Scienza di Firenze* 2.2 (1977): 45–66 (52–53).

13 See Turner, "A Critique," 549–551.

14 G.J. Toomer (ed.), *Ptolemy's Almagest* (New York: Springer-Verlag, 1984), 134.

equinox within the calendar scale is the sum of the single errors listed above, up to ±42 hours. Once transformed into years, the overall uncertainty on any calculated date could be up to ±230 years.

As an example, the vernal equinox of the Tomba-Koelliker astrolabe falls on March 11.5 (Figure 14.2).[15] According to Michel, the construction date associated with the instrument is the year 1449. The error is, however, ±230 years, and the astrolabe could have been made at any time between 1219 and 1679, an extremely large interval. A better estimate is easily obtained by considering other features of the instrument.

The Star Coordinates Method

The second widespread method of dating early astrolabes is based on measuring the star coordinates on the astrolabe's rete. The final procedure represents the evolution of the other three methods mentioned by Michel as better substitutes for the first point of Aries technique. The scientific phenomenon considered is no longer the apparent motion of the Sun along the ecliptic, but rather the precession of the equinoxes, which increases star longitudes at the

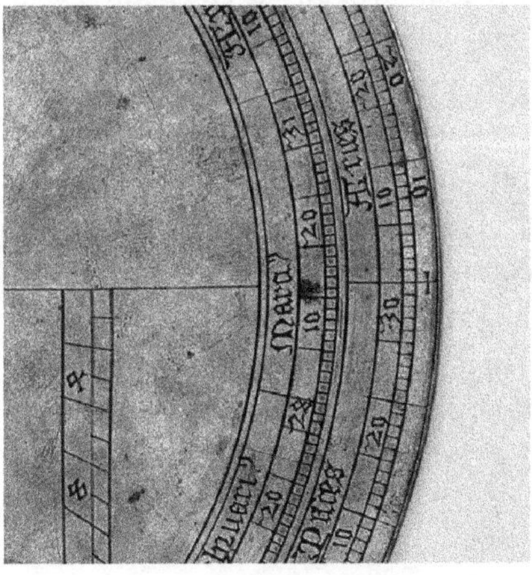

FIGURE 14.2 *The first point of Aries on the calendar scale of the Tomba-Koelliker astrolabe, detail.*

15 Tullio Tomba, "Poesia e scienza delle stelle nel Medioevo inglese: un astrolabio della 'Tradizione Chaucer,'" *FIMAntiquari: Arte viva: Pubblicazione della Federazione italiana mercanti d'arte* 5 (1994): 46–53 (48), cited the wrong date of 11 March.

constant rate of 1° every 72 years (present figure). The difference between the star coordinates on the astrolabe and those of the present sky is an indicator of the time elapsed between the epoch of the instrument and the present. According to Michel, the easiest way of dating by the stars involves their longitudes; there is, indeed, direct proportionality between the precession displacement and the time interval. In order to increase precision, it is preferable to consider only those few stars close to the ecliptic, such as Regulus (α Leonis) and Spica (α Virginis).[16] Their longitudes can be read directly on the astrolabe ecliptic ring and are affected by minimum random errors. If the distance from the ecliptic is great, the star longitude must be calculated and the random error increases.

The unreliability of this method is made apparent by a simple check. The longitude of Regulus, or Cor Leonis, in the Tomba-Koelliker astrolabe is 19° Leo. At the epoch 2000 Regulus was at 29° 50′ Leo. The precession difference of 10° 50′ corresponds to a time interval of 774 years. As a result, this particular Chaucer-type astrolabe—having the typical "V" frame of the rete—would have been made in 1226. Regrettably, the date is impossible in light of a couple of *post quem* indicators. The first is the year 1343, when Geoffrey Chaucer was born. The second lies in the star pointer for Alhabor, or Sirius. The pointer is in the shape of a dog head, which wears the so-called "Livery Collar of Esses," a symbol of the House of the Lancaster (Figure 14.3). The first documentary evidence of the collar is found on the gravestone of Sir John Swinford, who died in 1371, and is buried in the Church of St. Andrew, in Spratton, Northamptonshire.[17] The two indicators suggest that the astrolabe could not have been made before the second half of the fourteenth century.

Of course, the star pointer of Regulus might have been damaged over time, which could explain the erroneous result. In fact, if the calculation is repeated for the other suitable star Michel indicated, Alcimiek, or Spica, the result is different. In this case the calculation gives the year 1005, which makes even less sense!

To some extent Michel acknowledged the possibility of obtaining significant discrepancies,[18] and extended the analysis to the right-ascensions and

16 Michel, *Traité*, 142–143. Gunther did not expound any method of dating astrolabes based upon longitudes. He only cited the precession effect on star longitudes as a time indicator. See, for example, Gunther, *The Astrolabes of the World*, 1: 128.

17 Bennett and Strano, "The So-Called 'Chaucer Astrolabe,'" 226.

18 The third example mentioned by Michel, *Traité*, 142, closely recalls the case of the Tomba-Koelliker astrolabe. The French astrolabe of the Musée de la Vie Wallonne, Liège, inv. no. 400, revealed the date 1290 by using Regulus, 1160 by using Spica, and was nevertheless attributable to the fourteenth century from Gunther's stylistic point of view.

FIGURE 14.3 *Detail of the Tomba-Koelliker astrolabe pointer for Alhabor (Sirius) displaying the so-called "Collar of Esses".*

the mediations of the stars.[19] The three instruments on which he tested the different types of coordinates revealed the low reliability of longitudes and right-ascensions. Only in the case of star mediations, "notwithstanding the coarseness of the procedure, the results agree with each other remarkably, and allow one to determine, to within a few years, the age of these three astrolabes."[20]

In 1989 the very fact that different stars might indicate different dates inspired Roland Torode to generalise the star coordinates method. A single star pointer may be affected by a considerable random error, making the dating process unreliable. If, however, the scholar scrutinizes all the stars on the rete, the overall random error of the pointers would be reduced. From the statistical point of view, the random errors of a large data sample tend to have a null average (which does not mean that the statistical error of the system is null). The new method is, therefore, conceptually simple. It consists of calculating the date that best fits with the coordinates of each star of the rete and then averaging all the dates.[21]

---

19 Michel, *Traité*, 143–144 and 145–148, respectively.

20 Michel, *Traité*, 143, 144, and in particular 148.

21 Roland K.E. Torode, "A Mathematical System for Identifying the Stars of an Astrolabe and Finding its Age," *Astrolabica* 5 (1989): 53–73 (53).

The practical work is not elementary, even with the aid of a computer. In order to minimize the random errors in star positions, Torode outlined a specific procedure, whose principal steps can be summarized as follows: measure the position of the stars of the rete inside a Cartesian grid; transform the measurements into equatorial coordinates, roughly modernize them, and identify the stars by referring to a modern catalogue; transform the measurements into ecliptic coordinates; determine the precession displacement of each star from the present; calculate the date associated with the star; average all the dates; and determine the statistical error (standard deviation).[22]

When presenting the results obtained on a group of thirteen astrolabes, Torode strangely added another step. He discarded some of the dates calculated for each instrument on the guess that, "It is permissible to discard those dates which are more than one standard deviation from the mean."[23] By repeating the calculations on the remaining dates, he found that the statistical error was, in fact, remarkably reduced. Torode did not mention either a manual on statistics from which he derived such an additional step, or if the data selection was his own idea. He did not realize that, in the correct procedure, the experimenter is allowed to discard only the single datum affected by the largest apparent error before handling all the other data.[24]

Several sources of further uncertainty may however occur in the old astrolabes. As Torode indicated, the reference circles and points of the rete may be vaguely set; the ecliptic ring may contain divisions affected by random and systematic errors; the star pointers may be rounded, blunt or deformed; the rete itself can be warped or deformed; finally, an error of ±1° in the star positions leads to a time error of ±70 years.[25] In any case, Torode was convinced that his twofold method of measuring Cartesian coordinates and averaging a relatively large number of data would get rid of such errors. Also in the worst case, the method could at least indicate the year of the data adopted by the astrolabe maker.[26]

Torode would have been very surprised by how thoroughly his method was criticised. In 1992, Elly Dekker argued that, besides showing mistakes in

22 Torode, "A Mathematical System," 53–57. See also Torode, "A Study of Astrolabes," *Journal of the British Astronomical Association* 102.1 (1992): 25–30 (25).

23 Torode, "A Mathematical System," 61 t. 3.

24 On this problem, see Elly Dekker, "Of Astrolabes and Dates and Dead Ends," *Annals of Science* 49 (1992): 175–184 (178–179).

25 Torode, "A Mathematical System," 57; Torode, "A Study," 25–26.

26 Torode, "A Mathematical System," 58; Torode, "A Study," 26.

statistics, a scientific "method" never indicates the construction year of an astrolabe or the year its data were gathered.[27]

The Intermediate Approach to Star Data

The generous donation of the Tomba-Koelliker astrolabe to the Museo Galileo of Florence by its last private owner, Luigi Koelliker of Milan, furnished the occasion to perform a multifaceted study of this important artefact. The temptation to date the instrument by the scientific methods described above was strong. The puzzling and contradictory preliminary results, however, suggested that it was best to abandon this type of analysis.

There was, nevertheless, no reason not to gather and study star coordinates for another purpose than to find unconvincing and inconsistent dates of manufacture. Instead of checking the astrolabe against the sky (mediated by calculations to compensate for the present understanding of the celestial phenomena, like precession), the new purpose was to compare the gathered star data with other historical sources and, in particular, with data from different astrolabes belonging to the same tradition. In order to achieve such a goal, it was indispensable to sketch a surveying procedure very different from the scientific ones.

Determining Star Coordinates

According to the celebrated motto that every instrument "speaks its own language," the first indispensable step of the new approach to star data was, of course, to speak the same language. Cartesian, equatorial, or ecliptic coordinates had a null or marginal influence in how early astrolabes were made. Measuring the star positions on the rete requires one to adopt the same coordinate system used during the Middle Ages. In addition to being intrinsic to the astrolabe rete, this system is at the basis of several star tables. The system relies on two coordinates: mediation and declination. The mediation is determined by the intersection with the ecliptic of the great circle passing through the star and the celestial poles. It is given in degrees within the specific sign of the zodiac. The declination is determined by the angular separation of the star from the celestial equator measured along the above-mentioned great circle. It is given in degrees to the north or to the south of the celestial equator.

27 Dekker, "Of Astrolabes and Dates," 177–184.

There is little doubt that, in transferring the data from a star list, the instrument maker referred to the graduated scales on the astrolabe. On the one hand, it is therefore of marginal importance for the modern scholar to achieve great precision. On the other hand, while measuring, it would be wrong to neglect the instrument's own divisions. In fact, the goal of gathering data reliable at no less than ±0.01 mm, suggested by Torode,[28] is a scientific mirage which runs counter to the content of the instrument. It is hard to imagine that a medieval astrolabe maker could have been so precise and that, in placing the star pointers, he was free from any systematic error present in the graduation of the ecliptic ring or in the stereographic projection of the rete.

The modern scholar should be satisfied with measuring coordinates up to an average precision of ±1° with respect to the divisions of the instrument. This degree of precision is comparable to that of the maker and to the nominal precision—the smallest angle consistently mentioned, disregarding transcription mistakes—of the majority of the medieval star tables, which very seldom include figures up to 0.5°. In order to understand what the limit of ±1° implies, let us consider the Tomba-Koelliker astrolabe. The overall radius of the instrument is about 73 mm. Therefore, the limit means the ability to detect declination differences from about ±0.3 mm at a declination of 60° North up to about ±1 mm at a declination of 20° South. As for the other coordinate, the limit requires the ability to detect mediation differences from about ±0.2 mm at a declination of 60° North up to about ±1 mm at a declination of 20° South. Usually, due to the distortions of the stereographic projection, there are no stars placed on the rete outside of the range from 60° North to 20° South.

Star Identification

The second step of the new approach to the data consists, of course, in identifying the stars on the rete. The scientists consider the automation of this procedure to be the feather in their cap. Nevertheless, it is preferable to leave the computer switched off. This is not a form of snobbish pride by the historian. Only the programmer, who is seldom the same as the scholar, knows what the software is doing and, in particular, the weight it gives to each of the many variables on which star identification depends. Paradoxically, only a visual procedure involving a moderately sized planisphere, adjustable for precession, is reliable enough to perform the identification.[29] The planisphere does not include faint stars, which could fit any awkward combination of coordinates.

---

28 See respectively Torode, "A Mathematical System," 53; Torode, "A Study," 26.

29 Milton D. Heifenz, *Precession of the Equinoxes Historical Planisphere* (Somerville, Massachusetts: Learning Technologies Inc., 1997).

It only includes bright stars (up to the fourth magnitude), among which were those very few stars selected by the astrolabe maker.

It is, of course, imperative to repeat and recheck the identification a few times. About 75% of the represented stars create no trouble, but the identification of the remaining 25% requires, in addition to the planisphere, the consultation of historical documents. The most important document to identify the stars in the Chaucer-type astrolabes is the table based on the observations made by John of London in Paris in 1246 with an armillary instrument, which was revised in 1250. In turn, John's data influenced the other table included in the treatise on the astrolabe attributed to the Pseudo-Messahalla. As Paul Kunitzsch discovered, this star table was a source for many early English astrolabes.[30] Its best-known version is preserved at the Cambridge University Library (Table 14.1).[31]

It is remarkable to note that, when applied to the Tomba-Koelliker astrolabe, the identification procedure reveals the most interesting results with the hard-to-identify stars. As a first example, notwithstanding their shape, the two small metal bridges connecting the tail and the head of the dragon, which represents the tropic of Capricorn, to the ecliptic ring, are not star pointers, as was formerly believed (see Figure 14.1).[32] There are no stars, either significant or compatible with the Pseudo-Messahalla star table, in the areas where the bridges end.

30 Kunitzsch's star list type VIII. See: Paul Kunitzsch, *Typen von Sternverzeichnissen in Astronomischen Handschriften des Zehnten bis Vierzehnten Jahrhunderts* (Wiesbaden: Otto Harrassowitz, 1966), 51–58. See also: Owen Gingerich, "Zoomorphic Astrolabes and the Introduction of Arabic Star Names into Europe," in *From Deferent to Equant: A Volume of Studies in the History of Science in the Ancient and Medieval Near East in Honor of E.S. Kennedy*, ed. David A. King and George A. Saliba (New York, NY: The New York Academy of Sciences, 1987), 89–104 (96); reprinted in Gingerich, *The Eye of Heaven: Ptolemy, Copernicus, Kepler* (New York, NY: The American Institute of Physics, 1993), 81–101 (97).

31 Cambridge, Cambridge University Library, MS. Ii. 3.3, fol. 70v. The page of the manuscript is reproduced in Gunther, *Chaucer and Messahalla on the Astrolabe* (Oxford: University Press, 1929); facsimile reprint (London: Dawsons of Pall Mall, 1968), unnumbered plate between pp. 162–163. Transcriptions of the star table are in Walter W. Skeat (ed.), *A Treatise on the Astrolabe; Addressed to His Son Lowys by Geoffrey Chaucer, A.D. 1391* (London: N. Trübner & Co., 1872), xxxvii–xxxix; reprinted as: Geoffrey Chaucer, *A Treatise on the Astrolabe (1391)* (Amsterdam: Meridian Publishing Co., 1978), xxxvii–xxxix, which includes several mistakes, and in Kunitzsch, *Typen von Sternverzeichnissen*, 55–56.

32 Tomba, "Poesia e scienza," 48, discarded one of the two bridges, but considered the other as pointing to δ Scorpii.

TABLE 14.1 *The Pseudo-Messahalla Star Table from Cambridge University Library, MS. Ii.3.3, fol. 70v*

No.	Name	Identification	Mediation (°:′)	Declination (°:′)
1	mirach	Mirach, β And	(1) 0:07 Ari	32:30 N
2	bate[n]kaitoz	Baten Kaitos, χ Cet (2)	18:30 Ari	13:30 S
3	pate[n]kaitoz – vente[r] cethi	Baten Kaitos, ζ Cet	20:00 Ari	14:00 S
4	enif	Hamal, α Ari	22:00 Ari	23:30 N
5	finis flux[us]	Mira, ο Cet (3)	25:00 Ari	4:30 S
6	menkar – naris cethi	Menkar; α Cet	6:00 Tau	1:00 N
7	algenib – frons algonis	Algenib, α Per	10:00 Tau	49:00 N
8	algetenar	Zaurak, γ Eri	22:00 Tau	16:00 S
9	aldebaran – oc[u]l[u]s v[e]l cor taur[i]	Aldebaran; α Tau	29:00 Tau	14:30 N
10	alhaioh – hyrc[us] [ve]l hu[meru]s sag[itatoris]	Capella; α Aur	6:00 Gem	45:00 N
11	rigil – pes orionis	Rigel; β Ori	11:00 Gem	10:00 S
12	algeuze – hu[meru]s dext[er] orio[n]is	Betelgeuse; α Ori	15:00 Gem	8:00 N
13	alhabor – i[n] ore canis m[er]id[ionalis]	Sirius; α CMa	3:00 Cnc	15:00 S
14	razalgeuze – cap[u]d ge[m]i[n]o[rum]	Castor; α Gem	9:00 Cnc	33:00 N
15	algomeyza – in collo canis	Procyon; α CMi	13:00 Cnc	7:00 N
16	Markep	Tureis; ρ Pup	21:00 Cnc	22:30 S
17	Egregez	ι Uma (4)	24:00 Cnc	45:00 N
18	aldiran – in fro[n]te leonis	Alphard; α Hya (5)	6:00 Leo	6:00 S
19	alfart – equs v[e]l ci[n]gul[us]	– (6)	13:00 Leo	18:30 S
20	calbalezed – cor leonis	Regulus; α Leo	20:00 Leo	15:00 N
21	alrucaba – vrsa	μ Uma (7)	20:00 Leo	35:00 N
22	coruus	ν Hya (8)	1:00 Vir	11:00 S
23	dubhe – id e[st] vrsa	α Uma	2:00 Vir	(9) 6:00 N
24	denebalezed – cauda leo[n]is	Denebola; β Leo	15:00 Vir	19:30 N
25	algorab – in ce[n]tauro	Algorab; δ Crv (10)	22:00 Vir	13:30 S
26	alchimec – inhe[r]mis	Spica; α Vir	10:00 Lib	7:00 S
27	be[n]nenaz – filie f[e]r[e]t[ri] i[n] the[n]o[n]e	Benetnasch; η UMa	9:00 Lib	53:00 N
28	alramech – lanceator	Arcturus; α Boo	27:00 Lib	24:00 N

TABLE 14.1 *The Pseudo-Messahalla Star Table from Cambridge University Library* (cont.)

No.	Name	Identification	Mediation (°:')	Declination (°:')
29	alfeca – in corona ad[r]i[a]ne	Gemma; α CrB	16:00 Sco	29:00 N
30	alachil	Unuk; α Ser (11)	17:00 Sco	14:00 N
31	yed	Yed, δ Oph	26:00 Sco	3:00 S
32	calbalacrab – cor scorpii	Antares; α Sco	27:00 Sco	23:00 S
33	alhaue – cap[u]d dr[ac]onis	Ras Alhague; α Oph	13:00 Sgr	15:00 N
34	rahtaben – cap[u]d s[erp]e[n]tis	Ettanin; γ Dra	25:00 Sgr	52:00 N
35	wega – vultur cade[n]s	Vega; α Lyr	3:00 Cap	38:00 N
36	altair – vultur vola[n]s	Altair; α Aql	16:00 Cap	7:00 N
37	delfin	β Del (12)	29:00 Cap	12:30 N
38	alrif – in cigno	Deneb; α Cyg	29:00 Cap	42:00 N
39	addigege – cauda gallin[a]e	Deneb al-digege; α Cyg (13)	30:00 Cap	43:00 N
40	libedeneb – cauda capri	θ Cap (14)	6:00 Aqr	22:00 S
41	delfin – nubi[lior] et orien[tali]or	α Equ (15)	10:00 Aqr	6:00 N
42	aldiran	Alderamin; α Cep	10:00 Aqr	59:00 N
43	enifelf[er]az – m[us]ida eq[ui] pegasi	Enif, ε Peg	13:00 Aqr	7:00 N
44	denebalgedi – cauda capric[orni]	Deneb Algedi, δ Cap	14:00 Aqr	19:30 S
45	sceach – crus	Skat, δ Aqr	30:00 Aqr	19:00 S
46	alferaz – in pegaso (16)	Scheat, β Peg	6:00 Psc	24:00 N
47	mentichel – hu[meru]s eq[u]i alati	Sirrah – Alpheratz, α And (17)	18:00 Psc	25:00 N
48	denebkaitoz – cauda cethi	Deneb Kaitos, β Cet (18)	22:00 Psc	10:00 S
49	sceder	Schedir, α Cas	(19) 18:00 Psc	53:00 N

NOTES:

(1) Wrong transcription for 7:0°. See Skeat, *A Treatise on the Astrolabe* (cit. n. 31), xxxvii; Kunitzsch, *Typen von Sternverzeichnissen* (cit. n. 30), 56 n. 1; Bennett & Strano, "The So-Called 'Chaucer Astrolabe'" (cit. n. 3), 220.

(2) For Kunitzsch, *Typen von Sternverzeichnissen* (cit. n. 30), 56 n. 2, entries no. 2 and no. 3 are duplicates. The data suggest that the two entries represent two close stars, χ Cet and ζ Cet, both called Baten Kaitos.

(3) "Finis Fluxus" is Achernar (α Eri), but the star coordinates point elsewhere. This star cannot be θ Eri, as suggested by Kunitzsch, *Ibid.*, whose declination in 1250 was ca. 45° S.
(4) This star cannot be μ UMa, as suggested by Kunitzsch, *Ibid.* It is situated to the West, and not to the East of Alphard (α Hya) and Regulus (α Leo).
(5) This star, unidentified by Kunitzsch, cannot belong in Leo. The declination points to Alphard (α Hya). Apparently, this entry contains a transcription error. The name Alphard appears as the next entry, no. 19.
(6) This star cannot be Alphard (α Hya), as affirmed by Kunitzsch, *Ibid.* The coordinates point to an "empty" place.
(7) This star cannot be θ UMa, as stated by Kunitzsch, *Ibid.* It is most likely μ UMa, as suggested by Gingerich, "Zoomorphic Astrolabes" (cit. n. 30), 100, and Gingerich, *The Eye of Heaven* (cit. n. 30), 93. However, the declination of such a star is ca. 10° to the North.
(8) This star, unidentified by Kunitzsch, cannot belong in the constellation Corvus, which is placed about a full zodiacal sign to the East. The coordinates point to ν Hya.
(9) Dubhe is α Uma. Around 1250 its declination was ca. 65° N. A number is missing in the declination of the star table. See Bennett & Strano, "The So-Called 'Chaucer Astrolabe'" (cit. n. 3), 220.
(10) The name Algorab is given to δ Crv and not to γ Crv, as for Kunitzsch, *Typen von Sternverzeichnissen* (cit. n. 30), 55. The specification "in centauro" is a scribal mistake.
(11) This star cannot be Akrab (β Sco), as suggested by Kunitzsch, *Ibid.* The star declination is 14° N, and not ca. 20° S.
(12) This star cannot be ε Del, as suggested by Kunitzsch, *Ibid.*, 56. The brightest star of the constellation Delphinus is β Del, the position of which fits better with the declination in the list.
(13) This is a second entry for α Cyg, with slightly varied coordinates.
(14) This star cannot be Deneb Algedi (δ Cap), as suggested by Kunitzsch, *Ibid.* In 1250, the star δ Cap had about the same declination than θ Cap, but greater mediation. Moreover, Deneb Algedi is correctly listed at the entry no. 44.
(15) This star, unidentified by Kunitzsch, cannot belong in the constellation Delphinus, which is located ca. 10° in mediation to the West. The brightest star in the area indicated by the coordinates of the table is α Equ.
(16) The star names of entries no. 46 and no. 47 are inverted.
(17) Due to the inversion of entries nos. 46 and 47, this star is not a duplicate of β Peg, as instead suggested by Kunitzsch, *Ibid.*
(18) Deneb Kaitos is traditionally attributed to β Cet, and not to ι Cet, as suggested by Kunitzsch, *Ibid.*
(19) Possibly a wrong transcription for 28:00°.

More remarkably, the pointer labelled "corvus" does not indicate any part of the constellation Corvus.[33] It points at the lonely star ν Hydrae, about 20° in mediation to the west of Corvus. No other bright star can be found in the area

33 The wrong identification is in Tomba, "Poesia e scienza," 48.

marked by the pointer. Similarly, the pointer labelled "deyfin" does not indicate the small constellation of Delphinus.[34] It rather points at the relatively lonely star α Equulei, which is about 25° in mediation to the east of Delphinus.

While it helps to discard the role of the connections between the tropic of Capricorn and the ecliptic ring, it is noteworthy that the Pseudo-Messahalla star table includes double entries for both Corvus (no. 22: "corvus," and no. 25: "Algorab," the brightest star of Corvus) and "deyfin" (no. 37 and no. 41, both labelled "delfin"). The double entries are clues that the Tomba-Koelliker astrolabe lies in the tradition of the Pseudo-Messahalla star table. It is in this table that the two stars ν Hydrae and α Equulei are mislabelled. The erroneous labels were transferred from the table onto the instrument.

Data Comparison

The very centre of the analysis dealing with star coordinates resides in its third step: the comparison between different astrolabes within the same tradition. At the first stage of development of the new approach, the plan was to compare the star coordinates of the Tomba-Koelliker astrolabe with those of the other Chaucer-type astrolabes. It soon became apparent that to adopt this particular instrument as a main source of information was misleading. The different astrolabes have only a fraction of stars in common and the comparison gave limited results. Moreover, why should we assign special value to this particular astrolabe to serve as a touchstone or gold standard for the others?

The introduction of a relatively *super partes* term of comparison—the Pseudo-Messahalla star table—drastically changed things. For this purpose, the manuscript at the Cambridge University Library was again selected as representative of the star table for no other special reason than its widespread circulation in scholarly publications. Only three stars that are present in the group of six astrolabes selected for comparison are missing in the Pseudo-Messahalla star table.[35] As for the other stars, the mediation and declination differences emerging from each comparison indicate how close each astrolabe is to the data of the reference source. In particular, for each astrolabe it is possible to establish the percentage of star coordinate differences within the random error of ±1° involved in the measurements (Table 14.2).

34 The wrong identification is again in Tomba, "Poesia e scienza," 48.

35 β Persei in the Painswick and the Tomba-Koelliker astrolabes (perhaps to be re-identified with α Persei), ϰ Orionis in the Painswick astrolabe, and α Pegasi in the British Museum undated astrolabe.

TABLE 14.2 *Percentages of Star Coordinates Differences within ±1°*

Astrolabe	Stars	Mediation	Declination	Overall
Tomba-Koelliker, MG, Florence	24	63%	63%	63%
"Chaucer", BM, London	33	52%	61%	56%
Painswick, MHS, Oxford	21	38%	62%	50%
Parnel, SNM, Washington	13	38%	38%	38%
Undated, BM, London	23	22%	22%	22%
Gunther, MHS, Oxford	19	11%	26%	18%

The differences reveal that the Tomba-Koelliker astrolabe, with an overall 63% of coordinates within the limit of ±1°, is the one that depends most on the data of the Pseudo-Messahalla star table. It is followed by the so-called "Chaucer" astrolabe of 1326 at the British Museum, with 56%, and the Painswick astrolabe at the Museum of the History of Science, Oxford, with 50%. The other three astrolabes show a loose relationship with the star table's data: the Parnel at the Smithsonian National Museum of American History, Washington, with 38%, the undated British Museum astrolabe, with 22%, and the Gunther astrolabe at the Museum of the History of Science, Oxford, with only 18% of differences within ±1°.[36]

The comparisons yielded other results. A synoptic table, too large to be published, reveals other peculiarities of the six astrolabes. Relevant systematic differences in the mediation of Alpheratz (α Andromedae) show that at least four of the six astrolabes (the "Chaucer" and the Parnel have no such star) do not come directly from the data of the Pseudo-Messahalla star table as it appears in the manuscript at the Cambridge University Library. Such astrolabes depend on another source—a manuscript or an instrument—in which the problem of the Cambridge manuscript with Alpheratz was not present. In fact, the Cambridge manuscript, entry no. 46, reports the name "alferaz—in pesago [sic]" for the star Scheat (β Pegasi). The following entry, no. 47, "mentichel—hu[meru]s eq[u]i alati," is instead the one for Alpheratz.

From the same synoptic table, it is quite apparent that, as for stellar nomenclature, identification, and positioning, the six astrolabes belong to at

36 For detailed data see Bennett and Strano, "The So-Called 'Chaucer Astrolabe'," 214–219. Regrettably, due to a transcription mistake, all tables from 3a to 3f report the mediation of α Coronae Borealis in the sixth column as 26° Lib instead than 16° Sco.

least two different sub-traditions. On the one hand, the Tomba-Koelliker, the Painswick, the Gunther and the undated British Museum astrolabe contain the star ν Hydrae labelled as "Corvus," but do not have Algorab, the true Corvus. The Tomba-Koelliker and the Painswick also have α Equulei labelled as "delfin" or "deyfin." On the other hand, the "Chaucer" and the Parnel have Algorab for δ Corvi, but neither have the "false Corvus" nor the "false Dolphin."

Finally, the "Chaucer" astrolabe at the British Museum is extremely useful in confirming two transcription mistakes in the data of the Cambridge manuscript and in suggesting the possible corrections. This astrolabe alone among the six contains the two stars β Andromedae and α Ursae Majoris. As for the first star, the manuscript's entry no. 1, "mirach," reports the mediation 0° 7′ Aries. It has already been supposed that this odd datum could be a miswriting for 7° Aries[37] and this is confirmed by the "Chaucer" astrolabe. The second star, α Ursae Majoris, is listed in the Cambridge manuscript at entry no. 23, "dubhe – id e[st] vrsa." Its declination is, however, 6° North, a quite low figure for a circumpolar constellation. In this case, the astrolabe has a declination of 67° North. A number is missing in the manuscript, which perhaps should have read 66° or 67° North. These two remarkable entries confirm that the data of the "Chaucer" astrolabe come from an uncorrupted source other than the Cambridge manuscript.

The Catalan Astrolabe at the National Maritime Museum, Greenwich, London

The positive results expounded in the former paragraph encourage further research. They give support to the plan of extending the data bank and the confrontation of early astrolabes, in particular those that belong to the "Chaucer tradition." Within this plan, the Catalan planispheric astrolabe conserved at the National Maritime Museum of Greenwich, London, Caird Collection, appears puzzling. David A. King has already noted that, although the month-names on the back and some of the place-names on the plates are in the medieval dialect of Valencia, the instrument displays the same rete style of the Chaucer-type astrolabes.[38]

37 Skeat, *A Treatise on the Astrolabe*, xxxvii; Kunitzsch, *Typen von Sternverzeichnissen*, 56 n. 1.

38 D.A. King, "The Earliest Known European Astrolabe in the Light of Other Early Astrolabes," in King, *Astrolabes from Medieval Europe*, II/359–404 (II/381 and fig. II/14). Images of the astrolabe can be seen at the National Maritime Museum website: http://collections.rmg.co.uk/collections/objects/10714.html (accessed March 13, 2015).

Besides the superficial similarity and the comparable size (147 mm in diameter), the Catalan instrument differs from the other six astrolabes already examined. In addition to the "V" outline of the frame inside the ecliptic ring, the rete displays the colure of the equinoxes as an elaborate horizontal bar. The bar counter-changes six times and includes two semi-circular elements that may recall the relics of a quatrefoil decoration. The colure bar is missing in the other six instruments examined, and also in the retes depicted in the two best-known manuscripts of Chaucer's *Treatise on the Astrolabe*.[39] The rete of the Catalan astrolabe also displays a segment of the equator ring, which is present in the Parnel astrolabe only. Instead, the rete does not display any of the zoomorphic pointers that are a conspicuous characteristic of the Chaucer-type astrolabes. In addition, the pointers of the instrument have flame shapes usually associated with early French astrolabes.

These similarities and differences raise the following question: Did the maker of the Catalan astrolabe get the star data from a manuscript of the Pseudo-Messahalla star table? The availability of a couple of good pictures of the instrument made it possible to measure the mediation and the declination of the nineteen stars of the *rete* with the required precision of ±1°. The first five columns of Table 14.3 include:

(1) the entry number of the star in the astrolabe;
(2) the transcription of the star's name;
(3) the star's identification (its Bayer Greek letter, and the international three-letter abbreviation for the constellation);
(4) the mediation (μ) of the star in degrees within each sign of the zodiac;
(5) the declination (δ) of the star in degrees north or south of the equator.

Apart from spelling variants, the names engraved on the star pointers do not raise any particular problems. All the names but two are present in the Pseudo-Messahalla star table. The first exception, "Balkanu," unmistakably points to α Ophiuchi, which in the star table, entry no. 33, is named "alhaue – cap[u]d dr[ac]onis." The second exception, "Alhaz," is probably a corruption or contraction for Alpheratz, listed in the table, entry no. 46, as "alferaz – in pesago [sic]". Of course, the spelling variants and these two special cases would need further philological investigation.

39 Cambridge, University Library, MS. Dd.3.53, fol. 213v, and Oxford, Bodleian Library, MS. Rawlinson D. 913, fol. 29r.

Once the stars are compared to the corresponding entries of the Pseudo-Messahalla star table, additional information emerges. The five other columns of Table 14.3 include:

(6) the entry number of the star in the star table;
(7) the mediation (μ) of the star in degrees within each sign of the zodiac;
(8) the declination (δ) of the star in degrees north or south of the equator;
(9) the difference in mediation (Δμ) calculated in the form: astrolabe's less star table's coordinates;
(10) the difference in declination (Δδ) calculated as above.

TABLE 14.3 *The Catalan Astrolabe and the Pseudo-Messahalla Star Table*

Catalan Astrolabe					P-M Star Table			Differences	
No.	Name	Star	μ	δ	No.	μ	δ	Δμ	Δδ
1	Batencaytoc	ζ Cet	12 Ari	14 S	3	20 Ari	14 S	- 8.0	0.0
2	Aldebaran	α Tau	4 Gem	12 N	9	29 Tau	14.5 N	+ 5.0	- 2.5
3	Alioch	α Aur	13 Gem	45 N	10	6 Gem	45 N	+ 7.0	0.0
4	Rigil	β Ori	14 Gem	11 S	11	11 Gem	10 S	+ 3.0	- 1.0
5	Alabor	α CMa	4 Cnc	16 S	13	3 Cnc	15 S	+ 1.0	- 1.0
6	Algomeyza	α Cmi	13 Cnc	6 N	15	13 Cnc	7 N	0.0	- 1.0
7	Aldira[n]	ζ Hya	5 Leo	7 N	18	6 Leo	6 S	- 1.0	+ 13.0
8	Alfart	α Hya	13 Leo	8 S	19	13 Leo	18.5 S	0.0	+ 10.5
9	Algurab	δ Crv	28 Vir	13 S	25	22 Vir	13.5 S	+ 6.0	+ 0.5
10	Alchimeck	α Vir	15 Lib	6 S	26	10 Lib	7 S	+ 5.0	+ 1.0
11	Ursa	η UMa	18 Lib	53 N	27	9 Lib	53 N	+ 9.0	0.0
12	Alramech	α Boo	25 Lib	19 N	28	27 Lib	24 N	- 2.0	- 5.0
13	Corona	α CrB	22 Sco	29 N	29	16 Sco	29 N	+ 6.0	0.0
14	Balkanu	α Oph	15 Sgr	14 N	33	13 Sgr	15 N	+ 2.0	- 1.0
15	Uegua	α Lyr	3 Cap	37 N	35	3 Cap	38 N	0.0	- 1.0
16	Altayr	α Aql	15 Cap	7 N	36	16 Cap	7 N	- 1.0	0.0
17	Gallina	α Cyg	3 Aqr	40 N	39	30 Cap	43 N	+ 3.0	- 3.0
18	Libedeneb	θ Cap	6 Aqr	21 S	40	6 Aqr	22 S	0.0	+ 1.0
19	Alhaz	β Peg	6 Psc	24 N	46	6 Psc	24 N	0.0	0.0

The information associated with entry no. 19 of the Catalan astrolabe "Alhaz" for Alpheratz, forms evidence that the instrument lays in the Pseudo-Messahalla star table tradition. The coordinates of the star are exactly the same as those of the star table. There is, however, confusion among the table entries no. 46 and no. 47. As already noted in the previous paragraph, the coordinates of entry no. 46, "alferaz – in pesago [sic]," do not point to Alpheratz, but to Scheat. The coordinates of entry no. 47, "mentichel – hu[meru]s eq[u]i alati," are those pointing to Alpheratz. The maker of the Catalan astrolabe did not recognize the inversion of the two lines. He accepted entry no. 46, that is, Alpheratz with Scheat's coordinates. The mistake slipped from the star table onto the astrolabe.

More specifically, it is apparent that the Catalan astrolabe belongs to the same sub-tradition of the "Chaucer" and the Parnel astrolabes. On the one hand, entry no. 9 of the instrument is "Algurab" for δ Corvi. On the other hand, the astrolabe has neither the "false Corvus" nor the "false Dolphin."

The instrument contains two other interesting entries. When compared to the star table, they outline two remarkable mistakes of the Cambridge manuscript. Entry no. 7 of the astrolabe, "Aldira[n]," once compared with entry no. 18 of the Pseudo-Messahalla star table, reveals a mediation difference of only –1° and a relevant declination discrepancy of +13°. The subsequent entry no. 8 of the astrolabe, "Alfart," again reveals a null difference in mediation and another relevant discrepancy in declination of +10.5°.

A closer look at the manuscript reveals that entries no. 18 and no. 19 of the star table—respectively "aldiran – in fro[n]te leonis" and "alfart – equs v[e]l ci[n]gul[us]"—contain a significant blurb. Aside from their names, entry no. 18 points to Alphard (α Hydrae), and entry no. 19 points to an empty place. The pointers of the astrolabe, however, respectively indicate ζ and α Hydrae. The Cambridge manuscript is corrupt and the astrolabe—which received its data from another, uncorrupted, manuscript or instrument—suggests what would have been the original content of the Pseudo-Messahalla star table. It is very likely that entry no. 18 of the table, "aldiran," should have been ζ Hydrae, with star coordinates very close to those of the astrolabe. The scribe of the Cambridge manuscript wrote the correct mediation, 6° Leo, but exchanged 6° North with 6° South, therefore producing the declination discrepancy of +12°. As for entry no. 19, it should have indeed been Alphard, but, again, the scribe made another transcriptional mistake. He correctly wrote the mediation as 13° Leo, but added "1" before the original declination of 8.5°, therefore producing another discrepancy of +10°.

Such an explanatory potential of the Catalan astrolabe makes no surprise that the percentages of the differences in mediation and declination within

±1° are very high. The percentages are 42% in mediation and 74% in declination (without considering the two corrections for the star table entries no. 18 and no. 19). The overall percentage of 58% places the instrument between the Tomba-Kolleiker and the "Chaucer" astrolabes.

How can a Catalan astrolabe be so closely related to the John of London or Pseudo-Messahalla star table, and from which manuscript or instrument did the maker gathered the data? These are, for the moment, the unanswered questions.

Conclusions

The percentages of coordinate differences within ±1° outline what can be called "informational distance." Such a "distance" is an indicator of how close the relationship is between an instrument and the reference data source. This "informational distance" is not necessarily a distance over time. If the Tomba-Koelliker astrolabe is the closest to the data source, this does not mean that it is the oldest of the seven astrolabes studied. The data of the astrolabe represent a relatively faithful adaptation of those of the Pseudo-Messahalla star table, as do those of the Catalan or the "Chaucer" astrolabes. On this same basis, an instrument at a greater informational distance from the data source, such as the Gunther astrolabe, tell us that the coordinates of the Pseudo-Messahalla star table underwent several modifications before being measured (by me) on the rete. Modifications may include the intentional purpose by the maker to update the coordinates for precession, but they may also depend on many other factors, among which the ability of the maker and the story of the astrolabe. In particular, the rete of the Gunther astrolabe is badly damaged.

In any case, this new approach to the astrolabes' data, and especially to those of the instruments that belong to the same tradition, appears fruitful. While a scientific approach ends—or "dead ends," as Elly Dekker claimed—with an alleged year of fabrication, the new approach raises more questions than it solves. From which manuscript source does the data of the seven astrolabes examined come from? Is such a source a single one? What is the origin of the uncertainty between Corvus and ν Hydrae? What about the other uncertainty between Delphinus and α Equulei? Do the other four existent "V" English astrolabes belong to the same two sub-traditions delineated by these seven? What is the relationship between the Catalan astrolabe and the other English instruments? What is the data source for the other early English astrolabes, such as the "quatrefoil" ones? These and many other questions reveal that the new approach to the astrolabes data is not a dead end, but a long path of which only the first few steps have been taken.

Acknowledgements

Part of the present article (Sections 2, 3 and 5), has been presented at the *XXXIII Symposium of the Scientific Instrument Commission*, Tartu, Estonia, 25–29 August 2014 and at the *SISFA Symposium*, Florence, Italy, 10–13 September 2014. I would like to thank Karen Giacobassi for her kind revision of the English text.

Epilogue

Reconstruction of the Plate of Eclipses according to the Description by ʿAlī ibn ʿĪsā

Flora Vafea
Faculty of Languages and Translation, al-Azhar University, Cairo, Egypt

In his article *Astrolabes as Eclipse Computers: Four Early Arabic Texts on Construction and Use of the Ṣafīḥa Kusūfiyya*, Johannes Thomann presents, among others, two texts describing the plate of eclipses and its use, according to ʿAlī ibn ʿĪsā's description. The first text comes from the manuscript Istanbul, Topkapı Saray Kütüphanesi, Ahmet III 3509 ff. 304r–309v (MS T/ت) and the second one from Cairo, Dār al-Kutub, Zakī 706, ff. 20r–22r (MS Q/ق). The first part of each text describes the plate of eclipses, and in the second part it is explained how to use this instrument. Although the texts of the two manuscripts have some differences, they originate from the same treatise, as the parallel flowing of the two texts shows. Both of them have crucial omissions and deviations from the initial text, which do not allow us to understand completely how this instrument was and how it functioned, if we examine each text separately, but the combination of both texts can solve these problems.

Having the experience of reconstructing functional models of the plates of eclipses described by al-Bīrūnī and by Nasṭūlus,[1] I attempt here to reconstruct this instrument. The first necessary step is to merge the two texts, in order to fill the gaps that both of them include. However, there are some important differences in their overall look, such as:

i. In MS T excess wording has been added, such as "understand that", "know it" and "measure according to it", while the text in MS Q is terse, focusing on the process.

1 Flora Vafea, "Al-Bīrūnī: *The Plate of the Eclipses*", *Suhayl* 15 (2017), 297–354; Nasṭūlus' model is on p. 328.

 | DOI:10.1163/9789004387867_018

ii. MS T includes 9 numbered chapters on the use of the instrument, while MS Q only 8 unnumbered chapters; the first chapter of MS T has been omitted in MS Q. Also chapters 8 and 9 are separated at a different point in the two manuscripts.

iii. There are several repeated spelling errors in both manuscripts, e.g. "موري" in MS T instead of the correct "مري", i.e. "index", in MS Q; "انكشف" in MS Q instead of the correct "أنكسف", i.e. "to be eclipsed" in MS T; the term "محرق" (burnt) is always used in MS T, and "مخرق" (perforated) is used in MS Q, however, sometimes the former and sometimes the latter term is the appropriate one.

In the unified text that I present below, I tried to incorporate all information included in both manuscripts, in a few cases changing slightly the wording. When there were differences in the text, I selected the phrase that approaches better the construction of a working model. I have unified chapters 8 and 9 in order to avoid the problem of the different division of the chapters in the two manuscripts, and to present under the title of chapter 8 all that it announces. Although the term "خسوف", which appears in MS Q, corresponds to the lunar eclipse, I preferred to use the term "كسوف" (solar eclipse or eclipse) from MS T, in order to keep conformity, since the term "انكسف" (to be eclipsed) is mostly used here.[2] Al-Bīrūnī also uses the terms of the group "كسف" instead of "خسف" in the description of the plate of eclipses.[3] In all cases the initial text of both manuscripts is mentioned in the apparatus.

The purpose of merging the text of the two manuscripts was to construct an instrument according to the description presented in the first part of the treatise; this instrument should be able to operate according to the instructions given in the second part. The reader can actually make a copy of each one of the figures 1–4 and construct a working model of the device. Figures 3 and 4 must be attached on both sides of a circular piece of thin foam-board or paper-board to form the plate. Figure 2 should be first perforated cutting off the grey part; the rest can be laminated into a plastic transparent sheet in order to be strong enough and this part is the *rete*. Figure 1 may also be laminated and used as the limb. Then a thumbtack should penetrate these three parts through their centres, starting from the backside of the limb, then the plate and at the end the *rete*. The pin of the thumbtack may poke through an earring fixing, so that the three plates are fixed together. The construction of the alidade is not described in the text. After the construction, the reader may

2 The term "انخسف" that denotes that the moon is eclipsed appears only twice in MS Q.

3 Vafea, "Al-Bīrūnī: *The Plate of the Eclipses*", pp. 300, 303, 351–352.

perform all the uses described in the second part of the treatise and verify the functionality of the instrument, thus the correctness of the merged Arabic text and its translation into English.

This study could serve as a foundation for a formal critical edition of the treatise of ʿAlī ibn ʿĪsā on the plate of eclipses, accompanied by a comparison between the corresponding instruments described by Nasṭūlus and al-Bīrūnī, in order that the evolution of this instrument be researched and established.

Arabic Text

|ت/304|| رسالة علي بن عيسى الحاسب في معرفة العمل بالصفيحة القمرية والحُقّ الكسوفيّة. رحمه الله ورضي عنه[4]

|ت/304ب، بسم الله الرحمن الرحيم. ثقتي بالله.

ق/20|| قال علي بن عيسى، رحمه الله بعد حمد الله والثناء عليه لما هو أهله، والصلاة على نبيّه وعلى أهله: أمّا معرفة هذه الصفيحة والعمل بما فيها،[5] اعلم أنّ[6] حجرتها مقسومة بأربعة أقسام، نصفها مقسوم كح قسما، وهي ثمانية وعشرون يوما،[7] وابتداء الأقسام فيها،[8] إذا أخذتها بيدك وجعلت علّاقتها تلقاء وجهك، يكون ابتداء[9] أحد أرباعها عن يمينك صاعدا إلى العلّاقة. وهو[10] مقسوم ١٤ قسما[11]. ثمّ من الخطّ الذي نزل من العلّاقة في الربع الأيسر تمام كح يوما، وهو عدد ما يظهر للناس القمر. وأمّا النصف الآخر، فهو[12] أقسام ما يغيب عليه القمر، من[13] أول ليلة، من كسور الساعات والساعات[14]. وابتداءها[15] من الربع الذي يلي يسارك،[16] وهو[17] المكتوب فيه ه و، ثمّ

---

4 ق: --، ت: رسالة ... عنه

5 ق: باب العمل بالصفيحة القمرية، ت: ثقتي بالله ... فيها

6 ق: اعلم أنّ، ت: فإنّ

7 ق: أيّام، ت: ثمانية وعشرون يوما

8 ق: وابتداءها، ت: وابتداء الأقسام فيها

9 ق: بابتداء، ت: يكون ابتداء

10 ق: فهو، ت: وهو

11 ق: خطّا، ت: قسما

12 ق: فهو، ت: وهو

13 ق: القمر من ، ت: في

14 ق: والساعات، ت:--

15 ق: وابتداءها، ت: وابتداء

16 ق: يسارك، ت: اليسار

17 ق: وهو، ت:--

البيت الثاني فيه آ ه:[18] فالألف[19] ساعة، و ه كسر الساعة[20]، وهو[21] خمسة أسباع ساعة؛ حتّى تتمّ[22] أقسام هذا الربع من الساعات وكسورها[23] تمام ١٢[24] ساعة، التي هي ساعات الليل وكسورها،[25] وهو ما يغيب عليه القمر من أوّل ليلة يظهر، إلى ليلة البدر. ثمّ بعد ذلك إلى تمام كدَ، ما يغيب عليه بالنهار، بعد أن يلقي منه ١٢ ساعات الليل وكسورها،[26] حتّى يتمّ ما في حجرة القمر، وهي الحلقة المحيطة بصفيحة العنكبوت إلى أسفل العنكبوت، فافهم ذلك.[27]

باب في[28] معرفة العنكبوت وأقسامها ||305

فالعنكبوت[29] هي الصفيحة المشبكة،[30] التي في نصفها ثقب، تدور[31] حول القطب، عليه مكتوب [32] « فلك مري[33] نور القمر ». ومن هذا الثقب يُعلَم، إذا أدير العنكبوت[34] على الصفيحة المرقوم[35] فيها الفضّة والسواد، التي[36] تحت العنكبوت.

والصفيحة التي تحت العنكبوت فيها[37] دائرتان فيهما[38] فضّة محرّقة[39]، والتي فوق <من> هذه الدوائر <هي>[40] دائرة « فلك الشمس »، مما يحاذي العلّاقة. وبعدها[41]، إلى ناحية اليسار، إلى

18 ق: ة وَ. ثمّ البيت الثاني فيه آ ه؛ ت: عدد آ
19 ق: فالألف؛ ت: فألف عدد
20 ق: من ساعة؛ ت: الساعة
21 ق: وهي؛ ت: وهو
22 ق: حين تمّ؛ ت: حتّى يتمّ
23 ق: وكسورها؛ ت:--
24 ق: ١٢؛ ت: يب
25 ق: --؛ ت: التي ... وكسورها
26 ق: وهو ما يغيب ... وكسورها؛ ت: --
27 ق: --؛ ت: إلى أسفل العنكبوت ، فافهم ذلك.
28 ق: --؛ ت: باب في
29 ق: --؛ ت: فالعنكبوت
30 ق: --؛ ت: المشبكة
31 ق: تدور؛ ت: يدور
32 ق: القطب، عليه مكتوب؛ ت: الثقب
33 ق: مري؛ ت: يوري
34 ق: أنّ الذي؛ ت: إذا أدير العنكبوت
35 ق: المدفونة؛ ت: المرقوم
36 ق: السوداء، التي؛ ت: والسواد، الذي هو
37 ق: والصفيحة التي تحت العنكبوت فيها؛ ت: وفيها
38 ق: فيهما؛ ت: فيها
39 ق: قصة مخرقة؛ ت: فضّة محرّقة
40 ق: الذي يلي هذه الدائرة؛ ت: الذي فوق هذه الدوائر
41 ق: وبعدها؛ ت: وبعده

خطّ الربع[42] الأيسر، عدد ساعات طلوع القمر بالنهار، وفي الربع الذي بعده، عدد ساعات
|ق/ 20ب| طلوع القمر / بالليل.

وفي العنكبوت، في الجانب المخرق[43]، دائرة فلك الجوزهر[44]، وعليها مكتوب[45] « فلك الجوزهر ». وعليه[46] أيضا في هذا النصف المخرق[47] مري[48] الساعات، وهي الشظية التي عليها مرقوم « مري[49] الساعات »، وعلى الخطّ الذي يقطع[50] العنكبوت بنصفين، ويبتدئ من الخطّ الذي يقطع[51] فلك نور القمر،[52] ويمرّ بمركز العنكبوت وبفلك الجوزهر، وينتهي إلى آخرها[53]، حتّى[54] يتمّ الخطّ، عليه مكتوب مري [55] القمر. فاعرف كلّما وُصِفَ[56] لك، ليسهل عليك العمل بالصفيحة إن شاء الله. [57] وفي العنكبوت أيضا، في طرفها، أقسام على عدد أيّام الشهر القمري، في ناحية ٣٠ يوما، وفي الناحية الأخرى ٢٩ يوما، لما الشهر القمري كذلك يكون أبدا. [58]

|ت/305ب| وفي الوجه الآخر من الصفيحة، دائرة مقسومة باثنى عشر[59] قسما، يُعلَم[60] منها ما ينكسف من القمر، إذا كان له كسوف. وفي[61] الجانب الأيمن من هذه الدائرة عدد ما يكون بين[62] العقدتين والقمر عند الكسوف، فهذا[63] ما في الوجه الآخر من الصفيحة.

---

42 ق: خطّ الربع؛ ت: الخطّ الرابع
43 ق:في الجانب المخرق؛ ت: من الجانب المحرق
44 ق: الجوزهر؛ ت: الشمس
45 ق: وعليها كتابة؛ ت: وعليه مكتوب
46 ق: الجوزهر وفيها؛ ت: الشمس وعليه مكتوب فلك الشمس وعليه
47 ق: المخرق؛ ت: المحرق
48 ق: مري؛ ت: موري موري
49 ق: كتابة مري؛ ت: مرقوم موري
50 ق: يقطع؛ ت: يقع
51 ق: يقطع؛ ت: يقع
52 ق: نور القمر؛ ت: الشمس
53 ق: وبفلك الجوزهر، وينتهي إلى آخرها؛ ت: دائرة فلك القمر يؤدي نور القمر
54 ق: حين؛ ت: حتّى
55 ق: مكتوب مري؛ ت: مرقوم موري
56 ق: كلّما وصف؛ ت: علم ما وصفت
57 ق:--؛ ت: ليسهل عليك العمل بالصفيحة إن شاء الله.
58 ق: وفي العنكبوت ... كذلك يكون أبدا.؛ ت: --.
59 ق: ١٢؛ ت: باثنى عشر
60 ق: يعرف؛ ت: يعلم
61 ق: في؛ ت: إذا كان له كسوف وفي
62 ق: بين؛ ت: بعد
63 ق: فهذا؛ ت: فهذه معرفة

وفي ظهر هذه الصفيحة القمريّة عضادة مثل عضادة الأسطرلاب،[64] وفيها[65] خطوط يُعلَم منها ما مضى[66] من النهار وما بقي منه، إذا أقيس على ما أصف لك فيما بعد، إن شاء الله.[67] وفي ربع[68] من أرباعه أقسام الارتفاع [69]، كما يكون[70] في ظهر الأسطرلاب، لأخذ الارتفاع، وفي الربع المقابل له قسمة[71] الظلّ. وهو[72] جميع ما يحتاج إلى معرفة ما في هذه الصفيحة، فافهم ذلك وقس عليه إن شاء الله.[73]

الباب الأول: في معرفة أخذ الارتفاع. [74]

وهو كما يؤخذ بالأسطرلاب سوًى، وذلك أن تعلق الصفيحة بيمينك، وتحرّك العضادة بيسارك، وتحاذي بمنكبك الأيسر الشمس. وترفع العضادة أو تنزلها، حتّى يوافق أجزاء الارتفاع، وينفذ نور الشمس من الثقب الأعلى إلى الثقب الأسفل./ثمّ تنظر[75] رأس العضادة المحدّد ما وافق من ||306
الأجزاء، أعني أجزاء الارتفاع، فما كان فهو الارتفاع لذلك الوقت إن شاء الله.

الباب الثاني: في معرفة ما مضى من النهار من ساعات معوّجة، وهي الزمانيّة التي يكون بها النهار يبَ ساعة والليل كذلك، طال أم قصر. [76]

فإذا أردت ذلك، فاعرف ارتفاع نصف نهار يومك،[77] وضع رأس العضادة المحدّد[78] على غاية الارتفاع لنصف نهار يومك[79]. واستقبل بهذه العضادة[80] عين الشمس، حتّى ترى ظلّ هذه العضادة على باطن العضادة مستويا، غير منحرف إلى جهة من الجهات.[81] فإذا عرفت الظلّ أين وقع من العضادة، فعدّ[82] الساعات، وتبتدئ بعددها من الهدف الأعلى الذي أخذت

64 ق: --، ت: مثل عضادة الأصطرلاب
65 ق: فيها، ت: وفيها
66 مضى: ق:--، ت: يمضي
67 ق: للساعات، ت: يعلم منها ... إن شاء الله.
68 ق: ربع، ت: رفع
69 ت: الارتفاع ×٢
70 ق:--، ت: يكون
71 ق: قسمة، ت: قسم
72 ق: فهذا، ت: وهو
73 ق: إليه، ت: إلى معرفة ... إن شاء الله
74 الباب الأول لا يجد في ق
75 تنظر: ت: ننظر
76 ق: فصل في معرفة الساعات، ت:الباب الثاني ... طال أم قصر.
77 ق: --، ت: فإذا أردت ذلك، فاعرف ارتفاع نصف نهار يومك ذلك
78 ق: ضع مري العضادة، ت: وضع ... المحدد
79 ق: ارتفاع نصف نهارك، ت: غاية ... يومك
80 ق: بالعضادة، ت: بهذه العضادة
81 ق: يستوي ظلّ العضادة على باطنها غير منحرف، ت: ترى ... الجهات.
82 ق: فما بين وقع الظلّ، بعدّ، ت: فإذا ... فعد

منه الظلّ، <وقد> قُسم به الظلّ خطوطا قد قطعت باطن العضادة بالعرض،[83] فيما بينهما عدد الساعات بحروف الجمّل آ بَ جَ دَ هَ وَ.[84] الموضع الذي وقع عليه الظلّ <في> هذه العضادة، هو ما مضى من النهار من ساعات.[85] فإن كان القياس قبل نصف النهار، فعدّ الخطوط التي
|ت/306ب| على العضادة في الجانب المُوَقّع عليه بحروف/الجمّل آ بَ جَ دَ هَ وَ،[86] وإن كان بعد نصف النهار، فعدّ الخطوط من العضادة من الجانب المرقوم عليه عدد الساعات بحروف الجمّل هي زَ حَ طَ يَ يا يب.[87]

الباب الثالث:[88] في معرفة الظلّ.

إن[89] كان الارتفاع أقلّ من مه،[90] فانظر الرأس المحدّد من[91] العضادة على كم وقع من الظلّ، فما كان من الأجزاء المقسومة الظلّ[92] فاقسم عليه ١٤٤،[93] فالخارج هو الظلّ في الوقت الذي
|ق/ 21ا| أردت.[94] وإن كان الارتفاع مه،[95]/فالظلّ يب سوًى.[96] وإن كان الارتفاع أكثر من مه،[97] فالظلّ ما وقع عليه رأس العضادة المحدّد،[98] من غير حاجة إلى قسمة، فاعلم ذلك إن شاء الله.[99]

---

83 ق: وابتدائها من الهدف الأعلى التي آخذت منها الظلّ، من الخطوط التي على باطن العضادة؛ ت: ونبتدئ بعددها من الهدف الذي قسمت به الظلّ خطوطا قد قطعت باطن العضادة بالعرض
84 ق:--؛ ت: فيما بينهما ... آ بَ جَ دَ هَ وَ.
85 ق: فما كان فهو الماضي من الساعات؛ ت: بالموضوع، الذي وقع عليه الظلّ، هذه العضادة هو ما مضى من النهار من ساعة.
86 ق: إن كان العمل قبل نصف النهار؛ ت: فإن كان ... آ بَ جَ دَ هَ وَ.
87 ق: والباقي، إن كان بعده؛ ت: وإن كان بعد ... زَ حَ طَ يَ يا يب.
88 ق: فصل؛ ت: الباب الثالث.
89 ق: إذا؛ ت: إن
90 ق: ٤٥؛ ت: مه
91 ق: محدّد؛ ت: الرأس المحدّد من
92 ق:--؛ ت: من الأجزاء المقسومة الظلّ
93 ق: ١٤٤؛ ت: قسم طول دَ
94 ق: فالخارج هو الظلّ؛ ت: فما خرج من الظلّ فهو طول الظلّ في الوقت الذي أردت.
95 ق: ٤٥؛ ت: الارتفاع مه.
96 ق: فالظلّ ١٢؛ ت: فإنّ الظلّ في الوقت الذي أردت وإن كان الارتفاع مه فالظلّ يب سوي.
97 ق: أكثر؛ ت: الارتفاع أكثر من مه.
98 ق: العضادة؛ ت: رأس العضادة المحدّد.
99 ق: من غير حاجة إلى قسمة؛ ت: فالآن تقسم عليه قَ مَ دَ، فاعلم ذلك إن شاء الله.

الباب الرابع:[100] في معرفة ما يظهر من نور القمر[101] في كلّ ليلة من أيّام الشهر[102]. إذا أردت ذلك، ادر[103] العنكبوت، حتّى يقع[104] فلك الجوزهر على[105] فلك الشمس، ويقع فلك مري[106] نور القمر على الفضّة المحرّقة، ولا تبيّن[107] من الفضّة البيضاء شيء من ثقب مري نور[108] القمر. ||307/ فإذا فعلت ذلك، فادر العنكبوت جزءا واحدا من الأجزاء المقسومة في العنكبوت في طرفها[109]، من الأقسام التي بينهما آ بَ جَ وَ طَ يبَ يهَ يحَ، وهو[110] عدد أيّام الشهر القمري إذا كان الشهر ثلاثين يوما،[111] والجانب الآخر من الحاشية أيضا[112] إذا كان الشهر تسعة وعشرين يوما. فإذا[113] حرّكت العنكبوت جزءا[114] من أجزائها، حتّى يكون الجزء مع الخطّ الذي قد[115] أخرج من العلّاقة[116] إلى وسط الصفيحة، فحينئذ[117] يظهر لك ما في القمر من النور لليلة[118] واحدة. وكذلك تدير كلّ يوم لتنظر ما يزيد من النور في القمر، إلى أن يتمّ نوره في مقابلة الشمس ويكون بدرا تامّا.[119] ثمّ يبتدئ[120] بالنقصان، فلا يزال ينقص من نوره في البياض، ويزيد فيه السواد، حتّى لا يبقى من نوره شيء،[121] ويكون قد دار فلك مري[122] نور القمر حتّى طابق فلك الشمس في الاجتماع. فافهم ذلك وقس عليه، إن شاء الله عزّ وجلّ. وإذا شئت أن تعرف ما يكون في الشهر الآخر، فالعمل فيه مثل الأول سواء، وبالله التوفيق.[123]

---

100 ق: فصل؛ ت: الباب الرابع

101 ق: من نور القمر؛ ت: من القمر من النور

102 ق: الشهر؛ ت: أيّام الشهر

103 ق: ادر؛ ت: إذا أردت ذلك، فاردد

104 ق: يقع؛ ت: يقطع

105 ق: على؛ ت: و

106 ق: مري؛ ت: فلك موري

107 ق: يتبين؛ ت: تبيّن

108 ق: مري؛ ت: موري نور

109 ق: طرفها؛ ت: طرف

110 ق: يليها؛ ت: بينهما آ بَ حَ وَ طَ يبَ يهَ يحَ وهو

111 ق: ٣٠؛ ت: ثلثين يوما

112 ق: --؛ ت: أيضا

113 ق: ٢٩ فإذا؛ ت: تسعة وعشرين يوما فإذ أنت

114 ق: العنكبوت جزءا؛ ت: الشبكة مقدار جزء واحد

115 ق: قد؛ ت: --

116 ق: +ويبعد؛ ت: --

117 ق: --؛ ت: فحينئذ

118 ق: ليلة؛ ت: لليلة

119 ق: لكلّ يوم واحد، لينظر ما فيه من النور إلى أن يبدر؛ ت: كلّ يوم ... بدرا تامّا

120 ق: يبتدئ؛ ت: نبتدئ

121 ق: في النقص إلى المحاق؛ ت: بالنقصان ... شيء

122 ق: --؛ ت: موري

123 ق: لازق فلك الشمس؛ ت: طابق فلك ... التوفيق

|ت/307ب| الباب الخامس:[124] في معرفة ما في القمر من النور.[125]

فإذا أردت ذلك،[126] إذا كان طلوعه على ساعة من النهار، وأكثر من ذلك أو أقلّ، وأردت أن تعمل به،[127] فضع مري[128] الساعات على ساعة من ساعات النهار المقسومة في الربع الأيسر من الصفيحة،[129] أو على ما أردت من الساعات، فما وجدت[130] قد ظهر في فلك مري[131] نور القمر من الفضّة البيضاء، فهو مقدار ما يَرى فيه من النور.[132] وكذلك فاعمل إذا أردت علم مقدار ما يرى[133] فيه من النور في ساعات[134] الليل، فضع مري[135] الساعات على ما أردت من[136] ساعات الليل، وانظر[137] إلى فلك مري[138] نور القمر، فما يظهر فيه[139] من الفضّة البيضاء، فهو ما في القمر من النور، فاعلم ذلك وقس عليه.[140]

|ق/21ب| الباب السادس:[141] في معرفة على كم يغيب القمر[142] أول ليلة من الشهر.

|ت/308أ| وذلك بأن تضع مري[143] القمر -وهو الخطّ الذي يبتدئ من طرفي صفيحة العنكبوت/المرّ[144] بفلك الجوزهر-، حتّى تقطع[145] فلك الشمس[146] بنصفين، على ابتداء[147] العدد من كحَ، الذي[148]

---

124 ق: فصل؛ ت: الباب الخامس
125 ق: في الساعات نهارا أو ليلا؛ ت: --
126 ق: --؛ ت: فإذا أردت ذلك، فانظر
127 ق: --؛ ت: وأكثر ... تعمل به
128 ق: مري؛ ت: موري
129 ق: --؛ ت: من الصفيحة
130 ق: وجدته؛ ت: وجدت
131 ق: في فلك مري ؛ ت: من فلك موري
132 ق: من النور فيه؛ ت: فيه من النور
133 ق: وعلى هذا، إذا أردت ما؛ ت: وكذلك ... ما يرى
134 ق: في ساعات؛ ت: على ساعات من
135 ق: ضع مري؛ ت: فضع موري
136 ق: --؛ ت: ما أردت من
137 ق: وانظر؛ ت: ثمّ انظر
138 ق: مري؛ ت: موري
139 ق: ظهر؛ ت: يظهر فيه
140 ق: --؛ ت: فاعلم ... عليه
141 ق: فصل؛ ت: الباب السادس
142 ق: القمر على كم يغيب. إذا أردت؛ ت: على كم يغيب القمر
143 ق: فضع مري؛ ت: وذلك بأن تضع موري
144 ق: أي الخطّ المبتدئ من طرفي صفيحة العنكبوت، المارّ؛ ت: وهو الخطّ الذي يبتدئ من طرف الصفيحة، أعني صفيحة العنكبوت، وتمر
145 ق: يقطع؛ ت: تقطع
146 ق: +القمر؛ ت: --
147 ق: على ابتداء؛ ت: إلى أن ابتدئ
148 ق: الذي ؛ ت: إلى

عن يمينك في النصف الأعلى من المجرة. فإنّه يوافي عند ذلك مري[149] نور القمر من يسارك في مقابلته سواء.[150] ثمّ حرّك العنكبوت، حتى يقع مري[151] القمر على قسم « واحد »، وهو المكتوب فيه[152] آ، وانظر إلى مقابلة الخطّ من العنكبوت، فإنّه يقع على ما يغيب عليه القمر، وهو مكتوب في أحد يسارك <هَ> وَ:[153] ستّة أسباع ساعة.[154] وكذلك تعمل بأن تحرّك[155] العنكبوت، وتضع مري[156] القمر على عدد الأيّام وأنصافها،[157] فإن كان يوم تقسم المجرة بنصفين[158] لكي يكون اليوم نصفين. فإذا وضعت مري[159] القمر على ما أردت من الأيام وكسور الأيّام،[160] فانظر إلى مقابلة الخطّ من الجانب الآخر، على كم وقع من عدد[161] الساعات وكسورها، فما كان فهو ما يغيب عليه القمر من الليل، وبالله التوفيق.[162]

الباب السابع: في معرفة [163] على كم يغيب القمر بالنهار من الساعات وكسورها.[164]

3ب| فاعلم أنّه لا يكون ذلك إلا بعد انتصاف[165] الشهر. فإذا أردت ذلك،/فضع مري[166] القمر على ما مضى من أيّام الشهر، فإنّ الخطّ[167] المقابل له يقع على الساعات وكسورها، فاسقط منه يبَ ساعة، وما بقي فعليه يغيب القمر من النهار، فاعلم ذلك، إن شاء الله. فإذا كان[168] تمام كَح يوما،

149 ق: مري؛ ت: موري
150 ق: سواء؛ ت: سوآ
151 ق: مري؛ ت: موري نور
152 ق: عليه مري نور القمر؛ ت: فيه
153 ق: مكتوب في أحد يسارك و؛ ت: وَ
154 ق: --؛ ت: ساعة
155 ق: يعمل بأن يحرّك؛ ت: قائما كان لحرك
156 ق: ويضع مري؛ ت: وبضع موري
157 ق: وأنصافها؛ ت: وأيضا
158 ق: كل يوم قسم حرك المجرة بقسمين؛ ت: كان يوم نقسم المجرة بنصفين
159 ق: مري؛ ت: موري
160 ق: وكسورها؛ ت: وكسور الأيّام
161 ق: --؛ ت: عدد
162 ق: --؛ ت: وبالله التوفيق
163 ق: فصل في معرفة؛ ت: الباب السابع: إذا أردت أن تعلم
164 ق: من النهار؛ ت: بالنهار من الساعات وكسورها
165 ق: وذلك بعد نصف؛ ت: فاعلم ... انتصاف
166 فضع مري: ق: اجعل مري؛ ت: فإذا أردت ذلك، فإذا أنت وضعت موري
167 ق: فالخطّ؛ ت: فإنّ الخط
168 ق: ١٢ ساعة أبدا. تعمل كذا إلى ؛ ت: يبَ ساعة ... فإذا كان

يكون القمر قد دخل[169] تحت الشعاع، فلا يحتاج له إلى عمل[170] حتّى يفارق الشمس. ويبتدئ العمل فيه بعد ذلك على معنى ما مضى من القول المذكور فيه المنسوب إليه، فاعلم ذلك.[171]

الباب الثامن: (ت) في معرفة علم ما يكون من كسوف القمر، وكم مقدار ما ينكسف منه / (ق) في معرفة هيئة الخسوف وكمّيته.[172]

اعرف أوّلا من التقويم كم بين الجوزهر، والشمس والقمر من الدرج وقت الاستقبال،[173] |ق/22ا| فإن كان أكثر من يب درجة، فإنّ القمر لا ينكسف،[174]/وإن كان يب درجة أو أقلّ،[175] فإنّ القمر ينكسف، وبالله التوفيق.[176]

|ت/309ا| فإذا علمت أنه ينكسف، وأردت تعلم كم ينكسف منه،[177] فاقلب[178] الصفيحة التي فيها الفضّة المحرّقة إلى الوجه الآخر، وفيه[179] دائرة القمر مقسومة باثني عشر[180] قسما. فاطبق[181] العنكبوت على وجهها، ورد[182] القطب والفرس على حاله[183] الأول. وادر[184] العنكبوت، حتّى ينطبق دائرة فلك الجوزهر من العنكبوت[185] على دائرة القمر انطباقا لا يبيّن منه شيء[186]. ثمّ ادر العنكبوت، حتّى يقع طرف <فلك> الجوزهر على ما يكون بين القمر وبين[187] الجوزهر. فإنه عند ذلك تتنحى دائرة الجوزهر[188] عن دائرة القمر، ويظهر مقدار ما ينكسف من دائرة[189] القمر،

169 ق: ثمّ تدخل القمر؛ ت: يكون القمر قد دخل
170 ق: ولا يحمل له عملا؛ ت: فلا يحتاج له إلى عمل
171 ق: والعمل فيه بعد ذلك مضى؛ ت: ويبتدئ ... فاعلم ذلك
172 ق: فصل في معرفة هيئة الخسوف وكمّيته؛ ت: الباب الثامن: في معرفة علم ما يكون من كسوف القمر، وكم مقدار ما ينكسف منه
173 ق: اعرف ... الاستقبال؛ ت: فانظر في التقويم كم بين الشمس والقمر من الدرج في وقت الاستقبال، فإذا عرفت ذلك، وأردت أن تعرف الكسوف، فانظر إذا كانت الشمس والقمر وقت الاستقبال وبين الرأس والذنب.
174 ق: --؛ ت: فإن كان ... ينكسف
175 ق: فإن كان ١٢ درجة وأقلّ؛ ت: وإن كان أقلّ من يب درجة.
176 ق: فينخسف؛ ت: فإنّ القمر ... التوفيق.
177 ق: ولمعرفة كمّيته؛ ت: الباب التاسع: فإذا علمت أنه ينكشف، وأردت تعلم كم ينكسف منه
178 ق: اقلب؛ ت: فاقلب
179 ق: القصة المخرقة إلى الوجه الآخر، وفيه؛ ت: الفضّة المحرّقة، وفيها
180 ق: ١٢؛ ت: باثني عشر
181 ق: ويطبق؛ ت: فاطبق
182 ق: وترد؛ ت: ورد
183 ق: حالة؛ ت: حاله
184 ق: وتدير؛ ت: وادر
185 ق: فلك الجوزهر من العنكبوت؛ ت: الفلك الشمس
186 ق: لا يبيّن منه شيء؛ ت: حتّى لا يبان منه شيأ
187 ق: --؛ ت: وبين
188 ق: يتنحا دائرة الجوزهر؛ ت: عند ذلك تتنحا دائرة الشمس
189 ق: ينكشف من؛ ت: ينكسف من دائرة

ويكون ذلك مقدار ما ينكسف[190] منه، فافهم ذلك. وإن ظهرت[191] دائرة القمر من[192] تحت فلك الجوزهر كلّه، فإنّ القمر ينكسف[193] كلّه، وإن بقي منه شيء مغطّى،[194] فهو مقدار ما يبقى لا ينكسف[195] من القمر، فاعلم ذلك وقس عليه. وإن انكسف كلّه، فما زاد على مقدار[196] دائرة القمر، فهو مكث القمر في الكسوف.[197]

ولمعرفة[198] ما ينكسف[199] من الأصابع، انظر كم قطع فلك الجوزهر من أقسام دائرة القمر، وهي ١٢، فما كان فهو قدر الأصابع المنكسفة منه.[200]

وقد عرفت عند ذلك كمّية الكسوف والمكث، من غير حساب ولا تعب، إن شاء الله. وبه
30ب| الهداية والإرشاد/إلى قصد السبيل، وهو حسبنا ونعم الوكيل .

تمّ شرح العمل بالصفيحة القمريّة المنسوبة إلى علي بن عيسى الحاسب، رحمه الله تعالى. [201]

190 ق: فيكون مقدار ما ينكشف، ت: ويكون ذلك مقدار ما ينكسف
191 ق: فإن ظهر، ت: فافهم ذلك. وإن ظهرت
192 ق: --، ت: من
193 ق: ينكشف، ت: ينكسف
194 ق: مغطّي، ت: مغطا
195 ق: مقدار ما لا ينكشف، ت: فهو مقدار ما يبقى لا ينكشف
196 ق: وإن زاد على ، ت: فاعلم ذلك وقس عليه. وإن انكشف كلّه، فما زاد على مقدار
197 ق: مكث القمر في الخسوف، ت: ما يمكث القمر في الكسوف
198 ولمعرفة: ق: فصل في معرفة، ت: --
199 ق: ينخسف، ت: --
200 لمعرفة ... منه: ت: --
201 ق: انتهى والله أعلم، ت: وقد عرفت ... الله تعالى

English Translation[1]

Treatise of ʿAlī ibn ʿĪsā al-Ḥāsib on the knowledge of how to use the lunar plate and the receptacle for the eclipses.

May God have mercy on him and bless him.

In the name of God, the Merciful, the Compassionate—my trust is in God.

ʿAlī ibn ʿĪsā—may God have mercy on him—said, after praising God and extolling Him with what is due to Him, and prayer for his Prophet and for his family: As for the knowledge of this plate and how to work with its components, know that its limb (figure 1) is divided into four parts, and its [upper] half is divided into 28 parts, which correspond to twenty eight-days. [As for] the beginning of [numbering] the divisions on it, when you take it in your hand and you suspend it before your face, the beginning [of numbering] of one of its quadrants will be on your right side ascending to the suspension. [This quadrant] is divided into 14 parts. Next, [starting] from the line that descends from the suspension on the left quadrant, the 28 days are completed, being the number [of days] in which the moon is visible. As for the other half [of the plate], it consists of the divisions for the hours and fractions of the hours at which the moon sets, since the first night [of the month]. Its beginning is from the quadrant that is on your left, where "0, 6" are written; then, on the second compartment "1, 5": the "1" (*alif*) corresponds to one hour and "5" to a fraction of an hour -this [fraction] is five sevenths of an hour-, until the divisions of this quadrant in hours and their fractions are completed, as 12 complete hours, which are the hours of the night and their fractions.[2] This will be the time at which the moon sets, from the first night it appears, up to the night of the full moon. Subsequently after that, [you continue] up to the completion of 24 [hours], at which [the moon] sets during daytime after having encountered the 12 hours of the night and its fractions, until [the divisions] on the limb of the moon are completed;

1 Following the pattern of Johannes Thomann's article, the additions for the sake of clarity, are put into angular brackets.

2 The increase of the time of the moon setting is taken as 6/7 of an hour for each day, which corresponds to 24 hours/28 days. The number 6/7 of an hour on the first division is also confirmed in chapter 6 below. Starting from 6/7 and adding 6/7 of an hour per day, we obtain the arithmetic progression 6/7, 1 5/7, 2 4/7, 3 3/7, 4 2/7, 5 1/7, 6, 6 6/7, 7 5/7, 8 4/7, 9 3/7, 10 2/7, 11 1/7, 12 the first 14 terms of which are written on the lower left quadrant of the limb, and give the time of moon setting in seasonal hours after sunset. The next 14 terms of the same arithmetic progression, from 12 6/7 to 24, are written on the lower right quadrant of the limb and correspond to the time of moon setting in daylight, in seasonal hours after sunset. In figure 1, I have written the integer number of the hours and only the nominator of the fractions, since this way is described in the text.

the [limb] is the ring which surrounds the plate of the *rete* [situated] below the *rete*. Understand this!

Chapter on the knowledge of the *rete* and its divisions.

The *rete* is a lattice-like plate, with a hole in one of its halves (figure 2). It rotates around the pole, and "circle indicating the light of the moon" is written on it. Through this hole information is presented, when the *rete* is rotated on the plate with the silver and black markings on it, which is below the *rete*.

On the plate beneath the *rete*, there are two circles [plated with] burned silver, (figure 3) and the uppermost of these circles is the circle of the solar disc

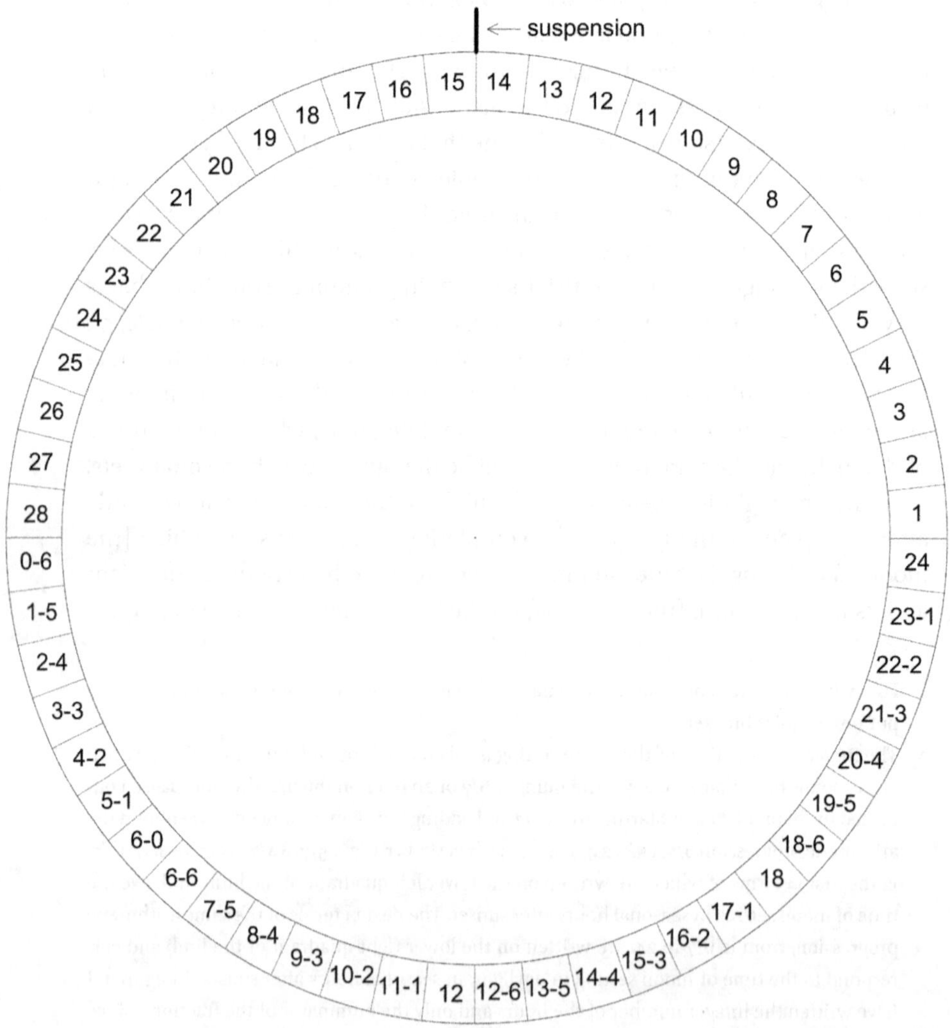

FIGURE 1 *The limb*

situated near the suspension. Next to it, on the left side, towards the line of the left quadrant, is the numbering of the hours at which the moon rises at day, and on the quadrant after it, the numbering of the hours at which the moon rises at night.

On the perforated side of the *rete* is the circle of *falak al-jawzahar*[3] (figure 2) and on it is written "*falak al-jawzahar*". On this perforated half, there is also the index of the hours; this is a splinter on which "index of the hours" is marked. On the line that splits the *rete* into two halves and begins from the line which intersects the circle of the light of the moon, goes through the center of the *rete* and through *falak al-jawzahar*, and terminates at its extremity, where the line is completed, on [this line] is written "index of the moon". Understand all that

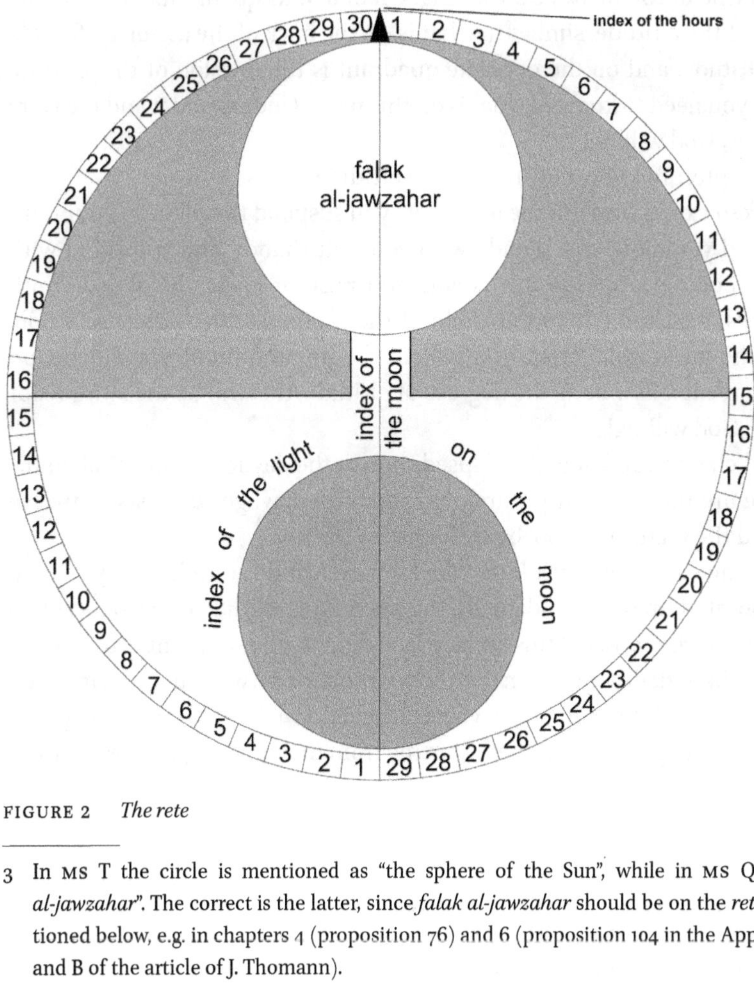

FIGURE 2 *The rete*

3 In MS T the circle is mentioned as "the sphere of the Sun", while in MS Q as "*falak al-jawzahar*". The correct is the latter, since *falak al-jawzahar* should be on the *rete*, as mentioned below, e.g. in chapters 4 (proposition 76) and 6 (proposition 104 in the Appendices A and B of the article of J. Thomann).

has been described to you so that the use of the plate is made easy for you, God willing! On the *rete*, on its rim, there are also divisions according to the number of days in a lunar month, on one side 30 days, and on the other side 29 days, because the lunar month is always like that.[4]

On the other side of the plate, there is a circle divided into twelve parts (figure 4), from which becomes known what is eclipsed of the moon, when there is a lunar eclipse. On the right side of this circle there is a numbering representing the distance between the two nodes and the Moon during an eclipse. This is what is on the other side of the plate.

On the back of this lunar plate, there is an alidade similar to the alidade of an astrolabe (figure 5). On it, there are lines from which becomes known what has passed of the day and what is left of it, when I measure according to what I will describe to you below, God willing! On one of its quadrants there are the divisions of the altitude, similar to what is on the back of the astrolabe for taking the altitude, and on the opposite quadrant is the division of the shadow. This is all you need to know of what is on this plate. Understand it and measure accordingly, God willing!

First chapter: To know how to take the altitude[5]

It is taken exactly as with the astrolabe: you suspend the plate by your right [hand], and you move the alidade with your left [hand], and you stand with your left shoulder opposite to the sun. You raise or lower the alidade until the degrees of altitude fit and the light of the sun passes from the upper hole through the lower hole. Then, you look at the pointed tip of the alidade, on which degree it falls among the degrees of altitude; this will be the altitude for that time, God willing!

Second chapter: To know the elapsed time of the daylight in unequal hours, and these are the seasonal [hours], by which the daylight consists of twelve hours, and the night-time too, being longer or shorter.

If you want that, you must know the [sun's] altitude at midday of your day. Put the pointed tip of the alidade on the maximum altitude corresponding to the midday of your day. Let this alidade face directly the sun,[6] until you see the shadow of the alidade straight on the body of the alidade, without being inclined in any direction. When you know where the shadow falls on the alidade, count the hours. You begin to count them starting from the upper sighting vane, from

4 The passage "On the *rete* ... like that." appears only in MS Q

5 The first chapter is not included in MS Q.

6 "directly the sun" lit.: the source of sun

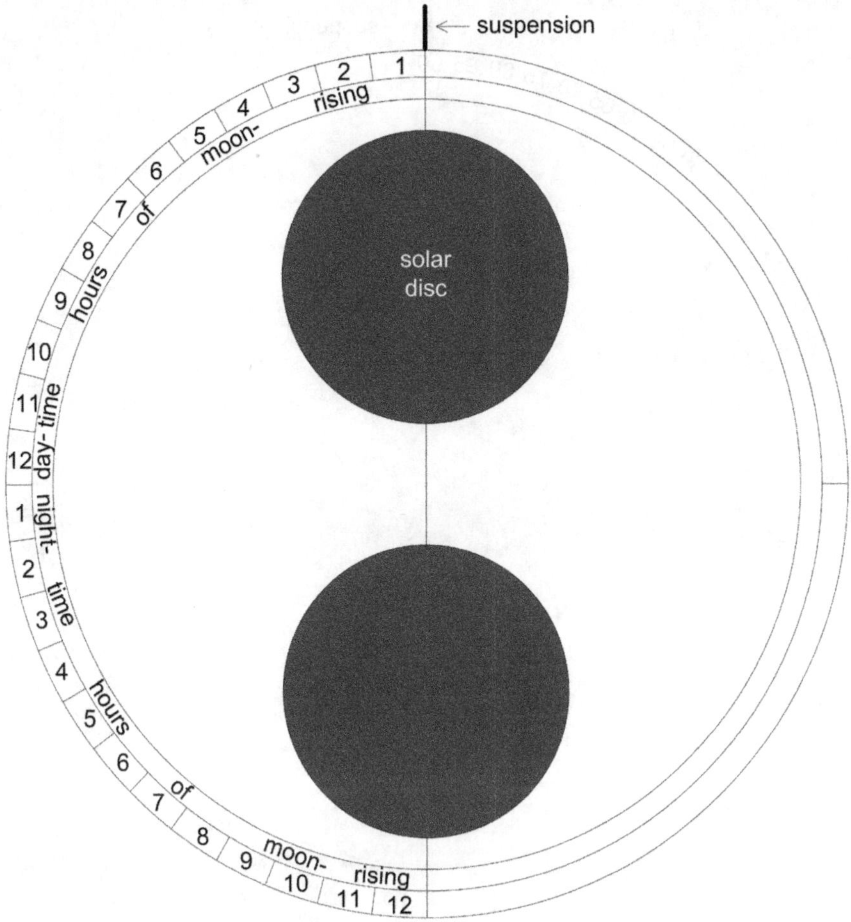

FIGURE 3 *The plate: Lunar phases side*

which you have taken the shadow, and which had been used for dividing the shadow with lines that cut the body of the alidade transversely (figure 5),[7] and

7 The construction of the lines of the unequal lines on the alidade is described in detail by al-Ṣūfī, in chapter XV.108 of his great treatise on the astrolabe (MS BNF, Paris arabe 5098, ff. 253ᵛ–254ʳ), see Flora Vafea, Les traités d'al-Ṣūfī sur l'astrolabe (Ph.D. Diss, University Paris 7—Denis Diderot, 2006), App. VII, 602–603, and by al-Bīrūnī, in *Istīʿāb al-wujūh al-mumkina fī ṣanʿat al-aṣṭurlāb* (MS Leiden, Or. 123b, ff.2ᵛ–3ʳ), see al-Ḥusaynī, Istīʿāb al-wujūh al-mumkina fī ṣanʿat al-aṣṭurlāb, allafahū Muḥammad Ibn Aḥmad Abū Rayḥān al-Bīrūnī, 1st ed. (Mashhad: Majmaʿ al-Buḥūth al-Islāmiyya, 2002), 114–115. For an English translation of al-Ṣūfī texts accompanied by explanatory figures and further texts concerning the hour lines on

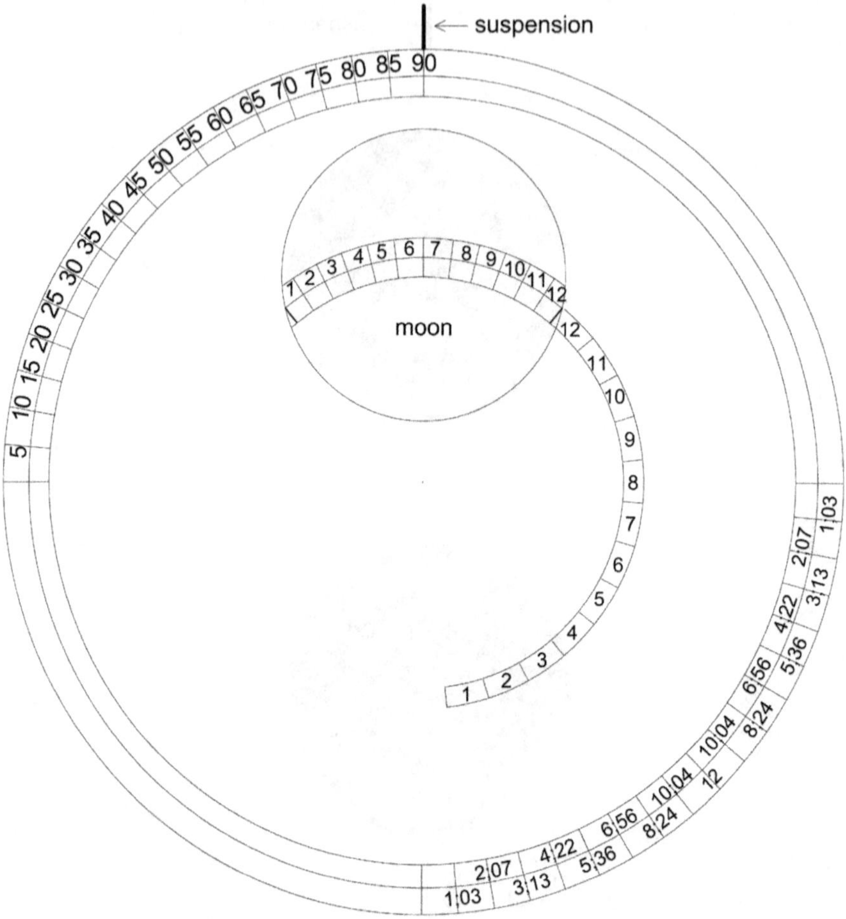

FIGURE 4 *The plate: Lunar eclipses side*

between them the numbering of hours are [written] with the *abjad* numerals [that correspond to the numbers] 1, 2, 3, 4, 5, 6.[8] The position where the shadow falls on this alidade [indicates] the elapsed time of daytime in hours.

---

the alidade, see Josefina Rodríguez-Arribas, "A Treatise on the Construction of Astrolabes by Jacob ben Abi Abraham Isaac al-Corsuno (Barcelona, 1378): Edition, Translation and Commentary" in *Journal for the History of Astronomy* 49(1), 27–82; here 76–79. The procedure for measuring the time is described by D. A. King in David A. King, In Synchrony with the Heavens: Studies in Astronomical Timekeeping and Instrumentation in Medieval Islamic Civilization, 2 vols., Vol. 2: Instruments of Mass Calculation (Leiden: Brill, 2005), 2:254.

8 These *abjad* numerals could be designated with the Greek numerals α, β, γ, δ, ε, ς.

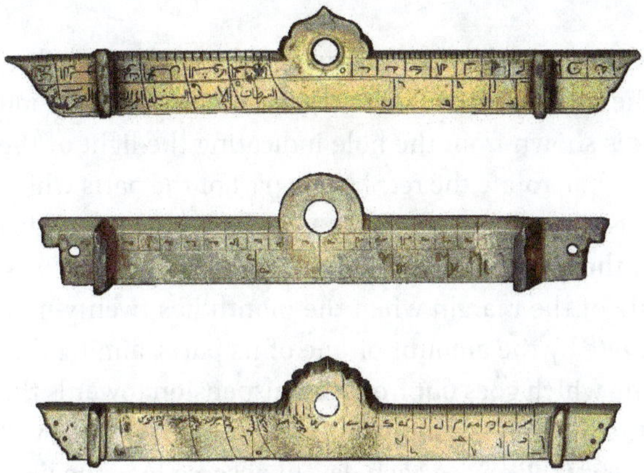

FIGURE 5 *The hour markings on three alidades from astrolabes (Inv. no. 53637, 33767, 42730) from the Museum of the History of Science in Oxford*
COURTESY OF THE MUSEUM OF THE HISTORY OF SCIENCE IN OXFORD

When the measurement is before midday, count the lines which are on the alidade on the side which is marked with the *abjad* numerals [that correspond to the numbers] 1, 2, 3, 4, 5, 6; and when it is after midday, count the lines on the alidade from the side on which the numbers of hours are indicated by the *abjad* numerals [that correspond to the numbers] 7, 8, 9, 10, 11, 12.[9]

Third chapter: To know the shadow (cotangent).

If the altitude is less than 45°, look at the pointed tip of the alidade at which indication within the shadow scale falls, and divide 144 by that indication—among the subdivisions of the shadow-; the result will be the shadow at the moment you want. If the altitude is 45°, the shadow will be exactly 12 [digits]. If the altitude is greater than 45°, the shadow will be the indication on which the pointed tip of the alidade falls, with no need of division.[10] Know that, God willing!

9 These *abjad* numerals could be designated with the Greek numerals ζ, η, θ, ι, ια, ιβ.

10 The procedure described in manuscript T is misleading, while that in manuscript Q shows that the shadow corresponds to the cotangent of the angle h of altitude, measured in digits. If $h=45°$, then $\cot h=12$ digits; if $h>45°$, then $\cot h<12$ digits and corresponds to the indication of the alidade; if $h<45°$, then $\cot h>12$ digits, and it can be a great number. For this reason the indication of the alidade is the tangent of the angle h of altitude ($\tan h$); the cotangent is calculated as $\cot h=144/\tan h$. The indications in the scale of shadow on the lower right quadrant of figure 4 have been computed on the basis of the above

Fourth chapter: To know the light of the moon appearing each night of the days of the month.

If you want that, rotate the *rete* until *falak al-jawzahar* falls on the solar disc, and the circle indicating the light of the moon falls on the burned silver, and nothing of the white silver is shown from the hole indicating the light of the moon. When you have done that, rotate the *rete* by one part of the parts which are allotted on the *rete*, on its rim, among which are 1, 2, 3, 6, 9, 12, 15, 18, which are the numbers of days of the lunar month when the month has thirty days, and also [on] the other side of the margin when the month has twenty-nine days. When you rotate the *rete* by the amount of one of its parts, aiming that the part is aligned to the line which goes out from the suspension towards the middle of the plate, then the light on the moon at one [particular] night will be shown to you. Rotate [the *rete*] likewise each day, to observe how the light on the moon increases, until its light is complete in the opposition to the sun, and it will be a complete full moon. After that, it starts decreasing and it continues to decrease its light [depicted] as the white [part], and the black [part] increases in it, until nothing remains from its light. The circle indicating the light of the moon may rotate until it covers the solar disc in the conjunction. Understand that and measure according to it, if God the Glorified and Sublime wishes. When you want to know what [the light of the moon] will be in another month, the work on it is exactly like that of the first one; and with God is the success.

Fifth Chapter: To know the illuminated [part] of the moon.[11]

If you want that, when it rises at an hour during daylight, or more or less, and you want to work with it, put the index of the hours on one of the daylight hours allotted on the left side of the plate, or [put the index] on what hour you want. The [part] of white silver that you have found to be visible, through the circle indicating the light of the moon, will be the extent of the illuminated [part] of [the moon]. Proceed similarly when you want to know the extent of the illuminated [part] of it [when the moon rises] during the hours of night.

---

interpretation; they are not included in the text of the manuscript. The phrase "now divide 144 by it" (فالآن تقسم عليه قَ مَ دَ) that appears at the end of proposition 73 in the Appendix A of the article of J. Thomann may had been a marginal note referring to the end of proposition 69 in an antecedent manuscript; this phrase may had been incorporated in a wrong position by a copyist.

11 In the title of MS Q, the additional phrase "in day- and night-time hours" appears. This could be understood as a remnant of a phrase that makes clear the difference between the procedures of the fourth and fifth chapters; this phrase could be "when the time of moon rising during day- or night-time is known" (من ساعات طلوع القمر بالنهار أو الليل).

Put the index of the hours on whichever hour of night you want, and look at the circle indicating the light of the moon. The [part] of white silver visible through it [depicts] the illuminated [part] of the moon. Know it, and measure according to it!

Sixth Chapter: To know when the moon sets the first night of the month.

That [can be found] by putting the index of the moon—and this is the line which begins from the two extremities of the plate of the *rete*, passing through the *falak al-jawzahar*- at the beginning of the numbering of 28 that is at your right, on the upper half of the limb, so that you cut the solar disc into two halves.[12] Thus, the index of the light of the moon comes to your left straight opposite to it. Next, move the *rete* until the index of the moon falls upon the compartment "one" on which 1 is written, and observe the other end of the line on the *rete*; it falls upon [the time] at which the moon sets: this is written on the first [compartment] at your left as <o> 6: six sevenths of an hour.[13] You will proceed likewise by moving the *rete* and putting the index of the moon on the number of the days and their halves, when there is a day that [a division of] the limb divides it into two halves, so that the day consists of two halves. When you put the index of the moon on the days and the fractions of days you want, observe on which number of the hours and their fractions the other end of the line falls on the other side [of the limb]; the indication will be [the time] at which the moon sets at night. The success is with God!

Seventh Chapter: To know when the moon sets during daytime in hours and fractions thereof.

Know that this occurs only after the middle of the month. If you want that, put the index of the moon on [the division corresponding to] what has passed from the days of the month. The opposite line falls on the hours and their fractions. Subtract 12 hours from it and the remainder will be [the time] at which the moon sets during daytime.[14] Know that, God willing! When the 28 days are completed, the moon has already gone under the rays [of the sun], and no work is needed for it, until it separates itself from the sun. The work starts after

12 The meaning is probably the following: You rotate the *rete*, so that the line that separates the perforated half of the *rete*, from the half with the hole, cuts the solar disc into two halves.

13 Number 6 is written as "و" overlined in MS T.

14 The numbers that correspond to the time of setting of the moon during daylight are written on the lower right quadrant of the limb; these are between 12 6/7 and 24. The number 12 should be subtracted from each one of these numbers, in order to obtain the time in seasonal daytime hours counted after sunrise.

that, according to the concept of the elapsed time, as mentioned in the speech attributed to him [i.e. ʿAlī ibn ʿĪsā]; know that!

Eighth Chapter: (T) To know whether a lunar eclipse will happen and what the magnitude of the eclipse will be. / (Q) To know the form and size of the lunar eclipse.

Learn first from the astronomical ephemeris how many degrees are between the lunar node[15] [on one side] and the Sun and the Moon [on the other side] at the time of the opposition. If [the distance] is greater than 12 degrees, the moon will not be eclipsed. If it is 12 degrees or less, the moon will be eclipsed. Success is with God![16]

If you know that [the moon] will be eclipsed, and you want to know how much of it will be eclipsed, turn upside-down the plate on which the burnt silver is, [and use] the other face, on which the circle of the moon is divided into twelve sections. Adjust the *rete* on [this] face, and bring the pole and the horse to their initial positions. Rotate the *rete* until the circle of *falak al-jawzahar* on the *rete* is adjusted in congruence onto the circle of the moon, so that nothing is visible from it. Then, rotate the *rete* until the rim of *<falak> al-jawzahar* falls on the distance between the moon and the lunar node.[17] In [the process of] that, *falak al-jawzahar* is moved away from the circle of the moon, and the magnitude of the eclipse is shown on the circle of the moon; this will be the magnitude of the lunar eclipse; understand that! If the circle of the moon appears entirely [i.e. is not hidden] below *falak al-jawzahar*, the moon will be entirely eclipsed. If something of it remains covered, this is the amount of what remains uneclipsed of the moon. Know that, and measure according to it. If it is entirely eclipsed, the magnitude, additional to that shown on the circle of the moon, corresponds to the duration of the totality during the eclipse.

To know how many digits will be eclipsed, look how many of the 12 sections on the circle of the moon *falak al-jawzahar* has traversed. The result will be the number of the eclipsed fingers of [the moon].[18] You have learned by that the

---

15 Or, according to MS T, the ascending or descending lunar node.

16 At this point in MS T, the eight chapter ends and the ninth one starts. However, only whether a lunar eclipse will wwoccur has been examined, without any hint for the magnitude of the eclipse, as the title announces. For this reason, I omit the title "Ninth chapter" that appears in MS T. In MS Q, next paragraph belongs to the previous chapter, and a new chapter starts as "To know how many digits ...".

17 The distance between the moon and the lunar node is measured on the arc-scale to the right of the circle of the moon. Similar scales appear on the plate described by al-Bīrūnī, but there the ecliptic latitude of the moon is the principal magnitude for the study of the lunar eclipse; see Vafea, "Al-Bīrūnī: The Plate of the Eclipses", pp. 318, 324.

18 The fragment "To know how many ... of [the moon]" appears only in MS Q.

magnitude of the eclipse and the [duration of the] totality without calculation and pain, God willing! The divine guidance and the guidance towards the right path are with Him. He is sufficient for us and the admirable attorney.

The commentary on the use of the lunar plate, attributed to ʿAlī Ibn ʿĪsā al-Ḥāsib—the Exalted God may have mercy upon him-, has been completed.

Index

www.ingramcontent.com/pod-product-compliance
Lightning Source LLC
LaVergne TN
LVHW011102300926
849160LV00008B/254
9789004383807